OPERATION PARAQUAT

The Battle for South Georgia

OPERATION PARAQUAT

The Battle for South Georgia

Roger Perkins

PICTON PUBLISHING (CHIPPENHAM) LIMITED

1986

BY THE SAME AUTHOR

THE PUNJAB MAIL MURDER
The Story of an Indian Army Officer
1st edition 1979, 2nd edition 1986

GUNFIRE IN BARBARY
Admiral Lord Exmouth's Battle with the Corsairs of Algiers, 1816
Co-researched with Capt K. J. Douglas-Morris, RN, 1982

THE KASHMIR GATE
Lieutenant Home and the Delhi VCs
1983

ANGELS IN BLUE JACKETS
The Navy at Messina, 1908
Co-researched with J. W. Wilson, 1985

First published in 1986 by Picton Publishing (Chippenham) Ltd
1 Frome Road, Beckington, near Bath, Somerset

ISBN 0 948251 13 1

Photoset in AM International Plantin
by Picton Print, Citadel Works, Bath Road, Chippenham
Printed in Great Britain
PP42194

Contents

Maps & Charts

Author's Introduction

For every Briton too young to remember former conflicts, 1982 was an extraordinary year, a year of totally new sensations and new experiences. The United Kingdom of Great Britain found itself, quite suddenly, plunged into the fervour of a popular war. After many years of gentle transition from Empire to Europe, the nation reverted to type. A great wrong had been committed. The armed forces of a foreign power had violated not only British sovereignty but also the accepted tenets of civilised behaviour. The Prime Minister evoked the nation's memory when, addressing the House of Commons on 8 April, she said: 'Throughout the Western world and beyond, there is a realisation that if this dictator succeeds in unprovoked aggression, other dictators will succeed elsewhere. We are fighting a battle against that type of aggression, and once again it is Britain that is fighting it.'

The despatch of the largest fleet to depart British shores since June 1944 aroused powerful emotions. The crowds lining the South Coast were almost painfully reminiscent of those which had waved farewell to the young men setting out for the Transvaal in 1899, their sons who sailed for France in August of 1914, and their grandsons who followed the same route in September of 1939. They did not sing the ballads of those earlier wars – *Goodbye, Dolly Gray*, *It's a Long Way to Tipperary*, or *Roll Out the Barrel*. They had a new song. The poignant melody and lyrics of *Don't cry for me, Argentina* were tinged as much with regret as with hope. The quarrel was with Argentina's policies and leadership, not with its common people. There was a fine irony to the fact that the nation's resources were being directed against a country for which the British felt no inherent animosity.

It was inevitable that a conflict which aroused such profound sentiments should, in its immediate aftermath, give rise to a number of publications describing those events. After a long and wasteful war, people prefer to forget. After a short and triumphant one, they wish to know more.

Operation Paraquat is the consequence of a decision, taken at Yeovilton in 1983, to commission a book describing the least known and least understood episode in the Anglo-Argentine conflict. The story which follows is unusual in many respects. South Georgia had witnessed the first direct confrontation between the two navies. The circumstances surrounding 'the Davidoff affair' would have been ludicrous if the principles at stake had not been so fundamental. The central character in the plot was not a great and powerful warship, but a workmanlike former Danish merchantman, HMS *Endurance*. On 3 April 1982, her detachment of Royal Marines fought an action unique in the history of the Corps. Little more than three weeks later, when the ships of the Paraquat group ended the ill-conducted Argentine occupation, there was another historic action; never before had a submarine been crippled by a strike force composed entirely of helicopers. The drama of Paraquat was highlighted by the fact that two of Great Britain's most secret military organisations – the Army's Special Air Service Regiment and the Royal Marines' Special Boat Squadron – received a degree of public exposure which they had not experienced for many years. These and other aspects of the South Georgia story had not previously enjoyed the analytical presentation which they merit, and it was for this reason that the author responded so positively to the original suggestion put forward by Commander Paul Madge, RN, of the Fleet Air Arm.

Although the originators of this book are associated primarily with naval aviation, it was apparent to them from the outset that it would be unjust and unrealistic to record the Fleet Air Arm's involvement without placing it in the overall context. The participants were drawn from nearly all services and branches of Her Majesty's armed and auxiliary forces. In short, Paraquat was a combined operation. It could not have succeeded if the constituent members of the team had failed to integrate their roles and functions. The following narrative is a tribute to everyone who took part – from Captain Brian Young, DSO, RN, who bore ultimate responsibility for the Task Group, to the Chinese laundrymen who helped to sustain its morale.

The book is a tribute also to the many other Britons who, over the years, have been involved in the story of South Georgia. Explorers, scientists, administrators, they are as important to any account of the island's history as the servicemen who went there in 1982. Indeed, the first two chapters of this book are devoted exclusively to the commercial, political and diplomatic background. Without that information, it would be impossible to understand how the conflict arose in the first place, or to appreciate the strategic and tactical considerations which shaped it.

The operational directive which first set in motion the British reoccupation of South George was codenamed Paraquet. The author has searched unsuccessfully for the origin of this word. It does not appear in any ornithological reference work known to him. Presumably it was a misspelling of Parakeet, that large genus of tropical and sub-tropical birds which constitute part of the 315-strong family of the *Psittacidae*. Alternatively, it may have been a mistaken attempt at the Old French name for these birds, Paroquet. Whatever the explanation, such an inappropriate title – blighted as it was – soon received the treatment it deserved. During the transmission of one of the earliest signals, it was corrupted to Paraquat, a commercial product for the elimination of weeds. This spelling provided, fortuitously, a clearer mental image regarding the purpose of the operation. It is also the spelling adopted by all of the participants and, as it is their choice, that is the version gladly adopted by the author.

It is appropriate to make reference to the way in which the author has approached his task. The prime requisite at every stage has been the pursuit of truth. This is an ambition more easily stated than achieved. What is truth? In essence, it is any statement which accords with fact. The difficulty in this instance has been that some of the participants in the events of 1982 perceive 'the facts' from diverse points of view. As Robert Browning said: 'The truth is within ourselves.' It was a United States Senator, Hiram Johnson, who said: 'When war comes the first casualty is truth.' In the case of Operation Paraquat, establishing the precise truth has entailed a search for consensus.

It is very rare that the military or naval historian has the privilege of being able to speak directly with the individuals who took part in the events which he is seeking to describe. Thanks to the good offices of the Fleet Air Arm, it has been possible on this occasion to make contact with many of the men who planned and executed Operation Paraquat. They, in their turn, have been unfailingly helpful and enthusiastic that a full account should be published. But the story which they tell inevitably varies from person to person because each one viewed those events from his own standpoint and within the framework of his own training and duties. The author decided at an early stage, therefore, to act as juryman rather than judge or counsel. He accepts that his verdicts on some aspects of the story will not meet with the unreserved approval of all concerned. Due allowance has been made for the natural healthy competitive rivalry which exists between services, ships, squadrons, branches and individuals. There is also the fact that the contributors were interviewed two and three years after the event. Memory does not necessarily fade, but opinions do become enshrined.

The author wishes to express his gratitude to the very many people who have given so generously of their time and expert knowledge. It has not always been possible to quote them by name. This is particularly the case with officers and men of the Special Air Service Regiment and the Special Boat Squadron, Royal Marines. Some are mentioned in the narrative by a name identifiable to their friends and comrades, but not to any readers who might have a less than benign interest in their future welfare. There is a sad exception to this rule. After the successful conclusion of Operation Paraquat, twenty-one members of 22 SAS were killed in a helicopter accident. It has been judged appropriate to show their full names, where relevant, in the following pages.

It is hoped that this book will be regarded as a definitive history, that it will stand the test of time. Apart from being a memorial to the strengths and weaknesses of the human spirit, there is much here to command the attention of future historians. Given the immense potential of Antarctica, 1982 may prove to have been an early milestone on a very long road. That which has happened once can all too easily happen again.

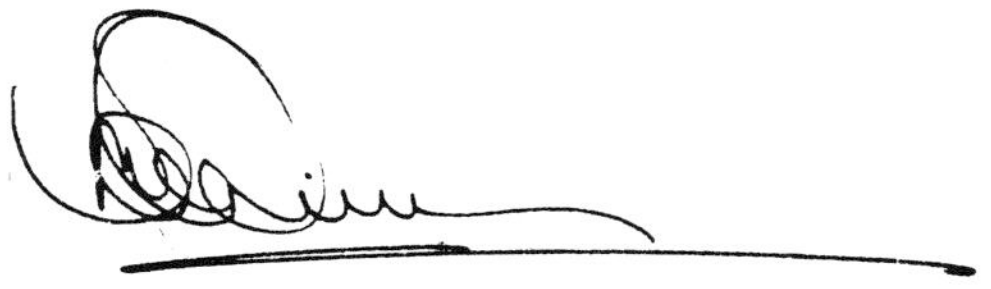

Roger Perkins
Newton Abbot
1986

Acknowledgments

The author wishes to place on record his appreciation of the part played by each of the following individuals in the compilation of the Operation Paraquat story.

Valuable advice and guidance has been received from numerous other individuals who, for one reason or another, cannot be mentioned here by name.

J. F. Ahearne, Esq
LA(Phot) N. A. 'Peter' Allen
B. Ainsworth, Esq (maps and graphics)
D. Ballington, Esq (Jane's)
Capt N. J. Barker, CBE, RN
Lieut A. C. Berryman, RN
Ch Offr A. Bilney, RFA
Lieut P. Blackman, RN
C. Blakeman, Esq
Cdr C. J. Blight, RN
Capt J. J. Bold
Miss A. W. Brown (DGST, MOD)
Maj C. C. Brown, RA
Lieut-Cdr B. W. Bryant, RN
MEA(H)1 S. Bugg
Lieut J. R. J. Burnell, RM
Lieut N. A. M. Butler, RN
Miss Cindy Buxton
The Lord Buxton
The Lord Carrington
Col F. W. B. Carter, RE
R. Cecil, Esq
Capt M. C. Clapp, CB, RN
Lieut-Cdr J. N. Clark, BSc, RN
D. S. Combes, Esq
Lieut S. G. Cooper, RN
Capt J. F. Coward, DSO, RN
WO G. S. Cox, BEM, RN
CPO P. Cutler
Cdr D. B. Davis, RN
Lieut-Col C. N. G. Delves, DSO
Lieut-Cdr J. R. Dransfield, RN
Lieut E. R. Duncan, RM
Cdre S. C. Dunlop, CBE, DSO, RFA
HRH The Prince Philip
Lieut-Cdr A. J. Ellerbeck, DSC, RN
Lieut-Col K. R. H. Eve, RA
Capt M. S. J. Farley, RFA
Admiral of the Fleet Sir John Fieldhouse, GCB, GBE, ADC
POACMN D. Fitzgerald
Lieut T. S. Finding, RN
Capt G. de Ferry Foster (Geest Line)
A. J. Francis, Esq (MOD)
Lieut I. A. Georgeson, RN
Lieut J. R. C. Gilbert, RN
Lieut I. A. Gibson, RN
R. Glenn, Esq (E. P. Barrus Ltd)
Cdr C. S. Gobey, RN
WO J. Graham, RM
Mrs M. Hallatt (BAS)
D. Hamilton, Esq, BEM
Sir Maxwell Harper Gow, MBE (Christian Salvesen plc)
LA(Phot) T. Harding
R. K. Headland, Esq
CAEA(M) D. J. Heritier
The late Vernon Hillier, Esq
Mne S. Holding, RM
Ms M. Hughes (DGST, MOD)
Sir Rex Hunt, CMG
Capt N. C. H. James, RN
K. Johnson, Esq (BBC Data)
WO D. L. Jones, BEM, RM
R. Jones, Esq (MOD)
D. Joy, Esq
E. P. Kinahan, Esq (NUS)
Mrs C. Kirk
M. J. Lacey, Esq (United Towing Ltd)
Miss J. Langton
Dr R. M. Laws, CBE, FRS
Admiral of the Fleet Sir Henry Leach, GCB
C/Sgt P. J. Leach, DSM, RM
A. M. Lee, Esq
Capt G. F. Liardet, CBE, RN
M. G. Little, Esq (RM Museum)
Lieut-Cdr A. J. Lockett, RN
POACMN A. Lomas, RN
Lieut-Cdr H. J. Lomas, RN
Col S. Love, MBE
The Rt Hon Richard Luce, MP
Mne I. D. McCallion, RM
Capt C. L. MacGregor, RN
B. McGrath, Esq
Lieut-Cdr P. A. McKay, RN
Lieut-Cdr I. S. McKenzie, RN
Sgt P. McLeman, RM
Cdr P. R. P. Madge, RN
Cdr H. K. P. Michell, RN
Rear-Admiral L. Middleton, DSO
Lieut-Cdr P. G. Miller, RN
Lieut K. P. Mills, DSC, RM
Capt J. J. Mitchell, RN
Maj-Gen Sir Jeremy Moore, KCB, OBE, MC
Capt R. Morgan-Grenville
Capt A. Morton, DSC, RN
Capt F. Morton (DMS)
Cdr D. C. Nairne, RN (DMS)
C/Sgt J. Napier, RM
C. E. Needham, Esq (Coalite Group plc)
Lieut-Cdr R. C. F. Nichol, RN
Maj G. R. H. Noott, RM
Sir John Nott
Maj C. J. Nunn, RM
Capt G. P. Overbury, RFA
Gp Capt G. J. Oxlee, RAF
Lieut-Cdr C. J. Parry, MA, RN
Sir Anthony Parsons
Miss M. Pauffley (British Red Cross Society)
PO J. Pearson
Capt D. Pentreath, DSO, RN
H. P. Playford, Esq (RM Museum)
Cdr F. N. Ponsonby, MVO, RN
A. Preston, Esq (Jane's)
Miss Annie Price
Flt Lt A. Pulford, RAF
S. Rabson, Esq (P&O Group)
Capt S. Redmond, OBE, RFA
Admiral Sir Derek Reffell, KCB
WO M. D. Richards, RA
P. M. E. Rooley, Esq (Heavylift Cargo Airlines Ltd)
D. Rootes, Esq
D. A. Rough, Esq
The Rt Hon Ted Rowlands, MP
CPO B. Savory
The Lord Shackleton
Lieut-Col J. M. G. Sheridan, OBE, RM
PO(Phot) P. Simcock
Cdr I. Stanley, DSO, RN
Cpl J. Stonestreet, RM
Capt I. B. Sutherland, RN
Surgeon Lieut-Cdr C. F. Swinhoe, RN
Lieut M. J. Tidd, RN
Maj-Gen Sir David Thorne, KBE
Miss A. Todd (BAS)
Lieut R. E. Veal, RN
M. D. Walker, Esq (DMS)
Capt G. F. Walwyn, CVO, RN
P. Watts, Esq
Flt Lt G. R. Weightman, RAF
Lieut D. A. M. Wells, RN
Lieut-Cdr M. D. Wells, RN
Rear-Admiral A. J. Whetstone, CB
Lieut K. P. White, RN
M. B. Wicks, Esq
Capt J. M. M. Wilkins, RFA
Sir Anthony Williams
Capt A. C. Wilson, RM
POACMN R. Wilson
Lieut C. J. Wood, RN
Cdr C. L. Wreford-Brown, DSO, RN
Capt B. G. Young, DSO, RN

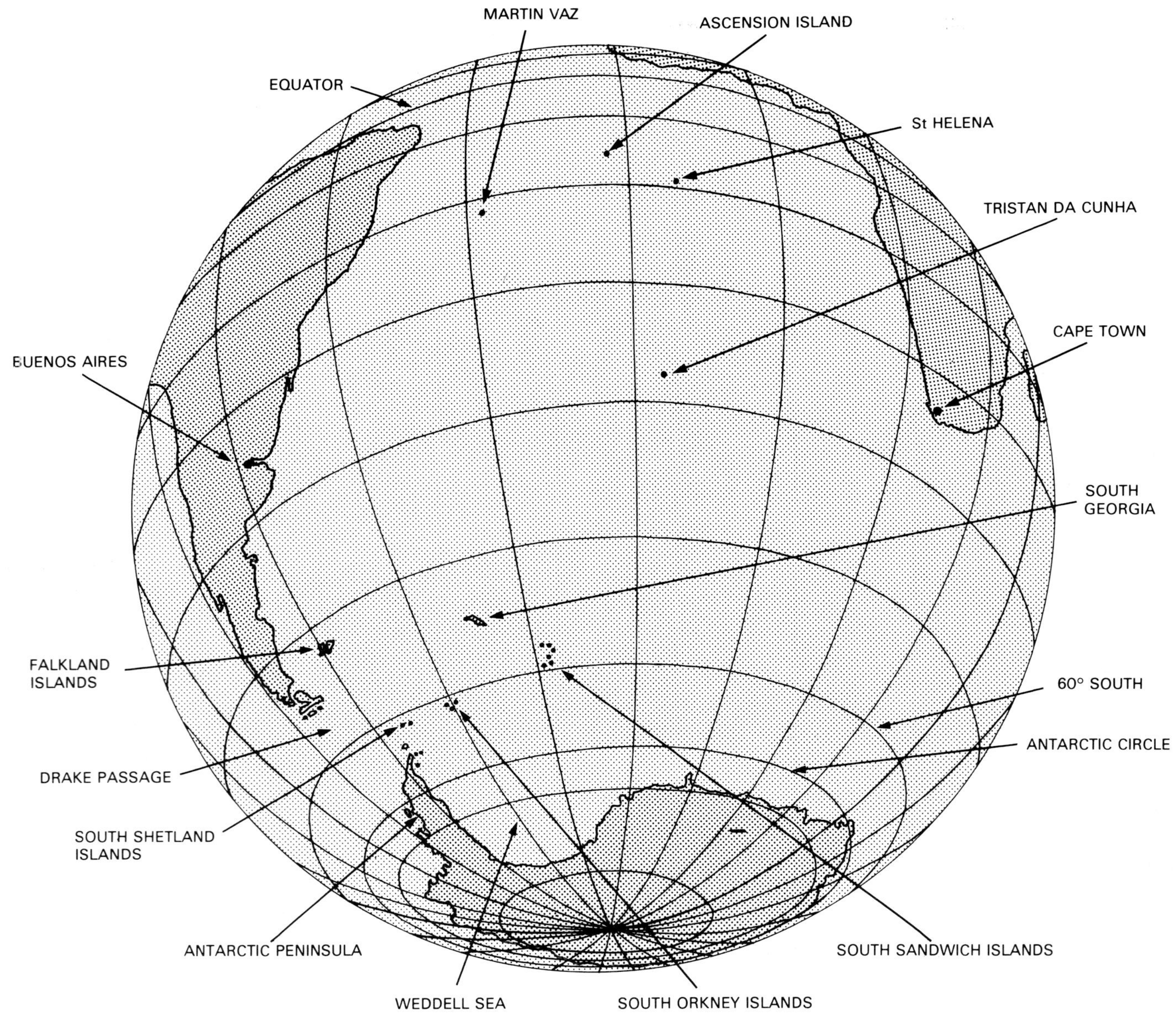

The South Atlantic, with a segment of the sub-Antarctic and the Antarctic Continent. This view of our globe emphasises not only the isolation of South Georgia but also the scale of the task facing the Royal Navy when, in April 1982, it was ordered to recover that island from the hands of an invader.

I | Backdrop to Conflict

On 17 January 1775, an officer of the Royal Navy formally declared a desolate windswept island, deep in the South Atlantic, to be henceforth a British territorial possession. The officer was Captain James Cook, RN, and he was the first man ever to set foot on this place. In honour of his king, he named it 'The Isle of Georgia'. He could not have imagined that it would become, two centuries later, the stage for the most southerly battle in all naval and military history. Still less could it have been conceivable to the great navigator that his wooden-walled brigs, the *Resolution* and the *Adventure*, would be succeeded in 1982 by steel vessels capable of cruising beneath the pack-ice of Antarctica or of destroying an enemy many miles beyond the horizon. Captain Cook entered the name Possession Bay on his scanty chart and, oblivious of the events which he had set in train, sailed on into history. Four years later he was murdered by the natives of the Pacific island of Hawaii.

James Cook was not the only officer of the Royal Navy to play a major role in charting distant waters during the eighteenth and nineteenth centuries. There were others, each contributing his part to the whole. Some of those best remembered were William Parry, Samuel Wallis, John Ross, John Franklin, Peter Heywood, Robert Fitzroy, Francis McClintock, and their doyen, Francis Beaufort. The detailed knowledge gathered during their cruises, and later compiled at the Admiralty's Hydrographic Office, not only ensured safer navigation for maritime trade but also assisted the scientific community in developing its understanding of the globe upon which we live and the forces which have shaped it. It was the combined efforts of many disciplines which produced, in 1960, general acceptance of the theory of continental drift. Known less lucidly as 'plate tectonics', it was the phenomena of the shifting land masses which created the setting for the war of 1982.

Even the briefest examination of the Atlantic map shows that Africa and South America were at one time a unified continental mass. For the past 200 million years, following the initial separation, the two have continued to move away from each other. Their movement is sustained by the seismic activity of the Mid-Atlantic Ridge, a major fault in the earth's crust which runs from Iceland in the north to the edge of Antarctica in the south. As molten rock constantly seeps up to the sea-bed from the earth's interior, so does it make larger the wedge forcing apart the plates on which stand the Americas and the continents of Africa and Europe. In the South Atlantic, the only visible evidence of these huge seismic forces are the volcanic islands of Ascension, St Helena and Tristan da Cunha.

It is between the southernmost extremities of the two land masses that the most rapid separation has occurred. At one time in close proximity to each other, the Cape of Good Hope and Cape Horn are now nearly 5,000 miles apart. The tectonic forces resulting from such a major movement have resulted in a massive rupturing of the earth's thin hard crust. In some places the layered folds of rock have been forced upward to such an extent that they now project above the surface of the surrounding ocean. Nearly two hundred such islands lay scattered to the east and south of Latin America's thin curving tail.

Discovered and charted mainly by British navigators, these lonely outcrops from the deep oceanic floor are the Falkland Islands, South Shetland Islands, South Orkney Islands, South Sandwich Islands, and South Georgia. Excluding the Falklands group, the islands form part of the Scotia Ridge which extends from the Andes to the Antarctic continent. In a region distinguished by its wild and spectacular seascapes, South Georgia is the most dramatic permanent example of nature's forces.

The island lies between latitude 53°56′ and 54°55′ South, and longitude 34°45′ and 38°15′ West. The total surface area is 1,450 square miles. Apart from a few offshore rocks and islets, it is a single land mass, 105 miles in length and 18.5 miles in width at its widest part. Superimposed upon the map of England, its two extremities stretch from Bristol to central London. In other words, and despite its slender crescent shape, South Georgia is a relatively large island. It is also very mountainous. The spine consists of the Allardyce and Salvesen Ranges, the former along the centre of the island and the latter in the south. The highest peak, standing at 9,625 feet, is Mount Paget, but there are five other substantial high points – Nordenskjöld Peak (7,725 feet), Mount Carse (7,649 feet), Mount Sugartop (7,623 feet), Mount Paterson (7,205 feet), and Mount Roots (7,483 feet). There are several other peaks which exceed 6,000 feet at their pinnacles and, with the offshore sea-bed plunging down beyond 1,000 fathoms, it is clear that this island is the tip of a truly huge isolated fragment of continental shelf.

The steep-sided mountains meet the sea in a series of high cliffs which are broken, particularly on the north-east coast, by numerous bays and inlets. Being protected from the prevailing westerly winds, these indents in the coastline provide a natural habitat for wildlife and excellent deep-water anchorages for shipping. Several bays are edged by steep sandy beaches. All of the valleys leading down to these places have been scoured by glacial activity in the past, or are still being deepened by the island's one hundred and sixty-one glaciers.

More than half of the island is covered permanently by ice. About 17,000 years ago the island was covered entirely – the glaciers extended eight miles out to sea and down to a depth of nearly 900 feet. With the ending of the ice age and a subsequent

A picture which captures the spirit of South Georgia's dramatic terrain and swiftly changing weather patterns. A stiff cold wind darkens the surface of King Edward Cove as yet another front brings rolling banks of cloud over the high peaks of the Allardyce Range. To the right, Mount Sugartop (7,623 feet) still reflects a brief spell of brilliant sunshine.

warming of the world's atmosphere, the glaciers retreated back onto firm land. As they melted, so they deposited millions of tons of moraine on the floor of the bays and surrounding ocean. This phenomena, which has occurred several times on a lesser scale, accounts in part for the irregular soundings to be found on the Admiralty charts.

For the past forty years the regional climate has been passing through a 'warm' period. The glaciers are presently much smaller than they were in the 1920s or in the late nineteenth century. Even so, glacial activity on the island is very powerful and, because many of the glaciers terminate at sea-level, they produce a flow of large ice blocks which break away to form icebergs or, if small, 'growlers'. The latter – known to the Navy as 'bergy bits' – constitute a significant navigational hazard because they are difficult to see or identify on radar in heavy swell or storm conditions. The danger is greatest during the spring and early summer when advances made by the glaciers during the winter are warmed and break free.

Apart from the icebergs calved locally, the waters of South Georgia are commonly littered with the much bigger offspring of the Antarctic polar ice cap. One particularly large visitor passed close to the Willis Islands, off the northern tip of the main island, in 1978. Measuring twenty-three miles by forty, it had been tracked by satellite for fifteen years before leaving the Weddell Sea to travel north into the South Atlantic.

Less dramatic, but equally hazardous for mariners, is the Antarctic pack-ice which covers the Antarctic Ocean in varying degrees of density. The northern extremity of the pack changes greatly between winter and summer and from year to year, but normally it is found to the south of South Georgia. In 1980, however, it advanced so far that it extended 125 miles *north* of the island, rendering the place ice-bound and unreachable by ships other than ice-breakers or vessels having ice-strengthened hulls. Even in a normal winter, the anchorages may become frozen over, or packed with drifting ice, until the next gale raises sufficient swell for the clutter to be carried out to sea.

Although South Georgia is nearly 800 miles north of the Antarctic Circle, it is well within the Antarctic or Southern Ocean. The southern limit of the relatively warm waters of the South Atlantic Ocean are to be found in a fairly stable position 220 miles to the north of the island – the point of encounter being known as the Antarctic Convergence – and there is a temperature difference of several degrees in the surface water between one side and the other. The colder Antarctic water not only ensures a longer life expectancy for the transitting icebergs but also creates a distinctly different environment for marine and bird life.

South Georgia's climate is in part dictated by the half-knot current which swirls in a clockwise movement from the Antarctic Peninsula, up to the South Orkney Islands, to South Georgia, eastward to the South Sandwich Islands, and then back down into the Weddell Sea. This very cold current is powered by the wind-driven masses of water which pour through the narrow Drake Passage separating Tierra del Fuego from the Antarctic Peninsula. Only 600 miles wide, this narrow stretch of water must

Only one member of the British Royal Family has ever been able to visit South Georgia. In January 1957, HRH Prince Philip, Duke of Edinburgh, went ashore at Leith Harbour, King Edward Point and Grytviken from Her Majesty's Yacht *Britannia* during her circumnavigation of the world. Upon entering King Edward Cove he was received with a twenty-one gun salute fired by whaler harpoon guns charged with blackpowder.

The Gentoo is one of eight species of penguin inhabiting South Georgia's coastline. The breeding season is March to November (the southern hemisphere's winter and spring).

accommodate the easterly movement of water which encircles the whole of the Antarctic continent.

It is through the Drake Passage that the westerly winds are forced to blow their hardest. Compressed by the Andes to the north and the mountains of the Antarctic Peninsula to the south, the winds are often accelerated to hurricane force. This is the one place on the surface of the world where wind and ocean can make an unobstructed circumnavigation. The Drake Passage is a constriction which generates the most violent weather to be found anywhere. The winter gales driving across the surface of the ocean create swells of a size unique to the area, as much as eighty feet in height and half a mile from crest to crest. It is easy to understand why Cape Horn carried such a fearsome reputation in the days of the square-riggers.

South Georgia is 1,200 miles due east of the Drake Passage. The island sits directly in the path of the weather emerging from that narrow channel. Fortunately it is sufficiently distant to escape the full effects but, even so, it is clear that the local climate is normally harsh. The average air temperature hovers only slightly above freezing. Winter or summer, the island is cold, wet and windy. Brilliant sunlit days of great beauty arrive but rarely.

Storms sweeping in from the west can bring winds of exceptional power (more than 140 miles per hour). Even on a quiet day, a steady 10 miles per hour wind is normal on the sheltered low ground. However, the weather found at sea-level on the protected north-east coast may be greatly different to that found, on the same day, on the upper slopes of the exposed south-west coast. The great height of the mountains generates sudden localised changes in atmospheric pressure, humidity and temperature which create *föhns* – high-speed warm winds – which sweep down into the valleys. They melt snow and ice, causing dangerous avalanches and the collapse of ice pinnacles on the high ridges.

Another feature are the 'williwaws' – violent whirlwinds or downdrafts – which descend without warning from the high ground to sea-level. Powerful unpredictable turbulence – especially around the north-east anchorages – is a hazard for mariners and aviators alike. The degree of violence of wind and sea multiplies rapidly as one moves away from the leeward shore of the island. Conditions at sea are frequently very different to those found at King Edward Point (where meteorological observations were recorded regularly until 1982).

The peculiarities of the island's climate were neatly summarised by the Duke of Edinburgh who went ashore at Grytviken, King Edward Point and Leith Harbour in 1957 while cruising the Antarctic in the royal yacht. In his subsequent book, *Birds from Britannia,* he wrote: 'It is difficult to believe that South Georgia, with its snow-covered mountains, treeless landscape and huge icebergs floating in its bays, is the same distance south of the equator as Manchester is north of it. The Gulf Stream does not always get the appreciation it deserves.'

The upper slopes of the steep-sided mountains are subjected to severe weathering. Scree and moraine are found everywhere in great quantities, the surface being unstable and treacherous underfoot. On the lower ground these screes have been further reduced to form gravel deposits and sandy loams which support areas of tussock grass, rushes, mosses, and a few perennial herbs. There are no trees on the island and former attempts to introduce them have failed.

Melt water from the glaciers feeds a number of freshwater streams and there are many lakes, ponds and patches of peat bog. The high precipitation sustains areas of pasturage, particularly on the lower slopes of the north-east coast. The most useful grass is tussock, its high nutritional value deriving from the guano of the large bird colonies and, in some places, from the faeces of the seals. Norwegian rats, imported accidentally by the early sealing vessels, have adapted to local conditions and survive largely upon a diet of tussock grass (although they do consume thousands of eggs and fledgelings during the sea-bird breeding season). There is even an account of a shipwrecked mariner surviving – on the Falklands – on a diet of tussock grass while awaiting rescue. The scanty grazing also supports two thousand reindeer, imported from Norway in 1911, which roam small areas of the island in three separate herds.

There are twenty-nine breeding species of bird, all but five of them being sea-birds. Apart from the spectacular Wandering Albatross, with it wingspan of nearly fourteen feet, the most

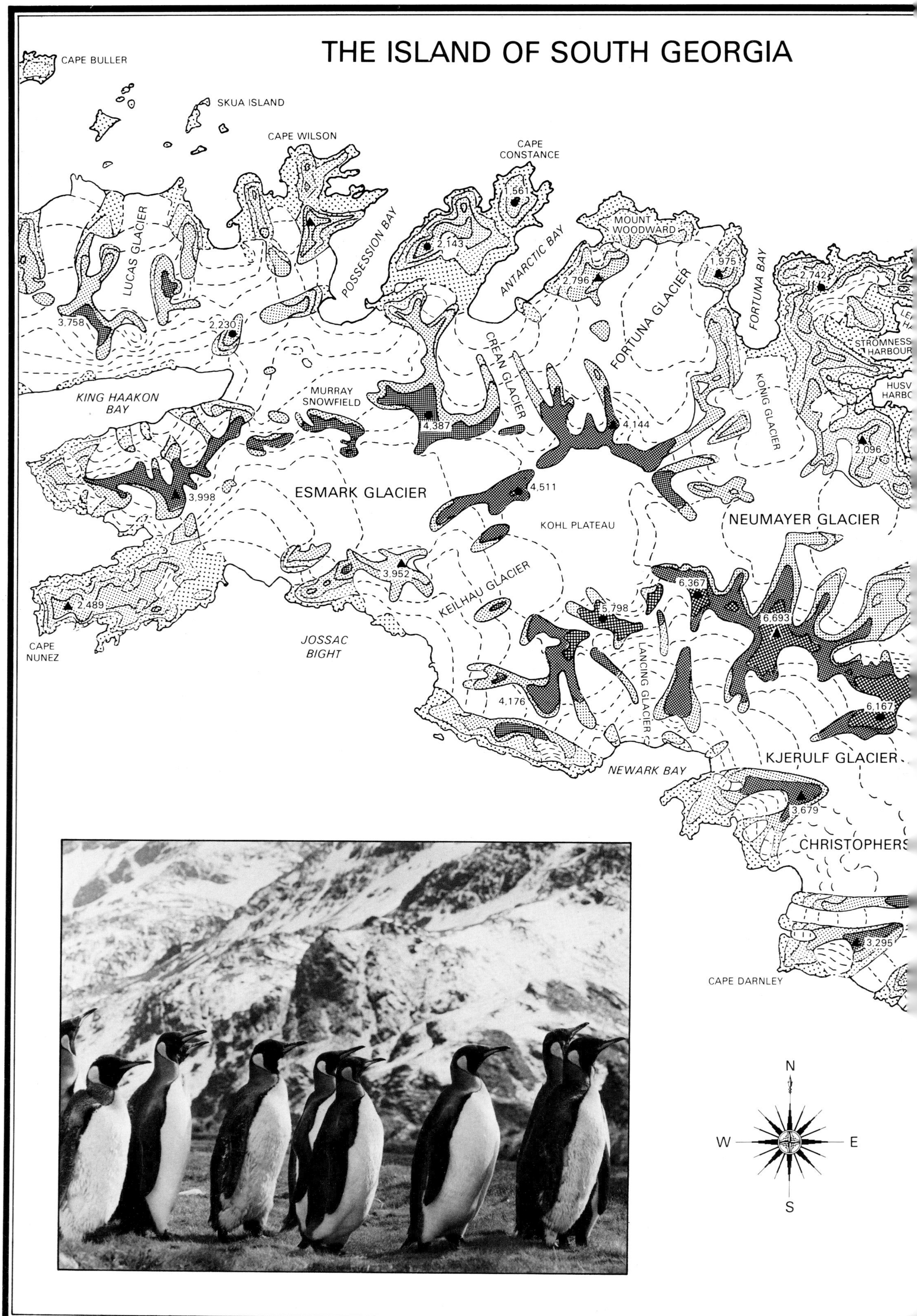
THE ISLAND OF SOUTH GEORGIA
CAPE BULLER
SKUA ISLAND
CAPE WILSON
CAPE CONSTANCE
1,561
MOUNT WOODWARD
LUCAS GLACIER
POSSESSION BAY
2,143
ANTARCTIC BAY
2,796
1,975
FORTUNA BAY
2,742
FORTUNA GLACIER
3,758
2,230
CREAN GLACIER
STROMNESS HARBOUR
KING HAAKON BAY
MURRAY SNOWFIELD
4,387
4,144
KÖNIG GLACIER
2,096
3,998
ESMARK GLACIER
4,511
KOHL PLATEAU
NEUMAYER GLACIER
3,952
KEILHAU GLACIER
6,367
2,489
5,798
6,693
CAPE NUNEZ
JOSSAC BIGHT
LANCING GLACIER
4,176
6,167
NEWARK BAY
KJERULF GLACIER
3,679
3,295
CAPE DARNLEY
N
W
E
S

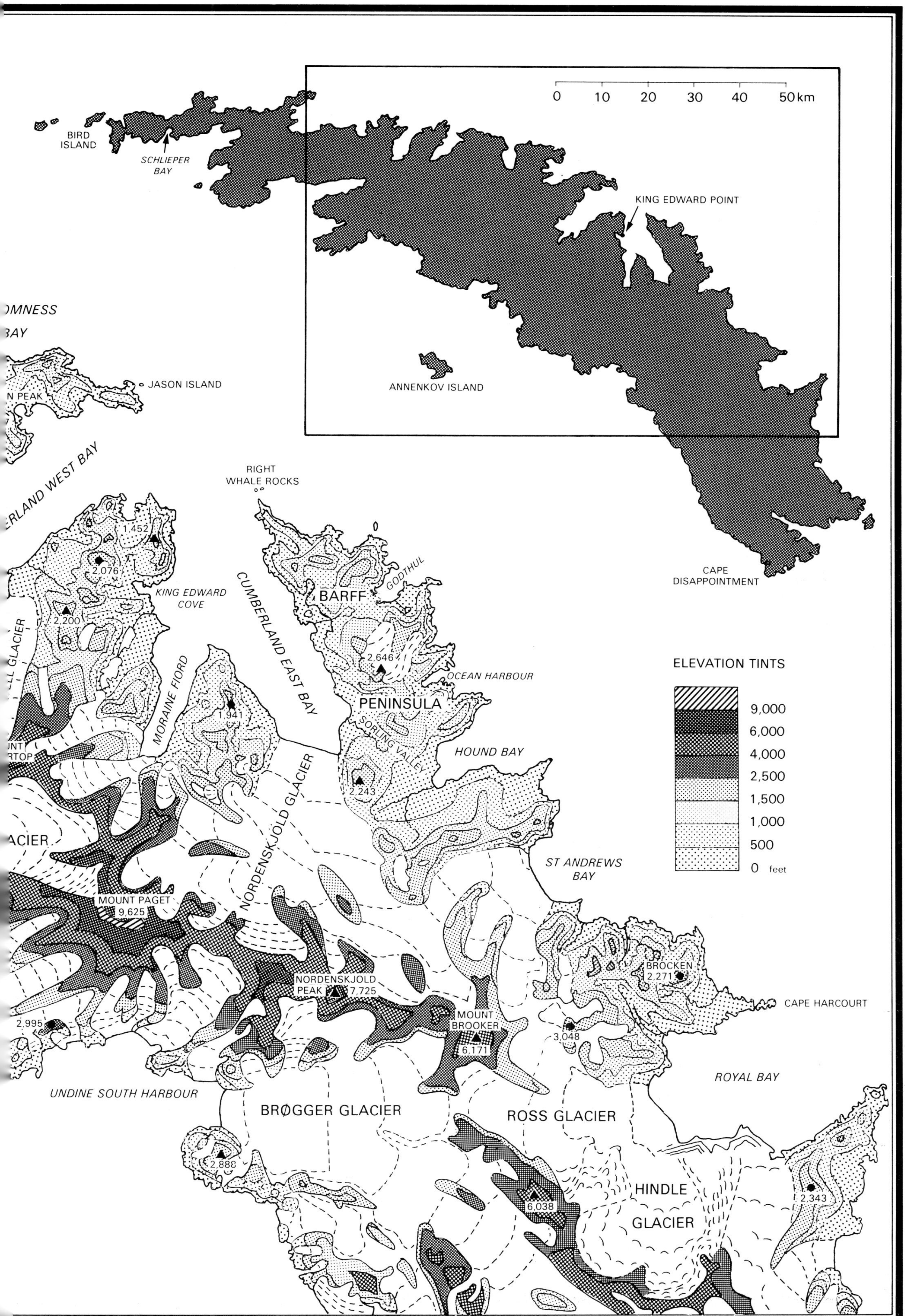
0 10 20 30 40 50km
BIRD ISLAND
SCHLIEPER BAY
KING EDWARD POINT
ANNENKOV ISLAND
CAPE DISAPPOINTMENT
JASON ISLAND
RIGHT WHALE ROCKS
KING EDWARD COVE
CUMBERLAND EAST BAY
BARFF
GODTHUL
PENINSULA
OCEAN HARBOUR
SORLING VALLEY
HOUND BAY
MORAINE FIORD
NORDENSKJÖLD GLACIER
ST ANDREWS BAY
MOUNT PAGET 9,625
NORDENSKJOLD PEAK 7,725
MOUNT BROOKER 6,171
BROCKEN 2,271
CAPE HARCOURT
ROYAL BAY
UNDINE SOUTH HARBOUR
BRØGGER GLACIER
ROSS GLACIER
HINDLE GLACIER
1,452
2,076
2,200
2,646
1,941
2,243
2,995
3,048
2,888
6,038
2,343
ELEVATION TINTS
9,000
6,000
4,000
2,500
1,500
1,000
500
0 feet

The elephant seal is the largest of all seal species. Some adult males are nearly twenty feet in length and weigh more than four tons. Weighing one hundred pounds at birth, the pups (above) quadruple their size in the first three weeks of life. Although almost totally inactive when ashore for the breeding season, elephant seals spend the rest of the year at sea and at least one individual tagged by scientists on South Georgia has been identified on the coast of South Africa, nearly three thousand miles away.

characteristic birds on South Georgia are the penguins. Eight species are present, but the commonest are the King, Gentoo and Macaroni penguins (the latter having five million breeding pairs in various colonies, particularly on the Willis Islands off the north-east coast).

The success of these birds depends upon the rich surrounding marine life. The sea contains large quantities of the small fish and krill which are essential for all of the local varieties of sea-bird, but there are in addition thirty-three larger types of deep-water fish (notably the Antarctic cod). East European trawlers and factory ships have exploited these stocks and, in the 1970–71 season, they took half a million tons of fish from local waters. Recent catches have been much lower and this resource is being seriously depleted.

South Georgia was at one time the world's largest whaling centre. Up until the turn of the century, the island's coastline and surrounding seas echoed with the voices of the great whales – Blue, Sei, Fin, Minke, Humpback, Southern Right and Sperm – but now they are almost all destroyed. The slaughter of the early twentieth century – conducted through to the 1970s by Russia and Japan – exterminated most of the local population. Lonely surviving families are still occasionally sighted, but the ocean is largely silent.

The decline of the whales has resulted in an increase in the krill population. These small crustaceans, resembling the shrimp, are the staple diet of the whale (excepting the carnivorous Killer and Sperm). As those great mammals are now much rarer in South Georgia waters, the quantity of krill available for other animals has multiplied rapidly and this in turn has permitted a swift recovery in the seal population. Specifically, the Antarctic fur seal was hunted and skinned for over a century but, with the introduction of protection laws and an abundance of krill, it has been able to re-establish itself on many of the island's beaches. It is estimated that there are now (1985) nearly one million breeding cows on South Georgia and the total population is thought to be increasing by ten per cent per annum.

When Captain Cook made his visit in 1775, he was accompanied by a geographer and two naturalists. They later published their observations, and their accounts of the enormous seal colonies attracted the attention of the hunters operating in the Magellan Straits. British sealers arrived on South Georgia in 1786, closely followed by the Americans. By 1800 the colonies were under full-scale attack. As late as 1812 there were still 3,000 men working on the beaches, clubbing and skinning the fur seals. A British sealer, James Weddell, estimated in 1825 that over a million had been killed. Sealing was to continue for many years more, but the devastation was such that the trade became uneconomic.

As the fur seals declined in numbers, so the hunters turned their attention to the elephant seal. These huge animals were killed for their blubber, the extracted oil having properties similar to those of whale oil. Large numbers were killed during the first half of the nineteenth century by gangs of men who went ashore armed with clubs and rifles and large iron pots for boiling the flensed blubber. Many of the pots were left behind and it was the Scandinavians who, upon finding these relics at Cumberland Bay in 1902, gave their name of *gryt* to one of the inlets. It became Grytviken – pot harbour.

The accounts which have survived of elephant seal hunting on South Georgia's beaches do not make pleasant reading. The participants added a 'sporting' flavour to their activities by blinding the frantic bulls with long pointed sticks before slaughtering their harems. The flailing and bellowing of the male seals, vainly attempting to defend their cows and pups, was described by one commentator as 'comical'.

The slaughter of the elephant seals recommenced in 1910. The modern technique was to drive the selected animals down to the beach, kill and flense them, and then tow the accumulated blubber to Grytviken for processing. Licences issued by the Falkland Islands Government set a 'limit' of ninety per cent destruction of the bull seals on any one beach. The overall limit of 6,000 animals to be taken in any one season was supervised by a

Forty per cent of the elephant seal's body is composed of fat and skin. As late as the 1960s, South Georgia's population of three hundred thousand was still being exploited for blubber under licences issued by the Falkland Islands Dependencies administration. Man's weapons are no longer a threat, and these two soldiers of the Royal Regiment of Fusiliers, serving with the post-1982 garrison, received no more than a mild warning from a 'beachmaster' when they approached his harem too closely.

Sealing Inspector and the trade produced about 2,000 tons of oil each year. The last licence for the hunting of elephant seals was issued in 1964.

The exploitation of seals and whales gradually came to an end in the mid 1960s for the simple reason that these activities were no longer profitable. Conservationist groups were becoming increasingly active and effective, but it was money rather than compassion which dictated the course of events.

The currency of the island is the Falkland Islands pound which is on a par with the pound sterling. In practice, however, there being no conventional settlements, the need for money in the usual sense does not exist. At the time of the events described later, financial affairs were handled by the Treasury in Port Stanley and supervised locally by the British Antarctic Survey. Revenue was generated by charges on postal services, port fees and the sale of fresh water to visiting ships.

Politically, South Georgia is a dependency of the Falkland Islands and is therefore a British Colony. Overall administration is the responsibility of the Governor in Port Stanley. Local administration was conducted, until 1982, by the Base Commander of the British Antarctic Survey stationed on the island. By authority of the Governor, this person was also legally empowered to act as postmaster, harbour master, receiver of wrecks, coroner, and in a number of other capacities, with the title of Magistrate.

It can be stated that formal *in situ* British administration of South Georgia dates from 1909 when the first Stipendiary Magistrate, Mr J. I. Wilson, was appointed, and the Civil Station at King Edward Cove became operative. This was however only one step in the history of the island, the logical outcome of previous events stretching back to the late seventeenth century.

Early records are incomplete, but it seems likely that 1675 was the year of the earliest sighting of South Georgia. The discoverer is stated to have been Antonio de la Roché, a merchant captain who, despite his French surname, was an Englishman. He and his crew may have been the first people to see *any* of the sub-Antarctic islands and they could be said, therefore, to have been the discoverers of Antarctica in general.

Some historians, particularly those supporting Argentina's claim to ownership, believe that de la Roché was wrong, that he had in fact sighted Beauchêne Island, 800 miles further west. They believe that it was a Spanish treasure ship, the *Leon,* which first discovered the island. In 1756, eastward bound from Lima to Cadiz with a cargo of bullion silver, she was blown far off course after rounding Cape Horn. The *Leon* first sighted the island early on the morning of 29 June. The feast day of St Peter is 1 July and, as his ship passed its southern tip, the Master gave the name of the saint to the island. Some Spanish and Argentine publications refer to the pace as 'Isla de San Pedro' in preference to the English name.

The next arrival, and the first to set foot on the island, was Captain James Cook, RN, in 1775. He and his people landed at three places around Possession Bay and partly charted the coastline. It was Cook who gave the name Cape Disappointment to the southern tip of the main island when he found that this was

Captain James Cook, RN, the man who first raised British colours on South Georgia. On 17 January 1773, he became the first navigator to cross the Antarctic Circle.

Maritime traffic increased rapidly after the establishment of Larsen's new whaling base at Grytviken in 1904.

A whale processing ship finds temporary haven in one of South Georgia's many deep inlets. Beds of tough kelp weed mark the shallow approaches to the surrounding beach.

not continental Antarctica. His subsequent account described his first cautious approach to the coast: 'the head of the bay . . . was terminated by a huge mass of snow and ice of vast extent . . . just like the side or face of an ice isle . . . pieces were continually breaking from them and floating out to sea. A great fall happened while we were in the bay: it made a noise like a cannon. Not a tree or shrub to be seen, not even big enough to make a toothpick. I landed in three different places, displayed our Colours and took possession of the Country in His Majesty's name under a discharge of small arms.'

The *Resolution* carried a party of Royal Marines – twelve other ranks of the Chatham division commanded by Second Lieutenant James Scott, RM. Scott was described as 'a man of unbalanced mind, suspicious and quarrelsome'. He left the Corps soon afterwards, but one of his party contributed a pleasing historical footnote to the South Georgia story. The ceremonial 'discharge of small arms' mentioned by Cook was the first recorded use of explosive on the island and one of the participants was Corporal Alexander Mills. In 1982, two hundred and seven years later, another Royal Marine by the name of Mills took part in a further display of firepower when he defended the title claimed by Captain Cook.

British and American fur sealers visited South Georgia every season from 1786 through to the 1820s, but the Russians also came for a few days when an Antarctic circumnavigation, commissioned by Tsar Alexander I and commanded by Thaddeus Bellingshausen, brought the *Vostock* and the *Mirny* to the island's shores in 1819. The Russians made a survey of the south-west coast and added more names to the charts. Thereafter the island was left in relative peace for the remainder of the nineteenth century, with only occasional visits by the few fur sealers still at work or by people engaged in research projects of one kind or another.

In 1877, an Austrian visitor, Heinrich Klutschak, made observations which he later published. He was followed in 1882 by a group of German scientists who, working in collaboration with the International Polar Year project, lived for a year at Royal Bay. They made extensive records of the island's geology, biology, meteorology and topography, and prepared detailed maps of part of the hinterland around Royal Bay. This was the first major attempt to define the island's natural characteristics.

In 1887, the Governor at Port Stanley, Thomas Kerr, received an enquiry from a retired naval officer, Captain C. D. Inglis, RN, who wanted to buy the island outright, or at least rent it for ninety-nine years, so that he might produce wool and mutton for export. Kerr passed the request to the Colonial Office with the comment that his administration knew very little about the place other than that it was 'covered in snow to great depths, surrounded by icebergs and fringed with glaciers,' but he saw no objection to renting the whole island to Captain Inglis for a nominal rent if he really did wish to graze his sheep there. In the event, the gallant captain lost interest in the project before London had had time in which to reply.

Fifteen years later another scientific party arrived. Sweden had mounted an expedition to explore the Antarctic Peninsula. In 1902, some of the Swedes made a winter visit to Cumberland East Bay and there made a mapping and geological survey. Commanding their ship was a Norwegian, Captain Carl Anton Larsen, a man with experience of whaling in Arctic waters. Larsen subsequently lost his ship in the Weddell Sea and was rescued by an

Argentine warship, but his short time on South Georgia had sown the seed of a plan to introduce whaling to the Antarctic.

At a dinner given in his honour in 1903 in Buenos Aires, Larsen floated the idea of forming a company - to be based on South Georgia - for the purpose of killing whales and producing oil. Several investors - Argentine, Swedish, German and American - backed the plan and the Compañia Argentina de Pesca was incorporated in Buenos Aires in February 1904. The event was reported in the press and must have come to the attention of the British Embassy in Buenos Aires, but Larsen was not contacted by the British and he did not apply for a permit or licence to set up his factory on the island. By June of 1904 he had assembled a flotilla of three ships, recruited a work-force, and purchased his equipment. By December he was established at Grytviken and was producing oil. The Argentine government meteorological service supplied him with instruments for the establishment of a weather station on the island.

Larsen's men wintered on the island and during this time received supplies from a chartered Argentine naval vessel, the *Guardia Nacional*. Reports of these activities appeared in newspapers and journals in both Argentina and the United Kingdom. Finally the Colonial Office in London started to show interest in the enterprise and correspondence flowed between London, Port Stanley and Buenos Aires. At this point, no doubt aware that the British government was less than happy with the situation, the directors of the Compañia Argentina de Pesca applied to the British Embassy in Buenos Aires, in November 1905, for a lease on the ground occupied by their whaling station.

In the meanwhile, beginning in 1903, another consortium of entrepreneurs was considering the commercial potential of South Georgia. In Chile, this group of sheep ranchers approached the British with the objective of acquiring the island's grazing rights. Later experience revealed that this was not a viable proposition, nor was their plan to search for mineral deposits, but the British did not reject the application out of hand. It was precisely the type of proposal which they had encouraged when, in 1900, they invited tenders for commercial leases on South Georgia's natural resources. After much debate, the Chilean group was granted a lease - on very favourable terms - and they formed the South Georgia Exploring Company (registered in Punta Arenas, Chile) in March 1905. Five months later their advance party, headed by a Yorkshireman, Ernest Swinhoe, arrived at Cumberland East Bay. Inevitably, he clashed with the manager of the Larsen group which, at that stage, had not yet obtained British authority to be there.

The Colonial Office now had a substantial problem on its hands. Larsen's people had been established on the island for many months - without a lease - but had not been told to remove themselves. Swinhoe had a lease, and had made temporary quarters at Hestesletten, but he wanted exclusive rights on the island. Larsen was interested mainly in whales, but also wanted to hunt seals. Swinhoe was interested mainly in pasturage, but he, too, wanted to exploit the seals. Friction was inevitable. The fact that men of so many different nationalities were involved did nothing to improve the prospect of an amicable solution.

Swinhoe complained bitterly to Governor Allardyce in Port Stanley that his carefully negotiated lease was worthless if he could not have the rights which it promised. Meanwhile, and presumably with tongue in cheek, the directors of the Compañia Argentina de Pesca were claiming that they had not realised that the British considered the island to be a sovereign possession and had not understood the importance of applying for a lease. Their formal application may have been prompted by rumours that the Royal Navy had been ordered to encourage their departure, by force if necessary. There was some substance in the reports.

The British government was anxious not to quarrel with the Argentines. Trade between the two countries was important to both sides. The tactics adopted were simple - the dilemma would be resolved best by tact and diplomacy, but the presence of a Royal Navy gunboat should certainly ensure a swift and satisfactory outcome to the discussions. The Commander-in-Chief of the South Atlantic Fleet was ordered to despatch HMS *Sappho* from Montevideo to Grytviken.

The *Sappho* arrived in Cumberland East Bay in January 1906. Her commanding officer, Captain M. Hodges, RN, was warmly received by Larsen's brother - acting as local manager -and shown over the whaling station. Prefabricated houses had been shipped down from Norway, some of the wives and children were in residence, the factory was working at full blast, the place was a hive of efficient activity. A hulk, the *Louise,* lay nearby, laden with a thousand tons of Welsh coal to feed the billowing factory chimneys. No fault could be found with the manner in which the company was conducting itself on the island. Captain Hodges' subsequent report clinched the decision - already made in principle by the Colonial Office - to grant a lease to the Compañia Argentina de Pesca and to persuade the directors of the South Georgia Exploring Company to accept the *status quo*. Reluctantly they did so, having understood too late that the only profitable activity was whaling and having won the right to add this concession to their original lease, but they never caught up with the pioneering Compañia Argentina de Pesca. It is ironic that Swinhoe's Chilean group, which followed all the correct procedures, was out-manouevred by the illegal Larsen enterprise and eventually failed.

The Larsen affair alerted the British to the fact that their posture as owners of South Georgia was based upon perception rather than legal title. True, Captain Cook had gone ashore, one hundred and thirty years earlier, and had raised the Union Flag, but this did not in itself constitute possession in perpetuity. There had been little attempt to administer the island, even less to colonise it. The British and American sealers who erected their shanties at the turn of the nineteenth century were a lawless crowd who owed allegiance to nobody but themselves. Then, from the end of the fur sealing boom, for eighty years, the island had been neglected by mankind in general and by the British in particular.

South Georgia was vaguely regarded throughout those years as being within the jurisdiction of the Falkland Islands government. The latter was established by Royal Letters Patent in 1843. Provision was made in the Letters for the administration of the Dependencies, but these were not specified by name. Ownership of South Georgia, in other words, had never been claimed in detail.

Fortunately, when the *Falkland Islands Gazette* of 1900 carried the offer of South Georgia leases, nobody had raised any serious objection. Similarly, when leases were granted to the two Chilean and Argentine companies, in 1905 and 1906, there was no protest by either of those governments (nor indeed by any other nation). There was further acknowledgement of *de facto* British title when, in 1908 and 1917, South Georgia, the South Shetlands, the South Orkneys and Grahamsland were declared, by Letters Patent, to be the Dependencies of the Falkland Islands.

It was not until 1927 that the British position was challenged. In that year, the Argentine government notified the International Postal Union that South Georgia and the South Orkney Islands should be regarded as Argentinian territory. This seems a curious method of raising such a claim, and, as Argentina had been an independent state since 1853, it was a very late stage in the game

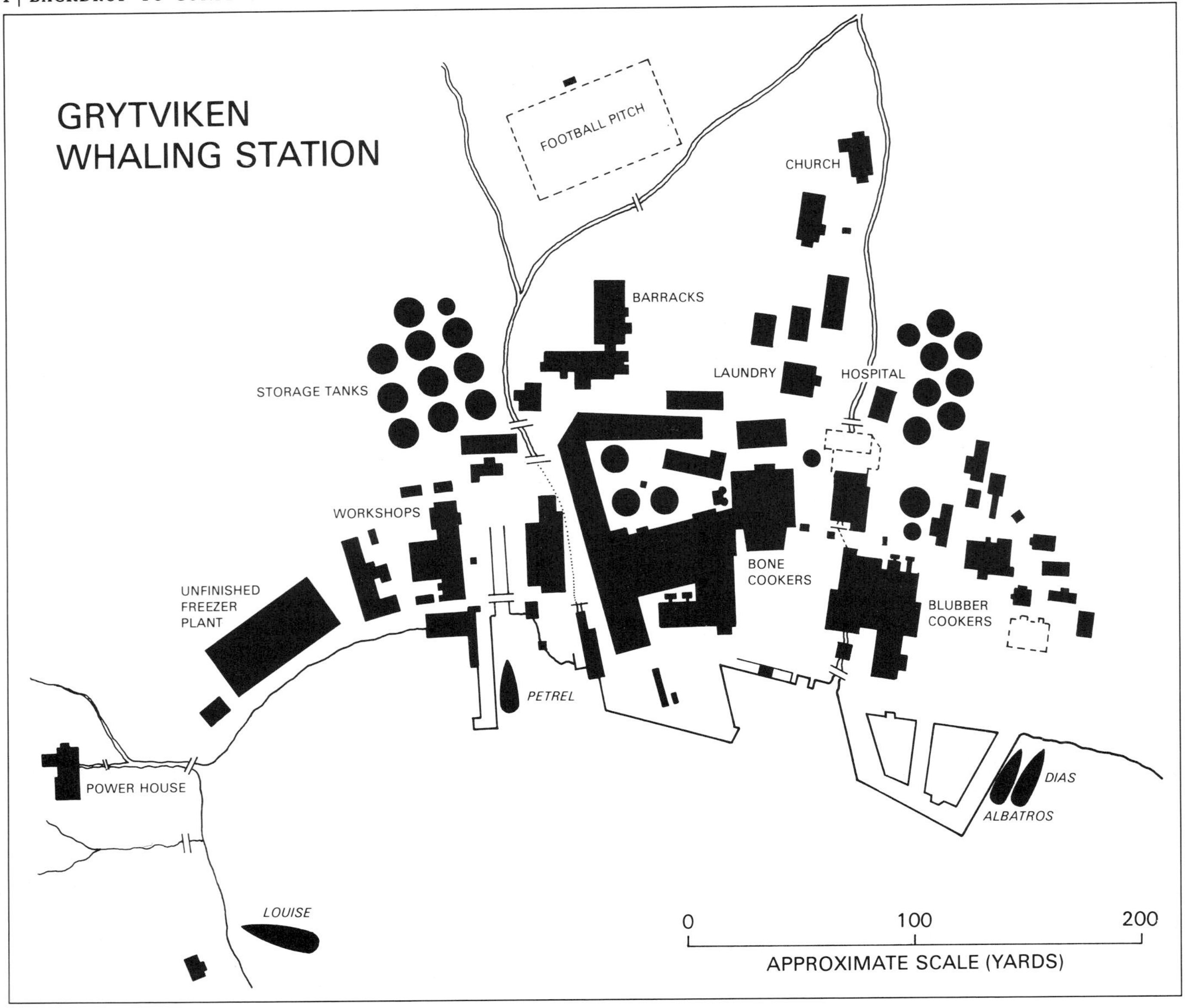

In 1982, the ghost town of Leith Harbour became a focal point in the dispute between Argentina and Great Britain. Forty-four years earlier, when this photograph was taken, the place had been a hive of activity. Hundreds of men and dozens of ships were engaged in the highly efficient and still profitable business of whaling. Most of the vessels moored at the jetties in this picture were the small handy 'catchers', numbers of which were pressed into service as convoy escorts and rescue ships during the war of 1939–45.

Sir Ernest Shackleton, portrayed by the artist R. G. Eves in 1921. Shortly afterwards the explorer departed for the south in his new ship, the *Quest*. On 5 January 1922, while she was moored in King Edward Cove, he suffered a major heart attack. He died almost immediately and his body was taken ashore to be embalmed in the Grytviken whalers' hospital. It was planned that his remains should be taken back to England for burial and the coffin had been conveyed as far as Montevideo, capital of Uruguay, when instructions arrived from Lady Shackleton that her husband should be interred on the island with which his name is forever associated. Shackleton was laid to rest in the Grytviken whalers' cemetery on 5 March 1922 and the present granite headstone erected in 1928.

at which to be raising it. In the meantime, much had been happening around South Georgia to explain the tardy Argentine enthusiasm for the place.

News of Larsen's success during his first whaling season attracted interest throughout the world. Applications for licences, issued by the Falkland Islands government, poured in as other countries sought to grab their share of the South Georgia spoils. The growth of the whaling industry accelerated rapidly as new methods were perfected and extensive use made of factory ships (otherwise known as pelagic whaling). The catchers were obliged to roam further afield as the stocks of whales in the local waters were depleted, but the fleets still needed shore bases for repair, maintenance and storage purposes. Between 1904 and the years following the Great War, bases were established and developed at Grytviken, Stromness, Leith Harbour, Husvik, Ocean Harbour, Prince Olav Harbour and Godthul. Further leases were granted in respect of Rosita Harbour and Jason Harbour, but these were never developed.

It is recorded that 175,250 whales were processed at South Georgia land stations between 1904 and 1965. This compares with the 1,432,862 animals taken from the whole of Antarctica between 1904 and 1978 (when hunting of the larger species ceased). It is indicative of the scale of industrialisation of this business that the work-force at Grytviken could flense and process a blue whale – 90 feet in length and weighing 150 tons – in little more than an hour. In the early years, the oil and by-products from such an animal made a profit of £2,500 for the whaling company. The potential for profits was reflected in the seventy per cent dividend declared by Larsen's company on its first year of trading.

The activities connected with the whaling fleets brought more and more maritime traffic to the island. The first official despatch of mail to Port Stanley from King Edward Point was made by the new Magistrate on 20 December 1909. Good communications and the presence of skilled men and workshop facilities made the place a useful port of call for expeditions headed to or from continental Antarctica. In 1911 a German party, southward bound, spent two summer months ashore, making scientific observations and leaving a memorial cross to one of their officers who vanished while off shore in a small boat. There were several other important expeditions during this period.

On 4 August 1914, Great Britain went to war against Germany. On that same day, as the troops marched to the docks to embark for France, a small ship sailed from England's shores for a far more distant destination. The ship was the *Endurance* and she carried a man whose name was to become synonymous with that of South Georgia. He was Ernest Henry Shackleton.

Born in Ireland in 1874, he was a mercantile marine officer with wide experience in sail and steam. Thirteen years earlier, in 1901, Shackleton had been with Captain Robert Scott, RN, when the latter made his first exploration of the Antarctic continent. In 1907, with the sponsorship of the Royal Geographical Society, Shackleton had returned with his own expedition. There were two main objectives. One party would make geomagnetic studies and seek to establish the position of the true magnetic South Pole, while Shackleton himself, with three others, would attempt to be the first men to reach the geographic South Pole.

The first party was successful, but Shackleton was forced to turn back, after a truly heroic effort, when only ninety-seven miles from his goal. His tantalising failure earned him a knighthood and blazed the trail for others to follow.

Two new major efforts were mounted. A British group led by Robert Scott and a Norwegian expedition led by Roald Amund-

Grytviken, photographed in late 1914 by a member of Ernest Shackleton's British Imperial Trans-Antarctic Expedition. The ship on the extreme right is the *Endurance*, a vessel fated to be lost in the Antarctic ice only a few months later. Alongside her is the coal hulk *Louise*, a relic which still occupies the same berth today.

Sir Ernest Shackleton's *Endurance*, caught in the ice of the Weddell Sea. Between 19 January and 27 October 1915, she was carried two thousand miles towards the Antarctic Peninsula before finally surrendering to the pack's crushing force.

sen competed for the prize. Amundsen was the victor, Scott and his companions the tragic losers.

With the title of 'first to the Pole' now held by the Norwegians, the only prize still unclaimed was 'first across the Antarctic'. In 1913, Shackleton announced his intention of sledging from the Weddell Sea to the Ross Sea, a distance of 1,800 miles. By December of 1914, the *Endurance* was ready to leave South Georgia for her final approach to the Antarctic mainland. The whalers warned the British adventurers that pack-ice conditions in the Weddell Sea were severe. They were correct. For nearly six weeks the *Endurance* tried to find a way through to Vahsel Bay before she became totally trapped in solid ice. Shackleton never even reached the shore. There then followed one of the most harrowing sagas in the history of polar exploration.

For nine months the *Endurance* drifted in the unbreakable grip of the Weddell Sea ice, swirling gently from east to west and then around to the north. Gradually her hull was squeezed to destruction. Shackleton and his people salvaged gear and food from their dying vessel and set up camp on the unstable floes. Most importantly they had with them three whaleboats taken from the *Endurance* before she finally foundered.

Six more months went by, the weary men surviving by sheer willpower on a shifting treacherous foothold of pack-ice. It was too dense for them to launch their boats and the transit of the pack too swift for them to walk across it. They were helpless, waiting upon events. Only on 15 April 1916 did they succeed in getting ashore on Elephant Island in the South Sandwich group. The physical condition of the party was by now much deteriorated.

Shackleton had so far not lost a single man, but he knew that their course was almost run. He decided upon the desperate venture of taking one of the boats, named the *James Caird*, and sailing the 800 miles to South Georgia (the only place of human habitation within reach). Choosing five men as companions, he set out, in a boat only twenty-two and a half feet in length, to cross this violent stretch of open sea. After sixteen days of storm and starvation, cold and exhausted, they landed on the north-west coast of South Georgia at King Haakon Bay.

The *James Caird* was upturned on the beach and two sick men left there in its shelter with a third to care for them. Shackleton, with the two fittest survivors, Thomas Crean and Frank Worsley, then climbed and stumbled over the icy plateaux and mountains which separated them from the whaling station at Stromness, more than twenty miles away. The ground rises to 4,000 feet and had never previously been seen by man. In less than thirty-six hours the starving trio reached safety.

Efforts to rescue the others at King Haakon Bay and on Elephant Island were immediately set in train. After many frustrations, Shackleton finally reached Elephant Island in the Chilean government steamer *Yelcho* on 13 August 1916. From the day of abandoning the *Endurance* to the day of their rescue, the Elephant Island party had suffered ten months of exposure to the most hostile environment on earth.

Despite all of Shackleton's previous experience in the Antarctic, despite the most careful preparation and pre-planning, and despite his superb qualities as a leader of men, he had been powerless in the face of the natural difficulties. Courage and fortitude brought him and his party through their long ordeal, but these qualities were not sufficient to ensure that they would achieve their objective. In microcosm, the same story was to be repeated during the campaign of 1982.

Like many others, before and since, Ernest Shackleton was enchanted by the vastness and pure beauty of the Antarctic. It was the only place that he wished to be. In 1921, having recovered his health and raised more funds, he sailed again for the South. But he had driven himself too hard. Arriving at Grytviken, he suffered a heart attack and died at the early age of forty-seven. He was buried in the cemetery near the whaling station, his body remaining on the island where he had completed one of the most remarkable feats of human fortitude.

After Shackleton, exploration of the region became less a matter of personal initiatives and more a question of national policy. In 1917, the Colonial Office set up a committee to study all aspects of the whaling industry, scientific research and commercial exploitation. Named after Robert Scott's ship, the Discovery Committee produced a large number of reports. They were based in part upon material gathered by ships of the Royal Navy, but the bulk of the work was done by the Royal Research Ships *Discovery*, *William Scoresby* and *Discovery II*.

Throughout the 1920s and 1930s a number of British, German and Scandinavian expeditions added to the fund of knowledge relating to South Georgia. Their work was interrupted by the outbreak of war in 1939, but the island still remained a focus of interest. With the war only a few weeks old, Great Britain scored a major success over Germany when the pocket-battleship *Graf Spee* was chased into Montevideo harbour, Uruguay, and there obliged to either scuttle herself or fight her way out. Captain Langsdorff chose to scuttle. The ships which drove him to this extremity were the cruisers *Exeter*, *Ajax* and *Achilles*. It was the *Exeter* which received most of the *Graf Spee*'s fire, sixty-one of her company being killed in action and many moore wounded.

The elderly whale catchers *Albatros* and *Diaz* still rest at Grytviken, symbols of a long-dead industry. Both remained afloat until 1980, but vandals from a visiting ship then scuttled them by opening the sea-cocks.

Following the battle, and after a brief call at Port Stanley, HMS *Exeter* limped eastward to the shelter of Cumberland East Bay. The shore facilities at Grytviken were superb and exceeded anything available at Port Stanley. This was in effect a complete and self-sufficient shipyard. Apart from a full-time staff of marine engineers and skilled tradesmen, there were machine shops, lifting gear, plate rollers and presses, stocks of engine parts, cable and chain, indeed everything needed by an injured ship trying to make good her wounds. There was also an excellent whalers' hospital where some of the *Exeter*'s casualties were treated.

In 1940, most of the whaling fleet operating in the Southern Ocean was mobilised for operational duty. Many of the ships were requisitioned by their national governments and employed in other waters as Allied naval auxiliaries. The majority were subsequently sunk while on active service. A small fleet remained behind, however, to maintain the production of whale oil. It was regarded as essential for certain industrial processes. This fleet worked from Grytviken and Leith Harbour, but the other South Georgia stations were closed for the duration of the war. The defence and supply of the 700-strong work-force on the island during the war years was the responsibility of the Royal Navy.

With the return of peace, and with an urgent need in Europe for all forms of oil and protein, the British, Norwegians, Russians and Dutch quickly built new fleets of catchers and mother factory ships to resume the hunt for whales. The new generation of vessels employed the technology developed during the war to hunt submarines: radar, sonar, echo-sounders and aircraft. In 1946, a voluntary association of whaling nations, the International Whaling Commission, agreed a self-imposed system of conservation (but without acting as a legislative body).

The modern fleets were self-supporting and no longer required the shore facilities of South Georgia to the same extent as in the past. Interest in the island therefore tended to veer increasingly towards pure scientific research. Indeed, soon after the Second World War, the Admiralty passed its responsibility back to the Colonial Office which then established the Falkland Island Dependencies Survey. In 1962 it was renamed the British Antarctic Survey, and this organisation was still very actively managing the day-to-day affairs of the island, on behalf of the Governor in Port Stanley, until March of 1982.

Apart from the British Antarctic Survey, various expeditions have visited South Georgia over the past forty years. Niall Rankin surveyed the bird population in 1946–47, the South Georgia Survey led by Duncan Carse conducted extensive mapping of the hinterland during the four summer seasons of 1951–57, Bernard Stonehouse and Nigel Bonner studied birds and seals in 1953–55, and George Sutton led the British South Georgia Expedition on a mountaineering visit in 1954–55. Sir Vivian Fuchs spent a short time on the island in 1956 in connection with his Trans-Antarctic Expedition and a laboratory was built to study glaciology, seismology, geomagnetism and similar subjects as part of the International Geophysical Year in 1957–58. There was another climbing expedition in 1961 when a party of Royal Marines from HMS *Protector*, led by Captain V. N. Stevenson, RM, made an attempt on Mount Paget.

In 1964, a Combined Services Expedition, led by Lieutenant-Commander M. K. Burley, RN, made a further attempt on Mount Paget and successfully retraced Ernest Shackleton's 1916 route from King Haakon Bay to Stromness. Burley's venture was to have an unexpected influence upon the conduct of Operation Paraquat in 1982. □

II | Overture to Battle

THE 1982 BATTLE for South Georgia was the consequence of a series of events stretching back over many years. In order to appreciate the factors - political, diplomatic and economic - which created the eventual conflict (and even in part determined the manner in which the battle itself was fought), it is appropriate to know something of the history of the aggressor nation.

Argentina is the second largest country in Latin America. It has its roots in the Treaty of Tordesillas of 1494, an agreement endorsed by the Spanish Pope, Alexander VI, and intended to ensure Spanish dominance in the new lands discovered by Christopher Columbus. It was followed by the voyages of discovery around South America by the Spanish and Portuguese navigators of the early sixteenth century - Cabral, Vespucci, de Solis, Magellan, Cano, Mendoza, and others. They were followed by the colonisers, men in search of treasure and territorial conquest. Portugal ocupied the central eastern coast (later expanded to become Brazil) while Spain conquered nearly all of the rest of the continent. In the south they found huge areas of savanna occupied sparsely by the Tehuelche Guarani Chulupi indians. There was no silver or gold here, no great centres of ancient civilisation to be sacked and plundered, so the region at first attracted little interest. Later, however, the land was settled and Spanish influence spread rapidly over the whole territory. The bulk of the European colonisers settled in the area of the River Plate, but huge *estancias* covered the hinterland. The native Indians were almost entirely wiped out during the sixteenth and seventeenth centuries.

The French and Napoleonic Wars of 1793–1814 brought great changes in Latin America. By late 1805, despite the triumph off Cape Trafalgar, the British found themselves facing the full power of the Napoleonic Empire in almost every quarter of the globe. Napoleon had established a strong presence in Egypt, and evidently hoped to throw the British out of India and the East Indies.

With the Middle East so strongly threatened, it was vital to the British that they should protect the sea route around the Cape of Good Hope. To this end, an expedition led by Commodore Sir Home Riggs Popham was sent from England to seize the Dutch settlement in Table Bay. The task force arrived on 4 January 1806 and, after a brief struggle, the Dutch capitulated.

Popham looked around for further ways of employing his force. Without waiting to consult London, he and General William Beresford settled upon Buenos Aires as being a legitimate target. It seems not to have occurred to them that they should first investigate the political implications of their otherwise purely military action. They apparently did not know that the population of the Spanish American Viceroyalty of the Rio de la Plata was looking for moral and material support in its ambition to break away from the control of Napoleon's puppets in Madrid. When Popham and Beresford landed at Buenos Aires in June, 1806, they and their 1,600 men should have been welcomed as allies and liberators. Instead, the British treated the local people as enemies. After a short sharp fight, in which the British lost one man killed and twelve wounded, the town was occupied. The Treasury was looted, Beresford was richer by £30,000, Popham by £6,000 and each private soldier by £18.50.

Two months later, Beresford and his men were forced to surrender to the enraged local Spanish and colonial levies and were imprisoned in various small towns in the interior. The British government was furious when news of all this reached London. Popham was recalled to England for court martial, but national honour demanded the sending of another task force to release Beresford.

A second and larger force, commanded by General Sir David Baird, arrived in October. It landed at Maldonado, in present-day Uruguay. The landing was contested by the Spanish, but they were quickly put to flight. Again, the political initiative was lost when the victorious British troops subjected the town of Maldanado to three hours of drunken pillage.

Encouraged by their local success, the British government sent a third force, led by General Sir Samuel Auchmuty, who captured Montevideo on 3 February 1807 for the loss of one hundred British and eight hundred Spanish dead. A few months later, on 5 July, Buenos Aires again was attacked, this time by a force under the command of General John Whitelocke. The assault was poorly planned and the British were beaten and forced to surrender. The subsequent negotiations led to an agreement for the total withdrawal of all British troops in the vicinity of Buenos Aires, Montevideo and Maldonado. So ended Great Britain's last attempt at conquest on the mainland of South America.

The fact that an ill-trained local militia army had defeated a professional European force lit the fuse of rebellion. There were less than six hundred thousand people living in the area of the Rio de la Plata at that time, and many of these were Amerindians and plantation slaves, but the powerful landowners were tired of paying taxes to the government in Madrid. They seized the opportunity provided by Europe's preoccupation with its own troubles to strike out for independence. Open revolt commenced in 1810 and within six years the long-standing Spanish link had been broken.

During the critical years, between 1807 and 1823, there was no direct diplomatic contact between Buenos Aires and London. After Spain became allied to Britain in the fight against Napoleon, the British found themselves in a curious position. As a colonial power in their own right, they had no wish to appear as supporters of any kind of revolutionary movement. In theory, they could have helped the Spanish to suppress the revolt, but the experiences of Beresford and Whitelocke had shown this to be impractical. A delicate balancing act was needed.

Following the final defeat of Napoleon at Waterloo and the restoration of Ferdinand to the Spanish Throne, Spain attempted to recover her position in Latin America. Fleets and armies were despatched across the Atlantic, but the task was too great, the continent too large, for any one nation to succeed. Spain could have regained her domination only if the British had aided her, and this they certainly would not do. By now they had a vested interest in the area. Ships of the Royal Navy had controlled the Southern Atlantic during the final years of the war and had allowed English merchants and bankers to gain a foothold in the commerce of the new nation. The United Provinces of Rio de la Plata needed capital and technology, the British needed hides, tallow and sugar. Trade prospects for both parties were excellent. In 1825, a treaty of friendship was signed by Great Britain and the United Provinces, affirming the communality of their interests.

Five remarkable officers of the Royal Navy succeeded each other during those years as commanders of the South American Station. The manner in which they walked the diplomatic tightrope is a prime example of the economic and diplomatic value of a global navy. They were Admirals Sir William Sidney Smith, Michael de Courcy, and Manley Dixon, and Commodores William Bowles and Thomas Masterman Hardy. With no resource other than their own commonsense and skills in advocacy, they fostered the climate in which a stable economy could evolve. Their ships protected the new republic from invasion and policed the growing maritime commerce. That the future nation of Argentina was able to survive its infancy and establish the trade needed for economic growth was due in large part to the Royal Navy.

Juan Domingo Perón, the army general who became nationalist dictator of Argentina in 1946 and ruled that country until exiled by his former brother officers in 1955.

The early years were marred by strife between rival internal factions, but stability was enforced when the first of several dictators, Juan Manuel Rosas, came to power in 1829. A federal constitution for the new country of Argentina was adopted on 25 May 1853 and, apart from occasional wars with her neighbours, she continued to build her strength throughout the nineteenth century. New farming methods produced huge surpluses for export. Many of the ranches and farms were owned or managed by British families and their descendents. Docks and railways were financed and constructed by British firms. Economic and cultural ties between the two countries were a model of their kind. Until quite recent times, in some smaller communities in Patagonia, Welsh and Gaelic were still to be heard, and many families in each country can count cousins in the other.

By the turn of the century the national population had grown to nearly five million, but wealth and power were concentrated in the élitist hands of the military, the landowners and the large companies. The majority – Amerindians, landless peasants, and a fast-growing number of Italian immigrants – had a low standard of living and little control over their own destinies.

Rising social aspirations, and a population expanding swiftly toward the current level of twenty-five million, led to growing unrest and clashes between authority and organised labour. The confrontation hardened during the 1920s and resulted in a series of military dictatorships which ran the country from 1930 until 1945. These men and their supporters admired the political philosophies of Hitler and Mussolini and, while declaring neutrality, were basically sympathetic to the Axis cause. Only in March 1945 did Argentina join the Allied cause by declaring war on Germany, but a significant number of Anglo-Argentines had served as volunteers with the British armed forces. Many fought with distinction – as had their forebears in the Great War of 1914–18 – and some were decorated for gallantry.

In 1943, after a revolution on 4 June, a former army officer, Juan Perón, emerged as the country's undisputed dictator. His success sprang from his ability to manipulate the military and the politicians, coupled with the charisma of his wife. A remarkable woman, Eva Perón helped her husband to win over the trade unions by championing the cause of 'the shirtless ones'. Perón did much to modernise Argentina and to broaden the economy so that it would be less dependent upon agricultural exports, but the price was high. To placate the working class, he sponsored overmanned and uncompetitive State industries which absorbed a large part of the nation's resources. Economic problems and the unbridled activities of his secret police led to his fall from power in 1955. He was exiled by the military, and control of the nation's affairs was taken over by a junta of generals.

The next thirty years were characterised by a pattern all too familiar in Argentina's earlier history – alternating periods of political instability and repressive authoritarianism. The tender flower of democracy found it difficult to take permanent root in Argentina's otherwise fertile soil.

It is a classic ploy, on the part of any government which seeks to divert popular attention away from its home-grown problems, to shine the limelight on topics beyond its own frontiers. In the case of Argentina, there have been periods when quarrels with her neighbours Chile, Uruguay, Bolivia and Paraguay have neatly fitted the bill, but the favourite diversions – particularly since the era of Juan Perón – have been her claims to sovereignty over the islands in the Beagle Channel, the Falkland Islands, the associ-

The German armed merchant cruiser *Kormoran*, typical of the well-armed raiders which roamed the South Atlantic and Indian Oceans and presented such a threat to British island possessions in those regions. In November 1941, the *Kormoran* fought and (apparently) destroyed the pride of the Australian Navy, HMAS *Sydney*, with the loss of all hands.

ated islands of South Georgia and the South Sandwich Islands, and a large part of Antarctica.

In the case of the Falkland Islands, there is long-standing resentment that a group of islands only 400 miles off shore should belong to a nation 8,000 miles to the north, and that Great Britain should be using these islands as a base from which to support her interests in Antartica. In Argentina, it is argued that the Falklands are a manifestation of the South American continental shelf and therefore integral to Argentina's natural boundaries. It is further claimed that the islands were settled by Spain and by the government of Buenos Aires during the second half of the eighteenth century and again between 1826 and 1831, thereby anticipating the establishment of formal British administration in 1833. The Falkland Islands have in fact witnessed the arrival of settlers from France, Spain, Great Britain and Latin America at various times since 1764, but the last of the Spanish-speaking colonists left the islands in 1831. The Falklands thereafter became a purely British domain.

The position with regard to South Georgia is even more clear cut, but it came to be one of the major bones of contention between the British and Argentine governments. In his definitive account of the Argentine standpoint, Rear-Admiral Laurio H. Destefani, Head of the Department of Naval Historical Studies of the Argentine Navy, stated in August 1982: 'In the South Georgias, Argentina has rights through her proximity, her neighbouring archipelagos, and because of her being the first occupant of the islands when, on 16 November 1904, the Compañia Argentina de Pesca started operations there, at which time the Georgias were uninhabited.'

The Admiral ignored the fact that South Georgia is 1,250 miles from the mainland, that the Compañia Argentina de Pesca sought and obtained a leasehold agreement from the Falkland Islands government, and that the island was occupied by British and American sealers long before Larsen's people landed at Grytviken in 1904. Further, he failed to mention the fact that, when the first load of South Georgia whale oil was unloaded from the *Rolf* in Buenos Aires docks in March 1905, it was classified and charged by the Argentine Customs Service as being a product imported from a foreign country. But basic facts were to a certain extent irrelevant to the core of the problem – the question of sovereignty came to rest upon a mixture of arguments determined by economic need, territorial expansionism, national pride, envy, factional rivalries, and Latin 'machismo'. Generations of Argentine children were taught that the Falkland Islands and South Georgia belonged rightfully to their nation and they grew to adulthood with this conviction planted firmly in their minds. It is against this background that the events of 1982 must in part be judged.

The quarrel between Argentina and Great Britain can be traced back over many years, but the early rivalry did not develop into something more serious until the Second World War. It was in 1942 that the British government began to worry about the military security of the Falkland Islands and the Dependencies. Germany had set loose a small but highly efficient force of armed merchant raiders – converted mercantile marine ships armed with six to eight 5.9-inch guns, four or more torpedo tubes, and one or two spotter aircraft. Their task was to raise alarm and despondency in every corner of the globe by pouncing on unescorted Allied merchantmen and either sinking them or taking them as prizes. They were eminently successful. Chosen for endurance rather than speed, the first group of seven such ships ranged the oceans and tied down a large number of Royal Navy units which attempted – for months at a time – to find them.

One of these elusive targets was the *Pinguin*. During her year-long cruise in the South Atlantic and Indian Oceans, which took her as far as the approaches to Sydney (Australia), the *Pinguin* sank or captured seventeen British and Norwegian freighters and sank eleven whalers. She replenished herself at sea from her prizes and spent some time sheltering in the Antarctic pack-ice off the coast of Queen Maud Land. She was finally caught and sunk, after a fierce duel with HMS *Cornwall*, in May 1941. The *Pinguin*'s course at the time of her sinking was taking her towards South Georgia. If her captain had raided the island he would have been able to destroy the rest of the whaling fleet and the facilities at Grytviken.

Several other German armed merchant cruisers were causing serious disruption in the South Atlantic and Indian Oceans and were thought to be making use of Kergulen, Crozet and Deception Islands as safe havens. Hiding in the wilderness of the Antarctic pack-ice, these ships were almost impossible to locate. They would have posed an even greater threat if they could have been reinforced and provided with an ice-free base. The Japanese sweep through the Pacific revealed the general vulnerability of all small islands and the difficulties of defending them.

There were several islands in the South Atlantic which offered obvious attractions to the Germans, and the most tantalising were the Falklands group. It would have been impractical to send an invasion force from Germany, but the Nazi Bund in Argentina and Chile had grown to a strength of 100,000. If the Argentine government – in furtherance of its own territorial ambitions – were to have turned a blind eye to the activities of these people, the British might have woken one morning and found that the Falkland Islands, and possibly South Georgia, had been seized. The invasion would have been made in the name of Germany, but the long-term ambitions of Argentina would have taken a giant

Port Stanley, Falkland Islands, 1944. The earlier invasion fears have started to subside and Force 122 is under orders to return to the United Kingdom. On 15 January a farewell parade swings past the saluting base in Ross Road, the salute being taken by Sir Alan Wolsey Cardinall, KBE, CMG, Governor from 1941 to 1946.

One of the 4-inch guns mounted during the Second World War to defend Stromness Bay, South Georgia. Pictured with this rusty relic of a former war is R.K. 'Bob' Headland, one of the British Antarctic Survey scientists serving on the island in 1982.

step forward. Once established, the invaders would have been very difficult to remove.

In June of 1942, Winston Churchill authorised the despatch of a military force to garrison the Falkland Islands. Known as Task Force 122, it comprised 11th Bn West Yorkshire Regiment, 359 Anti-Aircraft Battery RA, a Royal Engineers construction company, a field hospital, and numerous ancillary troops. By mid-August the area of Port Stanley was being turned into an armed camp, Bren-gun carriers were patrolling the 'Camp', and eight Bofors guns guarded the skies. Any immediate attempt at invasion – either by German or Argentine forces – had been thwarted.

On South Georgia, two 4-inch guns were placed on headlands overlooking Stromness Bay and Cumberland East Bay to provide some defence for the whaling stations at Leith Harbour and Grytviken. They were manned by volunteer Norwegian crews who were well accustomed to handling explosives and whaling guns.

Concurrently with the defence of the Falkland Islands, the British set in train a series of measures to watch and contain all German, Argentine and Chilean naval activity in the region of the Antarctic. Codenamed Tabarin, this operation involved constant patrolling by Royal Navy ships around the Antarctic Peninsula of Continental Antarctica and among the offshore islands. Despite the bitter battle being fought out in the North Atlantic – where Admiral Doenitz's U-boats had almost severed the lifeline between the United States and Great Britain – the Admiralty found ships for service in the deep south. Most of the workload was carried by the armed merchantmen HMS *Queen of Bermuda* and HMS *Carnarvon Castle*.

By 1945, three shore bases had been established at key harbours in the South Shetland Islands and on the Antarctic Peninsula. Their function was linked primarily to the provision of meteorological data for the Admiralty but, from the early months of 1944, the personnel serving at these bases were also conducting purely scientific research work. These three bases were to form the post-war nucleus of the Falkland Islands Dependencies Survey (FIDS). Seventeen more similar scientific bases were established in the ensuing years.

Signs of impending conflict with Argentina were apparent as early as 1943. While searching for German raiders along the coast of the Antarctic Peninsula, HMS *Carnarvon Castle* discovered that the Argentines were erecting bronze plaques on various islands, claiming possession for their country. Most of this work was being done by a vessel named the *Primero de Mayo*. She could not be physically prevented from making these landings, but the *Carnarvon Castle* followed on her heels and, whenever plaques were found, took them down again.

The game of cat and mouse was repeated in 1948. An Argentine naval task force set up a permanent base on Deception Island.

HMS *Sparrow*, a 1,350-ton sloop launched in 1946, searches for unauthorised Argentine incursions in Admiralty Bay, King George Island, in the South Shetland group, in 1949. Unlike some other countries, the United Kingdom has never designed and constructed a ship for the specific purpose of naval ice patrol deployments.

Further bases were being established by Chile. The President of Chile visited the area to inspect his country's investment of men and resources. The British government responded by despatching the 8,000-ton cruiser HMS *Nigeria* and the sloop HMS *Snipe* to show the flag. The *Snipe* was later replaced by another sloop, HMS *Sparrow*.

There was a much more serious incident in 1951. In December of that year, a force of six Argentine naval ships arrived with a construction team at Hope Bay, at the northern tip of the Antarctic Peninsula. There were already six Argentine bases on the Peninsula and on the South Orkneys and South Shetlands. The site of this new installation was close to that of a British base – at the time temporarily unoccupied following a serious fire – which had been built by the Falkland Islands Dependencies Survey in 1945. Three months later, when the *John Biscoe* arrived with a scientific party to repair and reactivate the British base, the Argentines drove them off by firing machine guns over their heads. The British were able to land only after a Royal Navy frigate, HMS *Burghead Bay*, was despatched from Port Stanley to Hope Bay, in February 1952, to defuse the situation. The Admiralty then withdrew the frigate from the South Atlantic, believing there would be no more trouble that year.

The generally aggressive conduct of the Argentine navy on this occasion was sufficient to set the alarm bells ringing again in Downing Street. Winston Churchill, newly reappointed Prime Minister, called for a review of plans for the defence of the Falkland Islands. He believed that Argentina was capable of pressing her 'Las Malvinas' claim to the point of direct military action. Having first attained the post of First Lord of the Admiralty in 1911, at a time when the Empire was at the pinnacle of its worldwide power, Churchill understood better than most the value of despatching a gunboat to concentrate the mind of a potential aggressor. Shortly afterwards, in April 1952, the frigate HMS *Veryan Bay* arrived at Port Stanley with a party of thirty Royal Marines. The Foreign Office did not share Churchill's view that there was a direct threat to the Falklands, but it welcomed his intervention because the display of naval power served to underline British interest in the region as a whole.

The Royal Navy's conventional modern warships – with their comparatively thin-skinned hulls – could not safely operate in the ice-covered waters of the deep south during the long winter months. Apart from the occasional despatch of ships such as the *Burghead Bay* and the *Veryan Bay*, there was little that could be done to assist the Colonial Office (responsible for the Falkland Islands and Dependencies) or the Foreign Office (dealing with Latin American affairs).

The pace of events quickened in the mid-1950s. Various overlapping territorial claims on Continental Antarctica contained the self-evident seeds of international conflict. It was fortunate that, at this time, the world's scientific community proposed a third International Polar Year (similar to those of 1882–83 and 1932–33). This proposal led to the International Geophysical Year of 1957–58 which covered the entire globe but which placed special emphasis upon the poles, especially the least known region, the Antarctic. Twelve nations – including Great Britain, Argentina and Chile – either used existing stations, or established new ones, for the venture. Its success led to the proposal that there should be an Antarctic Treaty to embody the new spirit of international co-operation. It was to apply to all land areas and ice shelves south of the 60th Parallel.

The Treaty was drafted in 1959 and, having been ratified by the twelve governments, came into force on 23 June 1961. The preamble declares 'that it is in the interest of all mankind that Antarctica shall continue forever to be used exclusively for peaceful purposes and shall not become the scene or object of international discord.' Under its terms, all territorial claims were to remain in abeyance for the duration of the Treaty. It did not, of course, affect the Anglo-Argentine disputes with regard to the Falkland Islands, South Georgia or the South Sandwich Islands. They are north of the 60th Parallel and therefore outside the Treaty area.

In the Letters Patent of 1908 and 1917, Great Britain had defined her claim, based upon prior discovery and exploration, to the Antarctic territory between latitudes 50 and 58 degrees South and longitudes 20 and 50 degrees West. Also specified was the territory south of latitude 58 degrees South between longitudes 20 and 80 degrees West. These territories were known as Falkland Islands Dependencies (FID), and were administered by the Governor of the Falkland Islands. The permanent scientific bases were manned by the Falkland Islands Dependencies Survey (FIDS), under the control of the Colonial Office.

In March 1962, to facilitate the implementation of the Treaty, that part of the FID which came within the Treaty area was renamed British Antarctic Territory (BAT). A few weeks earlier, on 1 January 1962, the title of the FIDS had been changed to British Antarctic Survey (BAS). This is the title still in use today and it applies to all British government-sponsored scientific bases in the region, including those located to the north of the 60th Parallel.

Although a signatory to the Antarctic Treaty, it is clear that Argentina has followed a consistent policy, over many years, of seeking to strengthen her claim to sovereignty over large tracts of the Antarctic continent and certain off-lying islands. Pregnant Argentine women have been flown down to Antarctic bases to have their babies delivered – thus establishing 'settlement rights' – and on one occasion the Marambio base was declared to be the capital of Argentine Antarctica and a Cabinet meeting held there.

In recent years, many nations have become interested in the economic potential of Antarctica but, to date, no economically exploitable minerals have been found. Although valuable expertise has been developed while working mineral deposits in Alaska and Greenland, conditions in the Arctic do not compare with the intensely hostile environment of the Antarctic. The southern land mass is covered almost entirely by an ice cap nearly 8,000 feet thick. Drilling through the ice is difficult because it is flowing slowly outwards to the coast. Nevertheless, as the rest of the world's mineral resources are exhausted, so will the economics of Antarctic exploitation become less daunting.

By 1966 so many former British colonies had attained independence that the Colonial Office was renamed the Commonwealth Office. The few remaining colonies – Gibraltar, the Solomons, Hong Kong and the Falklands – had no particular wish to become independent or to acquire a new patron.

Two years later, the Commonwealth Office was merged with the Foreign Office to produce a single office of State concerned with all overseas affairs. The new Foreign and Commonwealth Office (FCO) was increasingly preoccupied with the need to build bridges with emergent nations with which, in some cases, Great Britain had no particular historical affinity. There was increasing concern at the missed opportunities resulting from the earlier decision to remain outside the European Coal and Steel Community first formed in 1952. Prompted by a shift in the political climate, the FCO was working hard to facilitate the United Kingdom's entry into the European Economic Community (an ambition eventually satisfied in 1973). By comparison, the problems of the Antarctic and sub-Antarctic appeared very insignificant. They might have received scarcely any attention at all but for the increasing stridency of Argentine propaganda.

On the diplomatic front, the British had started to come under pressure as early as 1960. In December of that year, the United Nations General Assembly adopted Resolution 1514, a broad-ranging declaration on the granting of independence to all colonial countries and peoples. It specifically 'recognised the passionate yearning for freedom in all dependent peoples and the decisive role of such peoples in the attainment of their independence,' and expressed an awareness 'of the increasing conflicts resulting from the denial or impediments in the way of freedom of such peoples which constitute a serious threat to world peace.' This Resolution clearly did not apply to people such as the Falkland Islanders because they had never expressed a 'yearning for freedom', but the passing of the Resolution did give the Argentines a foundation upon which to build their diplomatic case.

Over the next six years the Argentines succeeded in keeping 'Las Malvinas' in the United Nations spotlight to the extent that the Special Committee considering the granting of independence to colonial countries and peoples recommended that the British and Argentine governments should 'proceed without delay... with a view to finding a peaceful solution to the problem.' This recommendation was debated by the General Assembly and approved under Resolution 2065 on 16 December 1965.

South Georgia was from the outset sucked into the vortex of the 'Malvinas' dispute by the misfortune of being labelled a 'dependency' of the Falklands. This was purely a matter of administrative convenience. The Falklands and South Georgia are different from each other in almost every way. However, the island was carried along on the tide of international debate which sprang from a declaration by the people of the Falkland Islands, conveyed to the General Assembly of the United Nations in 1964, that they wished under all circumstances to remain a British dependency. Lord Caradon made the ringing announcement from which the diplomatic wrangles of the next seventeen years were to follow: 'the interests of the inhabitants of these territories are paramount.'

While comforting to the inhabitants of East and West Falklands, the noble lord's message was met with silent indifference by the penguins and elephant seals of South Georgia. The only human residents were the temporary visitors of the British Antarctic Survey and a small force of Japanese and Norwegian whalers, and their reaction to the solemn discussion in New York has not been recorded.

Argentina continued to press the UN committee dealing with 'decolonisation' to rule that 'Las Malvinas' should be 'returned' to Argentine sovereignty. The claimant's cause was well served when, in 1965, Ian Smith's government in Rhodesia made its declaration of unilateral independence. The whole question of Great Britain's continuing rule in distant territories was brought into the full glare of international attention. If Mr Smith, who had fought loyally and bravely for the Crown as a 1940s Spitfire pilot, felt compelled to break away from London's control, then surely this was the moment for Spain to recover Gibraltar, and for Argentina to win her arguments over the Falkland Islands?

Buenos Aires made more noise than Madrid, so the British suddenly found themselves fighting a rearguard action in the South Atlantic. In the event, formal negotiations continued inconclusively for eight more years. The issue was once again dragged into the open for international debate in 1974 when the

United Nations General Assembly approved Resolution 3160. The text referred to 'grave concern that eight years have elapsed since the adoption of Resolution 2065 without any substantial progress having been made in the negotiations.' The text also 'expressed its gratitude for the continuous efforts made by the Government of Argentina ... to facilitate the process of decolonialisation and to promote the well-being of the population of the islands.' There was no reference to the United Kingdom's role in the affair, a sure sign that the British were losing the propaganda war.

In December of 1976, the General Assembly approved yet another Resolution, 31/49, calling upon the two governments to 'expedite the negotiations concerning the dispute over sovereignty as requested in Resolutions 2065 and 3160.' That the skids were being inserted under the feet of the British government became apparent – at least in the eyes of Buenos Aires – when a joint communiqué was issued by the two governments as a direct response to Resolution 31/49. To the joy of the Argentines, the British government agreed to 'hold negotiations from June or July, 1977, which will concern future political relations, including sovereignty [*sic*], with regard to the Falkland Islands, South Georgia and the South Sandwich Islands, and economic cooperation with regard to the said territories in particular, and the South West Atlantic in general.' For the first time, the British had admitted a willingness to discuss the issue for which, little more than five years later, they found themselves obliged to go to war.

The Royal Research Ship *Shackleton*, photographed off the west coast of the Antarctic Peninsula in the 1970s.

Over and above the question of sovereignty on the Falkland Islands, London was implying also that it was prepared to concede sovereignty – at least in part – over South Georgia and the South Sandwich Islands. The natural corollary of such action would be, in the view of some observers, an inevitable concession with regard to the claimed British Antarctic Territories. At the very least, it reduced the credibility of the declared British ambition to participate in future developments in Continental Antarctica.

The following years witnessed a procession of British politicians and FCO negotiators, each of whom for a short while plunged himself into the tangled story of sub-Antarctica. All had one basic dilemma – how best to sustain the principle enshrined in Lord Caradon's 1966 declaration and, at the same time, gracefully get rid of the 'colonial' connotation of the Falkland Islands and the so-called 'dependent' territories?

Gradually the Argentines came to the conclusion that the British had formulated a policy of withdrawing from the South Atlantic in the very near future. It was unfortunate that this impression should have been given by London or received by Buenos Aires. There was genuine concern within the FCO that the wishes of the Falkland Islanders should be respected but, because neither political history nor public opinion ever stand still, it was reasonable to assume that, with time, the islanders would come to view the Argentines in a different light. The large Anglo-Argentine community had thrived and prospered on the mainland for a century and a half. Thousands of families of purely British stock had been happy to live there. If people of British and Latin origins could learn to live and work together in the Argentine homeland, surely they could learn the same lessons on the offshore islands? It was never specified, but the FCO timescale seems to have been in the order of thirty years – a generation. In the words of Sir Anthony Parsons, one of the many British diplomats who became involved in the story: 'The long-term efforts were directed at reconciling the islanders to the idea that they were part of the South American community rather than a distant outpost of the United Kingdom.'

It was with this hope in mind that the United Kingdom government indicated to the Argentine government a willingness to negotiate a compromise on the question of sovereignty. Agreements were signed which granted to Argentina complete control of fuel supplies to the Falkland Islands, the construction of an airstrip near Port Stanley, and the operation of all scheduled commercial flights between the islands and the mainland. These were all loss-making activities, but the Argentines agreed to assume responsibility for them. They had two motives for so doing. Firstly, they wished to demonstrate their good faith and, secondly, they hoped to woo the people of the Falklands and convince them that the rulers of Argentina were not the bogeymen they were thought to be. It was for these reasons that Buenos Aires also committed itself to provide an advantageous programme of medical care and education for islanders willing to travel to the mainland.

As part of its own contribution to the new agreement, the British government agreed to initiate a regular sea ferry service between Argentina and the Falklands. This promise was never honoured, but Nicholas Ridley, the FCO Minister responsible for these matters, did try hard to convince the islanders that a compromise on the question of sovereignty was feasible. The idea which he was propounding was 'leaseback' – an arrangement whereby formal sovereignty passed to Argentina, but with all questions of nationality and constitutional rights remaining unaffected. It was to be a long-term transitional arrangement and perhaps it would have worked.

It was the FCO's hope that these developments might appease those pressure groups – in London, in New York, and in Buenos Aires – which wanted to see the Falkland Islands handed over to Argentina at the earliest opportunity, and the opposing factions – in Port Stanley and in Westminster – which wanted no truck with the Argentines at any price. The 'loyalist' lobby groups proved, in time, to be the most powerful and the most intransigent.

British resolve in the disputed area was tested to the full by two incidents in 1976. The first involved Lord Shackleton, son of the explorer. He had been appointed by the British government to assess the future development of the South West Atlantic region. When he visited South Georgia to evaluate the island's economic potential and to visit his father's grave, he did so as a passenger in HMS *Endurance*. The captain of an Argentine destroyer, the *Storni*, fired a shot across the bows of the Royal Research Ship *Shackleton* under the impression that the peer was on board and in an attempt to prevent the visit.

In the second incident, fifty Argentine service personnel landed on the frozen coast of Thule, an island in the South Sandwich group. This desolate and uninhabited place was one of the Falkland Islands Dependencies. It is of scientific interest to Argentina because it is an ideal location to observe meteorological conditions in the Weddell Sea. There appears to be a direct link between these conditions and the rainfall on the mainland's economically important farmlands. Obviously the Argentines have need for permanent bases on the edge of this sea, but the fact remained that this one was not authorised by the British and constituted a breach of international law.

The Argentines quickly and quietly erected their base, installed a small group of naval technicians and scientists, and sat back to see how the British would react. They had already given several twists to the tail of the ageing lion, now they had given him a really hard prod. The next move might be crucial.

Reaction in London was inscrutable. Several weeks passed before – following a routine visit by HMS *Endurance* – the establishment of the base came to London's attention. A formal protest was lodged with the Argentine government on 19 January 1977, but the matter was never pressed. The British Prime Minister, James Callaghan, and his Foreign Secretary, David Owen, conferred with the Joint Intelligence Committee and were advised that the base was not in itself a direct threat to British interests, but might well be the forerunner to other and possibly more adventurous sorties by the Argentine forces.

The following ten months slipped by without any further landings being reported, but there was evidence to suggest that their Southern Thule success might have whetted Argentine appetites for further encroachment on British territory. The latest round of negotiations between the two countries was due to begin in December. In anticipation of those meetings, and presumably acting upon the assumption that they might degenerate into a quarrel, the British government decided to establish a military presence in the South Atlantic. Orders were given to Admiral Sir Henry Leach, at that time Commander-in-Chief Fleet, to despatch a small force southward as a precautionary deployment. A nuclear submarine, two frigates and a Royal Fleet Auxiliary tanker departed from the United Kingdom in November. They sailed under a cloak of the greatest secrecy. The purpose of the small task group was not to act as a deterrent, but to loiter in a position from which it could respond in the event of a crisis. The submarine was placed one thousand miles north of the Falkland Islands, the two frigates and the tanker two thousand miles north. They remained on station until early January. It was then judged that the risk of armed conflict had subsided and they returned to their home ports. The deployment of this precautionary force was not announced by London to Buenos Aires – either officially or unofficially – at any time. It is a tribute to the security discipline of the four ships' companies that it remained a secret until revealed by James Callaghan following the invasions of 1982.

The Argentines did not in the event embark upon any fresh adventures after their successful occupation of Thule Island. The tide of diplomacy appeared to be running their way – there was much current talk of political and diplomatic accommodations which possibly would save face for the British but give the Argentines what they wanted. It had been announced, not for the last time, that the ice patrol vessel HMS *Endurance* – sole representative of the Royal Navy in the region – was to be withdrawn. Repeated announcements by the Ministry of Defence that Great Britain's naval, military and aviation strategy was geared primarily to the needs of NATO made it seem increasingly unlikely that Crown forces would in future be employed in any part of the world other than the North Atlantic and in Europe. In 1966, Denis Healey had declared that 'protection for island territories in the Atlantic, Indian and Pacific Oceans can readily be provided from our major areas of deployment', but this statement, in the view of some critics, appeared increasingly fatuous as the British government withdrew from Simonstown and Aden, scrapped the *Ark Royal*, and continued to build ships designed primarily for anti-submarine work in the NATO area.

By 1980 the 'signals' passing from London to Buenos Aires had become ever more encouraging for Argentina. Nicholas Ridley was working hard on his 'leaseback' proposals. Westminster approved the Home Office's new British Nationality Bill which withdrew full British citizenship from a quarter of the Falkland Islanders. The FCO was making it plain that, for the most laudable reasons, it wished to foster better relations with the 240 million people of Latin America. The British Antarctic Survey had announced the planned closure of its station at King Edward Point (South Georgia) through lack of funds. Many of the recommendations of the Shackleton Report had been ignored.

The total annual Budget of the Falkland Islands government was £1.5 million sterling. The annual Budget of the British Antarctic Survey for 1981–82 was £8.3 million. This included the cost of maintaining the Magistrature and scientific base at King Edward Point – little more than £200,000. These sums were miniscule compared with the money being invested on and around Antarctica by other countries, some of which were economically far less strong than the United Kingdom, but the Treasury evidently regarded any further expenditure in the South Atlantic to be an exorbitant luxury which Great Britain could no longer afford. Unfortunately, the financial exercises being conducted in London did not take into account the swiftly changing politics of the region.

Between 1976 and 1981, Argentina was ruled by a junta of officers of the three armed services headed by Lieutenant-General Jorge Videla. The tough economic policies followed by these people brought the annual rate of inflation down to under 100 per cent. Under President Isabelita Perón – the former President's widowed second wife – it had been running at 400 per cent. Success was accompanied by rigid controls. Political self-expression was suppressed ruthlessly by a secret police force employing every device of terror – kidnap, torture and murder. More than 9,000 Argentines are known to have disappeared during those years, many of them being 'interrogated' and 'eliminated' by serving officers of the Argentine armed forces.

In March 1981, Videla reached military retirement age and the junta elected as his replacement another soldier, General Roberto Viola. He came to power in a cloud of rhetoric. He spoke of an early return to democracy, free elections, and the like. Unfortunately for him, inflation once again soared to 400 per cent per annum and his swift removal was engineered by the three men of the junta – General Leopoldo Galtieri, Galtieri's good friend Admiral Jorge Anaya, and Brigadier Basilio Lami Dozo of the Argentine Air Force. They used the winter and spring months to plot the downfall of the President and, in December 1981, Viola was forced to resign on grounds of 'ill health'. The new leader emerging from these manoeuvres was Leopoldo Galtieri.

The handsome extrovert Galtieri was ambitious but unsophisticated in the sphere of international politics and diplomacy.

Admiral Jorge Isaac Anaya, head of the Argentine navy and the man whose personal drive was a key factor in the quarrel between his country and the United Kingdom.

General Leopoldo Fortunato Galtieri, former Commander-in-Chief of the Argentine army and, for a brief and tumultuous period in his nation's history, its President.

Argentine Air Force Commander Basilio Lami Dozo, the third member of the triumvirate which took Argentina down the road to war.

In the months prior to Viola's fall, Galtieri paid two visits to Washington where he was courted by the State Department. The Americans viewed him as a new anti-communist ally, easily manipulated, whose accession to power could be exploited to bolster America's influence over South and Central American affairs.

The State Department sought to extract a commitment from Galtieri that he would send some of his counter-insurgency specialists – who had long experience in fighting Argentina's own home-grown guerilla movement – to serve in Central America. Appealing to his intense national pride, they held out the offer of elevating the status of Argentina from the 'third world' category to that of a significant world power. The offer was overwhelmingly attractive. President Reagan's national security adviser, Richard Allen, publicly hailed Galtieri as 'possessed of a majestic personality'. It was all very heady stuff for any army officer with no previous exposure to American hospitality and good fellowship. Galtieri's ego became dangerously inflated in a very short period of time.

It is alleged that Galtieri was advised by the Americans to retain control of the army after his accession to power, but to cover himself by ensuring the total support of at least one other member of the junta. Admiral Anaya was the obvious candidate, but he had ambitions of his own. These centred upon his determination to advance the interests of the Argentine navy, a service which had never carried the same clout as the army within Argentina's closed ruling military circle.

Admiral Anaya was an energetic man of high professional competence. He did not like the British. He had served as Naval Attaché at the Argentine Embassy in London between January 1975 and January 1976. He had found it an unhappy experience, and it seems to have left a residual resentment. Those who knew him believed that he was the one man who possessed the ability and the personal commitment to mastermind a seizure of the Falkland Islands. This assessment was reported in detail – by the British Embassy – to the Foreign and Commonwealth Office in December 1981 and to naval intelligence in February 1982.

The structure of the junta was such that each of the three senior officers who controlled it was able, to a considerable degree, to exercise individual initiative with regard to his own branch of the armed forces without necessarily consulting the other members. The evidence suggests that Anaya was setting the wheels in motion at the end of 1981 for direct action against the Falkland Islands, and that he did not at that stage inform Galtieri and Lami Dozo. There was no immediate need to show his hand. The target date was January of 1983, the 150th anniversary of British settlement. Being a seaborne operation, it would be directed by the navy and would therefore reflect the greatest credit on that service. With twelve months in hand, Anaya could lay his plans quietly and thoroughly. It is most unlikely that the island of South Georgia came within the orbit of his scheme at that stage or, if it did, that it was of more than peripheral interest.

As Argentina entered more deeply into economic crisis and political unrest, so did the timescale shorten. It seems likely that Anaya revealed his thoughts to Galtieri and Lami Dozo in late January or early February of 1982. Galtieri was indebted to Anaya for having helped him to oust General Viola. He and Lami Dozo trusted each other and they jointly agreed – at least in principle – to Anaya's proposition. In the headlong rush of events within their troubled country, it offered the prospect of recreating national unity through national pride. It was a concept which subsequently earned the title of 'triumphalism'. The three agreed to work towards a seizure of the Falkland Islands at some date in June or July. It was a fatal pact which all three were to regret.

One of the appointments made by Galtieri upon becoming President was that of Dr Nicanor Costa Mendes, returning to the office of Foreign Secretary after a break of ten years. Costa Mendes was an anglophile and an old hand in the Falklands game. In 1968 he had negotiated with George Brown, then British Foreign Secretary, and believed at the time that he had won the argument over sovereignty. His successors – and Brown's also – had failed, despite considerable effort by both sides, to seal a compromise. The arrival of Galtieri at the supreme office now permitted Costa Mendes to try again, and this time his country had the valuable friendship of the United States. It came as a jolt to the Americans when, within days of taking office, Costa

Richard Luce, MP, Parliamentary Under-Secretary of State for Foreign and Commonwealth Affairs who took over the reins – in dealing with the Falkland Islands problem – at a time when the British government had already ceased to produce any new negotiating initiatives. It can be argued that the situation was already out of control some weeks prior to Luce's first encounter with the Argentines in New York.

Mendes bluntly reopened old wounds with Chile concerning the Beagle Channel islands and simultaneously renewed Argentina's claim to 'Las Malvinas'.

It is evident that Costa Mendes was fully aware that the junta might commit itself to occupy the Falklands by force. It is equally clear that, in common with his masters, he never seriously contemplated the possibility of a comprehensive military response by the British. With hindsight, it is astounding that the Argentines could have believed it possible to seize land regarded as British sovereign territory without provoking a violent reaction. It is equally surprising that the British intelligence services in London could have been so complacent. Both sides were equally guilty of self-delusion, but there is less excuse for the British than for their Argentine counterparts. There were plenty of straws blowing in the wind.

In February there was an important meeting in New York between Enrique Ros, the Argentine Deputy Foreign Minister, and Richard Luce, the junior British Foreign Office minister who had just taken over from Nicholas Ridley and who previously had specialised in African affairs. Luce had once visited the Falkland Islands, but his knowledge of the latest developments was based essentially upon briefings received from permanent civil servants in his department. Their general advice to him, as he set off for New York, suggested that the problem was chronic rather than acute.

Luce was joined in New York by Anthony Williams, Her Majesty's Ambassador to the Republic of Argentina. Williams did not share London's sanguine approach to the problem. Coming, as he did, straight from the eye of the gathering storm, he was deeply worried by the collapse of the Nicholas Ridley 'leaseback' initiative, and by the fact that, since then, the British government had run out of ideas. Williams knew that Galtieri and his colleagues were under increasing pressure to produce a popular national success before either handing power to a civilian administration or being displaced by yet another military junta. Their situation required fast and dramatic results. Williams assessed that, unless this New York meeting led to a new and clearly promising initiative, the Argentines would lose patience entirely. He did not believe that they would proceed immediately with an invasion of the Falklands, but he guessed that they would launch a major propaganda campaign and might duplicate their 1976 landing on Thule Island by seizing South Georgia or even one of the British possessions in Antarctica. If this led to an armed British response, then there could well be a clash on the high seas. Williams calculated that the crisis would culminate in June or July of that same year, 1982.

The British delegation at the New York meeting presented a working paper which proposed the creation of a 'permanent negotiating commission'. The purpose of this commission would be to 'identify all the elements in the dispute, to consider them in depth, and to recommend how they might be resolved within an overall settlement.' The period of operation of the commission would be for one year, at the end of which the participating Ministers would 'review progress and reach conclusions on whether the commission should continue its work.' According to the Committee chaired by Lord Franks which later examined these events, the proposals put forward by Richard Luce did not refer to the frequency with which these encounters would occur. The British recommended that either party could withdraw from the commission at any time, and that its existence should not be interpreted as prejudicing the sovereignty claims of either government. When viewed in the context of the previous many years of discussion, the latest British proposals were self-evidently flaccid and uninspired. To some Argentine commentators, they appeared blatantly cynical.

Luce and Ros issued a joint statement in New York on 1 March. This referred to the 'cordial and positive spirit' in which the meeting had taken place, and to 'their resolve to find a solution to the sovereignty dispute'. The statement was bland but fundamentally hopeful. The British were therefore dismayed when, on this same day, the Argentine Ministry of Foreign Affairs in Buenos Aires issued a unilateral statement which grossly distorted the sense of the New York communiqué. Buenos Aires was stating that the meeting had been intended to 'speed up to the maximum the negotiations in train to achieve recognition of Argentine sovereignty over the Malvinas, South Georgia and the South Sandwich Islands, and by this means to achieve substantial results within a time which at this advanced stage of the discussions will necessarily have to be short.'

There were other phrases in the Buenos Aires statement which made it very clear that the junta's patience was very nearly exhausted. The final lines were a classic instance of the steel fist being concealed within a velvet glove: 'The new system constitutes an effective step for an early solution to the dispute. However, should this not occur, Argentina reserves to terminate the working of this mechanism and to choose freely the procedure which best accords with her best interests.'

This communiqué was accompanied by much comment in the Argentine press. A government source was quoted as saying that, if the commission did not produce sufficient progress towards a solution, then Argentina might break off economic and political relations with the United Kingdom and once again raise the matter in the United Nations General Assembly. Whether it was officially inspired is not known, but there was further press speculation that the junta would seize the islands by *coup de main* sometime during the second half of the year if it felt that it had no alternative.

On 3 March, having seen the reports emanating from Buenos Aires, Richard Luce contacted Enrique Ros and expressed his concern at the apparent repudiation by the junta of their New York accord. He was even more concerned by the implied threat. On the following day, at a meeting between Anthony Williams and Dr Costa Mendes, the Argentines gave a conciliatory assurance that they were 'setting out a proposed programme which included only recourses contemplated in the United Nations Charter.'

Despite these soothing remarks, the United States government was requested by Richard Luce to use its good offices to 'encourage the Argentines to keep things cool in view of the dangers of confrontation if negotiations broke down.' Thomas Enders, the US Assistant Secretary of State for Latin American Affairs, had a prearranged meeting on 6 March with General Galtieri in Buenos Aires. The American repeated the British expressions of concern. The Argentine President's response was noncommital, but, according to subsequent reports, 'he did not give the impression that his government was about to do anything drastic.' It was during this period that the FCO was given a serious warning from a leading figure in Uruguay that 'Argentina might well take some military action if they did not get what they wanted.' Although prepared to talk softly when it suited them to do so, the exasperated Argentines were certainly carrying a big stick.

Warnings of increased Argentine pugnacity were coming through from Captain Nicholas Barker, RN, commanding officer of HMS *Endurance*. Since the beginning of the year, his operators had detected a sharp rise in the volume of Argentine naval radio traffic. As the man on the spot, he had more current knowledge of the region than any other Royal Navy officer of the same rank. Nick Barker was so disturbed by what he perceived to be the quickening pace of events that he continued to press his views to the irritation of some of his superiors. HMS *Endurance* was due to return to England and be withdrawn from further service. It may have been felt that some of his reports were deliberately alarmist in a last minute attempt to save his ship.

The sole regular bearer of the White Ensign in the southern hemisphere, the ice patrol vessel HMS *Endurance* had been intermittently threatened with the axe since 1975. Whenever the topic had been raised, the Foreign Secretary of the day had succeeded in persuading the current Secretary of State for Defence to keep her in service. Fresh rumours that she would be sold or scrapped started to circulate in the spring of 1981 but, when the annual *Statement on the Defence Estimates* appeared in April, her name did not appear in the list of ships scheduled for disposal. The only reference to the *Endurance* was a favourable comment regarding her usefulness as a hydrographic survey ship. It appeared, therefore, that the rumours were unfounded.

One sombre fact emerging from the 1981 Defence Estimates was the speed with which the nation's defence expenditure was growing. In 1950–51 it had been £780 million, in 1960–61 £1,604 million and, in 1970–71 £2,503 million. By 1981 the annual outlay had grown to £11,375 million, double the figure for 1976–77. Clearly there was a need to somehow contain expenditure at a level which the nation could afford. Responding to new guidelines from the Treasury, the Secretary of State for Defence, John Nott, drafted a White Paper – entitled 'The United Kingdom Defence Programme: The Way Forward' – which he presented to Parliament in June and which was fully debated in the House of Commons on 7 July. The White Paper made specific reference to certain ships earmarked for sale or disposal but, again, there was no direct mention of HMS *Endurance*. That she was due to be withdrawn from service was, however, by now widely known in Whitehall. The rumours during the spring and early summer had attracted some adverse comment, but not enough to effect the issue. Publication of the White Paper, with the accompanying well-informed story that the *Endurance* would be employed for only one more season, provoked heavier broadsides.

On 30 June, Lord Shackleton and Lord Buxton gained an opportunity to raise the matter in the Upper House. Each made a powerful speech, spelling out their reasons for counselling a change of heart. Shackleton described the decision as 'a disaster'. He and Buxton were politely rebuffed by Lord Trefgarne, a junior Minister who, curiously, normally spoke for the Department of Trade. He confirmed that the *Endurance* would be paid off in 1982 upon her return to the United Kingdom at the end of the season.

Neither Buxton nor Shackleton had expected an immediate retraction and were therefore not surprised by Trefgarne's reply. However, the announcement came as a disappointment to Lord Carrington, the Foreign Secretary, and to Nicholas Ridley, the junior Minister at that time still responsible for the Americas (including the Falkland Islands and Dependencies). The FCO had been pressing for the *Endurance* to be kept on station, and they believed that there was still hope of a reprieve. On 5 June, Lord Carrington had written the first of several letters to John Nott, stating the political perils implicit in her withdrawal. It was only when Lord Trefgarne made his statement in the House of Lords that they knew of Mr Nott's apparently final decision.

Over the years, the annual cost of running HMS *Endurance* was said to be somewhere between two and five million pounds, depending upon the year in question and the source consulted. It was, in any event, a trivial sum compared with the total national Defence Budget. Unfortunately, when obliged to seek areas of economy, the Ministry of Defence decided that the *Endurance* should appear high on the list of sacrificial lambs. The decision was based not only upon the need for economy but also a seeming lack of appreciation of her particular role and capabilities. The deficiency was revealed in a letter, dated 12 January 1982, addressed to one of several MPs who wrote to Mr Nott protesting against her proposed withdrawal. Two sections of his letter are particularly interesting: 'HMS *Endurance* has been a familiar sight in the South Atlantic and Antarctic waters for many years, and it is understandable that the news of her withdrawal from service should have met with some concern. During her annual deployments she has performed a wide variety of tasks. She will inevitably be missed and the decision to pay her off was by no means easy. That said, it is important to bear in mind that she has only a limited military capability. This decision should not be seen as in any way effecting our policy towards the Falkland Islands. We remain committed to their defence and we shall continue to maintain the Royal Marines garrison on the island at its present strength, as a visible and constant demonstration of that commitment. Your constituents can be sure that there is no question of our abandoning the islanders.'

The phrase 'limited military capability' was a reference to the fact that the *Endurance* carried only two 20 mm Oerlikon guns and two Wasp helicopters. But nobody had ever claimed that her value lay in her firepower, indeed the Antarctic Treaty specifically discouraged the introduction of armaments into the region, so this consideration alone could not justify the scrapping decision. His second point – that a Royal Marines garrison would remain at Port Stanley – was invalidated by the fact that, in general terms, it would be isolated from Royal Navy support and transportation. Apart from being too few to defend the Falkland Islands themselves, the marines, without the *Endurance*, would be unable to move to any other threatened British possession in the region. No other Royal Navy ship had an ice-strengthened hull, so her loss signified, in military terms and in the view of many critics, the effective abandonment of South Georgia, the South Sandwich Islands, and the whole of the deep south.

It must be said that John Nott, in common with most of his Cabinet colleagues, was under strong pressure to contain his Ministry's costs within strictly defined limits. The Prime Minister, Mrs Thatcher, and her Chancellor were following a tough line on all public expenditure. The cash limits were so stringent that, at one stage, several frigates lay idle at their berths, unable to put to sea for training because there was no more money left in the kitty to pay for the fuel oil.

The impact of the 1981 Defence Review is recalled by Admiral of the Fleet Sir Henry Leach: 'It is not the business of the armed forces to decide how much money should be spent each year on national defence. That is a political matter which must be decided by the Cabinet of the day. It is for the heads of the three Services to then advise the Secretary of State for Defence what the effects of his Budget will be upon the country's defence capability. Following the arbitrary decisions taken by Mr Nott in his White Paper, it emerged that Britain's naval capability, especially for overseas intervention, was due to be uniquely slashed. The cuts in the maritime element were twice as great as those affecting the Army and seven times greater than the cuts for the Royal Air Force. Additionally, it was decided that virtually the entire cost of our new national strategic nuclear deterrent - the Trident programme - would be financed from the Navy Budget instead of being spread across the Defence Budget as a whole, as had been the case on previous and similar occasions. I would have been quite happy to see the *Endurance* continuing in her usual role - particularly as her running costs were peanuts in the overall context - but the professional relationship between Mr Nott and myself had become so strained that any insistence on my part that she be retained would very probably have led to my losing some other, more valuable, asset. The price of the *Endurance* could easily have been a frigate. Indeed, matters had reached such a pitch that I had to contemplate the risk of losing both a frigate *and* HMS *Endurance*.'

It seems likely that similar dilemmas faced the FCO. Lord Carrington was arguing the diplomatic value of HMS *Endurance*, but over-insistence could too easily have led to the retort that, if the FCO regarded the ship so highly, then her running costs should be paid from the FCO Budget. Both the First Sea Lord and the Foreign Secretary were, to a degree, treading on thin ice.

It can be argued, with hindsight, that 1981 was one of the most traumatic years in living memory as far as defence expenditure - and naval expenditure in particular - was concerned. The background to the furious debate is recalled by Sir John Nott, the Minister about whose person the hurricane swirled: 'The reason why the Royal Navy were required to undertake larger savings in their future programme than the Army or the Royal Air Force was simply because they had expanded their programme out of all proportion to the sums of money that would be available in future years. This was partly a matter of tactics in attempting to acquire the maximum share of the total budget for the Royal Navy, but also because they were well aware that once something was in the programme, it was a political embarrassment to take it out! The Royal Navy was treated absolutely fairly relative to the other services, although I do not anticipate that Sir Henry Leach would ever agree with that. When the Royal Navy were required by me to bring their programme back into line with the sums of money actually available, two of the ships chosen by them for scrapping were HMS *Endurance* and the royal yacht, HMY *Britannia*. Of course, they wanted money for *all* their requirements, but it is interesting to note their *own* view of priorities which was actually to put *Endurance*, with *Britannia*, on the top of their list for cuts'.

While the cut and thrust of the *Endurance* argument continued within the discreet portals of government, other and noisier voices were being raised outside. There was a rising volume of qualified advice coming from a variety of sources during the second half of 1981 and the opening weeks of 1982.

One of the catalysts which led to a mounting wave of public debate was a visit to the South Atlantic by Bob Langley, a reporter with the BBC 'Pebble Mill' television team. Sent to the Falklands to produce a series of films entitled *Langley South*, he talked at length with Rex Hunt and others who knew the region well. Langley and his producer returned to Birmingham with an almost evangelical conviction that the government was on the wrong track in its dealings with the Argentines and with the Falkland islanders.

The 'Pebble Mill' management approached Lord Shackleton and invited him to take part in a televised debate on Langley's findings. The peer was an obvious first choice for such an event: he was a former Cabinet minister, his father had died on South Georgia, his daughter took a keen interest in the welfare of HMS *Endurance*, and Shackleton himself had chaired the major 1976 survey. Not only did he agree to appear on the programme, he also decided to use his political experience and personal prestige in backing the 'Pebble Mill' campaign.

Despite his seventy-one years, the donnish Shackleton had a razor-sharp mind and a patiently persuasive style. In early December 1981, having contacted a number of parliamentarians and other public figures whom he knew to be interested in the South Atlantic and Antarctica, he formed an unofficial pressure group with the pleasing acronym SWAG - the South West Atlantic Group. It became a focal point for the hundreds of letters and telephone calls coming in from members of the general public who had been alarmed by the *Langley South* reports, and acted as a well-informed lobby at Westminster.

There had existed, for many years, a group of MPs known as the Falkland Islands Committee. They kept an eye on any legislation which, directly or indirectly, affected the welfare of the islanders. Their efforts, combined with those of the new SWAG lobby, caused dozens of members of both Houses to be suddenly aware of matters to which previously they had given little thought.

It is an almost uniquely British tradition that anyone wishing to air an opinion or grievance can seek a platform for his views by writing a letter to the Editor of *The Times* of London. This was the next step for SWAG. Having carefully consulted among themselves, and having given due notice of their intention to the Ministry of Defence, they sent the following letter. It appeared in *The Times* on 4 February 1982, just eight weeks before the Argentine invasion.

'Your columns have already carried correspondence (Oct 26 *et seq*) regarding the retention in service of HMS *Endurance*, the Royal Navy's only vessel capable of carrying the White Ensign into the South Atlantic and the ice-filled Antarctic seas, thus maintaining, both symbolically and practically, Britain's traditional position in this potentially very important land and sea area.

Since the intention to pay off HMS *Endurance* at the end of the current summer season became known, considerable public and parliamentary concern has been expressed. We, the undersigned, who all have personal experience or knowledge of the present and developing situation in the area, share their anxieties.

While we appreciate the problems facing the Royal Navy in meeting essential defence commitments in a climate of economy, the decision to axe HMS *Endurance* is tantamount to the withdrawal of the Royal Navy from the hazardous waters where no other naval vessel can undertake all her various tasks. The saving is likely to be greatly outweighed by the consequences to Britain's future interests in what is expected to become a vital resource area.

With twenty-one nations from all quarters of the world now moving fast to establish or enlarge an Antarctic presence, in many cases at an expense considerably greater than this country seems willing to envisage, this is no time for Britain to pull back or be thought to be moving along a path of retreat. Besides the loss of the ship's special facilities, the withdrawal of HMS *Endurance* will be seen as a step in this direction. Thus Britain's influence will be diminished at a time when it will be greatly needed in the determination of conservation measures and the regulation of exploitation, as the successful Antarctic Treaty comes up for review, a Treaty which Britain played a leading role in instituting.

In anticipation of that review, and of the future beyond it, national self-interest will inevitably come to the fore. No country will then want to lose out on the prospects; least of all should Britain. It seems to us vital that the White Ensign should continue to fly regularly in the south-west Atlantic, and the Antarctic Ocean, for its implication of national interest goes far beyond that expressed by the presence of research ships. This would not only maintain general British interests in the area, but would also provide moral support for the totally British population of the Falklands Islands.'

There were nine signatories to this letter, and they represented a wealth of long-standing and hard-won experience in Falkland Islands and Antarctic affairs. If ever a government wished for a voluntary self-appointed 'think tank', this was it. Lord Shackleton appeared first, for obvious reasons, followed by Sir Peter Scott, conservationist and son of the explorer. Next came Sir Vivian Fuchs, deeply involved in Antarctic affairs between 1947 and 1973 and leader of the Commonwealth Trans-Antarctic Expedition of 1955. Then came Rear-Admiral Sir Edmund Irving, a Past President of the Royal Geographical Society and a former Hydrographer to the Royal Navy. Next was James Parker, former Governor of the Falklands and Dependencies, Rex Hunt's predecessor. Another name was that of Captain Tom Woodfield, RN, an Elder Brother of Trinity House and a former captain of the RRS *Bransfield*. The three other signatories were Lord Morris, Eric Ogden MP and Michael Shersby MP, each of whom had particular knowledge of the South Atlantic.

The British press carried other letters of a similar nature. As late as 25 March, *The Times* published an appeal, from Air Commodore B. G. Frow of the United Kingdom Falkland Islands Committee, which criticised the decision to withdraw the *Endurance* as 'madness'.

Encouraging though all this activity may have been to the *Endurance* supporters, it must be accepted that broad-based political and public interest in the future of the Falkland Islands and the Dependencies was conspicuous by its absence until the last twelve months prior to the Argentine attack. Apart from the sealers, whalers, farmers and scientists, only a handful of enthusiasts had ever shown any deep concern.

A number of British companies had acquired possession of tracts of land in the Falklands over a period of one hundred and fifty years. Many of them functioned as absentee landlords to the local share farmers. They tended to exploit the indigenous wool producing industry without making the long-term investments required for future growth and diversification. The most important influence was the Falkland Islands Company (FIC), a completely independent firm owning nearly one half of the utilisable land surface of the islands. It was founded upon a Royal Charter signed by Queen Victoria on 20 January 1852. Until the late 1970s, the pattern of exploitation was depressingly consistent over many decades. Far more money was taken out of the local economy than was ever ploughed in. To quote one well-qualified London source: 'If the islanders had been black, we would never have got away with it.'

The directors of the FIC were never involved in the management of South Georgia – that was the responsibility of the Falkland Islands Dependencies government – but they did at least have a fund of knowledge regarding the general ambiance of the South Atlantic territories. That knowledge had always been available for consultation, but the continuity started to crumble in 1972 when the company was taken over by Dundee, Perth & London Securities Limited, a financial holding group.

Only a few months later, in 1973, the DP&L group was absorbed by Charringtons Industrial Holdings Limited, a financial conglomerate which derived its main strength from the sale of fuels in the United Kingdom. The Falkland Islands Company had become just one more name in a portfolio of shareholdings, and South Georgia attracted even less attention than before.

In 1977, the Charrington group was taken over by the Coalite Group, yet another holding company, with an annual turnover (at that time) of £57.4 million sterling. Under the energetic chairmanship of Ted Needham, Coalite bought Charringtons in order to broaden its base in the area of fuel distribution services. Coalite's original financial strength derived from the production and sale of smokeless fuels and, although the company had expanded into shipping, warehousing, chemicals, vehicle manufacture and construction supplies, it remained essentially a United Kingdom trading organisation. Between 1977 and 1982, the Falkland Islands Company was the only firm in its long portfolio of holdings which operated outside Great Britain.

The book value of the FIC investment, as declared in Coalite's 1981 accounts, was four and a half million pounds. This was less than five per cent of total group assets. It had never been the Board's intention to assume the role of front-runner in a campaign to keep the British flag flying in the South Atlantic, and Ted Needham specifically warned his staff not to become 'too emotionally involved' in the affairs of the islands. For his own part, however, he found himself becoming increasingly interested in their unique society and economy. He made his first personal visit in 1979 to assess the situation at firsthand and, in the same year, agreed to the sale of 90,000 acres of Company land to local people. Split into freeholdings of approximately 15,000 acres each, the land was sold to six 'kelper' families.

The importance of these events lay in the manner in which they were perceived by the Falkland-watchers in Buenos Aires. Ownership of nearly one half of the land surface of a British sovereign territory had changed hands three times in little more than seven years. The shares which symbolised that land and the people who lived upon it had been bought and sold – more by accident than design – without any significant reaction in Whitehall, Westminster or Fleet Street. To any Argentine observer keeping track of these events, the developments on the London Stock Exchange suggested that the islands were available to the highest bidder.

It has been suggested that, if the Argentine government had possessed sound business acumen, it could have appointed nominees, created a British 'front' company, and attempted to purchase the entire stock of the FIC. If successful, this ploy would have established an Argentine presence on the Falkland Islands, via a totally legitimate route, and would have resulted in *de facto* control over most aspects of the island's affairs. Very probably it would have put the skids under any remaining British resolve to cling to their distant possessions in the South Atlantic.

There is evidence to suggest that something along these lines was attempted in 1975 and again in 1980. A leading entrepreneur, a well-known figure in the City of London, approached first Charringtons and later Coalite. Ted Needham accepted his invitation to lunch and learned that he wished to explore the possibilities of buying the Falkland Islands Company outright. Needham suspected that his lunchtime companion was acting as broker for other people. Bluntly he posed the question – 'are you representing Argentine interests?' The answer was affirmative and he terminated the conversation.

The more he learned of the South Atlantic, the more alarmed Needham became at the apparent complacency of the British government. He came to the same conclusion as that reached by others – it was vital to avoid a situation in which the Argentines might assume that the British wanted to be rid of the Falklands and the Dependencies. He was a businessman, not a politician, but he felt so strongly on the subject that he wrote directly, in January 1982, to three of the most senior members of the Cabinet. His main plea was for the retention of HMS *Endurance* and, just as the SWAG signatories had done, he stated the risks which he believed the British government to be incurring. The tone of the acknowledgements received from the three politicians showed that his plea had made no impact.

Although the campaign against the death sentence on the *Endurance* had failed, the heat and light generated by it had at least aroused Parliamentary interest to a greater extent than ever before. On 3 March, when Richard Luce carefully presented to an attentive House of Commons the official British version of his recent New York meeting with Enrique Ros, he was faced with a simple but potentially explosive question by the distinguished back-bencher Julian Amery: 'Will he [the Minister] assure us that all necessary steps are in hand to ensure the protection of the islands against unexpected attack?' Luce was in an impossible position. To give an affirmative answer would be both untrue and provocative. He knew that the Cabinet had not acceded to Carrington's request for the *Endurance* to be kept on station and that the only fighting troops further south than Gibraltar and Belize were the ship's own detachment of twelve Royal Marines and the token garrison of forty-two marines at Moody Brook, Port Stanley. On the other hand, no British government could ever reveal its defence arrangements by stating that it had decided *not* to protect any particular piece of British territory. The House of Commons is not always an ideal forum for the debate of such delicate matters.

The unhappy Luce took the only route open to him by concocting a non-committal reply which placated the House of Commons but which may have further encouraged the Galtieri regime to believe that the Thatcher government was lacking in resolution. With hindsight, it can be argued that this debate at Westminster had carried the two countries, unwittingly, one more step along the road to war.

Julian Amery was not the only Briton to express concern at the military vulnerability of the Falkland Islands. In October of the previous year, 1981, Colonel Stephen Love had written on the subject to the Ministry of Defence. Love was the Defence Attaché then serving at the British Embassy in Buenos Aires. As a professional soldier, and having held his appointment for two years, he was disturbed, even at that stage, by the belligerence of Argentine sentiment. It was a part of his task to report to his Ambassador and to London regarding Defence matters in general in Argentina, Uruguay and Paraguay, and on Argentine military and aviation developments in particular. Paradoxically, however, he did not cover the British islands in the South Atlantic. As he describes it: 'The accredited Defence Attaché in Buenos Aires could never have any official responsibilities in the Falkland Islands. Any such dual role would have been politically out of the question. However, as it was my job *inter alia* to offer military advice on the whole region, I felt that I must try to get to know something of the threatened British territory as well as the Argentine forces which posed the threat. To gain a balanced view, I needed an understanding of both, and I therefore asked for authority to make an official visit.'

On 10 November, a month after submitting his request, Colonel Love received a reply from his superiors in London which told him that the FCO had been consulted, but that the latter could 'see no particular advantage to your visiting the islands at this time.' The MOD letter went on to say that he could make a visit if he so wished, but added the provisos that he must not wear his uniform and must pay all his own travel expenses. This second consideration derived from the fact that there were no funds budgeted for the purpose. The return air fare from Buenos Aires to Port Stanley was, at that time, one hundred and sixty-five pounds. As Love recalls: 'Understandably, it was made clear to me that I must do nothing to draw attention to myself, that my visit must be as "low-key" as possible. But it was also explained that, if I went, it would be regarded as a private venture.'

Despite the lack of enthusiasm from London, Stephen Love remained convinced that it was necessary for someone having 'the Buenos Aires perspective' to draw up a military assessment based upon as much information as could be obtained during a short visit. His plan was endorsed by Anthony Williams and welcomed enthusiastically by Governor Hunt. 'By early January of 1982 it was clear that no one else would pay the fare, so I purchased three airline tickets at my own expense and flew down to Port Stanley with my wife, Robin, and our son, Tim. We stayed at Government House, but the pretence was maintained that we were there as private guests. Most local officials understood the real reason for the visit and Rex Hunt arranged a full programme of most useful meetings and visits. He also booked me onto a scheduled FIGAS Islander flight so that I could have a look at the ground outside Port Stanley and on West Falkland. There were several options open to the Argentine armed forces. They could have established a foothold on an outlying island, or conducted a demonstration exercise in a remote area for twenty-four hours, or built up the pressure with a series of overflights, or they could have conducted a full-scale invasion and occupation. What I was trying to do was to get some idea of the terrain and what sort of problem it might present to an invader and a defender.'

Colonel Love made an oral report to the Governor before flying back, after a week on the islands, to Buenos Aires. Rex Hunt had always had the military advice of the officer commanding the local Royal Marines detachment and the naval advice of Captain Barker of the *Endurance*, but this was the first time he had had an appraisal by a senior officer with firsthand knowledge of the Argentine army and air forces and their commanders. Hunt was able to question his visitor closely regarding their characteristics and capabilities.

On 2 March, Stephen Love completed a formal summary of his findings. He set out, in this document, his view of the courses of action open to the Argentine armed forces in the event that the junta decided to make a landing 'in the coming days [*sic*].' One of the options described by Love was an almost prescient forecast of the plan adopted by Admiral Anaya and his staff only four weeks later, but, as he explains: 'One of the most important points I made concerned the problem of giving advance notice from Buenos Aires of an impending invasion. The Defence Section of the Embassy – the two Attachés plus one civil service clerical officer – was not staffed or geared for the collection of tactical military intelligence. Argentina is a very large country, the distance north to south being equivalent to that between Oslo and Gibraltar. With things as they were, it would be entirely possible for important events – unit moves and deployments – to occur in areas remote from the capital of which we in Buenos Aires would know nothing until possibly too late. I felt that both Port Stanley and London should be reminded of this before things went any further.'

Stephen Love's assessment was discussed in detail in the

In the years following the Second World War, whaling developed into a highly industrialised business conducted mainly at sea. The shore bases of South Georgia became less important with the introduction of very large factory ships supported by flotillas of 'catchers'. The whale is alone amongst the world's non-domesticated animals in having been subjected to destruction and exploitation on such an intensive scale.

Embassy and then sent, as a letter, to Governor Rex Hunt. Copies were despatched to the MOD and FCO. Love had expected, at the very least, to be asked for further details or justification for his views. 'Although I received an acknowledgement from Rex Hunt, thanking me for the report and commenting that there was little they could do in Port Stanley if the worst did occur, from the other addressees I received no reaction. That was the pattern from then on. Silence. My report, and others which followed, had no discernible effect whatsoever at the other end. It was as though someone was trying to say, politely, "Yes, we know all that already, and we *are* frightfully busy." On my return to London, after the invasions, I found that this particular letter – my report of 2 March – had not been circulated above desk level in the MOD.'

Colonel Love's reconnaissance was a private and confidential affair, but other aspects of the continuing South Atlantic debate were being conducted in the full glare of publicity. The arguments and counter-arguments in the United Kingdom – during the second half of 1981 and the opening weeks of 1982 – were keenly monitored by the Argentine Embassy in London. Significantly, within two days of Lord Trefgarne's statement in the House of Lords regarding the withdrawal of the *Endurance*, Lord Shackleton received a telephone call from that Embassy, asking him whether the announcement indicated an effective end to British interest in Antarctica and the disputed islands.

Reports of British attitudes and opinions were channelled back to Buenos Aires where it must have seemed that, given just a little more time, the ambitions of several decades would soon be fulfilled. Unfortunately for the rulers of Argentina's unhappy people, time was a commodity fast running out. Events were about to unfold with a rush. There were two factors which brought matters to a head for the Galtieri regime. One was the rising tide of popular protest in Argentina caused by galloping inflation, the other was the old argument over sovereignty on South Georgia.

Both London and Buenos Aires were equally surprised when the simmering crisis erupted suddenly into a full-scale row. The opening scene of a plot which more resembled a Ruritanian comic opera than a serious international dispute was the landing on South Georgia of a previously obscure trader, of Balkan origins, by the name of Constantino Davidoff. This gentleman had been known for many years around the ports of Latin America as a businessman, never outstandlingly successful but always keeping an eye open for the main chance. That he, an Argentine, should have been able to insert himself into the scenario, with the full knowledge and initial approval of the British authorities, was to prove in time to be a particularly unfortunate coincidence.

The Davidoff affair was the trigger which detonated the South Atlantic war of 1982. The root causes of the conflict were complex and of long standing, and it is plain that this man's otherwise harmless business ambitions were not one of them. It was the 'scrap metal merchant' story, however, which first hit the headlines in the British popular press, which caused the first serious alarm in Whitehall, and which ultimately dictated the circumstances under which Great Britain was plunged into her biggest military campaign since 1945. For these reasons, it is appropriate to examine the circumstances in detail and, to do this, we must turn the clock back to the year 1907.

Following the initial success of the Larsen venture, seventeen more companies arrived in the sub-Antarctic between 1907 and 1912 with the express purpose of whaling. Of these, thirteen were Norwegian, one Chilean, and three British. Of the latter, one in particular already had a reputation for sound business management which it has never lost – Christian Salvesen & Company of Leith, Scotland. Founded in 1846 by a Norwegian of that name, the firm had for many years been involved in the shipping and whaling trades. The original whaling stations were in Iceland and

Officers and ratings of HM Yacht *Britannia* come ashore at Grytviken in 1957 to watch the dismemberment of a newly caught whale. The economics of shore-based whaling operations were already in decline.

at Sullom Voe, in the Shetland Islands, but declining stocks gradually reduced the attractiveness of whaling in northern waters. The directors decided to join the rush and head south. In 1907 they applied for a lease from the Falkland Islands government for a station at New Island, West Falkland. Trading as the New Whaling Company, Salvesens struggled to make a profit on the New Island operation but closed it down in 1915. They then concentrated their efforts on the station at Leith Harbour for which a lease was obtained in 1909 – after some difficulty – by a newly formed subsidiary, the South Georgia Company Limited.

Salvesens went on to become the largest whaling group in the world. According to the firm's own historian, 'man has hunted the whale mercilessly in every ocean of the globe ... none pursued it harder or farther than the Salvesens.' Between 1909 and 1963, this one firm killed an average of a thousand whales every year in Antarctic waters. In the peak season of 1915/16, they caught the huge total of 3,064 of these animals.

Salvesens were pioneers in the business of extracting maximum value from each catch. They protested against the wasteful habits of their competitors who simply flensed the blubber and abandoned the carcass. Equipment was developed for the total exploitation of the whale's bones and meat. After extraction of the oil, the residue was dried and ground into a meal for fertiliser and livestock feed.

Further bases were acquired at Stromness and Prins Olav in the 1930s, the former being used as a repair base and the latter

As late as 1961, whales were still being hauled up the slipway at Grytviken to be flensed and turned into oil, fertiliser and farm livestock feed. It was at this place that the story of South Georgia whaling operations had first started in 1904.

The former whaling station at the centre of the diplomatic wranglings in early 1982 was Leith Harbour. By that time the place was abandoned and semi-derelict but, when this photograph was taken in 1950, its billowing chimneys represented a period of intense activity for the whaling industry.

being stripped for removal to Leith Harbour. When Salvesens finally withdrew from whaling in 1963, the leases on Leith Harbour and Stromness were retained and most of the extensive plant left *in situ*. The other Crown leases still extant were those on Grytviken and Husvik, both held by the Argentine-owned Albion Star Company.

In the mid-1960s there was a brief flurry of activity when Salvesens sub-let the base at Leith Harbour to Japanese whalers for two seasons. The Japanese also rented the Albion Star base at Grytviken, but their venture was not successful and all of the island's stations were closed down entirely at the end of the 1965 season. Plant and buildings began to deteriorate under the combined depredations of the weather and the crews of passing ships coming ashore for souvenirs.

Thirteen years went by without any further active commercial interest in the island being shown. Then, in 1978, Constantino Davidoff approached Christian Salvesen with a proposition. He suggested that he should be granted a contract for the salvage of the large quantities of machinery at the old whaling stations. He would purchase the machinery and dispose of it as best he might, and at his own risk.

From Salvesen's standpoint, the proposal offered an interesting prospect. Although they had no immediate commercial plans in the area, they wished to retain a stake in the future of South Georgia. Ideally, they wished to eliminate any possible future competition. The Albion Star Company had ceased trading, but still held the Crown leases on Grytviken and Husvik. The liquidator of the company was seeking a buyer for these leases and it would be best if Salvesens could acquire them for themselves. This would cost money. Now here was Davidoff, offering to pay for the otherwise useless machinery on the island. There was an obvious attraction in the idea of taking money from Davidoff, using it to purchase the two residual leases, and simultaneously blocking the entrance for any other party which might try to establish a foothold.

In 1979 the directors of Christian Salvesen approached the Foreign Office in London and Governor James Parker in Port Stanley. They explained their plan and enquired whether there would be any objection from either the British or Falkland Islands governments. Governor Parker liked the idea that Salvesens should replace the Argentine presence as represented by the Albion Star Company, but was unenthusiastic about the deal with Davidoff. He was an Argentine national, and Parker

Like a flotilla of Great War destroyers, these 'catchers' lay at their moorings in King Edward Cove, ready for their next sortie to sea in 1953.

A Norwegian labourer prepares the spine of a large whale for the drying and grinding factory.

believed that this fact might not be palatable to the Falkland Islands Council. After some discussion, however, formal approval was granted and Salvesens became the leaseholders of all four stations. The way was clear for Salvesens and Davidoff to sign the scrap metal contract. Like all such contracts, it was drawn up according to the laws of the country in which it was to be carried out. In this case, the laws prevailing were those of the Falkland Islands and their Dependencies.

Davidoff was an opportunist, a freelance entrepreneur. He set about the task of trying to organise his new business venture, but it required time to do this with his relatively modest personal resources. It was not easy to mount an expedition capable of landing a work-force on an island hundreds of miles from home, sustaining that work-force over a protracted period, and then physically transporting the booty to Argentina.

Over the next two years he concentrated his efforts upon the financial aspect of the project. It appears that he obtained the backing of a wealthy Buenos Aires lawyer and of the Banco Juncal. One of the officers of the bank, Adrian Marchessi, was very conversant with the history of South Georgia and understood the economics of Davidoff's scheme. The contract anticipated a total payment to Salvesens of £115,000 for the purchase of the scrap metal at all of Salvesen's South Georgia whaling stations with the exception of Grytviken. It included the hulks of the former whaling and sealing boats at Grytviken, also a semi-submerged floating dock, but excluded the whale catcher *Karrakatta* on the slip at Husvik. Salvesen wished to preserve this vessel as an archaeological relic.

Davidoff and Marchessi gambled that the value of the scrap would be £7.5 million sterling at current prices. They calculated their direct and indirect costs at £3 million, so the theoretical profit of £4.5 million justified the effort involved. In the event, these figures were always highly optimistic and they became even more suspect when the onset of the world economic recession led to a fall in scrap metal prices. It is doubtful whether the financial return, even if the war had not aborted the whole scheme, would have been nearly as satisfying as Davidoff and his friends first hoped.

The powerful ice-breaker and Antarctic support ship, the *Almirante Irizar*, in her peace-time colour scheme of red and white (above). Following the outbreak of hostilities, she was registered with the International Red Cross as a temporary hospital ship and re-painted accordingly (above right).

Although Grytviken and Leith Harbour were the two former whaling bases named most frequently in contemporary reports of the Anglo-Argentine dispute, there are several similar sites scattered along South Georgia's north-east coast. One of the largest of these is Stromness (right), just three miles south of Leith Harbour.

Throughout this period Davidoff maintained contact with the British Embassy in Buenos Aires where the staff were aware of his intentions. It is also evident that he was talking with officers of the Argentine navy among whom he had numerous friends and acquaintances.

For a long time it seemed that Davidoff had bitten off more than he could chew, that his contract would lapse by default. Then, in the closing months of 1981, his luck seemed to change for the better. Informal conversations with middle-ranking naval officers resulted in an agreement that the Argentine navy would help him. A unit of the Antarctic Squadron, the *Almirante Irizar*, would take him to South Georgia so that he could reconnoitre the station at Leith Harbour and generally assess his next move. It would be the first time that he had ever seen the island upon which all his plans depended.

The *Almirante Irizar* was a Helsinki-built ice-breaker employed in round-the-year support operations to Argentine bases in the Antarctic. Displacing 11,800 tons, first commissioned only three years earlier, she had cost the Argentine government the impressive sum of 61.5 million US dollars. Equipped to carry helicopters and landing craft, fitted with two 16-ton cranes and a 60-ton towing winch, she had a ship's company of 133 and could accommodate 100 passengers. The question arises – why did the Argentine navy decide to divert such a valuable and sophisticated ship for such a mundane commercial venture as the Davidoff contract?

The evidence suggests that the decision was known to, and approved by, Admiral Anaya, but had no specifically military purpose at that stage. It was simply a useful way of obtaining up-to-date intelligence regarding the situation on the island and was one more gambit in Argentina's continuing policy – started forty years earlier – of testing British resolve. They were 'trailing their coat-tails' rather than 'throwing down the gauntlet'. Davidoff was a convenient pawn in the game.

The *Almirante Irizar* sailed from Buenos Aires on 16 December 1981 with Davidoff aboard. Before embarking, he wrote to the British Embassy, notifying the staff of his intention to land on South Georgia, at an unspecified future date, in pursuance of the plans which he had been discussing with them for the past two years. He stated that his landing would be in accordance with his contract with Salvesens, full details of which had been registered with the Foreign Office in London and the Falkland Islands government in Stanley in 1979. The ship's route, ports of call and ultimate destination were not declared, nor was diplomatic clearance requested in respect of the ship herself.

Apart from telling the British Embassy of his departure, Davidoff also sent an adivisory telex to Salvesen on 16 December and they relayed the message, on the following day, to Governor Rex Hunt in Port Stanley.

The *Almirante Irizar* reached Leith Harbour on 20 December. Captain Ceasar Trombetta wasted no time on passage – he covered 1,650 miles in four days. Throughout this time he maintained radio silence, a most unusual procedure in a region where ships' captains regularly broadcast weather reports and generally exchange news and greetings.

Trombetta and Davidoff remained only briefly at Leith Harbour. They went ashore – leaving ample evidence of their visit in the form of cigarette ends, discarded film packets and a childish message in chalk claiming possession in the name of Argentina – but they made no contact with the BAS.

The fact that a cladestine landing had been made at Lieth Harbour was established by Peter Witty, at that time the Base Commander. Having been alerted by the crew of a visiting yacht, he went there in person on 23 December. He noted the dated

Grytviken, photographed on a remarkably clear summer morning. The high ground to the right is Brown Mountain and the building in the right foreground is the old whalers' church. Both were to feature in the events of April 1982.

Apart from the obviously valuable stocks of non-ferrous metals still stored on the island, there were also thousands of tons of good quality steel and restorable machinery. Constantino Davidoff would have needed several years to remove it all, but it is evident that his initial plan was economically attractive. Only the decline in world metal prices and the political crisis in Argentina prevented him from clearing away these unsightly reminders of man's earlier commercial activities on the island.

chalk message and reported the facts to Rex Hunt in Port Stanley. The Governor in turn conveyed the information to the Foreign Office and it appears that his signal was received in London on 31 December.

In order to appreciate fully the subsequent events, it is important to understand the position in which Captain Trombetta and Constantino Davidoff had placed themselves. South Georgia was recognised throughout the world as being a British possession. Only Argentina disputed this fact. Every vessel calling at the island was obliged by law and by internationally accepted custom to report itself at King Edward Point, the official 'port of entry'. The purpose of this formality was to obtain permission to land, to explain the purpose of the visit, and to obtain Customs and Immigration clearance.

Precisely the same requirements exist at every port in the world at which a ship might call, or where an individual might wish to enter the country for the purpose of business or tourism. The circumstances prevailing on South Georgia were no different in this respect to the normal practices at the port of New York or indeed at Buenos Aires. In the case of South Georgia, where the Magistrate resided at King Edward Point, the correct course was to proceed in the first instance to the adjacent Cove. To do otherwise, to anchor within the three-mile limit or to go ashore, without permission, was an illegal act. Russian and Polish trawlers always respected the laws of the Falkland Island Dependencies, as did the various yachts – mainly French – which occasionally called while on passage between South Africa and South America. Only the *Almirante Irizar*, the first Argentine ship to visit South Georgia for several years, chose to flout the law. She did so despite the fact that Trombetta was a seasoned professional South Atlantic traveller, despite the instructions given to Davidoff by the British Embassy in Buenos Aires, despite the requirements of his contract with Salvesens, and

King Edward Point, 'capital' of South Georgia, the official port of entry, and main local base for the British Antarctic Survey. The cross stands on Hope Point and is a memorial to Sir Ernest Shackleton. The main accommodation building named after him stands nearby in the centre foreground. Grytviken can be seen on the far side of King Edward Cove.

despite the fact that naval ships calling at foreign ports require additional diplomatic clearance.

Governor Hunt drew attention to these facts in his signal to London. He proposed taking proceedings against Davidoff and recommended that a strong protest be delivered to the Argentine government. The Foreign Office responded by instructing Hunt 'not to institute proceedings (locally) which would risk provoking a most serious incident which could escalate and have an unforeseeable outcome.' He was further told to grant formal permission to land, if Davidoff asked for it, and to protest to the captain of the *Almirante Irizar* if he failed to ask for clearance. All of this was by now entirely academic – the *Almirante Irizar* had stayed only a few hours at Leith Harbour before sailing for Ushuaia, in the extreme south of Argentina.

Despite its initial reservations over Governor Hunt's recommendations, the Foreign Office soon decided that this was indeed a serious matter and should be so treated. On 4 January 1982, Ambassador Williams was instructed 'to deliver a formal protest in the strongest terms at this violation of British sovereignty.' He was told to say that, if any further attempt were made to land at Leith Harbour or elsewhere on South Georgia, without proper authority, then the British government 'reserved the right to take whatever action might be necessary.' These were brave words. The British simply did not possess the means, at that stage, of stopping the Argentines from doing whatever they pleased. A very small sabre was being rattled in a very large scabbard.

On 6 January, Ambassador Williams attended the Argentine Ministry of Foreign Affairs but, when the officials there denied any knowledge of the incident and asked for time in which to raise enquiries, he agreed. And there the matter rested for the next four weeks.

On 3 February, Williams was again instructed by the Foreign Office to deliver the protest and, six days later, he did so. On 18 February, to nobody's surprise, the Argentines rejected it.

At first glance, it might appear that the Davidoff visit to Leith Harbour, in December 1981, and the diplomatic exchanges flowing from it, was a storm in a teacup which was permitted to escalate to major proportions. Nations do not normally exchange 'strong protests' without good cause. Was the broadside delivered by Ambassador Anthony Williams on 9 February – seven weeks after the visit – really appropriate to the scale of the offence? The consensus view of those most closely involved states that it was. To have ignored the breach of law, to have treated it as nothing more than a naughty prank, would have been a serious loss of face for the British, a sign of weakness. The Foreign Office had decided to make a stand against Argentine encroachment.

Unfortunately it was all much too late. The British were no longer dealing with an Argentine government engaged in a game of diplomatic move and counter-move. The power structure in Buenos Aires had changed dramatically – President Viola was out, General Galtieri was in. It seems probable that the angry British response to the Davidoff landing at Leith Harbour served only to harden the resolve of those officers – particularly in the navy – who believed that their ships should be free to come and go as they pleased, without needing to seek the permission of a British civilian, even one vested with the cloak of Magistrate. South Georgia had not previously figured in the Argentine navy's plans, but now it attracted much greater attention.

Constantino Davidoff returned to Buenos Aires from Ushuaia in early February. His tail well up, he threw himself into the task of finalising his arrangements. He had seen enough at Leith Harbour – where he had gone ashore with several officers and crew from the *Almirante Irizar* – to know that there was still plenty of valuable machinery and metal to be harvested. Quantities had been stolen in former years by visiting trawlermen and yachtsmen. Indeed, the thefts continued after Davidoff's visit. On 13 February, a British Joint Services Expedition caught red-handed the crews of three yachts looting Leith Harbour. But the bulk of the salvageable material remained intact.

Davidoff's business associates and the Argentine navy had

The Royal Research Ship *John Biscoe*. Civilian-manned, registered in Port Stanley (Falkland Islands), she carries on her funnel the coat-of-arms of the British Antarctic Survey.

been hard at work during the early weeks of 1982. A work-force had been recruited and heavy plant purchased. The project was well advanced by the time he returned to Buenos Aires, so he presented himself at the British Embassy, on 23 February, to announce his current plans. He explained that he did not himself intend returning to South Georgia with his party of workmen. They would be in the charge of a man named Ricardo Cacace. Davidoff's own 19-year-old son would be a member of the group, the youngster being given the chance to broaden his experience. The expedition would sail in the 3,100-ton naval transport *Bahia Buen Suceso* with stores sufficient for a stay of four months. Davidoff offered to transport supplies for the BAS people at King Edward Point, and to make available to them the services of a doctor and a male nurse travelling with his party. Evidently Davidoff was trying hard to make himself agreeable. He had waited three years, but the profit to be made from the Salvesen contract was now almost within his grasp. It is unlikely that he knew, or even cared, that he had become nothing more than a stooge, or cover, for the Argentine navy's planned encroachment onto South Georgia.

Within hours of Davidoff's visit to the Embassy, full details of the conversation were reported to the Foreign Office. The signal was acknowledged without comment. Despite the furore earlier in the month, despite the Argentine outright rejection, on 18 February, of the British government's formal protest regarding the illegal *Almirante Irizar* visit, London raised no objection to this new visit. Presumably it was felt that he should be given the chance to redeem himself and to fulfil the terms of his contract with Salvesens.

The weakness in this line of reasoning was the fact that the work-force would not be under the personal supervision of Davidoff himself. Its movements would be entirely in the hands of an Argentine naval officer, Captain Briatore, captain of the *Bahia Buen Suceso*. Taking account of the behaviour of Captain Trombetta in December, and considering the mounting pugnacity of the Argentine navy in general, it was surely somewhat optimistic to suppose that Captain Briatore would in effect acknowledge British sovereignty on South Georgia by meekly asking the Magistrate for permission to land. Optimistic or not, this was the policy adopted.

On 9 March, Davidoff delivered formal written confirmation of his plans to the British Embassy in Buenos Aires. He said that his men were about to sail, that they would be at Leith Harbour for four months and would be transported there by the Argentine navy. He stated that he had informed Salvesens of the latest developments and, subject to payment of an additional sum, had obtained a further two years extension to his contract. Salvesens themselves sent confirmation to the Foreign Office, on 16 March, that an extension had indeed been granted and that Davidoff was acting with their full knowledge and consent. In the event, Davidoff never made the additional payment and the contract therefore lapsed on 31 March. However, it is difficult to imagine what other action Davidoff and Salvesens could have taken, at that stage, to keep the British government informed.

On 4 March the RRS *John Biscoe* arrived in King Edward Cove from Signy Island. Peter Witty embarked after handing over the duties of Magistrate and Base Commander to Steve Martin, a BAS boatman. Also leaving the island were the 'summer season' scientists and tradesmen. Staying behind were twenty-six BAS personnel who would be over-wintering: the majority at and around the Grytviken area, with four on Bird Island. The latter base is just off the northern tip of South Georgia and 1982 was the first season since 1963 that it had been planned to use it for winter observations. The *John Biscoe* sailed on 6 March for Port Stanley and, ultimately, the United Kingdom. Those remaining behind settled down to the prospect of the long dark silence of winter.

On 11 March, the British Embassy in Buenos Aires received one of the last communications which could be regarded as normal. It was a telephone call from a lawyer named Oliva who frequently acted as spokesman for Constantino Davidoff. He wanted to confirm the imminent sailing of the *Bahia Buen Suceso*. Michael Hickson, the First Secretary who usually dealt with Davidoff and who was known as 'Mr Falklands' in the Embassy, was not available at the time. The call was taken instead by the Consul-General, David Joy, who stressed to Oliva the importance of following the proper 'port of entry' procedures in order to avoid a repetition of the earlier diplomatic exchanges. David Joy recalls the conversation: 'I was essentially concerned to have details of the people involved and to make it repeatedly clear to Oliva that the ship should first report at King Edward Point for entry purposes. There was no reason to try to prevent Davidoff's people from sailing. That in itself was not illegal.'

The *Bahia Buen Suceso* sailed shortly afterwards for South Georgia and her specific destination of Leith Harbour. It is evident that her captain – and the senior Argentine naval staff officers under whose orders he was operating – never intended at any stage to obey the instructions conveyed by Hickson and Joy.

The main characters in the drama soon to be played out on the chill fringe of Antarctica were now assembling. They had no script, no inkling of the plot about to unfold. One player only had yet to appear, and she was a remarkable character destined to fill the role of heroine. She was HMS *Endurance*, a vessel whose distinctive paintwork explained the affectionate sobriquet used by all who knew her – 'the red plum'. □

III | Enter the Plum

In 1968, the Royal Navy took into service a ship which – in her own idiosyncratic way – was later to earn a reputation as familiar as those of the *Victory*, the *Hood*, the *Kelly* or the *Ark Royal*. At first glance she was an unlikely candidate for fame and glory. Her original name was the *Anita Dan*. Built by Krogerwerft of Rendsburg (Denmark) in 1956, she was employed in the business of carrying passengers and freight between Northern Europe and the ports of Iceland, Greenland and Spitzbergen. This involved transitting waters where icebergs and pack-ice are a common hazard, consequently her hull was designed and constructed for navigation in ice. Her bow was given extra framing and a steeply sloping curve, with steel plating an inch thick. She was not intended to work as an ice-breaker but simply as a polar merchantman.

For eleven years she went quietly about her lonely business, far to the north of all normal shipping lanes, until her owners, J. Lauritzen of Copenhagen, decided that she was no longer profitable. Taken out of service, she was left swinging at a mooring, her future uncertain until she was purchased by the Royal Navy.

Since 1955 the role of Antarctic ice patrol vessel and Falkland Islands guardship had been filled by HMS *Protector*, a converted net-layer of 2,900 tons built in 1936. This ship was reaching the end of her working life and a replacement was needed. The *Anita Dan* fitted the bill very well and she was cheap. Her purchase price of just under £300,000 was not much more than her scrap value and was arguably the most cost-effective acquistion the Royal Navy ever made.

Taken to Belfast, she was handed over to Harland & Wolff for a major refit. The most important structural change was the addition of a large hangar abaft the bridge superstructure and over the after hold. The new space was designed for the maintenance and housing of two helicopters. The deck of the hangar was extended aft over the former poop deck to create a helicopter flight deck. Essential though they were, these additions added 700 tons – above the centre of gravity – to the ship's original displacement of 3,500 tons fully fuelled and stored.

The high flat sides of the new structure made the ship less easy to handle when coming alongside in strong winds. The additional weight also gave her a tendency to roll in even the most moderate seas and, for reasons never fathomed, she listed to starboard when fully laden and to port when not. From the days of her first work-up, the Navy's newest recruit showed a determination to be different from any other ship in the *Navy List*.

Harland & Wolff ripped out a great deal of the *Anita Dan*'s freighting equipment and installed new accommodation for the twenty officers, twenty-eight senior ratings, seventy-eight junior ratings, and twelve Royal Marines who would constitute a normal ship's complement. She would be operating without break for months at a time, cut off from any Royal Navy support base, so self-sufficiency combined with a degree of comfort was a prime requirement. The cost of the conversion was approximately two million pounds.

One of the builder's original features retained by the Navy was the elegant wooden panelling in the officers' and senior ratings' accommodation and in the passenger cabins. Although the cause of much strong language – whenever the electricians needed to repair the concealed wiring – the polished wood provided a flavour of elegant living in what was otherwise a highly functional working ship. Another retained item of original equipment was the Burmeister & Wains two-stroke diesel engine. Producing 3,220 indicated horsepower at 170 rpm, this simple form of propulsion machinery had the great merit of reliability when run non-stop for very long periods. The single screw pushed the ship along at a modest maximum of thirteen and a half knots, but the task required range rather than speed. After refit, her fuel tank capacity enabled her to steam 4,000 miles without bunkering. The main disadvantage of the engine was that it was necessary to stop it before engaging reverse gear.

On 28 June 1968, the *Anita Dan* completed her metamorphosis and emerged as Her Majesty's Ship *Endurance*. The re-naming ceremony – in Harland & Wolff's yard – was carried out by Miss Alexandra Shackleton, granddaughter of the Antarctic explorer. As the bottle of champagne exploded against the red-painted bow, new life was given to the name which had distinguished Sir Ernest's ill-starred vessel on the Imperial Transantarctic Expedition of 1914–16. Taken from the family motto *Fortitudine Vincimus* – By Endurance We Conquer – the name had been remarkably apposite for the explorer's epic escape from the Weddell Sea to South Georgia. In 1982, it was equally appropriate to the historic ten months' deployment of his ship's successor.

The new *Endurance* followed a regular and well-ordered work pattern for fifteen years. In October of each year she sailed for the South Atlantic where the long winter nights of the southern hemisphere were ending. Six weeks later she would reach Port Stanley where the population always turned out to welcome 'our ship'. Her arrival was an important event in the colony's annual calendar, a harbinger of summer and a symbol of the close ties with the United Kingdom.

Facing page: The metamorphosis: the *Anita Dan* (above) photographed in her role as a polar merchantman sailing Arctic waters under Danish colours, and then seen four years later (below) carrying the White Ensign through the Lemaine Channel in Antarctica in 1971.

ANITA DAN

A171

Captain N. J. 'Nick' Barker, RN, the officer who commanded HMS *Endurance* throughout the most critical period of her long and unusual career. During the weeks preceding the Argentine invasions, his detailed reports to London were studied with interest but did not provoke the responses for which he had hoped. He was not the first officer in British naval history to incur the displeasure of Whitehall in having his warnings vindicated by subsequent events. At the conclusion of Operation Paraquat his name did not appear in the list of 'immediate' honours and awards. His appointment as a Commander of the Order of the British Empire was published in *The London Gazette* in the following October.

The following weeks were devoted to a busy schedule of visits to the scientific bases - operated by various nations - around the coasts of the Antarctic Peninsula and the islands of South Sandwich, South Orkney, South Shetlands and South Georgia.

There was regular and friendly contact with similar ships from the United States, Russia, Chile and Argentina. A typical season's programme included numerous movements of scientists and their equipment from place to place, thereby assisting the BAS supply ships *John Biscoe* and *Bransfield.* The *Endurance*'s own specialist officers conducted surveys and experiments connected with, among other subjects, the movement of pack-ice and icebergs. One of the most important functions was the mapping of coastlines - for which the two helicopters were needed in the aerial survey role -and charting the sea-bed. Large tracts of the Southern Atlantic have yet to be charted and the *Endurance* was therefore equipped with an extensive chartroom where the hydrographers - the 'droggies' - practised their craft for the Royal Navy's Hydrographic Office at Taunton, Somerset.

The *Endurance* was not simply a 'patrol' ship, a title which implies a passive role, but much more a mobile workshop, communications centre, laboratory and drawing office. She was also the sole regular bearer of the White Ensign south of the Equator. In the early 1980s the annual cost of operating the ship was less than five million pounds, not much more than the purchase price of one Sea Harrier. Whitehall and Downing Street tried several times to save this money by scrapping the *Endurance* and not replacing her. The old adage 'penny wise, pound foolish' was neatly illustrated when even the threat of scrapping became a contributory factor to the subsequent war with Argentina.

The ship's complement - most of whom were volunteers for Antarctic service - consisted half of officers and men who had already completed one commission in HMS *Endurance*, and half who had not. This practice ensured the handing on of specialist knowledge and experience peculiar to the ship's unusual role. A careful selection of her people was important in ensuring a 'happy ship' which, by inference, is an efficient ship.

Even more important to a ship's efficiency is the quality of the man who commands her. The *Endurance* had always attracted the services of captains of broad experience but, for this most important of cruises, she was commanded by an officer who was truly tailor-made for the job.

Captain Nicholas Barker, RN, was descended from a long line of military and diplomat forebears. His father was a naval officer - killed on 8 June 1940 when in command of the destroyer HMS *Ardent.* Off the coast of Norway, in company with another destroyer, the *Acasta,* Lieutenant-Commander J. F. Barker, RN, attempted to protect the aircraft carrier *Glorious* when their small force was attacked by the two German battle cruisers *Scharnhorst* and *Gneisenau.* To quote the official historian: 'The destroyers, in accordance with the heroic traditions of their class, made for their giant adversaries at high speed and laid a smoke-screen . . . having fired all her torpedoes the *Ardent* was overwhelmed by gunfire and sunk.' The unequal battle lasted two hours. There were very few British survivors from either ship to tell the tale, but the mighty *Scharnhorst* was crippled by a torpedo strike and forced back to dry-dock for repairs.

Some months later, a much-scrubbed best-suited little boy of eight was escorted to London from his preparatory school in Newton Abbot, Devon. He went to receive from the hands of King George VI the posthumous Distinguished Service Cross awarded to his father for previous gallantry during a convoy action. It was Nick Barker's first visit to Buckingham Palace, but not his last.

The example of the *Ardent* was an inspiration, and young Barker decided to follow his father into the Navy. Initially he did so by the unusual route of National Service, but two years later, in 1953, was accepted for a permanent commission. Over the next twenty-seven years he held a wide variety of staff and sea-going appointments, including command of the Type 21 frigate *Arrow* (1975–77) which he had earlier stood by while she was under construction at Yarrow Ship Builders. HMS *Arrow* was to see some sharp action, and receive battle damage, in Falkland Sound.

Nick Barker was appointed to command HMS *Endurance* in May of 1980. The first step was to familiarise himself with the very special circumstances surrounding his future duties. Two and a half months were spent on intensive briefings by experts at the BAS headquarters and the Scott Polar Research Institute at Cambridge, the Hydrographic Office at Taunton, various departments of the MOD (Navy), at Trinity House, and elsewhere. A captain of the *Endurance* was always much more than 'the driver' of the ship - he needed also to be sensitive to the historical, scientific, and navigational features specific to the Antarctic region. There could be few greater challenges for any officer in the Captains' List.

The 1980/81 deployment was completed with satisfactory results. The ship experienced the usual mixture of adventures, some happy, some less so, to be expected while operating in Antractica's hostile environment. Throughout those months Nick Barker spent much of his time in making an in-depth study of the region's political and economic climate. He was following the example of Bowles and Hardy who, between 1816 and 1823, sailed the same waters. He devoted himself to learning as much as he could of subjects which might normally be regarded as outside the responsibilities of a ship's captain. Although superficially the epitomy of the conventional Royal Navy officer, Barker was too inquisitive, too interested in the broader horizons, to content himself with pleasing his superiors and nothing more.

He came to know the staff of the British Embassy in Buenos Aires. The relationship was not an easy one. He formed friendships ashore which did not always meet with the approval of the Ambassador and his senior staff. There were clashes of personality between him and them, and there were other problems within the staff itself. Nick Barker, for his part, was not impressed by the arrangements made by the Embassy for the reception of his ship or the exploitation of her diplomatic value.

The Defence Attaché, Colonel Stephen Love, and the Naval Attaché, Captain Julian Mitchell, had as their main responsibility the promotion of British arms sales to the Argentine and Uruguayan governments, a responsibility which they shared with Michael Hickson, the First Secretary, and Michael Jackson, the Commercial Attaché. Two of these men – Mitchell and Hickson – were additionally responsible for general surveillance of the naval and political situation in Argentina and for keeping watch on developments with regard to the Falkland Islands. Stephen Love, in common with Julian Mitchell, was accredited to the British Embassy in Montevideo, capital of Uruguay, but his terms of reference did not extend to any of the South Atlantic islands.

As full-time residents of Argentina, the Embassy staff was well placed to keep a finger on the national pulse. Given the long-standing friendship between the two countries, there was no obvious need to cultivate a clandestine web of sources for the collation of military or political intelligence. Indeed, the very nature of Argentina's ruling military class was such that it was difficult to penetrate or assess, even if there had been a compelling need so to do. But at least the Embassy personnel were living at the centre of a city of seventeen million people where they could swiftly identify the changing patterns of power and public opinion.

The *Endurance,* on the other hand, visited the mainland on infrequent occasions. Captain Barker's assessments were based more upon his contacts with the ships and personnel of the many nations, including Argentina, then working in Antarctica and the South Atlantic. Inevitably, his views did not always agree with those of the Embassy. It would have been surprising if, under these circumstances, the opinions of the professional seafarer had always agreed with those of the professional diplomat. In broad terms, however, all agreed that the situation now developing in the region was far more worrying than the crises which had flared up from time to time in previous years.

There was little or no direct contact between Ambassador Williams in Buenos Aires and Governor Hunt in Port Stanley. Captain Barker was in some respects one of the men best placed to act as liaison between them. Barker, however, found himself more in sympathy with the views of Hunt than those of Williams, and this did not contribute to unity. There was a feeling of cordial wariness whenever the *Endurance* had direct contact with the Embassy, but the task of representing British interests was correctly fulfilled and Captain Mitchell succeeded in obtaining diplomatic clearance for the *Endurance* to visit the Argentine naval ports of Puerto Belgrano and Ushuaia, the first time she had been seen in either place for several years.

The relationship between the two navies was friendly and open. Apart from the traditional fellow feeling which exists between mariners of all nations, there was the fact that Argentina was modernising her navy with British assistance. Two of her most powerful units were the Type 42 destroyers *Hercules* (built by Vickers in the United Kingdom and delivered in 1977), and the *Santissima Trinidad* (built under contract at Rio Santiago with British supervision and just completing her trials in 1981). There was an encouraging prospect of further sales contracts in the future and the British Embassy was anxious to foster even closer ties.

An early but adverse note – and one which reinforced the serious doubts in the mind of Captain Barker – was the discovery of an Argentine vessel in a cove on the south-west coast of West Falkland. In February 1981, the pilot of a FIGAS (Falkland Island Government Aviation Service) civilian aeroplane reported having seen an unknown ship at anchor. She was later identified as the *Yehuin,* an oil rig tender owned by Geomatter SA of Buenos Aires. The *Endurance* was at the time visiting Port Stanley and Lieutenant-Commander Tony Ellerbeck, the Flight Commander, was despatched to investigate the sighting. A Spanish-speaking member of the ship's company was winched down to the deck of a vessel which subsequently became familiar as 'The Black Pig'. Her master was asked to explain his presence in territorial waters and his reason for not having reported his arrival in conformity with international maritime law. His excuses – that his radio was inoperative and that he had been blown four hundred miles off course while heading for Ushuaia – were regarded as unconvincing. Governor Hunt ordered him to sail immediately, which he did.

Barker and Hunt discussed the curious episode and debated whether the Argentines were testing the ability of the British to detect foreign ships in their local waters. It was possible that other intrusions would follow, either to provoke a response or to survey potential landing sites. Barker ordered the establishment of additional aviation fuel dumps at various sites around the Falkland Islands so that, if the need arose, his Wasp helicopters could swiftly reconnoitre the entire coastline without needing to return to the mother ship. Quantities of fuel were airlifted to a number of outlying settlements before the ship left the islands.

In May 1981, the *Endurance* returned to her home port of Chatham and was given a minimal refit. With the ship earmarked either for sale or for the knacker's yard, she was given sufficient refurbishment to complete just one more voyage to the south. On 13 October, she sailed from Portsmouth on the morning tide, unsung and almost unnoticed. The lack of official interest in her departure was epitomised by the fact that no crane-driver could be found in the dockyard to lift away the heavy brow or gangway connecting the ship to the quay. It was characteristic of her captain that he simply ordered the moorings to be cast off and sailed downstream, leaving the brow to fall into the water. Families and girlfriends watched from Southsea beach as the distinctive red hull, bearing the white code number A171 on its side, slipped quietly down the Solent.

After calling at Gibraltar, Funchal, Rio de Janeiro and Montevideo, the *Endurance* reached Bahia Blanca, an important Argentine naval base, on 26 November. The principal Argentine naval unit currently in port was the cruiser *General Belgrano.* The visit lasted only six days but, during that time, the British established a particularly happy relationship with the ship's company of the

Right: Governor Sir Rex Hunt lays a wreath at the monument on Port Stanley waterfront during the first Battle Day commemoration following the removal of Argentine forces from the islands.

Above: Four of the key figures in the events of 1982 (from left to right): Captain Nick Barker, RN; Governor Rex Hunt, CMG; Lieutenant Keith Mills, RM; and Sergeant Peter Leach, RM. Each, in his own way, was to play a prominent role in the drama of South Georgia between 25 March and 3 April.

big warship. Argentine ratings were invited to visit their opposite numbers aboard the *Endurance* and the Argentines, in their turn, organised various sporting and angling events for the two ships. The *General Belgrano* arranged a joint ships' barbecue with a display of folk dancing. In the words of Chief Petty Officer Geoff Cox: 'We were entertained to dinner in their homes and were always welcome when we felt the need for a retreat from shipboard life. Some of our lads made some very happy friendships during that time.'

Leaving Bahia Blanca on 1 December, the *Endurance* reached the Falkland Islands four days later. The first task was to lift, by Wasp, six members of the ship's Survey Department, with their equipment, to Green Patch (Berkeley Sound). They were tasked to carry out extensive survey work around a coastline which had last been surveyed by HMS *Beagle* in 1834.

'The red plum' dropped anchor in Port Stanley harbour on the following day and the usual flow of visitors – formal and informal – began at once. One group of matelots, however, had other matters on their minds. The ship's Royal Marines detachment, commanded by Lieutenant Keith Mills, RM, had performed their share of guards of honour and sunset ceremonies on the way south. It was now the turn of the sailors. They were to have the honour of representing the ship's company at the Battle Day Parade on 8 January. This annual event commemorates the first Battle of the Falklands when, in 1914, Vice-Admiral Sturdee caught and destroyed, in nearby waters, a German battle squadron commanded by Vice-Admiral Graf von Spee. During the previous days, with much hilarity, and suffering considerable ribaldry from the marines, the sailors sought boots of adequate fit, rehearsed their arms drill, and were exercised on the flight

deck by Sergeant Peter Leach. On the big day, the well-drilled contingent was complimented on its turnout by Governor Hunt, and the marines were suitably impressed. The ship had taken another step towards welding itself into a cohesive family unit.

Over the next five days, the two Wasp helicopters continued to loadlift survey equipment up the coast to Green Patch and to deliver thirteen tons of ammunition to the Royal Marines garrison at Moody Brook. Each aircraft completed forty hours of flying time during those five days, a fair indication of the workload to be expected throughout the coming weeks.

The two Wasps were in fact the most valuable asset at Captain Barker's disposal. Apart from being an important research and survey tool, and apart from having a useful 'lift' capability (three passengers or underslung freight), they were also the only medium range weapon of defence carried by the *Endurance*. In view of what was to follow, it is important to know something of the characteristics of these aircraft.

The prototype first flew in July 1958, the design being intended to meet the Royal Navy's requirement for a light anti-submarine strike helicopter operating from the decks of small ships. The first production Wasp, XS527, entered service in January 1963 and a total of ninety-eight were eventually built by the Fairey Division of Westland Aircraft Limited, Hayes, Middlesex. A training squadron was established at Culdrose, but the headquarters squadron for all operational Wasps was 829 Naval Air Squadron based at the Royal Naval Air Station Portland, near Weymouth in Dorset. No less than twenty-two frigates were equipped with these helicopters, normally as singletons, and they were known as Small Ship Wasp Flights.

The Wasp was a small aircraft, its basic weight of 4,350 pounds being not much more than that of a large family motorcar. It had a

A Wasp HAS1 helicopter launches a AS12 missile during a trial firing off Portland. The French-manufactured missile had a maximum range of four and a half miles and travelled at a speed of just under three hundred miles per hour. Developed initially as a lightweight anti-tank and anti-surface vessel weapon, the AS12 was already approaching obsolesence at the time of the Anglo-Argentine conflict.

maximum speed of 125 miles per hour at sea-level, a maximum climb rate of 1,200 feet per minute and a service ceiling of 10,000 feet. With a crew of two, the maximum range was 135 miles at 100 miles per hour, but the designers assumed that it would normally operate within thirty miles of its parent ship, armed with depth-charges, or the Mk44 anti-submarine homing torpedo, or air-to-surface missiles.

The type continued to be the Fleet Air Arm's standard small-ship helicopter for the next thirteen years until, beginning in 1977, it was gradually superseded by the Westland Lynx. As far as the fighting navy was concerned, the Wasp was no longer sufficiently large or powerful to carry the weight and diversity of sophisticated equipment needed for modern anti-submarine warfare. For patrol duties in the South Atlantic, however, it was still an excellent machine. Sturdy and simple in construction, the airframe was well proven and trouble-free. The Rolls-Royce Nimbus engine was likewise extremely reliable over long periods without major overhaul. Consequently the Wasp was still carried by HMS *Endurance* (in place of the earlier Westland Whirlwind) when the ship set off on her 1981–82 cruise to Antarctica.

The two examples on board were XS527 (call-sign 434) and XT418 (call-sign 435). Commanding the flight was Lieutenant-Commander Tony Ellerbeck. He usually flew 434, the first Wasp to enter Royal Navy service and at the time twenty years old. He

There have been three helicopter accidents on South Georgia and a number of near disasters. The first aviator to come to grief was Lieutenant Tim Finding, RN, of HMS *Endurance*'s Wasp Flight. This was the scene after XT418 was caught by a sudden change of wind at St Andrews Bay on 12 December 1981.

Right: Lieutenant Commander Tony Ellerbeck, RN, lifts the stripped remains of XT418 back to HMS *Endurance* from the crash site at St Andrews Bay.

had no reason to believe that his team would ever become involved in any form of armed conflict, but he had the comforting knowledge that his two aircraft were fitted with mountings for the external attachment of the Nord AS12 air-to-surface missile. This weapon was designed primarily to defend a parent ship from attack by small high-speed surface craft, and HMS *Endurance* carried sixteen of them. The Wasps also carried mountings for a general purpose machine gun (GPMG) to be operated from the rear passenger space.

The terms of the Antarctic Treaty prohibited the deployment of military weaponry within the Treaty area, but over the years this clause had been so openly flouted by other countries – most particularly Argentina – that it was an obvious precaution for a White Ensign ship to have the means to defend herself. In view of its limited range, the AS12 was regarded as a weapon of defence, as were the two obsolete 20mm Oerlikon heavy machine guns which could be mounted on the bridge superstructure at short notice. The other 'weapon' carried by *Endurance* was her contingent of one officer and twelve other ranks of the Royal Marines. In no sense of the word was the *Endurance* a warship – she was, to the contrary, tasked primarily to improve Man's knowledge of his own environment – but she had the capability to at least bare her teeth at any potential aggressor.

Leaving Port Stanley on 9 December, 'the plum' set course for South Georgia. On board she had His Excellency the Governor, Rex Hunt, and his wife Mavis. Hunt's parish extended far beyond the confines of the Falkland Islands – he was responsible for the administration and good government of the whole of British Antarctica and the associated islands. The *Endurance* was the ideal vehicle in which to make his rounds.

Other passengers included a sixteen-man Joint Services Expedition commanded by Lieutenant Bob Veal, RN. They were bound for Royal Bay, about twenty-five miles down the coast from Grytviken, where they planned to study glaciers and climb some of the local peaks. Accompanying them was a BBC team which had come south to film location sequences for a television account of Shackleton's epic journey. Another passenger was Commander Mario Fontanot of the Uruguayan Navy. He had joined the ship at Montevideo and was on board as a goodwill guest of the Royal Navy, to observe and report upon the prospects for international co-operation in Antarctica.

Dropping anchor off Molke Harbour at 1000, on 12 December, the *Endurance* ferried ashore, to eight widely scattered sites, the expedition party and their ten tons of stores. The ship then moved ten miles up the coast, in blustery weather, to St Andrews Bay. At 1400, the Governor and his wife climbed into Wasp 435 for the short flight to the beach where they intended to visit Cindy Buxton and Annie Price, the two civilian wildlife photographers living there in a small hut. The hazards of aviating around South Georgia were dramatically underlined when Lieutenant Tim Finding, the Senior Pilot, was caught by one of the notorious freak winds as he prepared to touch down. There was a strong wind blowing off the land but, at beach level, it was swirling in the opposite direction. Approaching the rocky beach at low speed, Finding discovered too late that he had a tailwind. He was swept forward, the wheels caught against the boulders, and the Wasp rolled onto its side. His passengers were shaken but unhurt. They continued with their programme, but the fuselage and machinery of 435 were badly damaged by the flailing rotor blades. This event led to a major change in the ship's programme.

Three mechanicians were left at St Andrews Bay to strip and salvage parts from the wreck while the *Endurance* completed her visits to other parts of the island. The Wasp was beyond repair with the limited resources available on board. A replacement was

essential if the summer programme was to be carried through. The ship returned to Port Stanley for Christmas and then made the thousand-mile passage to Montevideo to collect a new helicopter, XS539, flown out to Carasco Airport in a C-130 Hercules from RAF Lyneham. The replacement was soon assembled, flight tested and embellished with the number 435 and *Endurance*'s own penguin motif.

By 6 January, the ship was deep in Antarctic waters and the next two months were devoted to the second work period of the programme. A group of scientists from the Scott Polar Research Institute, Cambridge, had come aboard and they were airlifted onto various icebergs which they wished to study. A British Antarctic Survey team, carrying out geological surveys on James Ross Island, was assisted and embarked. Lord Buxton, a leading figure in the Westminster 'Falklands lobby', and father of Cindy Buxton, came aboard with his wife.

On 22 January, the *Endurance* made a courtesy visit to Ushuaia in Tierra del Fuego, only four hundred miles from the Falkland Islands. In complete contrast to the November visit to Bahia Blanca, the ship encountered a very unwelcoming reception by the local military authorities and naval staff. Barker had laid plans for various social activities with the Admiral commanding the district, and had been looking forward to the usual friendly contacts with the shore. Instead, he found that 'they had given us a gangway and a gash bucket, and that was just about all. The atmosphere was distinctly frigid and I couldn't understand why. I discovered that my Admiral chum had been recalled to Buenos Aires. I asked his Chief of Staff what was going on. He told me that we were now in a war zone and that the Admiral had been recalled to keep him from talking to me because we were on friendly terms. I was baffled, and asked what war he was talking about. Was he referring to a war with Chile? "Yes," he said, "but it is also the Las Malvinas war zone." This was said in a half-joking way, but conversations with other people during those three days made it plain that the armed forces were certainly thinking hard about the Falklands. The local civilian population, who seemed to be rather embarrassed by the whole thing, were as friendly as ever.'

At the end of the month, the *Endurance* docked for four days in the Chilean naval base of Punta Arenas, also in Tierra del Fuego. Captain Barker made a point of raising the 'war zone' conversation with some of the senior Chilean naval officers who were known to him. 'The sum total of their advice was that the Brits would have to take care because the Argentines were up to no good. Of course, I needed to take account of the current dispute over the Beagle Channel and the possibility that the Chileans had an axe to grind, but, even so, all the signs were there to indicate that something very unusual was going on. I reported all of this to the Embassy in Buenos Aires, as I had the episode at Ushuaia.'

In mid-February there was a welcome break of eighteen days, for maintenance and re-storing, at the Argentine naval port and holiday resort of Mar del Plata. Here the reception was very different to the frosty response at Ushuaia. Mar del Plata was always a 'good run ashore' for the *Endurance,* involving much socialising with the ships of the Argentine navy and a constant round of beach parties and formal dinners. On this occasion, the visit was even more glamourous than usual because it coincided with the arrival of dozens of ocean-going yachts taking part in the Whitbread Round the World Race. Many of the crews were British, and they were glad to see the *Endurance* and to invite her sailors and marines to their own parties.

The yachts filled the section of dock normally allocated to visiting foreign ships. The Argentine navy granted the privilege of permitting Captain Barker to berth inside the main naval base. His engineers were allowed to use the naval workshops for repair work and the ship was treated almost as though she was an Argentine vessel. Contacts with Argentine warships in harbour, which included a visit to the sail training ship *Libertad,* were as cordial as ever. Nick Barker and his officers particularly enjoyed the company – at dinner and a cocktail party – of a charming Argentine naval officer named Capitan de Fragata Horacio Alberto Bicain. He was the commanding officer of the submarine *Santa Fe.* He and Barker were destined to meet again, only a few weeks later, under very different circumstances.

It is now apparent that few, if any, of the Argentine officers with whom the *Endurance* made contact during this period knew of the plot being hatched by the junta in Buenos Aires. However, after six years of terrorist activity, they were extremely sensitive – almost to the point of paranoia – on all matters relating to security. This attitude led to a couple of unfortunate incidents.

The positioning of the *Endurance,* inside the naval base area, placed her in close proximity to the tightly guarded submarine base. Lieutenant Dick Buckland, one of the Wasp observers, cruised into this area on his sailboard and was promptly ordered out – with much blowing of whistles and pointing of rifles – by the sentries. Much more frightening was the experience of two leading rates who inadvertently wandered into an unfenced military area one night when returning to their ship. The *Endurance*'s berth placed her nearly a mile from the bright lights of the city and the route took the ship's company through a tightly patrolled security zone. It was easy, in the dark, to become lost. Tab Hunter and Nick Carter had enjoyed an excellent fish supper and were happily anticipating the comfort of their bunks when suddenly they were blinded by searchlights. Rifle shots cracked past their ears and they threw themselves flat. Armed guards quickly grabbed the two sailors, searched them thoroughly and

Despite the rising tensions between their two governments, the navies of Argentina and Great Britain continued to enjoy a cordial relationship in Antarctica during the opening months of 1982. Here the commander of the important Marambio base is delivered to HMS *Endurance*'s small flight deck in early February by a Chinook helicopter of the Argentine air force for a social visit. Open warfare is only two months away.

Captain Moya, commanding officer of the submarine base at Mar del Plata, makes a courtesy visit to HMS *Endurance* and is received with the guard of honour to which his rank and appointment entitle him. He is escorted by Captain Nick Barker, RN, (to the rear) and Lieutenant Keith Mills, RM, (left). Five weeks later, on 9 April, one of the submarines under Moya's command, the *Santa Fe*, will sail from Mar del Plata on her first and last war patrol under Argentine colours. HMS *Endurance* will be one of her potential targets.

emptied their pockets. Among the possessions taken from them were the ship's identity cards – printed in English and Spanish – known as the 'you can't shoot me' chitties.

After an hour in confinement, and having been interrogated by an officer waving a pistol in their faces, they succeeded in drawing attention to the chitties in their pile of personal effects. They were escorted back to the *Endurance* with a stern warning to be more careful in future.

Apart from these misunderstandings, there were no other untoward events during the eighteen days in port. Despite the growing anxieties at the diplomatic and political level, there was little evidence at Mar del Plata to suggest that direct armed conflict between the two countries was imminent. However, Nick Barker talked discreetly with his friends in the local Anglo-Argentine community, and they confirmed that the Galtieri regime was in desperate economic trouble and making increasingly bellicose statements with regard to the 'Islas Malvinas'. He reported these conversations to Ambassador Williams, to Captain Mitchell and to the Ministry of Defence.

The *Endurance* departed Mar del Plata on the morning of 22 February – a large crowd of shoreside friends waving her off in brilliant sunshine – and she reached Port Stanley three days later. She sailed again, after only a few hours, to commence her third and final work period in the true Antarctic. On 1 March, in foul weather, she reached the BAS base at Rothera, Four months earlier a violent storm had wrecked the two Twin Otter aircraft normally operated by the BAS at this station, so the *Endurance*'s Wasp flight was particularly welcome. Other bases around the Antarctic Peninsula were visited and serviced, and various evolutions carried out to satisfy the needs of a Royal Navy camera crew filming sequences for a projected public relations feature entitled '*Endurance*'. An important hydrographic survey was completed before she headed back to South Georgia to collect the

With the Argentine invasions only four weeks away, HMS *Endurance* continues the third period of her planned work programme in Antarctica. On 5 March, after a hydrographic survey and stores delivery visit to Brabant Island, she heads for the British Antarctic Survey base at Rothera. Lieutenant-Commander Ellerbeck, in Wasp 434, flies ahead for an ice reconnaissance and to give John Tippey's camera crew an opportunity to film the famous and photogenic 'Kodak gap' (so named for its appeal to past photographers).

Joint Services Expedition from Molke Harbour. On 16 March she anchored off King Edward Point.

The summer season was almost over. Soon she would be heading for home via Port Stanley, Montevideo, Buenos Aires, Barbados and the Azores. By 20 May, all being well, the ship's complement would be reunited with their families. In the event, HMS *Endurance* did not reach Chatham until 20 August, three months late, after a cruise of 311 days and having covered, almost without a break, 41,466 miles.

The root cause of the changed itinerary was contained in a series of signals now reaching Captain Barker from Port Stanley and London. Some days earlier, Constantino Davidoff had informed the British Embassy that his forty-one contract workers would depart Buenos Aires on 11 March aboard the naval auxiliary *Bahia Buen Suceso*. He did not declare an estimated date of arrival, but Captain Barker had to assume that it was imminent. If the *Bahia Buen Suceso* was currently making her way down to South Georgia, it would be reasonable to expect her captain to be making regular weather broadcasts – on the usual commercial maritime frequencies – as required by international convention. The *Endurance*'s operators, however, failed to detect any such transmissions, so it had to be assumed that Captain Briatore was

The *Bahia Buen Suceso*, built by Canadian Vickers in Halifax, Nova Scotia, and commissioned in 1951. First reports of her illegal landings at Leith Harbour caused some confusion in the Foreign & Commonwealth Office in London. It was believed that she was a mercantile marine ship having no connection with the Argentine navy. Reference to *Jane's Fighting Ships* (1980–81) would have revealed her to be what she was, a naval transport deployed under orders from Admiral Anaya's staff.

deliberately maintaining radio silence. This suspicious behaviour carried with it the echoes of Captain Trombetta's clandestine visit to Leith Harbour, in the *Almirante Irizar*, in December. Subsequent analysis suggests that Captain Briatore knew of the *Endurance*'s presence in South Georgia waters and was deliberately delaying his arrival until she was safely out of the way.

Nick Barker was unhappy with the situation, but he could not hang around South Georgia for days, achieving nothing. For all he knew, Captain Briatore might decide at the last moment to make a conventional landfall and respect the correct legal procedures. The *Endurance*, meanwhile, had several pressing reasons for a swift return to Port Stanley. The numerous passengers needed to be returned whence they came. The ship's stores and bunkers needed replenishment. Thirteen of the ship's officers and ratings were still surveying Berkeley Sound and had to be recovered. Finally, and although at that stage few people expected direct military action against the Falkland Islands, the noises emanating from the Galtieri regime suggested that a White Ensign flying in Stanley Harbour might be more useful than off Grytviken.

Before leaving, the *Endurance* performed one final service for the local BAS people. Although they sometimes jokingly referred to her as 'HMS *Encumbrance*', the scientists had a happy relationship with 'the plum'. The Navy was always willing to give a hand, particularly with loadlifting. On this occasion there were stores and a substantial wooden noticeboard to be delivered to Leith Harbour. The red and white lettering stated: 'British Antarctic Survey – Leith Field Station – Unauthorised entry prohibited – *Entrada prohibida sin autorisacion*'. The last line was repeated in Russian, Polish and French. Steve Martin, the Base Commander, was lifted to Leith Harbour on the morning of 16 March in one of the Wasps and he deposited the sign temporarily in the Villa (the former Station Manager's house). He checked the surrounding area but found no signs of any recent visitors.

The *Endurance* said farewell to Martin and his group on the afternoon of 16 March and sailed at 1630. The ship's company did not expect to return here within at least six months. Three days later 'the plum' arrived at Port Stanley and disembarked most of her recent passengers – Lord and Lady Buxton, Bob Veal's expedition personnel, the Uruguayan Commander Fontanot, and Lewis Clifton, the Falkland Islands Philatelic Officer.

Interesting guests are always welcome in any ship but, for the first time in many weeks, the ship's company had their home more or less to themselves. Apart from picking up the survey party from Green Patch, the season's work was almost done and it was time to think of relaxing and heading home to England. The last remaining task would be a visit to Montevideo to collect and then deliver to Port Stanley the new Royal Marines garrison detachment. But it was not to be. That night the first of Steve Martin's reports started to come through from King Edward Point, announcing the arrival of the *Bahia Buen Suceso* and reporting a direct encounter between his own people and the Argentine navy.

On Friday morning, 19 March, a four-man BAS party had left King Edward Cove by launch and rounded the coast to Carlita Bay. They planned to leave the boat there, walk across the Busen peninsula, and proceed around the edge of Stromness Bay to Leith Harbour. They were going to complete a task started earlier in the summer – moving the BAS stores out of the large Villa into the smaller and more convenient Customs House. The property needed to be generally repaired and put in good order before the arrival of the violent winter storms. It was intended also to erect the 'no entry' sign airlifted to the station three days earlier.

The party – Trefor Edwards, Brian Lockwood, Neil Shaw and Bob Banner – carried food and sleeping-bags for an overnight stay. They also had a VHF radio for contact with King Edward Cove. As they crested the high ground overlooking Stromness Bay, they heard several rifle shots and saw that a large ship was berthed at the deep-water jetty at Leith Harbour. The vessel was working cargo, slinging large metal containers ashore with her derricks. There were about fifty men in the harbour area, some in workmen's clothes, others in white 'Alpine type' uniforms of military style. A yellow mobile crane was shifting crates from the jetty to the warehouse buildings. It was scene of intense activity.

The scientists, who were unarmed, approached the settlement and were greeted by two men who smiled and waved vaguely at them. Neither side was conversant with the other's language, so the Britons continued towards the Villa. Behind them trailed a small retinue of curious Argentines.

Entering the house, they found twelve men sitting at the table and eating reindeer steaks. Propped in a corner was the 'no entry' board. The word 'British' had been scrawled over and 'Argentina' painted in its place. The language barrier prevented any conversation. Lockwood and Banner stayed where they were, while Edwards and Shaw went down to the jetty. On the way they passed an open-air barbecue and observed the Argentine national flag flying from the high structure of the station's generator station.

Escorted aboard, the two men asked for the Master and were relieved to find that Captain Briatore spoke English. They explained that Leith Harbour was not a port of entry, that he should proceed immediately to King Edward Cove, that the Villa was accommodation reserved for the BAS, and that his presence was illegal under maritime and Falkland Islands Dependencies

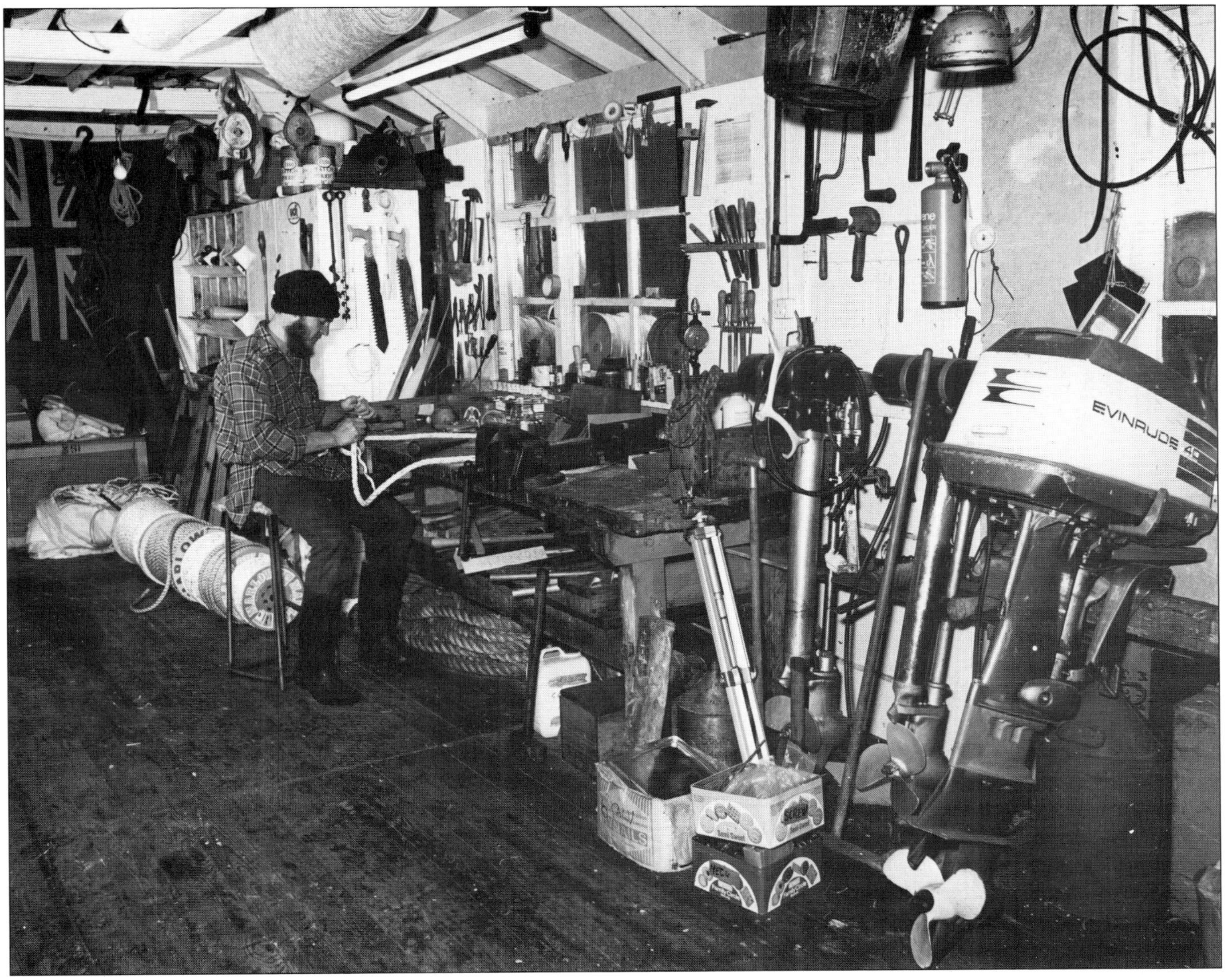

law. The Argentine responded by assuring his visitors, quite untruthfully, that he had been granted permission to land by the Foreign Office in London who, he claimed, had sent a radio message to him only two days before. This was obvious nonsense.

Captain Briatore was relaxed and cordial in his attitude and evidently felt confident. He offered overnight accommodation in his ship to the BAS party and said that he had a gift of fruit sent by Constantino Davidoff.

Both offers were courteously refused and the scientists rejoined their colleagues at the Villa. They then made their way to the Customs House where they found the front door smashed in and the furnishings disturbed. Six of the eight BAS emergency food boxes had been broken open, the contents pillaged and scattered about.

The Argentines were still going about their business and, after their initial curiosity, largely ignoring the British. There had been no attempt to examine or confiscate their belongings so, at 1630, Trefor Edwards was able to set up his radio and send the first detailed situation report to Steve Martin. The others went upstairs, where there was a good view, and unobtrusively took a number of photographs of the *Bahia Buen Suceso* and the busy jetty.

The BAS party maintained their watch throughout the evening and sent several reports to Steve Martin. He, meanwhile, was having difficulties trying to raise the Cable & Wireless operators at Port Stanley. They were expected to be available on an emergency listening frequency, but two hours were lost before they finally responded to calls by the *Endurance*'s operator. The ship, anchored off Port Stanley, intercepted Martin's signals and

The marine workshop in Shackleton House, King Edward Point. The British Antarctic Survey base was well equipped to sustain a round-the-year programme of scientific research, the results of which were shared with scientists worldwide.

took up the cudgels on his behalf. Martin, in the event, continued to talk with the *Endurance* until 0100 on the next morning, Saturday 20 March, before closing down for the night and after agreeing to reopen the channel at 0700.

Much of the radio traffic during the evening resulted from confusion caused by the BAS using one code and the Governor's office another. The codes were new and unfamiliar, and more time was lost. However, by late evening, Governor Hunt had before him a clear description of the scene at Leith Harbour. The next step was to inform London.

Rex Hunt sent a long signal to the Foreign Office that night, reporting the situation, suggesting a course of action, and requesting instructions. By 0730 on the following morning, 20 March, a detailed response had passed back through the complicated radio link-up and was being dictated to Trefor Edwards in the Customs House at Leith Harbour. Drafted by Rex Hunt, the message was a minor masterpiece of firmness and clarity, devoid of threat: 'You have landed illegally at Leith without obtaining proper clearance. You and your party must go back on board the *Bahia Buen Suceso* immediately and report to the Base Commander Grytviken for further instructions. You must remove the Argentine flag from Leith. You must not interfere with the British Antarctic Survey depôt at Leith. You must not alter or deface the notices at Leith. No military personnel are allowed to

Some of the telephoto photographs taken clandestinely by Robert Banner at Leith Harbour when, on 19 March, the illegal presence of the *Bahia Buen Suceso* was first detected by the British Antarctic Survey. The lower picture shows the Argentine flag flying from the tower-mounted generator.

land on South Georgia. No firearms are to be taken ashore. Ends.' By 0800, Trefor Edwards was climbing the gangway of the *Bahia Buen Suceso* to hand the message directly to Captain Briatore.

Despite the obvious communications problems, the British authorities had succeeded in presenting the Argentine navy's representatives with an official response less than sixteen hours after their illegal landing was first notified. In the circumstances, it was a notable achievement. Even so, all did not flow smoothly. Steve Martin was particularly anxious that his Director, then approaching Tierra del Fuego on board the RRS *Bransfield,* and the BAS Deputy Director, at the Cambridge headquarters, should be advised of what was happening that day. Later it transpired that neither Dr Laws nor Dr Adie were as fully informed as they would have wished.

The BAS was a purely scientific research organisation. On South Georgia, as a matter of administrative convenience, the Base Commander carried the title of Magistrate – a mantle of Crown authority which covered, among other matters, the non-scientific activities of selling water to visiting ships and selling the distinctive South Georgia postage stamps so highly regarded by collectors throughout the world. This latter activity yielded an income of several thousand pounds each year to the funds of the Falkland Islands Dependencies. However, the Base Commander was now engaged in diplomatic and para-military activities which went beyond those originally anticipated. Steve Martin was fulfilling the role of spokesman for Governor Hunt and therefore, in the final analysis, for Her Majesty's government. His colleagues were becoming increasingly involved in activities more appropriate to the specialist skills of reconnaissance troops. Martin several times asked for assurance that his superiors were aware of all this, but the evidence suggests that they were not. For the time being, the South Georgia scientists were out on a limb.

Having delivered his message to Captain Briatore, Trefor Edwards collected his small party and they left the Customs House to walk back to the launch at Carlita Bay. By early afternoon, on 19 March, he was giving an eyewitness account to Steve Martin of everything seen and heard at Leith Harbour. It was clear that some of the Argentines were genuine workmen engaged in demolition work. Equipped with oxy-acetelene equipment, lifting gear, hand tools, and other impedimenta of their trade, they were observed in various parts of the old whaling station. Their efforts were disorganised and unsupervised – indeed several seemed to be engaged in private looting – but there was no doubt that they were there to fulfil the Davidoff/Salvesen contract. The status of the men in white uniforms was less certain. The British scientists could not determine whether they were sailors or soldiers. They moved and worked in a manner which suggested that they were a disciplined force. It now seems probable that they were Argentine marines.

All of this was reported to HMS *Endurance* during the night of 19 March. Her radio operators, at that stage, were still receiving Steve Martin's signals, transmitted by the BAS radioman Peter Bayman, more clearly than the Cable & Wireless people. Governor Hunt had a meeting with Nick Barker and they pondered the various options open to them. In view of the rumours and counter-rumours emanating from Buenos Aires, it would have been ideal to retain the *Endurance* at Port Stanley. There was no reason to believe that the Argentines were about to invade the Falkland Islands, but the ship's White Ensign was an important symbol. On the other hand, the BAS people on South Georgia were in urgent need of support and could not long be left to fend for themselves.

Hunt and Barker were agreed that the landing by Davidoff's employees was nothing more than a cover for the establishment of an Argentine navy beachhead on South Georgia, leading later to assertions of sovereignty over the whole island. Their view was supported by Lord Buxton. He had had a meeting with Dr Costa Mendes in Buenos Aires on 23 February while awaiting passage in the *Endurance.* The Argentine Foreign Minister had made statements which caused Lord Buxton to conclude that the junta might be planning to make landings, provided that such incursions were not opposed, on South Georgia and elsewhere, and the peer had sent a personal message to Lord Carrington to this effect. The current events at Leith Harbour fitted perfectly into the 'creeping invasion' theory.

In his report to London, during the night of 19/20 March, Rex Hunt recommended that the entire Argentine group – civilian

A picture which summarises the cause of Governor Hunt's and Captain Barker's worries in the closing days of March. Robert Banner's secret photograph shows not only the civilian crew of the *Bahia Buen Suceso*, dressed in normal working clothes, but also a large group of unidentified men clad in white uniform jackets. Thought to have been Argentine marines, they are unloading numerous large steel crates. The quantity of stores disembarked is far in excess of the needs of Constantino Davidoff's relatively small workforce.

and naval – should be immediately ordered off the island. Further, in view of the presence of uniformed men, the landing of weapons and the shooting of wildlife, he recommended that the Davidoff/ Salvesen contract should be terminated forthwith. To enforce the eviction, he proposed that HMS *Endurance* should return to South Georgia as heavyweight bailiff.

In Whitehall, the arrival of Rex Hunt's signal set the telephones ringing in a number of different departments. It was a Saturday, but no time was wasted in arriving at a decision. Instructions were sent to Anthony Williams, the British Ambassador in Buenos Aires, instructing him to inform the Argentine Foreign Ministry that her Majesty's government regarded the incident as serious. The Argentine *chargé d'affaires* in London was called to the Foreign Office and told much the same thing.

The Argentines were informed that, if the *Bahia Buen Suceso* was not withdrawn immediately, the British government 'would take whatever action seemed necessary'. The Argentine diplomats, not surprisingly, shrugged their shoulders, raised their palms, and stated that they knew nothing about the ship. However, without making an open apology, they expressed the hope that 'the significance of the affair would not be exaggerated'. As on previous occasions, they were hoping to gain maximum advantage without provoking a full-scale row.

By the end of the day the decision had been made – by the Foreign Office and the Ministry of Defence – to follow Rex Hunt's advice by sending the *Endurance* back to South Georgia. The necessary order was signalled to Nick Barker by the Commander-in-Chief, Fleet, that evening. It stated that he should keep his destination confidential 'to avoid the appearance of escalating the incident'.

On that afternoon, Saturday 20 March, the *Endurance*'s football team had played the Stanley Football Club in the final of the Stanley Shield, winning it by the healthy margin of six goals to three. Leading Aircrewman Bob Nadin scored a hat-trick, so the Wasp flight was well pleased with itself. Several of its personnel stayed that night in the old schoolhouse at Green Patch, on Berkeley Sound, ready to commence lifting the last of the survey party's stores back to the *Endurance*. At 0615 on Sunday morning, however, they were awakened by an urgent radio call from the ship, telling them to re-embark immediately. They were sailing in three hours' time.

HMS *Endurance* weighed anchor at 0845 on 21 March. She recovered her Wasps and some of her survey personnel, but two 'droggies', Lieutenants Chris Todhunter and Richard Ball, were left behind at Port Stanley so that they could get on with the task of completing the records accumulated during the season. Subsequently they were captured when the islands were invaded and much of their findings were seized by the Argentines and lost. Another member of the survey party, Leading Seaman Steve Brook, was working alone in a remote part of East Falkland and he also was left behind. He returned to the United Kingdom many weeks later, having successfully evaded capture.

An important addition to the ship's fighting capability was the embarkation of nine more Royal Marines from the Moody Brook garrison. Nick Barker and Rex Hunt judged it prudent to bring the *Endurance*'s Marines detachment up to platoon strength before tangling with the scrapmetal workers and any Argentine troops who might be guarding them. Lieutenant Mills and

Sergeant Leach set about the task of planning the seizure of Leith Harbour. It was assumed that the Royal Marines landing party might encounter armed resistance. Mills prepared his men accordingly. They would go ashore fully equipped, with a generous scale of ammunition and, if it was forced upon them, ready to fight. The prospect of imminent action sent morale soaring.

Three full days would be needed for 'the plum' to retrace her track to South Georgia. What was required in the interim was a 'holding' operation – a breathing space during which the BAS could act as Captain Barker's eyes and ears, but without themselves becoming involved in anything resembling military conflict. They did have a number of elderly .303 rifles at King Edward Point, with a small quantity of ammunition, but these weapons were intended solely for the humane slaughter of the occasional injured seal or reindeer, and for self-defence against any ill-tempered bull elephant seal which might go on the rampage. Later, when invasion was imminent, Steve Martin was so concerned that these rifles might compromise the civilian status of his personnel that he gave orders for them to be dismantled (the bolts held in the base office and the rifles in Shackleton House). The BAS were effectively defenceless, but their intimate personal knowledge of the island made them a priceless asset to Nick Barker.

One of the decisions made at Government House on 20 March was to request the BAS to keep covert watch on Leith Harbour pending the return of the *Endurance*. Dr Laws was consulted by radio and agreed to the proposal. Steve Martin and his people made their plans accordingly and a two-man observation party was taken by boat to Carlita Bay early on Sunday 21 March. By 1130 they had established themselves on the slopes of Jason Peak, the main feature on the Busen Peninsula. This large hill, rising to 2,200 feet, dominates the southern side of Stromness Bay. Taking care to stay off the skyline, Bob Headland and Peter Stark huddled down among the rocks and tried to ignore the bite of the near-freezing wind. They were equipped with a powerful pair of binoculars which enabled them to see the whole of the bay and its sea approaches. They saw that the *Bahia Buen Suceso* had not moved from Leith Harbour, but the high surrounding buildings hid from view the activity behind the jetty.

Bob Headland had given much thought to the communications problem. A botanist by profession, he was also fluent in Spanish, a qualification which made him unique among the BAS personnel on the island. He had lived and studied in Spain and had conducted field work in Latin America. He had two tasks – to monitor any radio traffic from the Argentine ship, and to keep up a running commentary for the information of Steve Martin (and ultimately Nick Barker and Rex Hunt). This presented the obvious risk that his own transmissions might be intercepted by the radio operators on the *Bahia Buen Suceso,* with results which could be embarrassing if not fatal. To minimise this risk, he transmitted his VHF signals at very low power to the launch at Carlita Bay where a colleague relayed them, again at low power, to King Edward Point.

Although their main transmitter was silent, the Argentines were using UHF handsets to communicate between ship and shore, maintaining a regular flow of chatter which Bob Headland monitored. He soon discovered that more drums of fuel were being unloaded and that Captain Briatore hoped to complete this task and sail at 1600 that afternoon. The people on shore were heard to invite the Captain to join them for lunch – they were barbecuing more reindeer meat. Later in the day Captain Briatore asked for another reindeer to be shot – the skin presented to him earlier was not of sufficiently good quality.

The British radio channels were working well. Bob Headland's reports were reaching Nick Barker, pressing south-east at maximum speed, within an hour of being relayed via Steve Martin and Port Stanley. The *Endurance* did not acknowledge the BAS reports – she was herself maintaining radio silence.

At 1600, with the light failing, Bob Headland returned to base. His replacement, Neil Shaw, spent the night with Peter Stark in the hut at Jason Harbour. At 1900 they heard a ship's siren, presumably that of the *Bahia Buen Suceso* as she departed some three hours behind schedule. At dawn, on 22 March, the two men climbed stealthily back to their eyrie and saw that Leith Harbour was indeed empty of shipping. They were asked to remain on watch.

Steve Martin, in the meantime, was under great pressure. Apart from helping to man the radio, he was encoding and decoding signals, consulting with Rex Hunt, supervising his staff at Grytviken, and trying to keep the four field parties informed of local developments. In addition, bearing in mind that the *Bahia Buen Suceso* might appear in Cumberland Bay at any time, and might request entry permits for fifty men or more, he typed fifty questionnaire forms with appropriate wording. The brunt of all this fell upon the shoulders of Steve Martin, and those of Bob Headland when the Base Commander was sleeping, because the remainder of the BAS personnel around Grytviken could not abandon their scientific experiments or normal duties such as preparing food or maintaining the generator and other equipment. The normal life of the base could not be permitted to stop entirely.

Life was further complicated by the presence at Grytviken of a French yacht, the *Cinq Gars Pour.* She had arrived on 14 March with a crew which included a professional photographer and a ham radio operator. They had already caused trouble by going ashore with a .22 rifle and shooting a reindeer. Steve Martin had confiscated the weapon and warned them to respect the local laws. Now, on the morning of 22 March, they announced their intention of departing and asked for the return of the rifle. It was given back to them, but Martin warned them sternly to keep away from the area of Stromness Bay. To his irritation, they then sailed around the Busen Peninsula and proceeded directly to Leith Harbour.

It is unlikely that there was any sinister implication in the presence of the Frenchmen at that time. They were simply ocean wanderers in search of adventure. It is probable, however, that they had been in radio contact with the Argentines at Leith Harbour because they were seen by Neil Shaw and Peter Stark to pick up a small group of Argentines from the jetty and take them to a point on the shore below Jason Peak where a red-painted barge had driven ashore during the night. It was an Argentine craft which had lost its moorings in the strong wind and then drifted on to the beach. The scientists watched as the yacht towed the barge back to Leith Harbour where she berthed at the jetty.

Observation was maintained by Shaw and Stark throughout 22 March. Leith Harbour was five miles away, but their binoculars were of high quality and they could see that, since the departure of the *Bahia Buen Suceso*, there were far fewer men working around the jetty. None of them were in uniform. Increasingly cold and hungry as the day wore on, the scientists manned their post until the evening. They were then collected by the BAS launch and returned to King Edward Point.

Throughout this day the radio channels and telex lines were busy with a flood of signals between King Edward Cove, Port Stanley, London, Cambridge, Northwood and Buenos Aires. The *Endurance* was still 250 miles short of her destination. The *Bahia Buen Suceso* had disappeared. For a day, perhaps two or

Her landings at Leith Harbour were not the end of the *Bahia Buen Suceso*'s involvement in the Anglo-Argentine conflict of 1982. On May 16, at Fox Bay East (West Falkland), she was discovered by a Sea Harrier reconnaissance probe flown by Leiutenant Simon Hargreaves, RN, and Lieutenant Dave Smith, RN, of 800 Naval Air Squadron (HMS *Hermes*). Their report and photographs of the ship led to the decision to mount another Sea Harrier mission flown by Lieutenant-Commander Andy Auld, RN, and, again, Lieutenant Hargreaves. Late on that same afternoon the ship was hit with cannon fire which raked her bridge and set fire to a nearby diesel fuel tank. She was immobilised at this berth until after the cessation of hostilities.

three, the politicians and diplomats could count upon a breathing space in which to plan their next moves.

It was from this time onwards that the British governmental processes began to exhibit symptoms of distress. To comprehend the lack of cohesion, it is necessary to view the scene as it was perceived by the various people involved.

Seen from their end of the telescope, Lord Carrington and Richard Luce were looking at a problem which appeared, in some ways and in the very short term, to have resolved itself. The Argentine flag over Leith Harbour had been lowered and the few men remaining were apparently nothing more than labourers. True, the Argentines had been very naughty and had again broken the rules. Like children let off nanny's leash, they had made a lot of noise, had frightened the animals, and vandalised a notice-board. Certainly their wrists should be slapped, and contingency plans laid for the future, but was it a good idea to call in the Royal Navy right away?

From the other end of the telescope, Rex Hunt and Nick Barker viewed a scene which they judged to be far more menacing. To them, the cavalier behaviour of Captain Briatore was part of the continuum which had commenced four months earlier. The *Almirante Irizar* visit had been the first sign of impending trouble. In November, Captain Trombetta had lied to Nick Barker, telling him that he had no intention of visiting South Georgia (even though he had already laid his plans to do so). There had been three unauthorised flights over South Georgia by Hercules aircraft of the Argentine air force. One of these, on 11 March, was a low-level reconaissance along almost the entire north-east coast of the island.

The track of the aircraft was noted by the Joint Services Expedition and by BAS field parties who radioed their sightings to Steve Martin. The pilot's behaviour was consistent with an intelligence gathering mission – examining potential landing sites and searching for possible defences. Then there was the fact that the Argentines were delivering men and material to the island at the time of the year when summer was ending and the harsh winter commencing. Between April and October, the depth of snow covering Leith Harbour can be as great as six feet and there are only five hours of daylight in every twenty-four. The timing of the *Bahia Buen Suceso*'s arrival suggested that it was less influenced by the working requirements of the scrap metal contract than it was by the ambitions of the Argentine navy.

Censored by the Ministry of Defence.

These were some of the many clues which told Hunt and Barker that the departure of the *Bahia Buen Suceso* was not the end of the affair. They reported their views to the Ministry of Defence and the Foreign Office.

On 23 March, the Argentine government was told that HMS *Endurance* was proceeding to Leith Harbour to remove the scrap workers, this action being the inevitable result of Constantino Davidoff's failure to comply with the legal obligations, but that the eviction should not be viewed as an escalation of tensions between the two countries. In effect, the British were saying, 'we are going to throw your people out, but please don't take it personally.' Such an approach was doomed to failure from the outset. Three days earlier, after the usual Saturday night dance in Port Stanley Town Hall, an unidentified Briton had entered the adjacent LADE (airline) office, and had draped a Union Jack over the Argentine flag. It is alleged that the intruder left behind an insulting message. The Argentines, needless to say, were very angry and, after inflammatory reports of the episode had appeared in their national press, had been given a convenient stick with which to beat the British government.

At the request of Nick Barker, the BAS again manned their observation post on Jason Peak on 23 March, but they saw no change on or around Stromness Bay. They withdrew at midday when low cloud and flurries of snow blotted out the view. The lack of activity, combined with the advice of Anthony Williams and others who understandably wished to keep the diplomatic temperature as low as possible, resulted in a major change of plan. Orders went to Nick Barker to make his landfall at Cumberland East Bay, not Stromness Bay. He was to anchor off Grytviken and wait upon events. This he did, approaching the island under cover of darkness and dropping anchor at 0615 on 24 March. The idea of evicting the scrap men was abandoned, for the time being at least. Hunt and Barker were not pleased by London's apparent indecision. The *Endurance*'s Royal Marines were even less pleased.

The change of heart was prompted partly by the FCO's desire to contain the situation and partly by a stream of signals starting to arrive in London from the British Embassy in Buenos Aires. The feeling in the Embassy since the beginning of the month had been one of extreme gloom. As Ambassador Anthony Williams later expressed it: 'The New York talks on the future of the Falkland Islands had resulted in a fudge of the issue and none of us had any doubt that they merely represented a continuation of the UK government's line – talk and do nothing – which was starting to wear very thin. It was more disappointing than surprising, therefore, when the Argentine junta rejected the outcome of the talks and issued a statement saying that they reserved the right to take whatever action was necessary. The writing was on the wall and we waited with foreboding for an Argentine rejection of further talks and a shift to some radical action.'

For the Embassy, there was an immediate local problem to be tackled. On Monday 21 March, a message arrived from London, confirming the self-evident fact that the *Endurance* was now slipping ever further backwards in her programme. She was intended to be at Montevideo within forty-eight hours to embark forty Royal Marines arriving by air from RAF Lyneham, Wiltshire. This was the incoming relief garrison destined for Port Stanley. With the *Endurance* now heading back toward South Georgia, accommodation would be needed for the marines, in Montevideo, for an extended period.

The British Embassy in Montevideo did not have a resident Defence or Naval Attaché so, on 22 March, Captain Mitchell flew up from Buenos Aires to meet Major Mike Norman and his men. Fortunately the RRS *John Biscoe* was in port. She was due to return to the United Kingdom on the following day. The availability of the *John Biscoe* was an obvious solution to the transport problem and Mitchell sent a signal to the Ministry of Defence requesting her employment in that role. The MOD contacted the Foreign Office which, in turn, telephoned the BAS headquarters in Cambridge where Dr Adie forwarded the request to Dr Laws. He was currently at sea – off the coast of Chile – in the RRS *Bransfield*.

Dr Laws was asked if he would agree to the BAS ship making the passage to Port Stanley as a temporary troop transport. He immediately did so, and the well-laden *John Biscoe* sailed on the evening of 23 March. Her departure – and the purpose of her voyage – was swiftly misrepresented in the Argentine press and Richard Laws found his organisation under attack as being engaged in para-military activities. An Argentine military aircraft came sniffing around as the *John Biscoe* ploughed steadily southward. Laws had the uneasy feeling that the BAS was wandering into political territory where it had no wish to be. His doubts were heightened by the paucity of information reaching him from Cambridge and London.

Having seen the marines safely on their way, Julian Mitchell made the rounds of his local Uruguayan contacts. Their information supported the view that the Argentines were prepared to take very serious risks in achieving their objectives. Mitchell sent another signal to the Ministry of Defence, stating the estimated time of arrival of the *John Biscoe* at Port Stanley and suggesting that she should return empty instead of bringing away Major Gary Noott's tour-expired Naval Party 8901. Leaving Noott's men on the Falkland Islands would double the number of marines available to Rex Hunt as a defence force. The same proposition had already been put to the Foreign Office by Governor Hunt and it was agreed on the following day after consultation between the MOD and the Foreign Office.

While Mitchell was dealing with these matters in Montevideo, more signals were issuing from the Embassy in Buenos Aires. Early on 24 March, a meeting took place between the Ambassador and the Defence Attaché, Colonel Stephen Love. They reviewed the latest developments and drafted a telegram which Love despatched to the Ministry of Defence, with a copy to the Foreign Office. He renewed his earlier warnings that any attempt to remove the scrap metal workers and their armed escorts might be countered by force, either on the island or by interception of the *Endurance* at sea. Further, if the Argentine party was removed and taken to Port Stanley, this act might give the Argentine government an excuse to mount a 'rescue mission' leading to the occupation of the Falkland Islands. Colonel Love's warnings, endorsed by Ambassador Williams, underlined the military nakedness of the islands and the near impossibility of repelling even a limited invasion with the resources then available.

It seems likely that this and other signals of a similar nature –

The last consignment of mail to be processed through the Post Office at King Edward Point was franked on 24 March. This envelope was taken aboard HMS *Endurance* shortly before her departure for Port Stanley but did not reach its English addressee until several months later.

sent earlier by the Embassy – resulted in the decision to draw back from direct contact with the Davidoff party at Leith Harbour and to instead divert the *Endurance* to King Edward Cove. There is evidence also to suggest that this signal caused Lord Carrington to submit an urgent report to Mrs Thatcher and the Defence Committee, that same day, stating that the dispute had developed to a point at which an early confrontation with Argentina might need to be faced. He again asked John Nott to leave the *Endurance* on station. He also suggested that, in advance of the next Defence Committee meeting to discuss the Falkland Islands, the Ministry of Defence should circulate a consultative paper on military contingency planning.

Captain Mitchell arrived back in Buenos Aires on the afternoon of 24 March and was told of the telegram despatched a few hours earlier to London. Agreeing with its content, he sent his own latest assessment. This was based in part upon reports which had appeared in the Argentine newspapers during the past twenty-four hours, and in part upon information garnered from his own sources. According to Julian Mitchell: 'I had spent the first fifteen months of my appointment as Naval Attaché keeping my eyes and ears open. I had personal contacts not only with Argentines, but also with knowledgeable Chileans, Uruguayans and Paraguayans. I was talking to men who were privy – in one way or another – to the activities of the Argentine armed forces. One in particular was exceptionally helpful and well-informed. The advices received from these people made it clear to me that the Argentines were mobilising for active service.'

During the night of 24/25 March, Captain Mitchell despatched a number of signals to London, advising his superiors of the sailing from Mar del Plata of submarines and surface ships which ostensibly were to take part in joint anti-submarine exercises with the navy of Uruguay. He was not alarmed by this particular development because he knew that the exercises had been planned for some time. More disturbing, however, was the sailing from Mar del Plata of two small frigates with the stoutly Elizabethan names *Drummond* and *Granville*. According to Argentine press reports, their destination was South Georgia and their task was 'the protection of the civilian workers currently fulfilling a legitimate contract on that island.'

Mitchell was worried by the thought of two modern warships chasing after the *Endurance*, but he became even more alarmed on the next day. 'I spent several hours telephoning and meeting various contacts and they reported intense activity at the main naval base of Puerto Belgrano. The bulk of the Argentine navy had started to put to sea. It was accompanied by transport ships. Soldiers and marines had been seen embarking with full field equipment and armoured fighting vehicles. The unknown factor was whether or not – having formed an amphibious force – the junta would have the determination to commit it to action.'

Later that day, 25 March, British intelligence would have started to receive evidence which suggested that Admiral Anaya's ships might not be engaged in a routine peacetime exercise. The British were fortunate in having earlier broken Argentine cyphers – a fact subsequently confirmed in the House of Commons by Ted Rowlands, a leading spokesman for the Labour party and a former Minister at the Foreign and Commonwealth Office. The British had the advantage that a proportion of the Argentine radio traffic was transmitted on high frequencies in clear language and was reported by amateur radio operators.

Summaries of these signals would be passed to – among other recipients – the British Embassy in Buenos Aires. Williams, Love and Mitchell, with other Embassy staff, held a series of discussions on 25 March and collated their views and findings. On 26 March, having consulted again with his contacts during the night, Mitchell flashed a long signal to his superiors in Whitehall. No doubt his local findings and knowledge substantiated the information Whitehall was receiving from other sources. The main Argentine naval bases were almost empty of shipping.

He did not specifically express the personal view that an invasion was imminent, but the thrust of his urgent report made it plain that such an event was now probable. The recipients could have been in no doubt that the Embassy was deeply alarmed by everything which it had seen and heard. Like everyone else in the Embassy, Mitchell assumed that his reports, combined with those of Colonel Love and the information obtained from the broken cyphers, would provoke a fast reaction from the British government.

Captain Mitchell never received an acknowledgement to his signal, nor was he asked for further details. However, on that day, 26 March, a message arrived from the Foreign Office, suggesting that the Ambassador should start to prepare for the worst. Anthony Williams had already foreseen the possible need for this step. Twenty-four hours earlier, on 25 March, he had authorised the destruction of every sensitive document held in the Embassy filing cabinets.

All classified documents were stored in a secure area, within the Embassy, known as the Registry. Access to this area was restricted to authorised officials and clerks. Commencing on 25 March, they started to feed the most sensitive papers through their shredding machine. It was a huge task. A second machine was borrowed from a local branch of The Bank of London & South America. Those involved worked over the weekend of 27/28 March to ensure that the job would be completed quickly. During five days of methodical destruction, they shredded and then burned every document other than those which bore no security classification. The destruction was finished on Tuesday, 30 March, three days before the invasion of the Falkland Islands.

Having disposed of the classified material, the Registry clerks started to shred every other document in the Embassy except the signals cyphers. Anthony Williams was determined that, if and when the Argentines entered the building, they would find nothing which might embarrass Her Majesty's government or be useful to a protagonist. He began simultaneously to make plans for his British staff and their families to be quickly and safely transported out of the country when the time came. Very discreetly, some of the Embassy families started to pack suitcases and to crate precious possessions, all in readiness for the exodus. The word was passed quietly to representatives of the British Community Council, suggesting that it might be prudent if they too started to make their own preparations. There were at the time 17,000 registered British nationals in Argentina.

On 3 April, when the Embassy lost its diplomatic status, Williams and his people were almost ready. Colonel Love and Captain Mitchell, with their families, departed in a small convoy of Embassy and private vehicles on 5 April and took up residence in Montevideo (where they were also accredited). Three days later, after a final burst of shredding and burning over the weekend, Anthony Williams and the diplomatic and commercial staffs flew to London by scheduled airline.

Looking back at the events of that period, commencing with the arrival of the *Bahia Buen Suceso* at Leith Harbour on 19 March, the impartial observer might have difficulty in relating the developments in the southern hemisphere with those in the north. Between 24 and 27 March, when the Argentine invasion fleet was putting to sea, the British Secretary of State for Defence was to be found in Colorado Springs, attending a NATO conference. The Prime Minister and the Foreign Secretary were preparing to travel to Brussels for a Common Market meeting. The Foreign Secretary was planning then to travel on to Israel. The Chief of the Defence Staff was in New Zealand. Many units of Great Britain's armed forces were looking forward to the Easter leave period. Some of the Royal Navy's most powerful ships were about to have a weekend break in Gibraltar. Richard Luce, in London, was still trying to pour oil on troubled waters by announcing that 'further escalation of the dispute is in no one's interests – it is clearly right to pursue a diplomatic solution.' He added that 'the question of security in the Falklands area is being reviewed, although I prefer to say nothing in public about our precautionary measures.' In the event, the only concrete measures taken at that stage were the decisions to temporarily double the number of Royal Marines at Port Stanley and to send the *Endurance* back to South Georgia. Both developments were quickly reported in the Argentine press.

Instead of deterring the junta from invasion, the despatch of the *Endurance* back to South Georgia seems to have been a factor in stiffening its spine. There is evidence to suggest that, under the urging of Admiral Anaya, the Argentine government now decided unequivocally to bring forward by several months the plans which it had agreed for armed agression later in the year. The final decision seems to have been made on 26 or 27 March. The overall operation was codenamed Rosario and the units committed to the South Georgia occupation were designated Task Group 60.

Although politically attractive – to appease the restless crowds in the streets of Buenos Aires – the decision was militarily unsound. As later events were to show, the Argentine army and navy commanders had simply not had time to adequately think through the method and objectives of their adventure. But now we must return to the original flashpoint of the conflict.

Having made her high-speed return passage to South Georgia, the *Endurance* spent the next week cooling her heels off Grytviken. A certain amount of additional hydrographic work was done. The Wasps made reconnaissance flights to Bird Island and Gold Harbour. They collected Cindy Buxton and Annie Price from St Andrews Bay so that they could have two days of 'civilisation' aboard the *Endurance* before being flown back to their lonely hut. The Royal Marines rigged the ship's two elderly Oerlikon guns for action. They were given a detailed briefing by Bob Headland regarding local conditions ashore. They studied his photographs and maps, but time passed slowly for most members of the ship's company. There was only one important incident, and it was another portent of things to come.

Commencing on 24 March, the task of manning the observation point on Jason Peak was taken over from the BAS by Keith Mills' marines. They were flown up, two at a time, by the Wasps. On their second day, 25 March, they reported the arrival of another Argentine naval auxiliary ship. She had anchored in Stromness Bay, off Leith Harbour. This time it was the 9,600-ton *Bahia Paraiso,* the newest and best equipped ship of her type in Argentine service. She was flying the pennant of the Commodore Antarctic. Captain Trombetta had returned to the scene of his December visit.

It was difficult – and at times impossible – for the marines on Jason Peak to see exactly what Trombetta's people were doing. The decision was made to establish a second observation post very much closer to the scene of activity. At midday on 26 March, Tony Ellerbeck airlifted two Royal Marines forward to Grass Island, the island which sits less than three miles south of Leith and deep inside Stromness Bay. Ellerbeck succeeded in inserting these men covertly by flying very low behind the intervening cliffs and headlands. The operation was repeated on each of the following three days, the observers being withdrawn to the *Endurance* each evening.

The watchers saw nothing very remarkable during their time on the island. The *Bahia Paraiso* was working cargo, her derricks swinging backwards and forwards as she unloaded a large number of packing crates and netted freight into a barge, the *Fenix*. It was apparent that the ship had not come in response to British diplomatic pressure – to remove any of the men or gear delivered earlier by the *Bahia Buen Suceso* – but instead to further reinforce the Argentine presence on South Georgia.

The Grass Island observation post was able to report the Argentine activity in general terms, but it did not have a direct line of sight into the area of the buildings. On 27 March, therefore, it was decided that Keith Mills and Peter Leach would make a secret close-range reconnaissance.

They were taken by launch to Carlita Bay and put ashore, at first light, at the foot of the Olsen Valley. Carrying their side-arms, but with minimal equipment and no radio, the two men trekked through squalls of rain and sleet around the shore of Stromness Bay. They passed through the old abandoned whaling

The Carlita Bay field hut, built in 1971 as a refuge for persons travelling on foot between Cumberland West Bay and Stromness Bay. It was here, at the southern end of the Olsen Valley, that Lieutenant Mills and Sergeant Leach were landed from the launch *Albatros* to commence their reconnaissance of Leith Harbour on 27 March.

stations of Husvik and Stromness – where there was no sign of Argentine activity – until they came to the high ground south of Leith Harbour. Carefully ascending the 500-foot slopes of Harbour Point, they hid themselves in a rocky outcrop only six hundred yards from the settlement. They watched the busy scene on the jetty and counted the oil drums, crates and containers. The Argentines evidently intended to stay on the island for a long time and in strength.

Fortunately the Argentine commander had made no effort to place sentries on the high ground overlooking his new base, otherwise Mills and Leach might have been seen and caught. It was the first of several acts of military incompetence committed by the invaders during the coming days.

After forty-five minutes, satisfied that they had noted everything of importance, Mills and Leach withdrew from Harbour Point and started the long walk home. They covered fourteen miles that day, over exposed terrain, without apparently being detected. Shortly after leaving Harbour Point, however, they were forced to take cover in a patch of tussock grass. An Alouette helicopter had taken off from the *Bahia Paraiso* and it started to patrol the coastline. The two marines lay hidden for a long time as the aircraft cruised back and forth and, for a short while, dipped low over the track which they intended to follow. Eventually it returned to its mother ship and they continued their journey. Twenty minutes later they again went to ground when they saw three figures on the heights of Tonsberg Point.

Approaching the men cautiously, Mills was startled to recognise the crew of Wasp 434, Tony Ellerbeck and David Wells, and his commanding officer, Captain Barker. Subsequent conversation revealed that Barker had come to Stromness Bay to 'see for himself'. After a low-level approach over the Olsen Valley, Ellerbeck had parked the helicopter below the crest of the long finger of land which juts out into the Bay, and the three officers had then walked up the steep slope to observe the activity at Leith Harbour.

Tony Ellerbeck takes up the story: 'I was a trained commando pilot and had at one time taught "nap of the earth" techniques to the Royal New Zealand Air Force. By flying along the contours at between five and ten feet, and hiding behind the various cliffs and promontories, I reckoned I could put the Captain where he wanted to be without alerting the Argentines. I counted on the noise of the various generators and other machinery in Leith and aboard *Bahia Paraiso* to mask the noise of my engine. I hover-taxied up the final slope, found a hollow, and shut down. We climbed up to the ridge and spent about two hours observing the ship. Then we heard a jet engine starting up and an Alouette appeared out of Leith. It flew along the shoreline, crossed over Stromness and headed for Husvik. The pilot must have seen the bright dayglo paintwork of the Wasp because he suddenly circled back over us and came to the hover just fifty feet over our heads. I don't know whether the Argies had first spotted Mills and Leach, or perhaps us, but I suspect that it was just a coincidence that they made that flight when they did.'

The grinning face of Captain Trombetta appeared over the side of the hovering helicopter and he made the internationally acknowledged sailor's gesture to indicate that Nick Barker should 'go forth and multiply'. The two men had met several times previously and they instantly recognised each other. Nick Barker debated whether to return the gesture or simply to pray for the earth to open. A grenade dropped from the Alouette would have

When Captain Barker was flown north to Stromness Bay on 27 March to observe Argentine activity around Leith Harbour, the prominent red markings on Wasp 434's fuselage made the aircraft difficult to conceal. That same night both Wasps are resprayed in a camouflage scheme of black, grey and drab olive (above). The only remaining highlight is the ship's distinctive 'penguin' motif.

brought a swift end to an otherwise enjoyable career. Trombetta's pilot, fortunately, wheeled away and returned to his ship.

All four Britons returned safely to base, but the contact alerted the Argentines to the fact that they were being watched. On the following day, 28 March, the Alouette again searched the shores around the Bay and flew over Grass Island. The marines there had constructed two simple bivouacs to obtain protection from the elements while not on watch. These were located on the seaward side of the island, and were therefore invisible from Leith Harbour, but it seemed likely that they had been spotted by the pilot of the helicopter. The decision was made to abandon this observation post and to conduct all future surveillance from the Busen Peninsula. Later that day the watchers saw the *Bahia Paraiso* exit from Stromness Bay and head out to sea.

The next twenty-four hours passed without incident. Trombetta was maintaining radio silence and Nick Barker had no idea where he had gone. On 29 March, therefore, he decided to conduct a search at sea. He soon made radar contact with a large ship in poor visibility. Closing, he identified the *Bahia Paraiso* patrolling fifteen miles off shore. The ships came within a mile and a half of each other, the two captains exchanging courtesies by radio.

Barker wondered how he could discourage the Argentines from staying any longer in local waters without risking direct conflict. With his officers, he settled upon the idea of creating an illusion, of tricking Captain Trombetta into believing that there was a British hunter-killer submarine in the area. With Bob Nadin as his observer, and with the ship's doctor, Neil Munro, as passenger, Tony Ellerbeck launched in Wasp 434 and flew away over the horizon. The departure was conducted in full view of the Argentine ship and, as soon as he was airborne, Ellerbeck started to transmit on the distress frequency, Channel 16. At random, he and Barker had selected the call-sign of the nuclear-powered HMS *Superb*. Captain Trombetta was intended to believe that the submarine had a sick man on board and that the Wasp had gone to collect him for treatment aboard the *Endurance*.

Once out of sight, Ellerbeck went through the charade of conducting a radio conversation with the submarine while Nadin played the part of a submarine officer. 'We are ten miles from you. We shall winch your casualty up from your after casing.' 'Yes, very good, we are turning into the wind now. Speed eight knots. We are ready to receive your doctor.' Ellerbeck and Nadin went through their spontaneous dialogue and then returned to their mother ship with the 'casualty'. Whether or not the Argentines monitored the radio traffic, and whether they were deceived by it, the spoof did not have the desired effect. The *Bahia Paraiso* continued to patrol the same stretch of ocean as the *Endurance* turned back to King Edward Cove.

It was on 29 March that yet another ship appeared on the scene. The Russian ocean-going tug *Storki*, having requested landing permission and following all the correct procedures, came alongside Grytviken jetty to take on fresh water. She remained there until the following day and several of the *Endurance*'s people went aboard for a convivial evening. The timing of the *Storki*'s visit was coincidental. There were several Communist-bloc ships in the area. Only four days earlier, on 26 March, the Russian vessel *Spacatel* also had called at Grytviken for water. It is interesting to contemplate what the international repercussions might have been if any of these ships had been tied to the jetty four days later when the place became a battleground.

On 31 March, the *Endurance* made another search to seaward for the *Bahia Paraiso* and again found her patrolling about fifteen miles north-east of Cumberland Bay. Nick Barker did not close, but cruised back and forth, trying to keep the Argentine under visual observation in moderate visibility. Captain Trombetta had been pottering about in the same location for five days, so clearly

The big hangar and flight deck of the *Bahia Paraiso* made her an ideal assault ship for the invasion of South Georgia in early April. The same assets made her a natural candidate as a hospital ship when, in May and June, casualties from operations on East Falkland were being trans-shipped at sea under International Red Cross supervision. In this photograph, the *Bahia Paraiso* is seen sailing in company with the similarly employed Peninsular & Oriental SS *Uganda*. In the far distance is another British hospital ship, thought to be the hydrographic ship HMS *Herald*.

he was under orders to 'mark time' pending further developments. The obvious implication was that he expected more Argentine ships to arrive in the near future. Nick Barker accordingly kept his two Wasps busy with constant reconnaissance flights up and down the coast, but nothing was sighted that morning. These flights pushed the total flying time of the ship's two Wasps to a new record for any one month – 107 hours.

Early in the afternoon a signal was received from Northwood, ordering Captain Barker to return to Port Stanley with all despatch. The mounting volume of evidence arriving in London was indicating that invasion of the Falkland Islands was now a very strong possibility. Barker was told to put ashore his ship's detachment of Royal Marines at King Edward Point before sailing. In the same way that the Royal Marines on East Falkland gave some credibility to Great Britain's stance on the question of sovereignty there, a military presence was required on South Georgia for the same purpose. The military personnel could also provide protection – if required – for the unarmed BAS civilians. One platoon defending an island a hundred miles in length could be nothing more than a token force, but there was no alternative.

During the evening, Lieutenant Mills and his men were ferried ashore with a full allocation of platoon weapons and ammunition. This was the first occasion that armed troops had ever garrisoned the island. Steve Martin was not happy with this latest development. He would have preferred not to have had any troops on the island, particularly when they were billeted on him at only thirty minutes' notice. He was the Magistrate and therefore answerable to the Falkland Islands government for everything which happened on the island. He had received no advices from Rex Hunt, nor indeed from anyone other than Captain Barker. The BAS base was being turned into an armed camp, with all of the implicit perils, but Martin himself had no instructions as to how he and his personnel should conduct themselves in the event of serious trouble.

At a conference on board the *Endurance*, shortly before she sailed, Steve Martin voiced his anxieties to Nick Barker and to Commander Francis Ponsonby, RN. The latter officer was employed with the Royal Navy Directorate of Publicity. Originally he had come south with the Royal Navy camera team making the '*Endurance*' documentary, but he and his party remained on board at Nick Barker's invitation when the ship departed Port Stanley for Leith Harbour on 21 March. He and Nick Barker told Martin that the order to leave Keith Mills and his men at King Edward Point had come from London.

Steve Martin was not entirely reassured. He was a university graduate and boatman, not a soldier or diplomat. Everything was happening very quickly and he felt himself being carried along by circumstances which were totally unfamiliar to him. Who was now in charge of the island? Was he still the Base Commander, or had Keith Mills taken over? As Magistrate, and therefore legally delegated representative of Governor Hunt, he had considerable powers, but these appeared to have been in part overridden by the orders coming directly from London via Captain Barker.

After some discussion, his dilemma was resolved with a commonsense agreement – between Martin, Barker and Mills – that there would be no change to the usual administration of the

Base as long as the present situation continued. However, in the event of an emergency, and if fighting broke out, all BAS personnel would conduct themselves as non-combattants, keeping well out of the way, and Keith Mills would subsequently command.

Martin was not alone in feeling worried. Although elated at the prospect of independent command under novel circumstances, Keith Mills had the anxiety of not knowing exactly what was required of him. In the phraseology of his profession, what were the 'rules of engagement' (ROEs)? Nick Barker responded by spelling out the terms under which the marines could resort to force. Mills instantly recognised the parameters as being those defined in the Yellow Card rules.

It must have been an act of sheer perversity which caused the Ministry of Defence to choose yellow as the colour for the card upon which were printed the rules for troops engaged in 'aid to the civil power'. No fighting man, of any nation, wishes to be associated with a colour which symbolises cowardice. By custom and long-standing usage, however, all British troops came to accept the format and to understand the basic spirit of the wording.

The rules evolved as the years went by, but the need for them had first become agonisingly apparent in India. At Amritsar, in 1919, a mixed force of British and Gurkha troops was ordered to fire indiscriminately into a large crowd of unarmed civilians. More than four hundred people were killed and a thousand injured. The British subsequently introduced and developed the guidelines which later determined the conduct of troops operating in Aden, Cyprus, Northern Ireland and a dozen other trouble spots around the world.

The wording can be condensed thus: 'You are responsible for the safety and security of any persons, equipment or property which you are ordered to guard. You are not to use more force than is necessary to carry out your duty. You are to challenge anyone who is not known to you or who is acting suspiciously. Having given the warning, you may open fire against a person carrying what you can positively identify as a firearm, and if you think he is about to use it for offensive purposes and there is no other way of stopping him. At all times in carrying out your duty, and particularly if you are attacked, you must only use such force as is the minimum to achieve your purpose and is reasonable in the circumstances. If a warning is impracticable, you may fire without warning when any delay could lead to death or serious injury to people whom it is your duty to protect or yourself, and then only against a person carrying a firearm. Always fire single shots. Aim at the centre of the main part of the body exposed. Never fire warning shots. Fire only the minimum number of rounds necessary to achieve your purpose. If for any reason you are forced to use your arms on your own initiative, you are to continue acting independently only until the situation is in command of someone senior to you.'

Having already completed a tour of duty in Northern Ireland, Keith Mills was thoroughly conversant with both the terminology and the objectives of the Yellow Card. They did not exactly cover the circumstances on South Georgia because he and his men would not be faced with rioting crowds or stealthy bomb-carrying terrorists, but at least the Card authorised him to open fire if he judged that he had no alternative.

Steve Martin and Keith Mills agreed that everyone – scientists and marines – would live together in the large building known as Shackleton House. Another house, Quigley's, was chosen as the marines' military store and armoury. There was some apprehension among the BAS people at this disturbance to their normal routine. The main problems were space and food. The scientists needed peace and quiet to complete their 'end of season' reports, and they were alarmed by the appetites of their new guests. The marines brought a few 24-hour ration packs ashore with them, but were basically dependent upon the BAS for food and drink. A strict system of rationing was agreed and the two sides started to adjust to each other.

Nick Barker's next hurdle was to get his ship out of Cumberland East Bay without risking confrontation with the *Bahia Paraiso*. The Argentine had shadowed the *Endurance* back to the entrance to the bay and it was safe to assume that she was now hovering not far off shore. At first glance, the *Endurance* was a ship in a bottle and Trombetta was the cork.

Captain Barker and his Navigating Officer, Lieutenant-Commander Bill Hurst, studied the chart of the Cumberland Bays. The entrance to East and West Cumberland is approximately four and a half miles in width. Any ship leaving by a normal track would be spotted instantly by the *Bahia Paraiso*'s radar operators. The only answer would be to hug the shoreline and creep away, hoping that the radar scanner would fail to distinguish the ship against the steeply rising cliffs. The surface of the sea was littered with rocks and small icebergs and, as Nick Barker told his officers, 'we have got to pretend to be one of them.'

Announcing his intentions to the ship's company, he emphasised the need for secrecy. The navigation lights would be extinguished and all deadlights secured. The ship must be darkened 'as she has never been darkened before.' There would be strict radio silence. The radar would be used only in brief 'squirts' to confirm the ship's position (otherwise the impulses transmitted by the *Endurance* would be identified in the *Bahia Paraiso* and in any other Argentine vessel which might be in the area). The sole navigational aid – apart from the naked eye – would be the echo-sounder, but this would have limited value in such a poorly charted area. Bill Hurst was given the task of conning the ship away from King Edward Point and out to sea.

At 1900, on the evening of 31 March, the *Endurance* weighed anchor and moved very slowly across Cumberland East Bay until she was close under the high ground of the east headland. The wind was blowing from the west at half-gale force, kicking up the surface of the sea into waves three to four feet in height. Although the night sky was overcast, the horizontal visibility was six miles. This was much greater than Nick Barker would have wished. It was important not to let the ship's silhouette show against the horizon.

Bill Hurst maintained bare steerageway as he nursed the *Endurance* along the black jagged shore where, two hundred yards to starboard, a white line of bursting waves signalled the waiting rocks. At this slow speed, with the beam wind pushing hard against her high superstructure, 'the plum' was all too ready to drift to leeward.

Gradually the extremity of Barff Point came in sight. Here there is a narrow gap, four hundred yards wide, between the headland and the Right Whale Rocks. It is unlikely that this current-swept stretch of water, known as the Merton Passage, had ever previously been attempted by any vessel larger than a motor-boat. The charts show a minimum of eight fathoms, but the sea-bed had never been surveyed in detail and only extreme necessity could tempt a ship's captain to hazard his vessel in such a place. Bill Hurst quietly ordered the helm to starboard and eased the ship into the neck of the passage.

The *Endurance* had found herself in many curious places in the twenty-six years since she was launched, but her transit of the Merton Passage on this night was by far the most perilous. The men on her silent bridge watched as the cliffs and rocky shorelines, to starboard and to port, passed with almost painful slowness. The echo-sounder told them that the sea-bed was shoaling steadily closer to the surface. Soon there was a clearance of less than twenty feet between the rocky bottom and the *Endurance*'s keel. Even one large outcrop would have been enough to bring total disaster. But all went well, and she passed through safely.

Leaving the passage behind them, the people aboard 'the plum' were now faced with the task of getting completely clear of South Georgia without their departure being reported. It was known that there were several other ships in the area, including two Russian vessels. To avoid observation, Nick Barker chose the southerly route, heading for Cape Disappointment. Icebergs had reached the adjacent waters as the onset of the Antarctic winter carried them northward. He would move slowly down the coast, keeping close inshore, and hope that any Argentine radar operators would classify his ship as just one more iceberg.

For the first hour, it seemed that the worst was behind them. Then, approaching Royal Bay, the lookouts saw a brightly lit ship off the port bow. It was impossible to know her nationality or the reason for her presence in these waters. Captain Barker debated whether she was an Eastern-bloc ship, or whether she was another Argentine vessel coming to join the *Bahia Paraiso*. He ordered the course to be altered even closer to the coast. Accepting the hazards of offshore reefs, Bill Hurst conned 'the plum' slowly past the unknown ship until she disappeared into the darkness astern. Either she had failed to spot the *Endurance*'s blip on her radar scope, or had simply not been interested. The evidence later suggested that she was a Communist vessel.

By dawn, on 1 April, Barker had cleared the southernmost tip of the island and was pressing hard for the Falklands into a Force 10 gale. In the *Endurance*'s cavernous engine room, the five big pistons thundered up and down, making conversation impossible as the engineers coaxed maximum power from the elderly machinery. For Bill Hurst, it was the end of what he subsequently labelled 'a bloody gruesome night.'

The ship rolled like an empty barrel as she buffeted her way across a heavy south-westerly swell. Nick Barker's usually mild temper was strained to the limit by his current frustrations. Everything which he had forecast was coming to pass. The Argentine fleet was at sea and within striking distance of the Falklands and South Georgia. Reports spoke of major units to the north of the islands, with a second force approaching from the south; possibly two frigates, armed with Exocet missiles, were deployed somewhere between the Falklands and South Georgia, on the *Endurance*'s present track, and a submarine was rumoured to be *en route* to this same area.

To counter this display of power, Rex Hunt and Nick Barker had seventy-three marines on East Falkland, twenty-two marines on South Georgia, and the dear old 'plum' wallowing along halfway between the two. The nearest support was at Gibraltar, 5,500 miles and two weeks' steaming away. If ever there was a moment which exposed the poverty of the British government's policy in the South Atlantic, this was it. □

IV | In at the Deep End

Commodore Sam Dunlop on the bridge of the *Fort Austin*. His was the first surface ship to be ordered south in response to the storm clouds gathering over the South Atlantic.

STEVE MARTIN'S DISCOVERY of the Davidoff party at Leith Harbour on 19 March had given Lord Carrington the tool to finally extract an agreement from John Nott that the *Endurance* should remain on station for the time being. Although much criticised afterwards for its performance during the months preceding the war, the FCO had at least recognised the symbolic significance of the red ship with the White Ensign. Lord Carrington had submitted to the Cabinet Office several memoranda describing her purpose and value, and the risks inherent in her withdrawal. His recommendation had been ignored through to the eleventh hour but, at long last, on 24 March, it was made known that she would not be returning home on the planned date. This change of heart was announced with minimum publicity.

The decision carried with it the problem of how best to sustain the ship. Onboard supplies were running low and she would soon need an extensive replenishment. The problem was considered by Northwood and the ship selected to go the *Endurance*'s aid was the RFA *Fort Austin*, currently in harbour at Gibraltar. Commanded by Commodore Sam Dunlop, MBE, the senior master of the Royal Fleet Auxiliary, the *Fort Austin* was homeward bound after a five-month deployment in the Indian Ocean. Her own reserves were running low when, on 26 March, Sam Dunlop received his new orders. The next three days were devoted to re-storing and generally preparing for the long passage down to the South Atlantic. The civilian crew were not entirely enthusiastic at this interruption to their domestic plans – at least one wedding would now be postponed – but Commodore Dunlop soon convinced them that the diversion was important. News of the invasions – received as the ship sailed down the west coast of Africa – created an entirely new sense of purpose.

On 29 March, while Captain Barker and Captain Trombetta were playing their game of cat and mouse off the Cumberland Bays, Lord Carrington had another opportunity to awaken Mrs Thatcher to the peril in which she stood. During a flight from London to Brussels to attend a Common Market meeting, he pressed the Prime Minister to take further action. He told her of Ambassador Williams' latest encounter with Dr Costa Mendes. The Argentine had stated that 'the door is now closed on all further negotiations on South Georgia.' Disturbed by the intransigence of the Argentines and their refusal to recognise the British administration at King Edward Point, Carrington succeeded in impressing upon the Prime Minister the fact that diplomacy might need to be backed by a show of force. Having concentrated her mind upon the subject, she reacted with speed and determination. When their aircraft touched down at Brussels, she and Carrington found a telephone and called John Nott in London. He was instructed to contact the First Sea Lord, Admiral Sir Henry Leach, and to set the wheels in motion for the immediate despatch to the South Atlantic of at least two nuclear-powered submarines. Capable of very high speeds over prolonged periods, needing no replenishment at sea, accustomed to operating independently for weeks on end, they could reach the sub-Antarctic sooner than any other vessel.

Nuclear submarines (SSNs) are valuable and expensive vessels. Their operational deployments are planned long in advance and they cannot lightly be diverted to new tasks. The immediate implications of the Prime Minister's decision are best

Sir Anthony Parsons, the United Kingdom's Ambassador to the United Nations, addresses the Security Council in emergency session on 3 April. An expert on Islamic history and Middle East affairs, he had served as Britain's Ambassador in Teheran throughout the Iranian revolution before moving to New York in 1979.

described by Admiral Sir Henry Leach: 'At the time we had a number of major exercise and important trials commitments, with barely sufficient SSNs to meet them. Nevertheless, a need had arisen to deploy a deterrent or retaliatory capability to the South Atlantic against the possibility that the Argentinians really would do something stupid this time. For this, one SSN would suffice but, in view of the distance from base and the advantages derived from mutual support, it was clearly prudent to send two. At that stage it was not necessary to send a third – there were other and important operational reasons for not doing so. But after the Falklands had been invaded, then it *did* make sense to deploy a third SSN as soon as other operational tasks permitted, and this was done.'

The first of the nuclear submarines to be deployed was HMS *Spartan* (Commander J. B. Taylor, RN). She was operating off Gibraltar when, on 29 March, the call came. She began loading stores for a protracted cruise and departed for the south within forty-eight hours. HMS *Splendid* (Commander R. C. Lane-Nott, RN) sailed from Faslane, Scotland, on 1 April, and, three days later, she was followed from the same base by HMS *Conqueror* (Commander C. L. Wreford-Brown, RN). Two weeks would pass before any of these powerful units could expect to be on station, but at least something positive was being done to sustain Great Britain's claim to sovereignty.

The machinery of collective government was starting to stir. The Prime Minister had returned to London from the Brussels meeting. Commencing at 1900 on the evening of Wednesday, 31 March, a long and crucial meeting was held in her office in the House of Commons to discuss the latest batch of intelligence summaries. Her Foreign Secretary was not present – he had travelled on to Israel from Brussels – so the FCO was represented at this meeting by Richard Luce. Among the others present were John Nott and Sir Anthony Acland, head of the Joint Intelligence Committee.

The Prime Minister was totally, if belatedly, aware that she had a full-blown crisis on her hands. What must now be done? For an hour, those present each offered his and her personal view. The general consensus veered in the direction of caution. All were agreed that the British response must be framed in accordance with international law and custom. It was decided to instruct Sir Anthony Parsons, the United Kingdom's permanent representative at the United Nations, to request an emergency meeting of the UN Security Council.

Parsons had been in regular contact with the FCO and Downing Street regarding the South Atlantic since Monday, 29 March. The substance of these exchanges related to the illegal status of the Davidoff party at Leith Harbour and the much more serious developments which might flow from that otherwise relatively insignificant affair. It was apparent to Parsons that there was an increasing anxiety in London at the reports of the Argentine fleet putting to sea and the bellicose statements issuing from the Galtieri regime. He was asked to advise Downing Street and the FCO what the reaction would be if he were to request an emergency meeting in the Security Council (which at the time was debating the crisis in Nicaragua).

Sir Anthony Parsons replied that he would be more likely to obtain a sympathetic response if he could reveal some hard evidence with which to substantiate British fears. This advice created a dilemma for London because any attempt to place detailed factual evidence before the Council might compromise British intelligence sources. A complicating consideration was the fact that the President of the Security Council for the month of March was the United States representative, Mrs Jean Kirkpatrick, a lady known to be sympathetic to Latin American problems and less sympathetic to British affairs.

The formal instruction to lay the whole matter before the Security Council was received by Sir Anthony Parsons at 1230 on 1 April. With a time difference between London and New York of five hours, this signifies that the telephone call from the FCO to Parsons' office was made at 1730 (London time) on the afternoon following the Prime Minister's Wednesday evening meeting in the House of Commons. There is no obvious reason for this apparent delay. However, despite the fact that it was lunchtime in New York, Parsons was able, after only five minutes of hurried telephone calls in the United Nations building, to set the wheels in motion. He was able to reach the new (April) President of the Security Council, Mr Kamanda wa Kamanda, the representative of Zaire, and, an hour later, had delivered to him the following letter: 'I have the honour, on instructions, to address you urgently about the situation in the South Atlantic. The Government of the United Kingdom has good reason to believe that the armed forces of the Republic of Argentina are about to attempt to invade the Falkland Islands. In these circumstances, I request your Excellency to call an immediate meeting of the Security Council.'

This letter produced a statement, that evening, from the President on behalf of the whole Council. In it, Mr Kamanda called upon both sides to exercise restraint. On the following

Born at Yarner, in the maritime county of Devon, Henry Conyers Leach was the son of a Captain, Royal Navy, and seemed almost destined to enter the same profession. Having followed the time-honoured route into the Royal Naval College at Dartmouth as a 13-year-old Cadet in 1937, he specialised subsequently in gunnery. He served at sea throughout the Second World War in the Atlantic, Indian and Pacific Oceans. Later in his career he was appointed to the major NATO command of Allied Commander-in-Chief, Channel, and nationally, Commander-in-Chief, Fleet. Knighted in 1977, he became First Sea Lord in 1979, and thereby professional head of his Service. In the early 1980s, he was one of a handful of serving officers who could still argue the case for a balanced global Navy from a position of personal experience.

morning, news of the Port Stanley invasion caused Parsons to return again to the Security Council where his representations led to the adoption, that same day, of Resolution 502: 'Recalling the statement made by the President of the Security Council at the 2,345th meeting of the Security Council on 1 April 1982 (S14944) calling on the Governments of Argentina and the United Kingdom of Great Britain and Northern Ireland to refrain from the use of force in the region of the Falkland Islands (Islas Malvinas); deeply disturbed at reports of an invasion on 2 April 1982 by armed forces of Argentina; determining that there exists a breach of the peace in the region of the Falkland Islands (Islas Malvinas); demands an immediate cessation of hostilities, demands an immediate withdrawal of all Argentine forces from the Falkland Islands (Islas Malvinas), calls on the Governments of Argentina and the United Kingdom to seek a diplomatic solution to their differences and to respect fully the purposes and principles of the Charter of the United Nations.'

From the very outset, Great Britain was seizing the diplomatic initiative. Resolution 502 was a priceless asset in the war of words. It would be further argued that Article 51 of the United Nations Charter – the inherent right to self-defence – applied to military actions aimed at removing foreign troops from British sovereign territory. In the following weeks, Anthony Parsons never wavered in his pursuit of his Argentine opponent at the United Nations or in his brilliantly successful rallying of support by other nations. Equally important were the activities of Sir Nicholas Henderson, the British Ambassador in Washington, who succeeded in winning over the pro-Latin American pressure groups and calming American suspicions that the British might be embarking on some sort of 'colonial' adventure.

We must return now to the evening of 31 March at the House of Commons. Having dealt with the initial legal and diplomatic aspects of the crisis, Mrs Thatcher and her colleagues were left with the military option. What action – if any – could be taken to display British resolve and, if necessary, to eject any invaders by force of arms? It was only when the group was joined by the dynamic First Sea Lord that serious consideration was given to the mobilisation of a major British naval and military response.

Admiral Sir Henry Leach had been keeping his finger firmly on the pulse of Argentine activities for several days past. On his own initiative, he had set his staff to the task of studying the immense planning and logistical problems to be overcome if the government did decide to despatch a naval task force to the South Atlantic. Earlier in the day he had been in Portsmouth on official business, with a Sea King helicopter on standby to rush him back to London in the event of a sudden worsening of the situation. Completing his programme as planned, he returned to his office in Whitehall in the early evening. The latest – and radically different – intelligence assessments were waiting on his desk. Quickly he scanned them and then went in search of the Secretary of State for Defence.

Told that the Prime Minister had called Mr Nott to attend an emergency meeting at the House of Commons, Leach made his way there. Although kept waiting for fifteen minutes by the Police Sergeant in charge of security, Leach was able eventually to track down his quarry. His entrance into the Prime Minister's private room came as a breath of fresh air. The only professional serviceman present, he understood exactly the anxieties of the civilians. The fact that he was still in uniform made him automatically a focus of attention. His positive replies to Mrs Thatcher's searching questions were a major factor in dispelling much of the previous uncertainty. The meeting dispersed with no final decision made, but with a general agreement that the armed forces should study urgently the feasibility of mounting a comprehensive expedition within the next seven days, if necessary.

Admiral Leach had been presented with a golden opportunity to display the continuing importance of a global and balanced navy. For years, the politicians had been telling Leach and his predecessors that they would never again be called upon to operate on a significant scale outside the NATO area. Great Britain's naval contribution to European defence would be confined, in their view, to the hunting and destruction of Russian submarines in the North Sea, the Western Approaches and the North Atlantic. Air cover, they argued, could be provided, whenever necessary, by the giant carriers of the United States Navy. The Royal Navy would no longer need carriers of its own, nor would it need an amphibious capability. The old carriers, such as HMS *Hermes*, could be scrapped, and the first of the new breed of small carrier, HMS *Invincible*, could be sold off. The

HMS *Hermes* was in dockyard hands when the first warning signal for a South Atlantic deployment was received onboard. The spider's web of scaffolding around her superstructure was dismantled in record time.

elderly specialised amphibious warfare ships, HMS *Fearless* and HMS *Intrepid*, were irrelevant to anti-submarine warfare and too costly to be replaced.

Although they contained an element of hypothetical logic, these theories depended upon an opponent who dutifully followed the script written for him. As author of the current scenario, John Nott had made no allowance for the appearance on stage of a new character. Despite the emergence of several totalitarian regimes in various parts of the world during the previous decade, the Treasury and Mr Nott had overlooked the ambitions of a Leopoldo Galtieri.

The First Sea Lord seized his opportunity with both hands. The following day, Thursday 1 April, was packed with planning conferences. At 1600, just ten hours before Argentine special forces first set foot on East Falkland, a signal was received in HMS *Hermes* and HMS *Invincible*: 'Prepare for prolonged operations in the South Atlantic.' A few hours later, Leach attended a second meeting with the Prime Minister and received authority to despatch a fleet of sufficient force to recover the Falkland Islands in the event of their occupation by Argentina. At 0227 on 2 April, the aircraft carriers were ordered to come to four hours' notice to sail. Operation Corporate had started.

Far to the south, at Gibraltar, a considerable force of Royal Navy ships was engaged in a large-scale maritime exercise. Code-named Springtrain, this assembly of frigates, destroyers and Royal Fleet Auxiliary ships was under the command of Rear-Admiral John 'Sandy' Woodward. Promoted to flag rank only ten months previously, forty-eight years of age, Woodward was a complex and highly intelligent officer. Articulate, at times introspective, possessed of a quirky sense of humour, his intellect demanded challenge and stimulation. He yearned for a more active role. His wish was shortly to be fulfilled beyond any normal expectation.

On Monday 29 March, three days before his first meeting with the Prime Minister at the House of Commons, the First Sea Lord had contacted the Commander-in-Chief Fleet, Admiral Sir John Fieldhouse, and had alerted him to the probability that the Royal Navy would be committed to extensive and prolonged operations in the South Atlantic. Fieldhouse was at this time observing Exercise Springtrain from the comfort of the guided missile destroyer *Glamorgan*. He immediately summoned Woodward from his flagship HMS *Antrim*. and the two men discussed contingency plans for a swift response, if and when the call came.

Admiral Fieldhouse returned to London where his staff were starting to assess what would be the Royal Navy's largest and most difficult operation since 1945. Woodward and his small staff continued with the Springtrain programme and simultaneously examined their own problems. By 2 April their plans had taken shape. The order to launch Operation Corporate led to a busy day of refuelling and re-storing by those ships which would be going south.

The Springtrain force was divided into two groups – the ships which were to return to the United Kingdom and those selected for Corporate. The Task Force Commander would be Sandy Woodward. He was already in the driving seat, he knew the captains under his command, and he was in any event the nearest seagoing officer of flag rank to the scene of impending action. The scale of the problem was underlined by the 6,000 miles which separated him from his objective.

Favoured by fine weather, the embryonic Task Force com-

Governor Rex Hunt relaxes in the gardens of Government House, Port Stanley. The stretch of lawn leading up to the large conservatory became part of the battleground when the building was attacked by Argentine special forces on the morning of 2 April.

pleted its replenishment at sea and turned away from the ships returning home. Apart from HMS *Antrim*, Woodward had under his immediate command Her Majesty's Ships *Arrow*, *Coventry*, *Glamorgan*, *Glasgow*, *Plymouth* and *Sheffield*. To the subsequent astonishment and dismay of General Galtieri's advisors, the Royal Navy was embarking upon the one course of action which they had always rejected as impossible – the deployment of a powerful fleet.

Although they had been caught flat-footed by the crisis, those politicians most directly involved – the Prime Minister, the Secretary of State for Defence, and the Foreign Secretary – each responded positively in their own individual ways when the moment came. The Prime Minister backed the Navy to the hilt while insisting that the search for a diplomatic solution should be pursued to the bitter end. John Nott stood back and allowed his professional civil servants and the naval staff to get on with the jobs for which they were trained. Lord Carrington headed back to London from Tel Aviv with the knowledge that his colleagues and senior staff were already taking all possible measures to enlist the support of the government of the United States of America and within the General Assembly of the United Nations. On the face of it, everyone was making the best of a calamitous situation. There was only one astonishing oversight – the Governor of the islands at the centre of the impending storm had been left in blissful ignorance of the feverish activity now developing in Northwood, Whitehall, Gibraltar, Portsmouth, Devonport, Rosyth, and elsewhere.

The first warning of an invasion of the Falkland Islands was received by Rex Hunt at 1530 local time on Thursday, 1 April. Allowing for the time difference between London and Port Stanley, this signifies that the FCO did not get around to contacting the Governor until very nearly twenty-four hours after the Prime Minister's House of Commons meeting with John Nott, Richard Luce and Admiral Sir Henry Leach.

On the evening of Wednesday 31 March, while that meeting was in progress, Rex and Mavis Hunt had been entertaining fourteen guests to dinner at Government House. Those present were all islanders and they had assembled to celebrate the fiftieth anniversary in Falkland Islands government employment of Harold Bennett, a popular and much respected member of the island community. Lord Buxton, head of Anglia Television, had earlier given Hunt a video copy of his company's documentary 'More British than the British,' an account of 'kelper' history and current political status. The happy group chatted and watched the film until midnight, oblivious to the frantic activity in the United Kingdom.

Rex Hunt was up at 0630 to talk by radio with Steve Martin at King Edward Point. By 0800 he was at his desk and that fateful Thursday morning passed with the customary round of routine meetings and government business. He lunched with his wife and son and returned to his office with the intention of clearing the rest of the day's correspondence. At 1530 a telex arrived from the FCO via the nearby Cable & Wireless office. The contents came as a bombshell: 'We have apparently reliable information that an Argentine task force could be assembling off Cape Pembroke by dawn tomorrow [2 April]. You will wish to make your dispositions accordingly.'

Governor Hunt had been aware for some days that part of the Argentine fleet was putting to sea, ostensibly for anti-submarine exercises in conjunction with the Uruguayans. He had not been told that the Argentines had subsequently sailed at least one amphibious warfare ship, and that a full battalion of marines had been seen embarking with armoured personnel carriers (APCs). Like most of his ex-Colonial Service contemporaries, Rex Hunt was a down-to-earth pragmatist. In the light of the limited information which earlier had been given to him, it had seemed most improbable that the Argentines would yet strike directly at the Falkland Islands. He had assessed the risks in a different light. He believed it much more likely that some or all the Argentine frigates would proceed to South Georgia (to support the Leith Harbour force), or form a cordon to intercept the *Endurance* at sea if she arrested the Leith Harbour people (and tried to bring them to Port Stanley).

Hunt took a deep breath and reached for his telephone. Having during his career served in various trouble spots around the world – Uganda, Vietnam, Borneo and Malaya – he had no doubt where his priorities lay. Military defence was the first. By 1600 he was in conference with Major Mike Norman, RM, commanding the new Naval Party 8901 which had arrived aboard the RRS *John Biscoe* three days earlier. Norman had assumed local command from Major Gary Noott, RM, the officer commanding the outgoing NP8901, within hours of his arrival on the island. Norman's men had taken over the Royal Marines accommodation building at Moody Brook while Noott's party moved in with various local families with whom they had become friendly

Port Stanley, photographed in 1984. The large building in the foreground is Government House. The initial Argentine assault was launched through the enclosed gardens at the rear of the building. Here the attackers lost an officer and two other ranks killed by the accurate fire of three Royal Marines: Corporal Sellen, Corporal Fleet (a cook), and Marine Dorey (armourer).

during their tour. Hunt showed the new commander the FCO telex and said, 'it looks as if the buggers really mean it this time.'

The two men quickly assessed the likeliest landing point and decided that it would be the broad sheltered beach at Yorke Bay on the north shore of Cape Pembroke and close to the airfield. Major Norman decided to cover that beach with a machine gun section while covering the eastern approach road to Port Stanley with further sections. More marines would guard Government House.

The two Royal Marines majors had received no warnings or orders from the FCO or from their own headquarters in the United Kingdom, but they mobilised their men while Rex Hunt called in the heads of department of the Falkland Islands government. By 1630 he was issuing orders for the FIGAS Islander aircraft to be flown from the airfield to the racecourse, for obstacles to be placed on the airfield runway, for the Falkland Islands Defence Force to be mobilised, for the hospital to be brought to full-scale alert, for the civil defence volunteers to report for duty, and for a hundred other jobs to be done with inadequate resources.

Darkness prevented the Islander from getting airborne to reconnoitre the area of sea off Cape Pembroke, but Captain Jack Sollis of the Motor Vessel *Forrest* sailed with orders to wait off Port William and to report any approaching ships. At 0330 his radar showed numerous vessels approaching Port Stanley from the direction of Cape Pembroke and he hurried back to report his findings to Governor Hunt.

Hunt's small secretarial staff was working hard to destroy all classified documents in Government House. Having witnessed the chaotic fall of Saigon, Rex Hunt had developed the habit of 'weeding' his files regularly so that they could be shredded and burned quickly in the event of an emergency. Even so, there was an immense volume of paper to be put through the shredding machine. The hydrographic party from HMS *Endurance* was called in to help with the task before being armed by the Royal Marines and sent off, under the Chief Secretary, Dick Baker, to arrest the twenty-eight Argentine technicians living in the town. These men had come over from the mainland to work on the fuel depôt, but they were all young and fit and Hunt suspected that they were not what they claimed to be. Dick Baker, escorted by Lieutenants Todhunter and Ball, collected them from their lodgings and, by 0500, had them all under armed guard in the Town Hall.

Instead of making an initial amphibious landing at Yorke Bay and advancing from the east, the Argentine spearhead landed south of the town, at Mullet Creek. It seems likely that they came ashore soon after 0400, splitting into two groups. One group advanced over Sapper Hill and approached Government House from the rear. The other party went directly to the Royal Marines accommodation building at Moody Brook, one and a half miles to the west. Unaware that it had been evacuated, the Buzo Tactico special forces assault group quietly surrounded the building and then, without warning, raked it with automatic fire. Kicking

down the door, they raced through the interior, hurling phosphorus grenades into each room and spraying the bunks with their sub machine guns. This murderous attack was intended to kill each and every British serviceman on the island in a single blow. Their lives were saved only as a result of Major Noott's and Major Norman's precautionary moves, and by the coincidence of the date. If the Argentines had launched their invasion on the previous night, at which time Hunt and the marines had no indication whatsoever that they might be in danger, the plan would have succeeded. Given the almost total lack of warning, the marines could count themselves extraordinarily lucky to be alive. Rex Hunt and his wife and their fourteen guests were similarly fortunate.

Shortly before 0600 the people in Government House heard firing from the direction of Moody Brook. Soon afterwards fire was opened on Government House itself. Argentine marines forced their way into an outbuilding and a fierce firefight began as the Moody Brook contingent joined the first group. The attackers were held at bay for two and half hours, three of their number being captured and five, including one of their officers, a captain, being killed.

The main invasion force, meanwhile, had come ashore to the west of York Point, unwittingly avoiding the Royal Marines machine-gun section covering the bay to the east. The Argentines advanced in strength toward Port Stanley along the airfield road in their armoured personnel carriers. Here they ran into the hastily arranged line of Royal Marine defences. Despite their overwhelming strength, the Argentines received a very bloody nose. Lieutenant Bill Trollope's party scored a hit on the tracks of the leading APC with a round from their 84mm Carl Gustav rocket launcher. A second hit was then scored against the hull with a 66mm launcher. These vehicles normally carried ten armed soldiers. The hollow-charge missile punched a hole through the steel plate of the vehicle's flank, spraying the interior with small chunks of molten metal. Nobody got out. Whether they had all been killed, or whether the survivors guessed that they would be shot at should they try to escape, has never been established.

Sheltering under the desk in his small office – equipped with a small transmitter and now his 'operations room' – Rex Hunt continued to broadcast a running commentary on the battle as bullets cracked through the windows, walls and roof of the handsome Victorian building. Half a mile away, Patrick Watts, Broadcasting Officer of the Falkland Islands Broadcasting Service, performed the bizarre function of playing a selection of recorded light music interspersed with information on the invasion and the battle raging outside. Reluctantly he broke off his broadcast when Argentine troops burst into his studio and thrust a rifle muzzle against his spine.

In Government House there was a sense of relief that the initial attack had been successfully repulsed. The Argentines had failed to break into the main building and had failed to capture the Governor. After the first thirty minutes of heavy firing, the battle gradually reduced to sporadic sniping. As reports came through of the approaching armoured units, however, Rex Hunt faced the inevitable decision to surrender. He knew that the grossly outnumbered Royal Marines were willing to fight on, but he was deeply concerned for the civilian population in the town. Having had such inadequate notice of the enemy fleet, he had had no opportunity to evacuate the women and children and the elderly.

The officer commanding the new Naval Party 8901, Major Mike Norman, RM, is greeted at Government House by Governor Hunt. Between them stands Norman's predecessor, Major Gary Noott, RM. During his final briefings before leaving the United Kingdom, Major Norman had received no indication that he might be entering a potential battle zone. Four days after this photograph was taken, Norman and Noott found themselves once again standing outside Government House, but this time under radically different circumstances.

A prolonged firefight could only cause unnecessary loss of life without altering the final outcome.

It is said that the Argentines broadcast a pre-recorded ultimatum from one of their ships, demanding the surrender of the islands, but Governor Hunt did not receive this message and he wondered how best to establish contact with the commander of the invasion forces. The telephone system was still working and Hunt was able to make a call to the one Argentine officer whom he knew to be on the island. This was Vice-Commodoro Hector Gilobert. For two years, this air force officer had administered the local LADE office. He had then handed over to a new incumbent, but had returned to Port Stanley to investigate the affair of the vandalism on the previous Saturday night. According to Rex Hunt: 'I had been glad to see him back with us. I knew him well, he was well liked, and I was as anxious as he that our one regular weekly lifeline to the outside world, the Tuesday Fokker F28 flights, should not be cancelled by the junta. I knew he was staying with his successor at a nearby house, so I called him up and asked if he would act as mediator with the invaders.' Hunt persuaded Gilobert to join up with Dick Baker and make his way along the waterfront to the Town Hall. Gilobert and Baker, who were on good terms, were authorised to contact the advancing armoured units and negotiate a ceasefire. At great danger to themselves, the two men, Argentine and Briton, set off through the bullet-swept dawn toward the town.

The Governor sent orders for the Falkland Island Defence Force to make their way back to their homes and to hide their uniforms and weapons. They had done what was asked of them, had guarded key positions in the town, and had mounted guard over the Islander aircraft at the racecourse, but dead heroes were of little use to the future of the islands. Hunt then sent orders to the *Endurance* party at the Town Hall to get rid of their weapons and to offer no resistance when approached by the invaders.

At 0920, Admiral Busser, commanding the invasion force, was

Pictures which the triumphant Argentines flashed around the world: Buzo Tactico marines search the defenders of Government House before marching them away to their brief captivity. A family of local islanders can only watch and wonder what their future now holds.

escorted into Government House. He complimented Hunt on 'the bravery and professionalism of his soldiers', but urged him to give the order for the Royal Marines to lay down their arms 'in order to avert further bloodshed.' The unhappy Governor agreed, and arranged with Majors Norman and Noott for a ceasefire to be broadcast. There was no formal declaration of surrender, no signing of documents, no handover of power, but, by 0925, it was all over. The Falkland Islands had fallen and, as the day wore on, the news was received around the world with stunned disbelief.

In Argentina, the crowds which only the day before had been shouting abuse at their leaders were given the message for which they had waited a lifetime – their national flag was at last flying over *nuestras Malvinas*. The whole country went wild with pride and joy. The people were told that the occupation had been achieved peacefully, the British garrison putting up no more than a token resistance. The British had in fact fired 6,450 rounds of 7.62 mm small-arms ammunition, four 84 mm rockets and eight 66 mm launchers. The number of troops killed and injured in the armoured personnel carrier is not known but, upon the evidence observed, the invading force lost five killed and seventeen wounded. The occupation was certainly not the walkover for which the junta had been hoping.

In the British Embassy in Buenos Aires, the worst of the earlier fears had become a sickening reality. The shock of the invasion was perhaps more severe for the local staff than it had been for Rex Hunt and his colleagues because, as one of them explains: 'We knew that an amphibious force was at sea and we were pretty certain that something very serious was about to happen, but we were not told that the Argentine ships had been detected off Cape Pembroke. The warning signal sent to Governor Hunt by London on the Thursday afternoon was never copied to us. The first we knew of the Port Stanley attack was when we heard it announced by special newscast on the Argentine radio. That was mid-morning on Friday. Rex Hunt may not have been given much advance notice, but at least he was better served than us.'

Apart from the sudden rupture in their previously amicable relationship with numerous Argentines whom they had come to know well, each member of the Embassy staff was now facing a major unheaval in his or her personal life. The Argentine Ministry of Foreign Affairs had made no move to declare them *personna non grata*, but it was evident that it would very soon be necessary to leave the country.

On Saturday, 3 April, when the FCO formally broke diplomatic ties with Argentina, the Embassy building became the 'Swiss Embassy, British Interests Section'. At the invitation of the British government, Swiss officials entered the building, made an inventory of the contents, and accepted responsibility for its safe custody. The red and white of Switzerland was hoisted up the flagpoles in the garden and above the residence. The British staff were now simply temporary guests in the country.

Everyone worked hard throughout the weekend to complete the business of shutting down the Embassy and simultaneously making their domestic arrangements. Sir Anthony Williams describes the final days: 'Prior to the invasion, the task of shredding and burning had been going forward with urgency. When the news came through, urgency turned into blitz. My wife and I not only joined in, but also put up in our house – across the garden from the office – any burners who wanted to doss down or snatch a quick meal.'

Colonel Love's files, like those of Captain Mitchell, had been destroyed several days earlier, but the two officers assisted their colleagues over the weekend before loading their cars and departing with their families on Monday 5 April, for Montevideo. Stephen Love remembers the startling contrasts of his final days in Argentina: 'The road convoy up to the Uruguayan border was very tightly controlled by the authorities. They were anxious to see that we left the country without any untoward incident. Our

Lieutenant Keith Mills, DSC, RM, photographed in 1985. Still a junior officer, already he has an entitlement to four medals: the Distinguished Service Cross, the Campaign Service medal (with clasp 'Northern Ireland'), the South Atlantic medal with rosette, and the United Nations medal for peace-keeping services with 40 Commando RM in Cyprus.

escort consisted of police, military and civil officials from the Ministry of Foreign Affairs. We had often driven to Montevideo and we knew the road, but our escort kept getting lost and we had a problem persuading them to follow our instructions. It was extraordinary to think that only six days earlier – on the evening of 31 March – my wife, Robin, and I had entertained three Argentine generals to dinner at our house in Acassuso. The main topic of conversation was the civil unrest, particularly the big trade union demonstration at Mendoza the previous day when fifteen hundred people were arrested. We sat there – discussing their nation's affairs – while their fleet was poised to invade our islands and the British government was drafting instructions for Anthony Parsons to go in to bat at the UN Security Council. None of that information ever came through to us, of course. The irony of the dinner-party was the fact that the most senior guest was General Mario Menendez, their Chief of Operations. Presumably he already knew then that he was to be the next Governor of the Falkland Islands. He was sworn in – as Rex Hunt's successor – on the following Wednesday, 7 April.'

Returning now to South Georgia, the Royal Marines detachment had spent their first full day ashore, Thursday 1 April, trying to make themselves comfortable in the overcrowded quarters of Shackleton House. There was little that Mills could do in the way of warlike preparations. He was at pains to respect Steve Martin's request that the BAS personnel should not be distracted from their usual work. The pretence was sustained that the marines were engaged in nothing more than an 'adventure exercise'.

At first glance, the orders given to Keith Mills by Nick Barker were simple and clear. Firstly, it was his duty to maintain a British military presence on the island. Secondly, he was to protect the BAS personnel at King Edward Point in the event of an 'emergency'. Thirdly, he must maintain a continuing surveillance of Leith Harbour. In practical terms, such orders – particularly the first two – were open to a dozen different interpretations, with everything depending upon the future actions of the Argentines. Initially there was little scope for Mills to use his own initiative. He was bound to wait upon events and then use his training and commonsense as circumstance dictated. A great weight of responsibility had descended suddenly upon the 22-year-old shoulders of this cheerful slender young man.

Keith Paul Mills was born in Abingdon, Oxfordshire, and grew up in Amlwch, on the island of Anglesey. The son of a nuclear engineer, he attended Sir Thomas Jones' School before entering the Royal Marines, in September 1978, as a Probationary Second Lieutenant. After training in England and Brunei, and a tour of duty with 41 Commando RM in Northern Ireland, he attended a Professional Studies Course at the Royal Naval College, Dartmouth. At the suggestion of one of his troop sergeants, Barry Nelms, he applied for the appointment of Officer Commanding Royal Marines Detachment, HMS *Endurance*, and was mildy surprised when his application was approved in the spring of 1981.

Of medium height, lightly framed, he was physically well adapted for the various sports which interested him. A sub-aqua diver, skier and mountaineer, he was also a light middleweight boxer and judo black belt. The angular face and deep-set brilliant eyes reflected great vitality and a driving ambition to live life to the full. Some older officers occasionally found his boisterous spirits and obvious self-confidence mildly irritating, but readily acknowledged his professional competence and enthusiasm. Perhaps the ultimate tribute to Keith Mills was the comment – made after the war of 1982 – by one of his contemporaries: 'We were bloody lucky in the character of the bloke who happened to be in command down there.'

Like any other young officer ever given the chance to show his mettle, Mills was fully aware of the opportunity which suddenly had presented itself. Apart from any fresh instructions which might come to him by radio, he was free of any constraining influence by officers of higher rank. It did not require a vivid imagination to understand that his reputation in the Corps might be made or marred by his personal decisions in the coming hours and days. His situation was in every respect unique. It is not possible to recall any other episode in British military history when such a small force has been placed in such an isolated position and with so much hanging on its conduct.

At the meeting in Nick Barker's cabin on the evening of 31 March, it had been agreed that, while the Argentines were certainly making very pugnacious noises, they were unlikely at that stage to commit themselves to open warfare. From Mills' point of view, the Argentine force at Leith Harbour was the most immediate concern. If it became active, if it advanced toward King Edward Point and threatened the base there, then he might find himself involved in a head-to-head conflict. But the idea that the Argentines might mount an invasion from the sea, that they might launch a frontal attack, was so outlandish that neither Barker, Martin nor Mills raised it during their short discussion.

Despite all of this, Keith Mills was determined to be ready for every contingency. One of his assessments was the need for adequate weaponry and ammunition. Before leaving Port Stanley, he had obtained from the Moody Brook garrison a quantity of anti-tank missiles: twenty 66 mm rockets and a Carl Gustav launcher with twelve rounds. In addition, he had two general purpose machine guns (GPMG) and two Bren guns (LMG). Each marine was armed with the self-loading rifle (SLR), the standard individual infantry weapon. He himself carried a Sterling sub machine gun (SMG) and Sergeant Leach, a qualified sniper marksman, was armed with the L42 sniper's rifle. All this weaponry would be useless without ample reserves of ammunition. Mills went to Lieutenant Dave Hutchings, RN, the officer responsible for the ship's ordnance.

Hutchings, in common with everyone else in the ship, still did not believe that the Argentines would attempt an armed assault. 'What are you talking about, Royal? There just isn't going to be any war.' He told Mills that he could have only 2,000 rounds of small-arms ammunition, this being the remaining balance of the 3,000 rounds allowed for training purposes (of which 1,000 rounds had been fired off in practice shoots earlier in the season).

Mills agreed with Hutchings' assessment of the war risk, but training and instinct told him to back his own judgement. He appealed directly to Captain Barker who authorised the issue of the detachment's full war allowance of 15,000 rounds (plus the 2,000 rounds of training ammunition). In the event, the marines fired off approximately 8,500 rounds during the subsequent battle. At their own level, Barker and Mills ensured that the means to resist were available, if required.

On their second night in Shackleton House, the marines settled down to a quiet evening of relaxation with the BAS civilians in the recreation room. A few men sat at the bar, others played cards or snooker. One of the BAS men wandered over to the radio and switched it on. The voice of Rex Hunt announcing the probability of an imminent Argentine invasion of the Falklands cut through the chatter. In silent amazement, they all listened to the news which, for Keith Mills, had previously been only a vague possibility, but which, for the BAS men, was a very disturbing shock. Hunt completed his announcement and the room immediately burst into a babble of raised voices. The marines crowded around their officer while the BAS personnel pressed their Base Commander with the obvious questions. 'Steve, what's going on? Did you know about all this? Why weren't we told? What happens now?'

After the initial confusion, a tense calm fell on Shackleton House. Some people drifted off to their bunks, others stayed close to the radio, listening to the friendly voice of Pat Watts in Port Stanley. The flow of relaxing light music was interspersed with official messages, asking the civil population to remain indoors, mobilising the civil defence organisation, and ordering the Falkland Islands Defence Force to report for duty. By the early hours of Friday morning, 2 April, only a small group remained gathered around the radio at King Edward Point. At 0425 they heard the Governor announce a formal State of Emergency. Two hours later, he began his commentary on the battle then raging around Government House. All doubts were now swept aside. The Galtieri regime had committed itself to military action. It could be only days, even hours, before they made their move against King Edward Point. Mills and Martin swiftly agreed that they must prepare to receive unwelcome visitors.

There were three immediate priorities: to evacuate the civilians to a place of safety, to prepare defensive positions, and to maintain a comprehensive radio listening watch. Before first light, Shackleton House was a hive of intense activity. The scientists began to pack their valuable equipment and irreplaceable records. Mills and Leach started to draw up a plan of action for the marines. The BAS radio operators continued to monitor the Port Stanley transmissions and, while listening to Rex Hunt, were shocked to hear gunfire in the background. The signals ceased at 0830 when the Argentines seized Pat Watts in his studio and occupied the Cable & Wireless station. Shortly afterwards, HMS *Endurance* broke radio silence. Still steering for the Falklands, but soon to reverse course, she confirmed the invasion reports and informed Martin and Mills that intercepted radio transmissions indicated the presence of another Argentine naval force, in South Georgia waters.

Steve Martin quickly decided how best to safeguard his own people. Three of them – Peter Stark, Tony North and Miles Plant – departed for Saint Andrews Bay in the launch *Albatros*. They were to make their way to the Sörling Valley and then overland to the hut occupied by Cindy Buxton and Annie Price, remaining with the two young women for as long as necessary. Orders were sent to the outlying BAS parties at Bird Island and Lyell Glacier to stay out of sight and avoid any provocative acts if Argentine forces appeared in their areas.

Two more BAS men – Trefor Edwards and Damien Sanders – set off from King Edward Point to join their two colleagues – Campbell Gemmell and Ian Barker – occupying the survey hut at Lyell Glacier. Both were physically very fit and knew the route well. Steve Martin was trying to disperse his people, but the onset of winter made it likely that they would be obliged to give themselves up if the Argentines gained control of the island and its sea approaches. The BAS staff could only watch and wait – they did not have either the training or weaponry to engage in open warfare. Indeed, if any of them had done so, acting as *francs-tireurs*, they would have had no protection in international law.

The bulk of the BAS personnel, thirteen in number, were obliged to remain where they were, on the coast. Martin and Mills agreed that they should be removed from the vicinity of the marines' main defensive position. The best alternative accommodation was the old whalers' church behind Grytviken, half a mile west of King Edward Point. It had been maintained in excellent condition over the years and, as a holy place, it offered the vague prospect of being a sanctuary.

Most of the scientists and other BAS personnel took their essential kit and sleeping-bags and, with a generous stock of food and heating equipment, moved into the church at 0900. They took with them a transmitter/receiver so that they could maintain contact. Only Steve Martin and Bob Headland remained at King Edward Point. Headland took over from the BAS operators, Peter Bayman and Brian Stanswood, in assisting with the cryptographic work and keeping a listening watch. He cleared out the administrative files, collecting all the official cyphers and government seals in a weighted sack and hurling it far out into the deep waters of the Cove. He later joined his colleagues in the church.

Steve Martin intended to stay until the last possible moment. He understood the excellent Racal radio equipment better than any of the marines, but it was in any event his duty, as Base Commander and Magistrate, to remain at his post until such time as any shooting obliged him to step down in favour of Keith Mills.

Mills knew that he could not remain for long on King Edward Point if faced with a superior force, but his men were all trained in Arctic warfare techniques and were extremely fit. He had plenty of ammunition and plastic explosive. Each man had his full issue of Arctic warfare clothing and other equipment. With a little luck, they could sustain themselves for weeks in the empty wastes of the hinterland.

He ordered his men to pack their Bergen rucksacks with spare clothing, food and ammunition. The packs were loaded onto the BAS tractor and trailer and hauled around the edge of the Cove to Grytviken where they were dumped at the church. Having resisted the initial landing, the marines would make a fighting withdrawal, under cover of darkness, recover their packs, and then move inland. The intended route would take them up the steep 2,000-foot slopes of Mount Hodges and northward to Maiviken. Mills hoped to hide his small force among the rock caves and snow pinnacles and conduct a guerilla campaign until such time as a relief force might appear. He also counted upon being able to make use of the BAS hut and emergency food dump at Maiviken. As Peter Leach said later: 'It would not have been easy, but at least we would have stayed on the island and could have given the Argies a bad time.'

The next step was the local defence of the key feature, King Edward Point. An obvious objective for an Argentine invading force would be the jetty. Mills discussed the problem with Marine Les Daniels, a trained assault engineer and therefore the detachment's explosives expert. The slim energetic Cockney was asked if he could improvise some means of making the jetty inoperative if a ship tried to lay alongside it. His solution was to sling a 45-gallon drum of petrol under the working deck where it was lashed to the supporting girders. A necklace of plastic explosive charges was secured around the drum and connected to an electrical cable laid back to a central control point hidden in the buildings of King Edward Point.

Then there was the question of the beaches. It was known that the *Bahia Paraiso* was carrying three, possibly four, landing craft. It seemed probable that she would be able to deploy a force of at least fifty marines or other specialist troops. The ship normally operated one or two Alouette III helicopters, but these had a limited lift capability. It was reasonable to assume, therefore, that any attempt by the Argentines to assault King Edward Point would be seaborne. There were two beaches over which the attackers could come ashore. The first was inside King Edward Cove, curving westward from the jetty towards Grytviken. The second and most likely beach stretched along the eastward edge of the promontory between the jetty and Hope Point. From the attackers' viewpoint, the second beach was not ideal because there are large beds of kelp nearby, and these could foul the propellers of approaching small craft. However, it provided a direct line of approach to the front door of Shackleton House and avoided the risks of trying to go around the tip of the Point into King Edward Cove itself.

Keith Mills lacked the strength to cover both beaches with effective fire, so he concentrated most of his efforts on the east beach. Again he consulted Marine Daniels and asked him if he could concoct some means of defending it. Daniels searched through the buildings and collected a quantity of scrap metal, nails, bolts and old whale harpoon heads. This material was packed into ammunition boxes, each having a core of plastic explosive. It was planned to bury these home-made mines at intervals along the length of the beach. The charges would be connected by cables running back to the central control post and detonated, if necessary, by electrical impulse.

Helping Daniels were two other marines, Porter and Church. At this stage, Keith Mills had no idea whether the Argentines would make a full-blooded assault or whether they would first offer terms for a capitulation. Conscious of the need to avoid unprovoked violence, he told the three-man explosives party that, once they had completed their work, they must remain hidden in the nearby buildings. They were to keep the jetty under close observation and, if appropriate, leap out and arrest anyone coming ashore unarmed.

The well maintained whalers' church at Grytviken in which the British Antarctic Survey personnel took refuge when the imminence of an Argentine attack was finally recognised.

While Daniels, Porter and Church prepared their charges, in worsening weather, the other marines began to dig a line of slit trenches along the low plateau on which stands Shackleton House. The line commenced just in front of the big red building and curved away westward on a front of one hundred and fifty yards. There were seven trenches, each accommodating two or three men. The BAS staff produced picks and shovels and, until they were evacuated to the church, helped the marines to hack through the tough tussock grass into the saturated peat beneath.

The gale strengthened to Force 9, bringing with it more squalls of sleet and rain which filled the trenches almost as quickly as they could be dug. Conditions were miserable and more than one marine grumbled quietly to Peter Leach: 'Sergeant, just what exactly are we supposed to be doing here?' They had already suffered one false alarm when told, two weeks earlier, that they were to arrest the scrap metal workers at Leith Harbour. Having spent many hours rehearsing that operation aboard the *Endurance*, they had felt deflated and let down when the operation was subsequently cancelled. It was now Leach's task to impress upon them the fact that this latest episode was indeed important and might well lead to a shooting contest. Despite the news from Port Stanley, many of them remained sceptical and did not fully accept the seriousness of the situation until literally the last moment – when they had their first sight of the Argentine invasion force.

By 0930 on that Friday morning the weather had deteriorated even further. The wind increased in strength, gusting to Force 10. Visibility was very limited. The marines were struggling to dig their trenches and Daniels had great difficulty in fixing the petrol drum under the jetty. He had not yet even started to lay his beach mines. In tactical terms, the detachment was extremely vulnerable. It was at this point that a large ship was seen to emerge from the curtain of driving rain at the entrance to Cumberland East Bay. It was the *Bahia Paraiso*. 'Stand to, stand to!' Everyone threw themselves flat or tried to crouch in the half-dug water-filled trenches. Mills left Peter Leach in charge and hurried down to the jetty to receive possible visitors.

Captain Trombetta announced himself on the radio and, speaking in good English, informed Steve Martin that he should stand by for an important message which would be passed to him on the following morning. Trombetta stayed long enough only to satisfy himself that the *Endurance* was not anchored in the bay. The ship then turned and disappeared into the murk. On shore, everyone drew their own conclusions from this brief visit and work went forward at an even sharper pace.

With hindsight, it is evident that the Argentines intended originally that their seizure of South Georgia should coincide with their invasion of the Falkland Islands. Only the foul weather prevented them from keeping to the timetable. The wind was much too strong for the launching of helicopters and the seas too violent for the operation of landing craft. Thanks to the storm, Keith Mills was granted twenty-four hours of additional breathing space in which to complete his preparations. As Peter Leach commented later: 'If it hadn't been for the lousy weather we would have been caught with our pants at half mast.' Inexplicably, the invasion warning sent by London to Governor Hunt at 1530 on 1 April was not repeated to Martin and Mills at King Edward Point. That warning referred only to Cape Pembroke, but the information would have been helpful to the South Georgia people also.

Before leaving the *Endurance*, Mills knew that he would be able to talk by radio to the ship's communicators and thus maintain two-way contact. The *Endurance* herself was naturally very well equipped with communication systems of various kinds. She could conduct live voice conversations via satellite with the Commander-in-Chief's staff at Northwood in complete security, but there was no such equipment at King Edward Point and the BAS cyphers were not appropriate for naval use. During the half hour in which the marines were busily preparing to go ashore, steps were taken to devise a way of transmitting short simple messages to the *Endurance* that carried only a slight risk that they might be decyphered by any Argentine eavesdroppers who might have been listening.

At 1030 on 2 April, Mills used this *ad hoc* device to signal the *Endurance*. He reported the visit to Cumberland East Bay of the *Bahia Paraiso* and asked for instructions. Forty-five minutes later the reply came through: 'When the Argentines again make contact you are not to co-operate.' It required fifteen minutes for Mills to encode his second signal: 'Your last message ambiguous. Please clarify'. There was now a long pause before the *Endurance* again responded.

At approximately 1300, Mills was told: 'When asked to do so you are not to surrender.' Mills and Martin studied the words and looked at each other. 'Christ! What does that mean?' They debated the matter for a few minutes and Keith Mills then decided that – allowing for the limitations of the home-made code – this signal more or less permitted him to use his own on-the-spot judgement. The question of surrender had not previously surfaced in the orders given to him.

Reasonably happy with this interpretation, he left Steve Martin alone in the radio shack and went up to Shackleton House to talk with Peter Leach. A short time later, at 1320, Martin received another message from the *Endurance*: 'The Officer Commanding Royal Marines is not repeat not to take any action which may endanger life.'

Martin called Mills back to the shack and they stared in amazement at the slip of paper. This was something entirely new. If followed, this instruction did not allow Keith Mills even the modest latitude of the Yellow Card rules. Literally, it denied him the right to open fire under any circumstances. He concluded that 'nobody wanted to take the rap for dead men' and tried to dismiss the message from his mind. It seemed that he had been 'dumped in at the deep end' and could expect little official support if things went wrong. Uttering an appropriate marines' curse, he got on with the job in hand.

What he could not know was that Nick Barker was himself unhappy with the text of the third (1320) signal. He had been talking throughout the morning with Whitehall, passing to them the information reported by Mills and asking for decisions. When told by a very senior officer on the naval staff that the marines on South Georgia 'must not endanger life', Barker did not take it kindly. 'Good God! The Argentines have invaded the Falklands and shot at our people there. We have a war on our hands. What more do you need to be convinced?' Whitehall's reply has not been recorded.

Nick Barker was now very worried for the safety of the marines. Apart from his natural professional concern, there was also the fact that Keith Mills was one of his ship's officers and most of the other ranks were members of the ship's company. They needed support and they needed it quickly. 'The plum' continued to pound south-eastward – for the third time in two weeks – still three hundred miles short of her destination.

It was during this return passage that Captain Barker thought he might have an opportunity to take direct action against the Argentine navy. Reports showed that the two small frigates already mentioned in Argentine newspapers as being on their way to South Georgia – the *Drummond* and the *Granville* – were heading in his direction. In addition, there was the possibility that they had been joined by two other ships, destroyers – elderly former United States Navy ships renamed *Piedra Buena* and *Commodoro Py* – and by the 16,000-ton fleet support tanker, the *Punta Medanos*. It was also reported that the tanker had suffered storm damage and was turning south toward Beauchêne Island. Presumably she intended to shelter there and repair the damage before proceeding. In that case, the thirsty destroyers – which needed to refuel from her every two or three days – might turn away and for a while leave her unescorted. If the *Punta Medanos* could be sunk, the frigates and destroyers might be forced to return to the mainland.

Nick Barker requested authority to go in search of the *Punta Medanos* and attack her with his two Wasps, armed with AS12 missiles. If they failed to sink her, he proposed to take advantage of the *Endurance*'s ice-strengthened bow by ramming the Argentine ship. The proposal was considered briefly by Northwood, but then rejected. Barker was told to stay on course for South Georgia. The first priority was to support the isolated marines and BAS civilians.

With the limited number of men available, Mills could not indulge in any fancy 'in depth' defence works. The left flank trenches had a good view of the beach and their arcs of fire

The setting for one of the most extraordinary small battles ever fought: King Edward Cove, South Georgia. To the left, King Edward Point and the shoreline track which was the Royal Marines' only possible escape route. In the foreground, the abandoned whaling station of Grytviken with the tiny white-painted church closest to the camera. To the right, the coal hulk *Louise*, the square shape of the cemetery, and the lower slopes of Brown Mountain where the Argentine Puma helicopter crash-landed.

covered the area where Marine Daniels intended to plant his home-made minefield. The field of vision for the centre and right flank trenches was restricted because it was partly blocked by the roofs of the buildings in the area of the jetty. However, the weapons in these trenches could be brought to bear upon the track leading up from the jetty, upon anyone attempting to occupy the buildings, and upon anyone advancing across the open ground, near the radio shack, normally used as the helicopter landing area.

Concealment of the trenches was aided by the fact that they were set back from the edge of the Shackleton House plateau. This ground is approximately thirty feet higher than that on which stand the other buildings of King Edward Point. Furthermore, the thick tussock grass still had its summer foliage and this helped to camouflage the marines' positions. The only isolated trench was one of the Bren gun positions. It was sited on higher ground, two hundred yards away, on Hope Point (where Ernest Shackleton's memorial cross forms a prominent landmark). The Bren's accuracy over long ranges made it ideal for this dominant feature. It covered the sea approaches to Hope Point, could fire down onto King Edward Point, or could rake the beach in front of Shackleton House.

The home-made mines were buried in the sand and pebbles of the beach and the final connections made. Daniels still had some explosives in hand, so he proceeded to booby-trap Quigley's House and several other buildings. These were clearly very unconventional measures, but Mills believed that it was his prime duty to protect the lives of his men and to sustain their fighting capability. To achieve this aim he used whatever materials were available.

Mills inspected Daniels' handiwork and felt satisfied that he had created an effective killing ground. If necessary, any hostile force coming ashore on the beach could be shredded with scrap metal. Alternatively, any craft coming alongside the jetty could be engulfed in a fireball. He had done everything possible to protect the BAS personnel and their property. His methods clearly breached the spirit of the instructions received that morning via HMS *Endurance*, but he was determined to ensure a hot reception for any Argentines who might insist upon making an armed landing.

The third element in his orders – to maintain a watch on Stromness Bay – had been covered continuously since the first hours. The *Endurance*'s marines had taken over the surveillance task from the BAS on 25 March. During the last six days of the month they were ferried to and from Grass Island and the Busen Peninsula by Wasp helicopter, returning to the ship at the end of each watch. Then, following the departure of the *Endurance* on 31 March, a four-man team had been stationed around the clock on Jason Peak. With the Wasps no longer available to lift them back and forth, they had been taken to Jason Harbour at dawn on 1 April, by Steve Martin and Peter Leach, in the BAS launch *Albatros*.

The party consisted of Marines Poole, McCallion and Combes under the command of Corporal Nigel Peters. They worked a 'two on, two off' routine, one pair watching the Bay and its sea approaches, the other pair resting in the Jason Harbour refuge hut. For communications they relied upon a BAS transceiver. Normally they would have used their own Clansman radios, but all of the batteries were found to be exhausted. The marines had

Above: Corporal Nigel Peters poses beside the Jason Harbour refuge hut during a short break in the otherwise foul weather.

Below: A face to be reckoned with. Sergeant Peter Leach, RM, HMS *Endurance*'s Detachment Sergeant-Major, and by far the most experienced member of the defending force.

left the *Endurance* in such a hurry that nobody checked to ensure that they were properly charged.

Corporal Peters' team did not have a pleasant or easy task. Off watch, they could sleep and prepare food in the hut, but the observation point itself was just a patch of exposed rocky ground. Driving sleet frequently blotted out the stretch of water which they were trying to watch. However, on their first day, they noted the presence of the *Bahia Paraiso*, anchored half a mile off Leith Harbour. Landing craft were moving between ship and shore. The ship's Alouette helicopter made several reconnaissance flights over Leith Harbour, Stromness and Husvik, but did not approach the Busen Peninsula. Apparently the Argentines did not know that their arrival had been observed and were hoping to avoid detection.

Corporal Peters sent an initial situation report to Keith Mills, but thereafter stayed silent. Like Bob Headland before him, he did not wish to reveal his position to the Argentine operators by excessive use of the radio. Mills also kept his transmissions to the minimum, but he ensured that Peters was informed of the developments on the Falklands. Peters was not at first too alarmed by the news from Port Stanley, but he started to feel anxious when Spike Poole jokingly asked: 'And what are we supposed to do if the Argies come up here looking for us?' Peters had no orders as to how or whether his party should defend itself, nor did it have sufficient ammunition for a prolonged firefight.

The storm continued to blow throughout Friday. The watchers on Busen Peninsula saw very little of Stromness Bay. At King Edward Point, the task of digging in continued until last light. Wet and hungry, the marines retired to the comfort of Shackleton House. For the first time since the *Endurance* sailed from Portsmouth, six months earlier, the Royal Marines found themselves entirely alone. Keith Mills decided to follow Napoleon's maxim that, 'an army marches on its stomach.' He told Marine Stonestreet, an expert cook, to take full advantage of the BAS food stocks. The freezer was raided and Stonestreet prepared a gargantuan meal, thereafter known as 'the last supper'. The building shook to the horseplay and laughter of eighteen tough young men. Each, in turn, every thirty minutes, slipped out into the violent night to take his turn as sentry on Hope Point. Mills doubted whether anything would happen in such weather, but it was prudent to have a man watching and listening on the shoreline.

With the coming of darkness on that Friday, 2 April, the two sides settled down to their final night of rest and contemplation. What would tomorrow bring? Would there be a battle? Who would win and who would lose? In Stromness Bay, the men aboard the *Bahia Paraiso* slept snug in their cabins as the ship yawed uneasily at her anchor chains. Twelve miles to the south, at Grytviken, the BAS civilians tried to settle in the draughty discomfort of the whalers' church. At King Edward Point, chinks of light shone from the windows of Shackleton House where Keith Mills and his men were finishing their memorable supper. At Jason Harbour, Nigel Peters and his party lay curled in the cocoon of their 'green slug' sleeping-bags. All around, the ancient glaciers and mountains looked down on these four tiny outposts of humanity. Out in the cold darkness and howling wind, the only movement came from colonies of penguins and seals whose occasional domestic squabbles continued late into the night.

In the way which is so typical of the region, the weather changed completely during the small hours of Saturday morning. By sunrise, the wind had died away to a whisper, the sea was almost calm and there was not a cloud in sight. South Georgia's spectacular snowfields glared white against a brilliant blue sky. Peter Leach went to his officer and suggested that it was time to retrieve the observation party from the Busen Peninsula. They had very little fighting value where they were, but they represented a fifth of the total Royal Marines force on the island. If it came to a fight at King Edward Point, their firepower would be badly needed. Mills agreed and authorised Leach to bring them in.

A heated argument then followed between Peter Leach and Steve Martin. With unlimited visibility, this was precisely the wrong sort of day for a secret evacuation by sea. Leach wanted to depart immediately in the *Albatros* and, with two other marines, motor across Cumberland West Bay to Jason Harbour. He calculated that he could complete the round trip in little more than two hours. Martin was opposed to the plan because he did not believe that the sergeant could handle his boat, especially if its diesel engine failed, and he rightly regarded the venture as extremely hazardous. The launch was a valuable BAS asset and Martin was loathe to place it at risk.

Apart from the possibility of colliding with the chunks of ice which continually break away from the nearby glaciers, there was real fear that the launch might be attacked. A radio report had just been received from Nigel Peters, announcing the arrival in Stromness Bay of an Argentine frigate. In the words of Keith Mills, 'this threw a whole new light on the situation.' The warship was reported going alongside the *Bahia Paraiso* and taking on

As the ice-topped mountains start to catch the first rays of the rising sun, Corporal Peters' observation party on the Busen Peninsula watch the movements of the Argentine invasion force. These historic pictures, taken early on the morning of 3 April, show the arrival in Stromness Bay of the small frigate *Guerrico* and her rendezvous with the *Bahia Paraiso* (anchored off Leith Harbour).

fuel, but that task would soon be completed and she would then be free to take the offensive.

The powerfully-built Leach informed the Base Commander that he intended to take the launch, with or without his permission, and Steve Martin reluctantly agreed to show him how to operate the controls. The *Albatros* departed King Edward Point shortly after 0630. Astern, she towed a Gemini assault boat. It was Leach's plan that, if the *Albatros* was attacked and damaged, he would abandon her and try to make his escape in the smaller boat which was much faster and highly manoeuvrable.

Peter Leach stood holding the wheel as the heavy launch plodded northward along the coast at a steady eight knots. There was plenty of time to think, to ponder the curious circumstances in which he found himself. He was the father figure of the *Endurance*'s marines detachment. Thirty-seven years of age, a Liverpudlian, married with two children, he could look back over nearly eighteen years' service with the Royal Marines. Like many other members of the Corps, he was a man of many parts: air photography instructor, qualified helicopter observer, weapons instructor, and unarmed combat expert. In the 1960s he had seen action as a leading scout in Borneo, had later served a tour in Northern Ireland, and had witnessed the Turkish invasion of

Guerrico, the warship so nearly destroyed by a handful of Royal Marines. This picture was taken before the removal of her Mark 54 rocket launcher and the installation of a 40 mm gun on her stern. The four Exocet missile launchers can be seen abaft the funnel. It was the 100 mm gun mounted on the bow which swung the balance of the battle against Lieutenant Mills and his men.

Cyprus. It had been a long and varied career, but he could recall no other episode remotely comparable to this one. His immediate worry was the possibility that one of the *Bahia Paraiso*'s helicopters might see and attack him. Initially he hugged the cliffs of Sappho Point, but then he had nearly six miles of open water, crossing Cumberland West Bay.

Peter Leach's other anxiety was the lack of medical facilities on the island. Unlike the rest of the detachment, Leach had been in battle before. He had seen the horrendous injuries inflicted by modern high velocity ammunition. Each man had been given a morphine ampoule by Surgeon Lieutenant Neil Munro when they left the *Endurance*, and they had a small first aid bag and the usual field dressings, but none of the group was medically trained. Given the isolation and severe climate of the island, there was every chance that any man wounded by gunshot would subsequently die from lack of treatment.

One factor which gave confidence was the quality of the men. Their average age was twenty-four, and this meant that they were unusually experienced by Royal Marines standards. The men of the *Endurance*'s detachment he knew very well indeed, but he had also formed a favourable opinion of the nine extra marines attached from the Moody Brook garrison (Naval Party 8901). They had just completed their tour on the Falklands and had been hand-picked by Major Gary Noott. Leach had no doubt that they would all give a good account of themselves.

Peter Leach's determination was rewarded when he collected his very chilled passengers from the beach at Jason Harbour and brought them safely back, at 0900, to King Edward Point. They were quickly debriefed by Keith Mills. He wanted full details of the warship seen entering Stromness Bay three hours earlier. He knew that his detachment would be helpless if subjected to sustained gunfire from a ship standing off beyond the limited range of his own weaponry.

By coincidence, one of the marines in the observation party, Paddy McCallion, was an expert on warships and weaponry. As other men might study horse-racing form, or the performance of a football team, so had Marine McCallion devoured everything published on the equipment of the world's navies and armies. Raised in Carrickfergus, Ulster, he had served an apprenticeship as a fitter in Harland & Wolff's shipyards before joining the Royal Marines eight years earlier. He instantly recognised the warship for what she was – a small but powerful frigate of the French A69 class. She was the *Guerrico*.

Displacing 1,200 tons, with a complement of five officers and seventy-four ratings, she had been well designed for her main anti-submarine role. Originally she carried a Mark 54 rocket launcher and two triple torpedo tubes for attacking submerged targets but, additionally, she had a 100 mm semi-automatic gun on her forecastle and twin MM38 Exocet rocket launchers aft. Completed at the Lorient Naval Dockyard in 1978, she was first intended for the South African Navy. She was named *Transvaal* while under construction, but the sale was cancelled as the result of a United Nations embargo on arms sales to South Africa. The French navy renamed her *Commandant l'Herminier*, but then sold her to Argentina while still fitting out. At a later stage her armament was changed from the specialist anti-submarine warfare (ASW) configuration to give her a general purpose role. The Mark 54 ASW rocket launcher on her stern was replaced by a 40 mm gun.

McCallion explained that, like many modern warships, the *Guerrico* carried little armour. Her superstructure was largely of aluminium. The slender hull, 262 feet in length, was propelled by twin diesel engines producing 11,000 brake horsepower, and they could push her to a maximum speed of 24 knots, but she was vulnerable at close range. The Exocets would be useless against a land target and the anti-submarine equipment was irrelevant to the task of seizing King Edward Point. Mills realised that the 100 mm and 40 mm guns were a serious threat but, if the confined waters obliged the frigate's captain to come close to shore, she could be penetrated by small-arms fire.

Mills was much relieved by McCallion's expert opinion. The

Guerrico was certainly an additional problem but, as she was the only vessel to have joined the *Bahia Paraiso* in Stromness Bay, it looked as though this was the total invasion force. A few radio signals had come in during the night, but the most important originated from Nick Barker. Using the agreed code, he announced that he hoped to make landfall in Cumberland East Bay at 1500 that afternoon. His arrival might serve to balance the scales.

By 0930 the activity around King Edward Point had started to subside. Apart from Steve Martin, all of the BAS personnel were safely settled in Grytviken church. Martin and Keith Mills were in the radio shack, maintaining contact with the BAS bases on Signy and Faraday and listening for the *Endurance*. The marines pottered around their trenches, bailing out rainwater, covering the spoil, and laying out their reserves of ammunition. It was the classic calm before the storm.

Suddenly the morning stillness was broken by the sound of a distant engine. Several voices shouted together: 'Chopper!', 'Freeze!', 'Take cover!', 'Don't move!' In the event, everyone leapt into his trench except Sergeant Leach. He had been too busy to dig one for himself, so he lay down in a fold in the ground just in front of Shackleton House. 'Watch your front and report. Don't shoot. Don't point your weapons. Pass the word.' Leach shouted his orders to the nearest marines and they shouted to the others on the flanks. Everyone watched as a helicopter circled over Cumberland East Bay and then came lower to make several fast passes over King Edward Cove.

The aircraft was Alouette III/0699, assigned to the *Bahia Paraiso* some weeks earlier for reconnaissance purposes. Flown by Teniente de Navio Remo Busson (pilot), Teniente de Fragata Guillermo Guerra (observer) and Sub Segundo Julio Gatti (aircrewman), it was this aircraft which had overflown Wasp 434 when Captain Barker landed on Tonsberg Point on 27 March. Busson's task this morning was to inspect the area of King Edward Cove prior to the arrival of a second helicopter carrying twelve Argentine marines. They would be landed by the Puma assault helicopter AE-504 which had been sharing the *Bahia Paraiso*'s hangar with Busson's Alouette since both aircraft were embarked at Ushuaia in late February. As usual, the Puma would be flown by Primer Teniente Juan Villagra (pilot) and Primer Teniente Eduardo Leguizamon (second pilot).

On the ground, everyone watched as Busson circled the area for nearly ten minutes. He did not hover or touch down and, to the British, seemed to be distinctly nervous. He did not fly directly over Shackleton House and Keith Mills assumed – correctly – that he had not seen the trenches where the marines lay hidden. Eventually he turned away and flew northward at low level.

The *Bahia Paraiso* now came into sight, steaming slowly into Cumberland East Bay. Captain Trombetta called Steve Martin on Channel 16 VHF and delivered his promised message: 'Following our successful operation in the Malvinas Islands the ex-Governor has surrendered the islands and dependencies to Argentina. We suggest you adopt a similar course of action to prevent any further loss of life. A ceasefire is now in force.' Martin briefly acknowledged the message and asked for five minutes to consider his response.

Trombetta's ultimatum contained two blatant lies. Governor Hunt had not surrendered the dependencies, only the Falkland Islands. Martin and Mills quickly agreed that they were not being told the truth on this point. The use of the word 'ceasefire' was, however, a different matter. They now knew that the United Nations Security Council had been in emergency session since Thursday night. All manner of peace initiatives were in the air. The BBC World Service was broadcasting regular reports on the diplomatic developments. Was it possible, wondered Mills, that the British government had agreed to a formal ceasefire in the past few hours? If so, what would be the consequences, for himself and for his country, if he opened fire? Of all the pressures placed upon him, this was the one which caused him the greatest anxiety.

Trombetta at this point was a perplexed man. With the British all having taken cover, he was viewing a shoreline which appeared entirely lifeless. He came through again on Channel 16 VHF, ordering Steve Martin to call his people out into the open so that they could be seen and counted. Martin ignored this instruction and consulted with Mills. 'What now, Keith?' 'Tell him that there is a military presence on the island and it will be defended if he tries to land.'

More than anything else, Mills needed to stall for time. He and Martin both hoped that mention of a military presence might cause Trombetta himself to delay while he consulted his own superiors. If talks could be dragged out until mid-afternoon, the arrival of the *Endurance* might possibly discourage the Argentines from pressing their point. Martin carefully and slowly pronounced the words suggested by Mills, and there is no reason to doubt that the Argentine understood exactly what was being said.

Judged by their subsequent actions, the Argentines simply did not believe Martin's warning that there were British troops in the vicinity. Trombetta at first thought that he had HMS *Endurance* trapped in the bay but, when he discovered that she had slipped away, he assumed that his landing would be unopposed. He may have contemplated the possibility of resistance, but only by a few men who could be easily overawed and frightened into submission. If this was his reasoning, he made a very bad error of judgement.

Keith Mills realised that the *Endurance* might not be able to monitor these exchanges. He suggested to Martin that he should switch to a high frequency channel. On the pretext that the VHF reception was not good, Steve Martin changed to a different channel and slowly read back Trombetta's ultimatum 'to be certain that he had understood it correctly.' The *Endurance*'s operators did indeed pick up this transmission. They also heard, shortly afterwards, an announcement by Trombetta that he intended to send troops ashore by helicopter, and Steve Martin's repeated warning that any such illegal landing would be resisted by force.

With the situation now coming to the boil, Captain Barker decided to release Lieutenant Mills from the previous restraints ordered by London. Again breaking his self-imposed radio silence, Barker tried to talk directly with King Edward Point. Radio propagation conditions were not good, and the *Endurance*'s operators failed to make contact. Paradoxically, however, their calls were heard by almost every other station in the region. The BAS operators on Signy Island were among those monitoring the radio traffic and they tried to act as intermediaries. They were quickly joined by the BAS people at Faraday, hundreds of miles to the south in Antarctica.

Signy: 'Grytviken, Signy. We have a message from the *Endurance* for you. Over.'

Endurance: 'Grytviken, Grytviken, this is *Endurance*. Copying you loud and clear. How me? Over.'

Faraday: 'Grytviken, Grytviken, this is Faraday. Do you copy? Over.'

Signy: 'Grytviken, Grytviken, Signy, Signy. I have a message from the *Endurance*. I have message from the *Endurance*. Over.'

Endurance: 'This is *Endurance*. You are very distorted. Go ahead.'

King Edward Point as seen by Teniente de Navio Remo Busson when he approached the tip of King Edward Point with the first serial of marines from the *Bahia Paraiso*. Lieutenant Mills was standing on the jetty in the foreground, waiting to receive an expected envoy. Beyond is the track which he and his escort covered at speed when he saw the Alouette disembarking its passengers and decided to rejoin his men near Shackleton House.

Signy: '*Endurance*, *Endurance*, Signy, Signy. You are barely readable. Can you connect me via *John Biscoe*, via *John Biscoe*, please. Over.'

Endurance: '*Endurance* over . . .'

Signy: 'Grytviken, Signy, over ... *Endurance*, Signy, *Endurance*, Signy, very faint here, very faint. We are still trying to contact Grytviken. Over.'

Endurance: 'Grytviken, *Endurance*, Grytviken, *Endurance*. How do you copy me? Over.'

At this point, ironically, an English-speaking operator aboard the *Bahia Paraiso* joined in the desperate race to aid Captain Barker in his attempt to make contact with Steve Martin and Keith Mills.

Bahia Parasio: '*Endurance*, *Endurance*, *Bahia Paraiso*, *Bahia Paraiso* calling. Do you read me? Over. This is the *Bahia Paraiso*, *Bahia Paraiso*, go over with your message, go over with your message to Grytviken. He is listening on 11255. I repeat, go over with your message to Grytviken, to Grytviken. He is listening on 11255. Over.'

Signy: '*Endurance*, Signy, *Endurance*, Signy.'

Faraday: 'Signy, Faraday, Signy, Faraday.'

Signy: '*Endurance*, Signy, *Endurance*, Signy. *Bahia Paraiso* is calling you on 11255, calling you on 11255. Over.'

And so the crackling voices continued, fading and strengthening unpredictably. At St Andrews Bay, Cindy Buxton and Annie Price sat by their own radio, following the exchanges with fascination and preserving them for posterity on their tape recorder. They heard no response from King Edward Point (Grytviken), but they could hear Signy Island, the *Endurance* and the *Bahia Paraiso* still calling vainly to each other and to Mills and Martin.

Nick Barker now tried again to send his fresh instructions, regardless of the lack of response.

Endurance: 'Grytviken, this is *Endurance*. Over . . . I assume you can hear me. Rules of engagement are as stated earlier this morning. Rules of engagement are as stated earlier this morning. Defend if you are provoked. Over.'

This transmission was monitored by several other stations – including the *Bahia Paraiso*, presumably – but it did not get through to the man for whom it was intended, Keith Mills.

Another voice now joined the clamour. It was the RRS *Bransfield*, a thousand miles to the west, with a long message from Dr Richard Laws. He was trying to instruct the BAS personnel on South Georgia to co-operate with the new Argentine Military Governor in Port Stanley and to agree, if necessary, to be evacuated. In a confused situation, and isolated from any hard news, Laws was trying to reasssure his colleagues that everything possible would be done to keep them out of personal danger. The *Bransfield*'s operator failed to raise Steve Martin, but Signy Island offered to re-transmit the message. In the event, Signy's repeated transmissions of the Director's message were a wasted effort. In the past ten minutes, events at King Edward Point had made irrelevant the instructions which Barker and Laws were trying so hard to convey.

As they stood in the radio shack, debating what should be done next, Mills and Martin glanced out of the window. There, on the far side of Cumberland East Bay, was the *Bahia Paraiso*. Around her, circling low over the water, was her Alouette helicopter. It was tracking backwards and forwards at a range of about 5,000 yards from the British position. Suddenly, moving slowly, and very much closer, appeared the grey bows of the *Guerrico*. She had rounded Hope Point and was approaching the confined waters of King Edward Cove itself. Her two guns were trained on the shore. Keith Mills had the distinct impression that their point of aim was him, personally. Her arrival had been observed by Marines Steve Parsons and Steve Chubb, at the Bren gun post, but the lack of radios prevented them from reporting it to their officer.

With hindsight, it is evident that the *Guerrico* was trailing her coat-tails to see if she could provoke a response from King Edward Point. Mills' belief at the time, however, was that she had come close inshore to launch a small boat and send ashore an envoy to discuss surrender terms.

The likeliest landing point would be the jetty. Leaving Steve Martin to stand by the radio, Mills left the shack and walked down to the jetty, two hundred yards away. With him he took the explosives party – Daniels, Church and Porter – as his personal escort. While he stood waiting on the jetty, the *Guerrico* entered the Cove, turned, and started to make her way out again. Puzzled, Keith Mills gazed at the ship, looking for any signs of a boat being lowered overside. There was no such activity, but the ship's guns had been relaid and were again pointing at the King Edward Point buildings.

At this point he heard the clatter of helicopter blades to his left.

The Royal Marines' main field of fire: King Edward Point viewed from the Shackleton House plateau. The Alouette landed at the end of the Point (beyond the radio masts), the Puma hovered over the helicopter pad (left centre), and the *Guerrico* twice cruised around the Point into the confined waters of the Cove (right centre).

There, only fifty yards away, was the Alouette, already on the rough ground of the Point, with armed men jumping out and running for cover. To this day, none of the Royal Marines can explain exactly how it came to be there. They had been concentrating upon the warship to the exclusion of everything else. Remo Busson had collected his passengers from the flight deck of the *Bahia Paraiso* and then skimmed in over the waters of King Edward Cove to make a swift and skilful landing on the Point.

Mills turned and faced the new arrivals, his SMG slung across his shoulder. With the idea of letting them know that he was a soldier, to confirm that there were British troops on the island, he waved to the Argentines and pointed at his combat jacket. The fifth man – just leaving the Alouette – saw Mills and his face registered total surprise. After a moment's hesitation he reached forward and tapped his nearest companion on the shoulder, shouting and pointing at Mills. The second soldier swung around, spotted the British officer and started to swing his rifle up to his shoulder. His stance was clearly that of a man about to fire.

Keith Mills took to his heels and ran. Ducking and weaving, making use of dead ground, he covered the three hundred yards up to Shackleton House in record time. At any moment he expected a bullet through his back, but in fact no shots were fired at him. As he ran, he took Les Daniels and the two other marines with him. 'Come on, get back.' There was no point in leaving the engineer at the mines control post. The Argentines were not landing by boat, as expected, so that part of the original defence plan was now useless.

It was at this stage that Steve Martin wisely decided that the situation had become a matter for the military. Dressed only in slacks and a shirt, he ran from the radio shack and started to head toward Grytviken church. He never arrived there. He had gone only a short distance when firing broke out and he dived into a shallow gulley for protection. For more than two hours he lay in this ice-cold place – shot and shell cracking past only feet above his head – and developing all the classic symptoms of hypothermia. Another peril was the possibility that a panicky elephant seal might choose the gulley as an escape route to the sea. Two tons of seal would certainly prove fatal. In the event, his only visitor was a Norwegian rat.

On the other side of King Edward Point, immediately in front of Shackleton House, was a section of five marines commanded by Lance-Corporal George Thomsen, a huge Northumbrian noted for his fighting spirit. He and the rotund Marine Jock Hunter lay in a natural fold in the ground, concealed by the long tussock grass. To their right and rear, in a large L-shaped trench, were Marines Holding, Hare and James with a GPMG. This section had been positioned by Keith Mills with the objective of putting enfilade fire across the beach if anyone tried to land on it. They had been told to be prepared for an assault by landing craft, but to stay out of sight until ordered otherwise. One of those crouching in the bottom of the trench, Brasso Hare, an eccentric, otherwise known as 'the mad axeman', heard the throb of marine engines. Overcome by curiosity, he peered over the parapet. There, four hundred yards away, was the long low grey shape of the *Guerrico*. 'Hellfire! That ain't no bloody landing craft!' The others stood up and, easing themselves into firing positions, prepared for whatever was to come.

Up on the plateau, Peter Leach and the other marines viewed with alarm the scene unfolding before them. Entirely cut off from their officer, they had watched the *Guerrico* cruising into the

Cove. They could see the *Bahia Paraiso* on the other side of the Bay. They watched the small helicopter patrolling offshore and saw its sudden arrival on the Point. Armed troops were already on the ground and advancing towards them. Then Keith Mills came running over the crest and dropped, panting, into the trench occupied by Marine Knocker White. 'What's the score, Sir? What's happening?'

Mills had no time to reply or explain. As he turned and looked back to the shore, he saw a Puma troop-carrying helicopter closing fast. White and several other marines shouted a warning. They had all spent much time, during their service careers, in and around helicopters. It was obvious to them that this one was preparing to land. Hovering thirty feet above the flat ground of the shoreline, the pilot started the usual pre-landing manoeuvres – yawing the nose from side to side, checking for overhead cables or other obstructions. The side doors were open. Two machine guns were pointing out of the door and a number of men could be seen inside, holding rifles at the ready.

Corporal Al Larkin, one of the steadiest men in the detachment, shouted anxiously: 'Sir, what are we going to do?' It was a good question. The thought flashed through Keith Mills' mind: 'If this thing lands, we're in the shit.' He hesitated only a few seconds before bellowing the order: 'Hit it!' Instantly, every weapon opened up and raked the helicopter at a range of between fifty and one hundred yards. Even Mills himself opened fire, loosing off half a magazine (fifteen rounds) at the hapless machine. They were the only shots he fired during the entire battle. Nearby, concealed in the tussock grass, Peter Leach did not need his telescopic sight to score repeated hits on the huge target with his sniper rifle. Marine Andy Lee joined in the barrage with a discharge from a 66mm rocket launcher. The missile passed a few feet below the Puma, otherwise it might have disintegrated there and then.

In their excitement and inexperience of battle conditions, the marines made the error of firing at the troops within the hull of the aircraft. Hundreds of rounds were pumped into it, killing or wounding nearly all of the passengers, but nobody thought of firing into the cockpit. One burst would have been sufficient to kill the pilot and the Puma would have crashed immediately. In the event, the machine wobbled around the sky, emitting thick clouds of black smoke from the engine compartment, and staggered across King Edward Cove to the far shore. Almost miraculously, Juan Villagra succeeded in staying airborne long enough to avoid ditching and made what was later officially described as a 'heavy landing'. The marines saw the machine drop suddenly into semi-hidden ground, its blades flailing vertically and clearly having rolled over. No further movement was observed from the site of the crash.

According to a subsequent statement by the Argentine armed forces, the Puma 'was hit by fifty rounds of small-arms fire, two men were killed and several wounded.' In the circumstances, these figures would appear to be no more reliable than any other Argentine account of events on South Georgia. It cannot be disputed, however, that Villagra and Leguizamon displayed fine airmanship. The casualty roll would have been much higher if they had allowed their stricken aircraft to drop into the deep and frigid waters of the Cove.

During this engagement, the small party of Argentine marines which had disembarked from the Alouette was moving cautiously along the narrow pebble beach towards Shackleton House. They were in dead ground as far as most of the Royal Marines were concerned, and they in any case were concentrating upon the Puma. Their route was leading the Argentines directly towards Lance-Corporal Thomsen's position.

When the Puma had appeared overhead and hovered eighty yards in front of them, Thomsen's whole section blasted it with rifle and machine gun fire. Tracer from Steve Holding's GPMG joined that of the other automatic weapons, creating a wide cone of fire rising from all parts of the area and meeting at the focal point of the helicopter's fuselage. As the machine started to smoke, Brasso Hare had grabbed Holding's arm and, jumping up and down, laughed and shouted: 'Brum, Brum, we've got it, we've got it!' More yells and Indian war-whoops could be heard from the trenches to their right.

Despite the intense excitement of having such an unexpected target at point-blank range, Thomsen's men had not lost sight of the beach, their main area of responsibility. They now saw the seven Argentine soldiers from the Alouette advancing along the shoreline. Holding shouted up to Sergeant Leach who was some yards to his rear. 'Pete, look at those daft buggers in the open down there.' Leach bellowed back: 'Well, don't just look at them – shoot the bastards.' Holding squinted down the barrel of the GPMG and, at a range of less than two hundred yards, fired between sixty and a hundred rounds in one long burst. Three of the Argentines dropped to the ground. The survivors disappeared into the nearby buildings, tracer bullets flashing past them as they ran.

Keith Mills looked around and assessed the situation. So far, so good. The big Puma had been driven off. The landing party had been forced into cover and made to keep their heads down. The Alouette had taken off and presumably returned to the *Bahia Paraiso* (nobody saw it depart). Spasmodic small-arms fire was coming in from the buildings on the Point, but the situation was basically encouraging.

At this moment he was startled to see the *Guerrico* coming back toward the shore. At no more than three knots, she was retracing her track of ten minutes previously. This time, however, her guns were not simply trained on the British position, they were firing. The marines ducked down in their trenches as 100mm high explosive shells screamed over their heads, from left to right, to explode harmlessly on the scree slope one hundred yards to the rear.

The 40mm quick-firer at the frigate's stern came into action, also firing slightly too high, but tearing holes in the ground all around the area. The salvation of the marines was the fact that the *Guerrico* was so close to her target that the 100mm gun was obliged to fire in the horizontal trajectory. She could not 'lob' her shells. With the trenches being thirty feet above sea-level, the shots were passing just over the parapets. Like gamebirds flushed from cover, George Thomsen and Jock Hunter ran to the trench nearby and jumped in with Holding, Hare and Jesse James.

Overcoming a natural inclination to remain at the bottom of his own trench, Keith Mills stood up and watched the advancing ship. Her menacing grey shape and low speed seemed almost to say, 'We are the Argentine navy, we are too powerful for you.' Mills felt irritated. He also realised that the *Guerrico*'s captain was placing his head in a noose. The dangerous Hobart Rocks, at the mouth of the Cove, were obliging him to sail closer to the Point than he might have wished.

Mills shouted the order 'Stop firing, pass the word,' and waited to see what might happen next. To his delight, the frigate came straight on. It was a golden opportunity. 'Nobody fire 'til I give the word. We'll wait 'til he's committed himself.' Obligingly, the Argentine cruised gently into everyone's gunsights.

'Fire!' Mills screamed the command and every weapon on the Point burst into action. At that range, the Argentine was a sitting duck. Hundreds of rounds of rifle and machine gun fire smashed through her hull and superstructure.

The narrow stretch of water separating King Edward Point and Grytviken. It is difficult to understand why the Argentines should have decided to place the *Guerrico* in such an obviously vulnerable position in daylight without adequate previous reconnaissance of the shore defences.

One of the first men to fire at the ship was Marine Dave Combes. He and the detachment's cook, Marine John Stonestreet, were respectively number one and number two on the Carl Gustav 84 mm rocket launcher. The effective range of this potent weapon is normally 450 yards, but the *Guerrico* was about 500 yards away. Combes took careful aim and, hoping for the best, pulled the trigger. The missile streaked across the open waters of the Cove, then plunged down toward the midships section of the *Guerrico*'s hull. Striking just above the waterline, it punched a surprisingly large hole through the grey-painted steel. A column of smoke and spray shot up high above the ship's deck and the marines heard a deep booming explosion. Loud cheers and whoops of triumph came from the line of trenches, to be repeated moments later as Corporal Peters launched a 66 mm rocket which flew over the water towards the frigate. But the range was too great – the missile fell into the sea 100 yards short of its target.

Rifles and machine guns sent an incessant storm of fire at the Argentine vessel, the tracer rounds from the automatic weapons disappearing through the ship's flanks. Apart from killing and injuring several sailors, this small-arms fire caused havoc within the hull. Equipment was damaged, fuel pipes punctured, electrical cables cut. It seems likely that the cables serving the semi-automatic 100 mm gun were damaged by this fire. Within a few minutes, the turret ceased to traverse to starboard as the frigate moved further into the Cove. It remained jammed in the same position and ceased firing. As Keith Mills later commented: 'One minute she was a complete fighting ship – the next she was just about dead in the water. Just us alone, we wrecked that ship.'

Only the stern 40 mm gun was still firing. A Bofors-type weapon, it was manned by five sailors wearing bright yellow life-jackets. Protected by a steel front shield, they continued to pump short bursts in the direction of the Royal Marines. As the ship progressed deeper into King Edward Cove, however, they became exposed to Marines Steve Parsons and Steve Chubb, both quiet steady men, totally reliable. Parsons squinted down the barrel of the Bren, holding his fire, waiting for the Argentine guncrew to be seen clearly. The *Guerrico* was moving away toward Grytviken and was six hundred yards from Hope Point before Parsons judged the moment to be right. Squeezing the trigger, he sent a series of tightly grouped bursts toward the distant cluster of yellow. Two Argentines fell to the deck and the survivors ran forward to shelter behind the ship's superstructure. The gun remained silent throughout the remainder of the battle.

The frigate now had the problem of turning around and escaping from the Cove. This kept her occupied for several minutes. Her manoeuvres could not be followed from the marines' trenches because she was partly obscured by the tussock grass in front of their positions and by the King Edward Point buildings, so Peter Leach ran into Shackleton House and went upstairs to the first floor. From here he could see almost the entire sweep of the Cove. Moving to the extreme right-hand room, he bashed out the window with his rifle butt, dragged a table to the centre of the room and, laying down on it, adopted the prone firing position. Taking a firm grip of his L42, he carefully adjusted his telescopic sight for a range of 500 yards. Slowly the *Guerrico* turned a half circle, her bows facing toward the sniper. He was presented with a perfect frontal view of the ship's bridge. Steadily and methodically, he fired the five rounds in the magazine, reloaded, fired five more, and then five more again. His shots went directly through the windows of the bridge housing. The effect of this fire – upon the people who were attempting to con the stricken vessel – must have been devastating. Later reports suggested that Leach's shots killed the captain and severely wounded another officer.

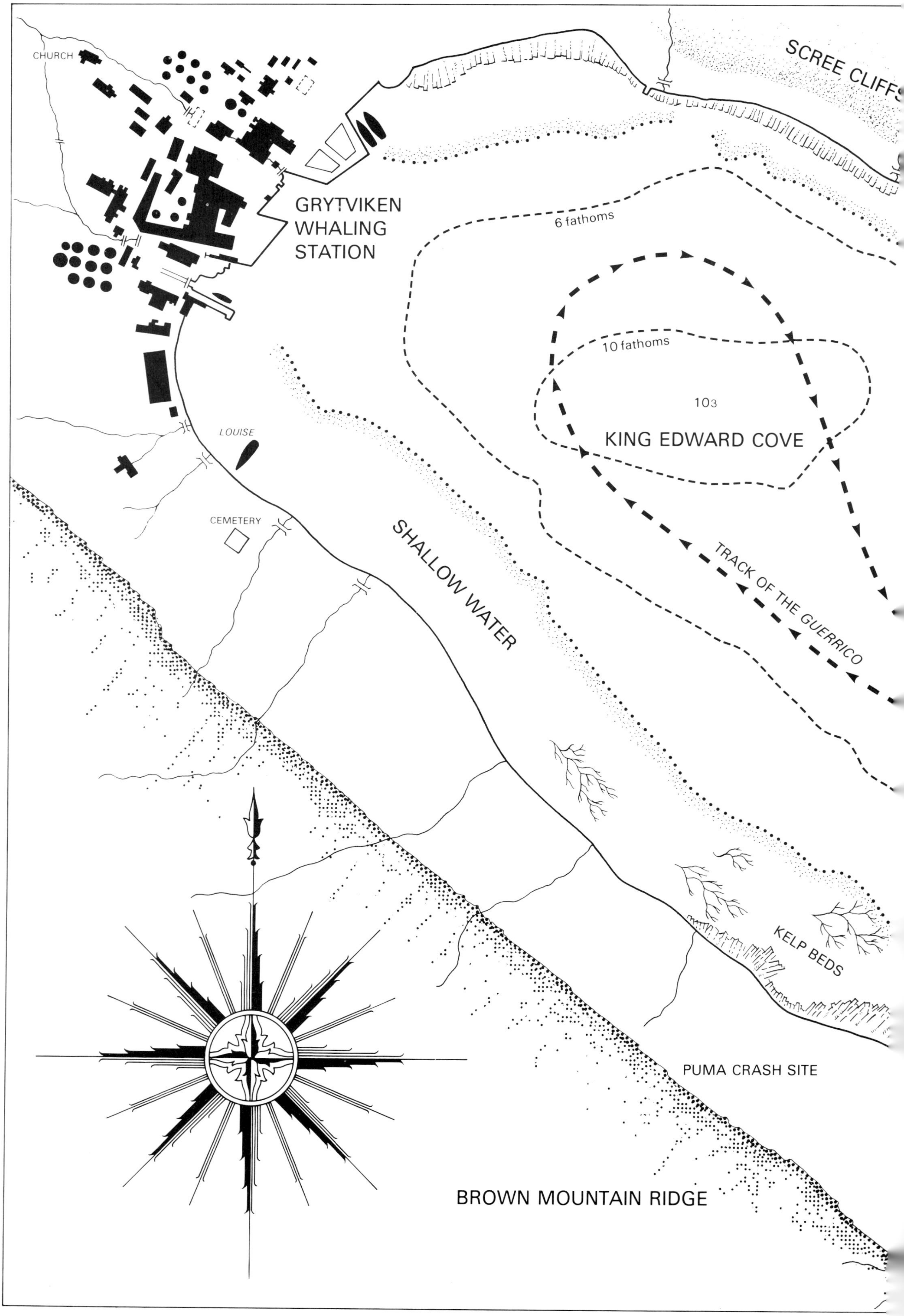
CHURCH
SCREE CLIFFS
GRYTVIKEN
WHALING
STATION
6 fathoms
10 fathoms
103
KING EDWARD COVE
LOUISE
CEMETERY
SHALLOW WATER
TRACK OF THE GUERRICO
KELP BEDS
PUMA CRASH SITE
BROWN MOUNTAIN RIDGE

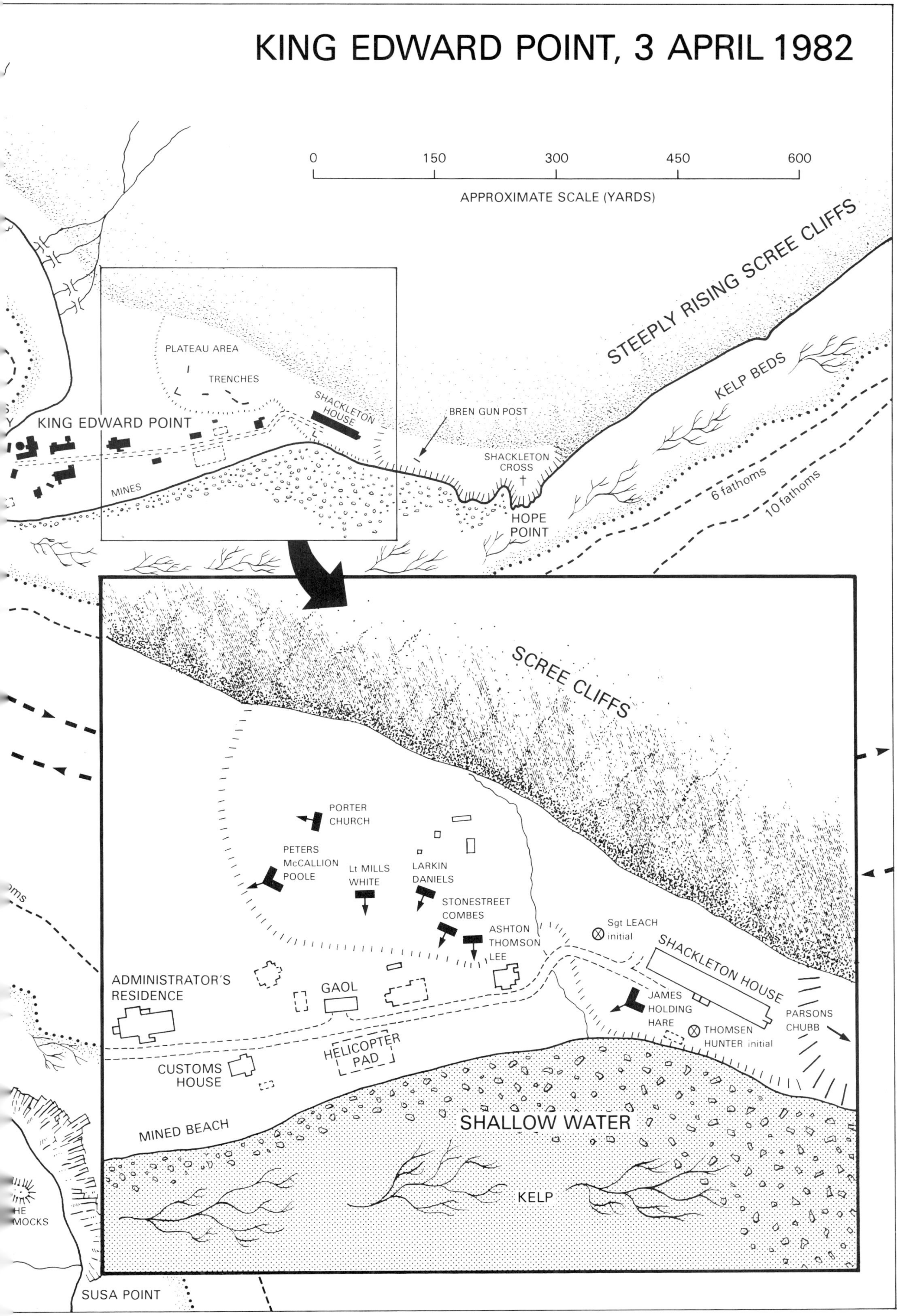
KING EDWARD POINT, 3 APRIL 1982
0
150
300
450
600
APPROXIMATE SCALE (YARDS)
STEEPLY RISING SCREE CLIFFS
KELP BEDS
PLATEAU AREA
TRENCHES
SHACKLETON HOUSE
KING EDWARD POINT
BREN GUN POST
SHACKLETON CROSS
MINES
HOPE POINT
6 fathoms
10 fathoms
SCREE CLIFFS
PORTER
CHURCH
PETERS
McCALLION
POOLE
Lt MILLS
WHITE
LARKIN
DANIELS
STONESTREET
COMBES
ASHTON
THOMSON
LEE
Sgt LEACH initial
SHACKLETON HOUSE
ADMINISTRATOR'S RESIDENCE
GAOL
JAMES
HOLDING
HARE
THOMSEN
HUNTER initial
PARSONS
CHUBB
CUSTOMS HOUSE
HELICOPTER PAD
MINED BEACH
SHALLOW WATER
KELP
SUSA POINT

Having finally turned, the frigate belched a stream of black smoke from her funnel and rapidly gained speed. Her stern dipped lower in the water as the propellers bit. She started to exit from the Cove with a haste which was in marked contrast to her original entry. Once again she was obliged to run the Royal Marines' gauntlet and again she was subjected to a barrage of small-arms fire. The sound of the bullets punching through her hull made the noise of a dozen rivetting machines. This sound was heard clearly on shore despite the roar of weapons being fired.

As the *Guerrico* moved from right to left, so did Peter Leach race from room to room, smashing windows and loosing off two or three shots from each before moving on to the next vantage-point. The reserve ammunition dump was sited under one of the windows. Marine Brasso Hare was collecting fresh supplies here when a shower of glass splinters fell on him. Startled, he looked up and demanded to know what was happening. Leach poked his head out of the empty frame: 'Sorry, pal.' The moment of unintentional humour almost ended in tragedy. As Leach hurried to the next window, a burst of Argentine machine gun fire blasted glass and timber work to the rear of the room. He flung himself to the floor, resolving to be more careful in future.

Keith Mills and Knocker White had a similarly narrow escape. Standing shoulder-to-shoulder in their trench, they were shaken by a bullet which cracked past between them, plucking at their sleeves. Startled, they fell to the floor. 'Shit, that was close!' The two young men burst into loud relieved laughter.

The marines operating the Carl Gustav launcher were experiencing problems of their own. Only one 84mm rocket had been launched successfully when the *Guerrico* came into the Cove. More rounds could have been fired, but they refused to leave the launcher. The rounds had been removed some hours previously from their protective casings so that they would be available for instant use. Unfortunately, it was not appreciated that the electrical circuitry was so vulnerable to humidity. When Combes attempted to fire three more rounds, they all failed to ignite.

According to the drill manual, he should have waited several seconds before unloading the launcher and reloading with a fresh round. At great risk to themselves, Combes and Stonestreet ignored the rules and removed the faulty missiles by hand immediately it was obvious that something was wrong. It was not unknown for faulty rounds to ignite after an initial failure. If that had happened on this occasion, the two marines would have been killed or, at the very least, severely injured. However, they did succeed in launching one more missile as the *Guerrico* made her escape. This one struck the Exocet launching gear, damaging it badly. Subsequently there was some debate regarding what the results might have been if Combes had instead hit one of the warheads of the Exocets. The resultant explosion could have destroyed the *Guerrico*. That might have stopped the Argentine invasion in its tracks, and the final outcome might have been very different. By such narrow margins are battles lost or won.

Corporal Peters now had another 66mm launcher on his shoulder, ready to fire, but he was denied any further chance of scoring a hit. Standing head and shoulders out of his trench, he was suddenly knocked off his feet and flung to the bottom of the trench. Two sub machine gun bullets had struck his upper left arm, breaking the bone and throwing the arm behind his back at an unnatural angle. He lay there, shocked and stunned, while Marines McCallion and Poole crouched anxiously over him. 'Sir! Corporal Peters has been hit.' Mills barely heard the shout, he was too absorbed in controlling his battle. 'Sir, Corporal Peters has been shot in the arm.' The response was brusque: 'Yes, alright, stick a field dressing on it then.' It was a surprisingly mature response from a young officer who had just received his first battle casualty.

Spike Poole started to deal with the injured corporal – winding a first field dressing around the mangled limb – but Paddy McCallion's Irish temper flared. He was very angry that the Argentines should have hurt his friend. Seizing the GPMG, he raked the wooden building to his front with a series of long sustained bursts. Windows caved in and pieces of wreckage flew into the air as – firing long streams of ball and tracer – he blasted the shed in which he believed the snipers were taking cover. It is doubtful that he hit anyone, but the effect must have been terrifying for the Argentine landing party.

Despite the action on and around King Edward Point, Keith Mills did not lose sight of the activity on the far side of the Cove. The *Bahia Paraiso*'s Alouette was maintaining a constant shuttle between ship and shore. Remo Busson – who made fifteen flights during the battle – was evacuating the wounded from the crashed Puma and ferrying fresh troops to the strip of hidden ground near the old radio station. At least one machine gun position had been established by the Argentines in that area and it was bringing harassing fire to bear on the Royal Marines' position. The fire was not particularly accurate, but it might have inflicted further casualties at any time. A greater worry was the number of riflemen who could be seen moving around the shoreline toward the whaler's cemetery, the football pitch, and Grytviken settlement itself. They were advancing steadily, making skilful use of cover. These troops would eventually reach the church, capture the BAS people, and cut off the marines' only possible escape route. To counter this threat, Mills passed the order along the line of trenches that fire should be brought down on the fleeting figures, a thousand yards away across the water. Marine Laurie Church, the tall quiet signaller, and the cheerful Paddy Porter, maintained a steady rifle fire in that direction, but the advance continued.

With so few solid targets now available, Keith Mills gave the order for fire to be directed at the Alouette as it flitted to and from the far shore. The range was extreme – up to fifteen hundred yards – but he hoped to discourage the Argentines from reinforcing the troops already ashore. Several hundred rounds were directed hopefully in the direction of this difficult target. Surprisingly, the fire was effective. The helicopter's emergency flotation gear was punctured and numerous holes were made in the fuselage.

The battered *Guerrico* now made her presence felt once again. Having moved two miles out toward the open sea, she turned around and, almost stationary, again started to fire her 100mm gun. Four shells exploded on the cliffs, wildly off target. It was apparent that the detachment's combined firepower had wrecked the elevating and training mechanism of the gun. In order to aim, whoever was now commanding the ship (and he must have been a very determined man) was obliged to manoeuvre the whole ship. Only by swinging the bows from side to side, and going slowly ahead and slowly astern, could he gradually correct his aim. The maximum effective range of the weapon was ten miles, so even at two miles he could still fire effectively with the barrel jammed in a near horizontal position.

The marines continued to fire spasmodically at the Alouette and at the troops on shore, but the firefight was starting to die away on both sides. The *Guerrico* was the threat and, rather depressingly, there was nothing Mills could do about it.

With practice, the Argentine was steadily improving his fall of shot. The shellbursts crept ever closer to the trenches and to Shackleton House. Shells were arriving in series of four or five –

at twenty-second intervals – followed by several minutes of respite as the *Guerrico* changed her heading and range. Keith Mills now had a chance to review his situation and decide upon the next move.

The battle had started sometime after 1030, two and a half hours earlier. On the positive side, his command was still intact and holding its ground. He had suffered only one casualty. He still had plenty of ammunition and his men were in excellent spirits. Fire discipline had been exemplary. He had given his enemy a very sharp blow and, in theory, he could fight on. The *Guerrico*'s shellfire was becoming increasingly effective, however, and it could be a matter of only ten or fifteen minutes before the Argentine mastered the knack of pointing his ship in the right direction. A few rounds of high explosive impacting on this small plateau might kill half of the defending force. Furthermore, Mills realised that the Argentine troops in the Grytviken area would eventually approach his right flank and he would be unable to fire at them. His 'overs' would imperil the civilians in Grytviken church.

With his escape route already cut, he decided that the sands of time had run out. He had done his best to protect the people and property on the island. His men had done everything, and more, that he had asked of them. If his superiors were to decide later that he had acted improperly, that would be just bad luck. His own conscience was clear.

He waited for the next lull in the firing and poked his head cautiously over the parapet: 'Well, guys, that's it. We've made our point, that's enough. I've decided to surrender. Does anyone have any violent objection?' His announcement was greeted with silence. This is not surprising as many of his men were too far away to hear him. He waited a moment, then: 'Good. Because that is what I have decided.'

Seventy yards to the rear, high in Shackleton House, Peter Leach could hear Mills shouting but did not understand what he was saying. Then the word was passed that his officer wanted to speak with him. Calling for covering fire, Leach ran downstairs, sprinted across the open ground and threw himself down beside Mills' trench. Despite his initial forebodings, prior to the battle, Leach's blood was up. 'What's the problem, Sir?' Mills told him of his decision to surrender. 'What the hell for? We're winning hands down.' Mills briefly gave his reasons and confirmed his decision. 'OK, whatever you say. Personally I would like to carry on. But you're the boss.' Sergeant Leach felt highly incensed by this reversal of their fortunes, but he darted back into the building while the order to stop firing was shouted along to the trenches on the flanks.

Mills wondered how best to communicate his intentions to the Argentines. For the first time in more than three decades, a British officer was about to surrender his command to an enemy. It was an unpleasant sensation, but he had no sensible alternative. Taking off his Arctic waterproof, he turned it inside out to expose the white facings. Using Knocker White's SLR, he poked the jacket high in the air. Its appearance was met by a burst of fire from the King Edward Point buildings. Thoughtfully he took it down and waited a few moments before repeating the process. This time there were no shots.

After a short debate with himself, Mills decided that the moment had come to take the initiative. Drawing in a deep breath, and hoping that his sphincters would continue to function normally, he climbed out of the trench and stood in the open. Nothing happened. The only sound was the whirr of the distant Alouette. After a brief pause he started to walk down the track toward the jetty, his SMG slung across his back. It was a long and lonely journey. Then, passing through the buildings, he saw a strange uniform. It was an Argentine officer.

Placing his weapon on the ground, Mills approached the man: 'Hello, do you speak English?' 'Yes, I do.' Much relieved, Mills launched into the speech which he had rehearsed on the way down from Shackleton House: 'Look, you are in a difficult position. We are well dug in and can go on fighting for a long time. We shall all get killed, but we don't care. You will lose more and more of your own men. To avoid needless casualties on both sides, I am prepared to surrender now if you will guarantee good treatment for my men.'

The Argentine officer reached out, seized Mills' hand, and shook it fervently. A smile of delight on his face, he promised that there would be no reprisals. Later in the day, he admitted that he and his fellow officers had fully expected the defenders to fight on to the death.

Mills was now feeling the reaction of battle and did not fully comprehend the speed with which Argentine troops seemed to appear from every direction. He was told to call his men forward. Gradually they emerged, unarmed, and walked down to the beach. As they arrived, they were lined up and counted. Suddenly there was tension in the air. The Argentines could account for only twenty-two prisoners. Uneasily they fingered

Later during the conflict, the *Bahia Paraiso* was re-deployed as a hospital ship and allocated another Puma helicopter to replace AE-504 lost on 3 April. The replacement, painted in accordance with International Red Cross requirements, is seen here coming in to land on the flight deck of the British hospital ship SS *Uganda*.

The Royal Marines force which defended King Edward Point on the morning of 3 April 1982. Back row, (left to right), Cpl Al Larkin, Lieut Keith Mills, Mne John Stonestreet, Mne Les Daniels, Mne Steve Chubb, Mne Paddy Porter, Mne Laurie Church, L/Cpl George Thomsen, Mne Jesse James, Mne Paddy McCallion, Sgt Pete Leach, Mne Dave Combes, Mne Spike Poole, Cpl Nige Peters. Front row, (left to right), Mne Steve Parsons, Mne Knocker White, Mne Steve Holding, Mne Jock Hunter, Mne Andy Lee, Mne Bob Ashton, Mne Jock Thomson, Mne Brasso Hare.

their triggers, looking up at the cliffs, searching for the bulk of the British force. They suspected that Mills had tricked them into an ambush. Surely this tiny force of less than two dozen men could not have held off – for more than two hours – two naval vessels, two helicopters and eighty special assault troops?

Only with difficulty did Mills convince his captors that the muddy little group was indeed his complete force. Thereafter the Argentines seemed almost respectful in their treatment of the marines. They became even more thoughtful when they searched – as they did the entire party – the person of Marine Brasso Hare. Hidden under his combat jacket they discovered two hand-axes and, in his trench, an enormous red BAS fire-axe. Possessed of the true Samurai spirit, Hare had intended to use these weapons in personal combat in the final defence of Shackleton House.

Teniente de Navio Alfredo Astiz, the officer commanding the military element in the invasion, came ashore by Alouette with his personal staff. All were immaculately dressed in smart combat gear. Mills was impressed by his bearing and air of authority. He spoke perfect English and congratulated the marines upon their defence. He acknowledged Mills' advices regarding the mined beach and booby-trapped buildings. Marine Daniels was called forward and invited to dismantle the devices which he had planted the day before. Although the terms of the Geneva Convention absolved him from such duty, Daniels agreed to make the area safe. Several times he had to tell the curious Argentine troops to go away when they stood around – expressions of mild horror on their faces – watching him at work on the mines which had been intended to kill them.

The deeply shocked Corporal Peters was brought down to the beach. Peter Leach injected morphine into the back of his thigh and covered him with spare clothing. Later he was examined by an Argentine doctor who came ashore. Mills and Leach were told that they could return to Shackleton House, under escort, to collect the small valuable possessions of their men. They did so, selecting everything of obvious financial or sentimental value. Several cameras went into the kitbag. These were first opened by the guards, and the film ripped out and destroyed. When they reached Keith Mills' own camera – an Olympus OMIO – the inexpert Argentine failed to discover the knack of opening it. He gave up the struggle and threw the camera into the bag. Everything was later returned intact. Hence, by pure chance, the historic photograph which appears on this page survived. By an equally lucky fluke, Nigel Peters' photographs of Stromness Bay – taken from Jason Peak – also survived the search. Before leaving, Peter Leach scribbled a note – addressed to the BAS – apologising for the broken windows and promising to return soon 'to clear up the mess'. It was a prophetic pledge.

All of the Royal Marines, with the exception of Keith Mills, were ferried out to the *Bahia Paraiso* by landing craft. Crossing the bay, they passed the *Guerrico*. The guards ordered them to lay down in the well-deck so that they could not admire the damage which they had inflicted on the Argentine warship. Despite their feelings of relief at having survived, and their professional pride in what they had achieved, the marines were quickly learning that they were now prisoners. The future course of events, at least in the short term, would be dictated by their captors.

Similar sentiments were being experienced by the BAS personnel on the other side of King Edward Cove. Since the firing of the first shots, more than two hours earlier, they had continued to shelter inside the church at Grytviken while the firefight raged outside. The explosions caused by the *Guerrico*'s

Even in September, five months after the battle, the *Guerrico* was still bearing obvious scars from her encounter with the Royal Marines. Fresh daubs of paint on her hull and superstructure are evidence of the repairs made necessary by small-arms fire, while the black patch on the mid-section waterline confirms the point of penetration by Stonestreet and Combes' Carl Gustav missile.

100 mm rounds had been particularly disturbing, especially the wild shots fired from offshore. The surrounding cliffs and mountains magnified the noise, the echoes giving the impression that a whole battery of artillery was at work. When the frigate made her first entrance into the Cove, Chris Bales climbed up inside the church steeple and took several photographs of her as she made her turn. Unfortunately they were spoiled by the grime on the windows and by the fact that, in his excitement, he neglected to focus the lens correctly. After that, Bales and the others lay on the church floor and hoped for the best.

At around 1300, after the firing ceased, Bob Headland and some of his colleagues got up and looked out of the window. They watched as a section of ten Argentine marines moved silently and carefully through the factory buildings. They were ignoring the church and seemed unaware that it was occupied. Headland, as the only Spanish-speaker in the party and in the absence of Steve Martin, went out and approached the Argentines. He introduced himself and explained the presence of the unarmed British civilians. He was told to stay indoors and the marines then continued their advance along the shore toward King Edward Point.

An hour later, accompanied by Mills and twenty of his own men, Astiz arrived on foot at the church. Headland formally surrendered his party to the Argentine who ordered them to come out, one at a time, to be searched. They were then gathered into a group and marched, under guard, along the beach to Shackleton House. There was great anxiety over the missing Base Commander, but Steve Martin emerged from his gully in response to shouts from Keith Mills. Although shaken and very cold, Martin was immensely relieved to discover that none of his colleagues had been killed or injured in the noisy battle which he had heard but could not witness.

Some of the kit in the church was loaded on the tractor and trailer and carted back to King Edward Point by Dermot Hopkins. After several requests, Astiz agreed that each of the BAS men should be allowed time to enter Shackleton House to collect personal valuables, but this concession was curtailed by the impatience of the guards and the approach of dusk. Some of the party never had an opportunity to rescue any of their belongings and they lost everything. Claims for compensation were to be settled more than a year later after painstaking negotiations with the National Environment Research Council.

With dozens of Argentine troops and British civilians milling around the King Edward Point buildings, several opportunities were exploited to salvage anything of value. Bob Headland demonstrated the intricacies of a scintillometer to an interested naval lieutenant and, having distracted his attention, managed to smuggle cash and postage stamps to the value of £10,000 sterling into his personal baggage from the Post Office. Headland also rescued all of the photographs taken at Leith Harbour, on 19 March, by diverting the attention of his guard with an 'art' magazine.

The scientists tried to ensure that their delicate research equipment would not be damaged. Astiz was warned of the location of dangerous chemicals and radioactive materials, and he was advised of the presence of other BAS parties in various parts of the island. It was emphasised to him that these people were all non-combatant civilians. He promised to make contact with them but, in the event, they were left to their own devices throughout the short Argentine occupation.

As darkness filled the deep valleys of the South Georgia coast, so the wind strengthened yet again and thick cloud blotted out the early evening stars. Captain Trombetta urged his officers to complete the task of landing stores for the new garrison and the embarkation of the BAS prisoners and the bodies of the dead and wounded. By 1800 he was ready to sail. Leaving the perforated *Guerrico* to lick her wounds before attempting the long passage back to the mainland, Trombetta turned his ship toward the open sea. He could claim to have beaten the British in a fair fight, and to have evicted most of them from the island. The Argentine newspapers would tomorrow be able to boast of yet another triumph. But it was a hollow victory. The cost in men and matériel was too high for the short-lived propaganda advantage gained. The manner in which the Argentine soldiers and pilots had conducted themselves suggested that they were lacking in neither skill nor courage, but the tactics adopted by their senior commanders placed in doubt their ability to hold on to the territory which they had seized. Many of the newly landed troopers must have wondered how long they were likely to remain undisturbed in this desolate place □.

V | The Waiting Time

WHILE KEITH MILLS and his men were fighting their battle on the morning of 3 April, Nick Barker and his engineers were pressing the *Endurance* as hard as they could to cover the last remaining miles to the threatened island. On a good day, under ideal conditions, the old lady could be pushed to thirteen and a half knots (and she needed thirty minutes to build up to even that modest speed). There was nothing they could do to make the miles reel past more quickly, but steps were taken to prepare for military action once they arrived off Cumberland Bay. Barker's intention was to round Cape Disappointment, sweep northward up the coast and launch his AS12-armed Wasps against the *Guerrico* and the *Bahia Paraiso*. If he arrived in time, he might possibly prevent the invasion. On the other hand, if the Royal Marines had been already killed or captured, he could avenge their defeat by inflicting damage on the Argentine ships. Barker was under no illusions regarding the probable outcome of such action. The life expectancy of a man immersed in those frigid waters is very limited.

By midday the *Endurance* had passed the southern tip of the island and was heading north. Barker started to call King Edward Point, but there was no reply. The silence was ominous. He launched Wasp 434, piloted by Lieutenant-Commander Tony Ellerbeck, RN, with orders to make a covert reconnaissance of Cumberland East Bay.

In deteriorating weather, Ellerbeck flew low and fast up the east coast past St Andrews Bay. Keeping close to the high cliffs to hide his radar reflection in the land 'clutter', he reached the Barff Peninsula undetected. Turning in through the bay at Godthul, he landed near the small lake in Reindeer Valley. Ellerbeck and his observer, Lieutenant David Wells, shut down their helicopter and walked forward to find a hidden vantage point. As they did so, they heard two heavy explosions – the final rounds fired by the *Guerrico*.

Through the gathering gloom, they saw the whole Argentine force laid out in front of them. The damaged *Guerrico*, only a mile away, was moving slowly in toward the shore. Ellerbeck noted a hole in her side. The *Bahia Paraiso* was lowering a landing craft full of armed men and the Alouette was lifting passengers from her flight deck. The far shore, around Grytviken and King Edward Point, appeared to be deserted.

Hindsight suggests that Ellerbeck and Wells arrived on the scene just a few minutes following Keith Mills' decision to surrender. The question must be asked – what would have been the outcome if Mills had fought on into the afternoon, holding the Argentines at bay pending the arrival of the *Endurance*? There is no doubt that the three ships involved would have fought a fierce engagement there would have been heavy loss of life. However, even had she survived, the *Endurance* might easily have suffered damaged to her communications systems and this was an asset which the C-in-C Fleet was desperately anxious to preserve. With the Falklands and South Georgia under Argentine occupation, the last remaining source of purely naval intelligence, both electronic and human, anywhere in the South Atlantic, was HMS *Endurance*. Captain Barker was ordered to withdraw to a less dangerous position.

To hide a 4,000-ton ship, her hull painted bright red, is not the easiest of tasks. The only hope was to turn back to the south and find a quiet bay which Argentine forces would be unlikely to enter. For the next thirty-six hours, the ship hugged the shoreline, drifting slowly in a field of icebergs making their way north with the advancing Antarctic winter. As a blip on a radar screen, she was no more than just another iceberg. She was safe from visual identification during that first long night, but dawn brought the risk that she might be spotted by patrolling Argentine aircraft. It was an uncomfortable day. For the time being, Captain Barker and his ship's company could only wait upon events.

The British public had been outraged by the Friday evening reports of the fall of Port Stanley. It was not difficult to identify with the plight of the 'kelpers' because they were so clearly fellow Britons. The loss of South Georgia, announced on the following evening, seemed somehow less dramatic, less significant. The very name of the island was still associated with the recent and apparently almost ludicrous arguments about scrap metal merchants. Prior to the emergence of the Davidoff story, only a very small proportion of the British public had even heard of this fly-speck on the South Atlantic map. During the first week of April, as the attention of the world shifted to the Central Atlantic and to the intense activity at Chatham, Portsmouth, Rosyth and Devonport, the island disappeared from the headlines and slipped back into temporary obscurity. Very few people were aware of the continuing British activity on the island and these were mainly the administrators of the British Antarctic Survey.

When Argentina invaded the Falklands on 2 April, it was immediately apparent to Dr Richard Laws, Director of the BAS, that his communications had been compromised. With the local Cable & Wireless office at Port Stanley now in Argentine hands, he could no longer rely upon a continuing flow of information from his staff in the field, nor could he depend upon its accuracy. He wasted no time in searching for other methods of maintaining contact. He was anxious to keep open the channel of communication not only with King Edward Point, but also with the BAS bases in Antarctica. If the Galtieri regime had screwed up its courage to the point of occupying the Falklands, then there was every possibility that it might seize or harass other places where the Union Flag flew. Specifically, Laws was deeply concerned for the welfare of his people on South Georgia, at Faraday and Rothera on the Antarctic Peninsula, at Halley on the ice shelf of continental Antarctica, and at Signy in the South Orkney group of islands. Within hours of the fall of Port Stanley, Richard Laws and his staff were investigating alternative means of communicating with those distant stations.

One of his first steps was to contact Dr E. P. Todd, Director of Polar Programs, National Science Foundation, Washington, DC. It took several days to decide that, if necessary, the Americans would handle limited BAS radio traffic through their facility at Palmer, on the Antarctic Peninsula. Laws also called his opposite number in Germany, Professor G. Hempel, Director of the Alfred Wegener Institute for Polar Research, in Bremerhaven. An arrangement for unlimited traffic via the sophisticated German communications satellite facility was quickly agreed. The BAS radio operator at Halley, Steven Williams, also talked with the German scientists at their Neumayer base and made local arrangements for them to forward any traffic which he could not transmit by other means. The reaction of the Germans and Americans was not necessarily based upon nationalistic senti-

HMS *Endurance*'s 20mm Oerlikons had been taken out of storage and mounted, ready for action, as soon as the crisis started to come to the boil in late March. These two guns were the only ship-mounted weapons available to Captain Barker and, fortunately, he never needed to fire them in anger. Here the two-man crew of one of the guns prepares to conduct a practice shoot at a flare.

Halley, the British Antarctic Survey geophysical observatory on the Brunt Ice Shelf, Antarctica. First constructed on the surface in 1956, the base has since been reconstructed three times as it has been gradually buried by snow and carried horizontally towards the ice edge.

ments – pro-British or anti-Argentine – but simply reflected the need to aid fellow scientists in difficult circumstances.

Simultaneously with these negotiations, the BAS was trying other radio routes which did not depend upon the now silent station at Port Stanley. Test transmissions were made via Signy Island, Rothera, and the RRS *Bransfield*, to Bermuda and to Ascension Island. They were not successful. Richard Laws then told the various stations to make their own tests to establish which of them should assume the role of control station (allowing for the very variable radio propagation conditions in that part of the world).

Halley was immediately eliminated because conditions there, for transmission and reception, were the least favourable. It was soon confirmed that Signy base – more or less equidistant from South Georgia, Halley and the Antarctic Peninsula – was the best suited as a control station. Commencing late on 2 April, Signy became the new linchpin in the BAS communications network in the South Atlantic and Southern Oceans. Signals received there were re-transmitted to the RRS *Bransfield* and she, in turn, passed them on to Cambridge via the British Telecommunications ship-to-shore telephone receiving station at Portishead, near Bristol. The final link was the normal BT land line to Cambridge.

The opening days of April were characterised by numerous minor communications problems but, in general terms, Laws and his colleagues had already created an effective alternative radio network by the time the King Edward Point transmitter fell silent at 1045 on the morning of 3 April. Consequently, despite the loss of the principal local base, he was soon receiving fresh reports from the four field parties remaining on South Georgia. Each was equipped with a Squadcall transceiver and, with one exception, a generous supply of batteries. Commencing on the afternoon of 3 April, they started to transmit a stream of reports. They were able to confirm the fact – simultaneously reported by the *Endurance* – that King Edward Point had indeed been occupied by the Argentines and that they had no information regarding the fate of Steve Martin's party or the Royal Marines.

There were twelve huts scattered along the north coast of the island, of varying age, size and primitiveness, intended as temporary quarters for scientists engaged in field work during the summer season. Typically, they were constructed of corrugated iron over timber framing, were divided into two small rooms, and could accommodate three or four people under cramped conditions. The largest and most elaborate buildings (apart from the main base at King Edward Point) were on Bird Island, just off the northern tip of the main island. This station had been enlarged and modernised during the previous season so that it could house up to nine scientists throughout the year. Working there at the time of the invasion were two biologists, Paul Goodall-Copestake and Julian Hector, and two construction workers, Paul Humphries and Barry Wheeler.

Facing above: Rothera, the British Antarctic Survey earth sciences research station on the eastern flank of the Antarctic Peninsula. This was one of the British bases in the region which played an unexpected role in maintaining contact with the United Kingdom in early April.

Facing below: The BAS biological laboratory at Signy Island, South Orkney Islands, with Coronation Island in the background. The flat area in the middle distance is a frozen bay, with several icebergs trapped for the duration of the winter. This was the spectacular natural setting for an intensive period of radio communications work by the local staff. Signy Island became the regional co-ordinating centre for all British Antarctic Survey communications throughout April 1982.

Below: The new seal and sea bird observatory at Bird Island. Located at the northern extremity of South Georgia, it provided a useful lookout point to watch for over-flying Argentine aircraft.

A few miles to the south, at Schlieper Bay, towards the northern extremity of South Georgia, were two more biologists, Bill Doidge and Paul Martin. In the centre of the island, only five miles west of King Edward Point near the foot of the Lyell Glacier, was a very small hut (seven feet square) occupied by Ian Barker, the mountaineering expert, Trefor Edwards, a carpenter, Campbell Gemmell, a glaciologist, and Damien Sanders, the group's diver. Fifteen miles further south, at St Andrews Bay, in a new two-room hut, were Cindy Buxton and Annie Price, filming for the Anglia Television 'Survival' series. With them, sleeping in tents, were the three BAS men – Tony North, biologist, Peter Stark, meteorologist, and Myles Plant, builder – who had been sent there in anticipation of the Argentine occupation of King Edward Point.

Most of these people were accustomed to moving around the island on foot, but the obstacles provided by the mountains, glaciers and treacherous scree slopes placed a severe limitation upon the distance and direction in which they could travel. In most respects, they were now isolated from each other and from the outside world. The only form of communication available was that provided by their radios, and these were soon put to work as each party tried to find out what had happened at King Edward Point and checked to establish that the others were safe and well.

The St Andrews Bay party was further reassured when, on the evening of 3 April, Tony Ellerbeck flew low over their heads in Wasp 434 and dropped a message with the latest news. They had another visit on the following morning when Wasp 435 swooped in across the huge colony of King penguins. Tim Finding and Bob Nadin shut down their helicopter and spent fifteen minutes with the party, explaining what had happened and what was likely to happen next. Lieutenant Finding offered to evacuate the five civilians to the *Endurance* – the 'red iceberg' as she was now jokingly calling herself – at her secret hiding place further down the coast. The two young women discussed the situation with the BAS men, and collectively decided to stay where they were for the time being. There was every possibility that the *Endurance* might be engaged in armed action, and their presence on board could be an unfair hindrance to Nick Barker and the ship's company. Furthermore, the RRS *Bransfield* was scheduled to collect them all later in the month. They thought that, as an unarmed civilian research vessel, she might be allowed by the Argentines to go about her lawful business in the usual way.

Finding and Nadin departed quickly and at low level after warning the party to exercise great care in the use of their radio. The operators aboard the *Endurance* were monitoring some of their signals and it was very probable that the Argentines were doing the same. Nick Barker did not want his opponents to be given any gratuitous intelligence regarding the location of his ship or the reconnaissance flights by her Wasps.

Beginning on the evening of 3 April, the reports of the South Georgia field parties were collated by Goodall-Copestake and Hector at the Bird Island station. They transmitted a summary every hour (or, later, every two or three hours) to Signy Island, 650 miles to the south-west. The three operators at the Signy base were David Rootes (Base Commander), Dennis McCrohan and Roland Lewis. During the following three weeks, despite severe storm damage to the radio installation which by their own resource they managed to repair, these three men sustained a round-the-clock schedule and transmitted more radio traffic than they would have handled during a normal nine-month winter.

Apart from talking with Bird Island, they were in regular contact with the other BAS bases in Antarctica and, less frequently, with the Russian, Chilean, German, South African and United States bases. The business of reporting meteorological and other scientific information went forward as usual. Dennis McCrohan even called up the radio operators at the Argentine bases on Laurie Island, in the South Orkney group, and on Thule Island, in the South Sandwich group. He asked if they could throw any light upon the current status of the Royal Marines and civilians captured at King Edward Point, but the Argentines could not, or would not, supply any details.

As soon as the initial problems had been resolved, the bulk of all BAS communications from Signy was forwarded to Portishead and Cambridge via the RRS *Bransfield*. Her sole radio operator, Hugh O'Gorman, acted as a vital link in the chain and, throughout most of the month of April, was one of the hardest-worked men anywhere in the South Atlantic.

The *Bransfield* was one of the finest vessels operating in the Antarctic. Launched in 1970, designed and built specifically as a support ship for BAS stations, with laboratories for 'on board' research work, she was in some ways similar to the *Endurance*, having an ice-strengthened red-painted hull, and displacing 4,816 tons. Unlike the *Endurance*, however, the *Bransfield* was a purely civilian-manned ship. She carried no armaments and did not operate under Royal Navy orders.

On 2 April, the *Bransfield* had been sailing in Argentine territorial waters, off the coast of Tierra del Fuego, eastbound for Faraday base and having just completed a routine visit to the Chilean port of Punta Arenas. Her captain, Stewart Lawrence, was aware of the tensions between Buenos Aires and London and, like many other mariners in the region, knew that a large part of the Argentine fleet was at sea. Even so, the news of the landings at Port Stanley came as a shock, particularly so given the proximity of his ship to what might now be regarded as an enemy coast.

Captain Lawrence soon received orders from Dr Laws to follow his planned itinerary, to continue eastward, between the Falklands and Staten Island, and to then turn south for Faraday. He made the usual end of season visit and then sailed north-east for Signy Island. After being held there on an extended call, he was instructed to proceed further eastward, to the south of the South Sandwich Islands. He was told to stay well below the 60th Parallel – the northern limit of the Antarctic Treaty area. The *Bransfield* was, in effect, hiding herself in the icy wastes of the Southern Ocean until such time as Argentine intentions had been clarified and a decision could be made regarding her next move.

Hugh O'Gorman had more than enough work on his hands, but the majority of the *Bransfield*'s crew were deeply unhappy. They wanted to take an active part in relieving the field parties on South Georgia or somehow aiding the Falkland Islanders. Several members of the crew were themselves native-born 'kelpers'. Captain Lawrence reported to Cambridge, 'they are as close to mutiny as I have ever seen them,' but the decision to keep her out of harm's way, agreed between the BAS, the Ministry of Defence and the National Environment Research Council, was the only sensible step in the circumstances. She could not defend herself if attacked, she possessed no offensive capability, and she was completely isolated from Royal Navy support and protection. On the other hand, her radio was a vital link in the BAS communications chain, and would be even more important to British interests if anything happened to the *Endurance*.

Throughout the middle of April, while the *Endurance* was far to the north and meeting up with the Task Group coming south from Ascension, the lonely *Bransfield* continued her cruise along the edge of the icefield. Her position was dictated by three considerations. Firstly, she needed to be as far away as possible from the threat of Argentine ships or aircraft operating from the Falklands or from the mainland. Secondly, the northernmost limit of the Antarctic Treaty was the 60th Parallel, and there was

The Royal Research Ship *Bransfield*, 4,816 tons gross, diesel-electric powered and class '100 A1 ice strengthened' at Lloyds. Normally engaged in supply and relief duties around British scientific bases in Antarctica, she was a major element in sustaining radio-telephone communications between Cambridge and the deep south during the mid-April absence of HMS *Endurance*.

some notional protection to be gained by staying within the Treaty area. Thirdly, the British had declared a 'maritime exclusion zone' of 200 miles around the Falklands and South Georgia, signifying that any ships found within that radius might be treated as hostile. The Argentines had declared a similar 'military zone' around the same islands and also around the South Sandwich Islands, the latter being a clear violation of the Antarctic Treaty.

It was not until 17 April that Captain Lawrence finally was told to head north and make for Ascension Island. He and his crew did so with very mixed feelings. They wished that they could have done more for their friends and colleagues and, in the case of the Falkland Islanders aboard, for their relatives. But, apart from maintaining the radio link, they had made their own gesture against Argentine aggression. On 6 April, when the invaders were broadcasting on Port Stanley radio an announcement banning the use of civilian transmitters (and detailing the penalties for disobedience), Hugh O'Gorman tried to drown the message by playing on the same wavelength the theme music from the film 'Born Free'. The Argentine announcer followed the decree with his country's national anthem. O'Gorman responded with 'Land of Hope and Glory'. He was heard by radio operators throughout the region, and some, who previously may have doubted British resolve, began to realise that the story was far from over.

On the island of South Georgia, throughout this period, the four field parties had initiated what was, in effect, a voluntary 'coast watching service'. Those based at Schlieper Bay and on Bird Island were obliged, by the geography of their locations, to remain where they were. However, they could observe the air and sea approaches to the island. Argentine ships and aircraft coming down from the Falklands, or from their mainland bases, were likely to choose this route, over-flying or making landfall at the island's north-eastern tip. In the event, neither party noted any Argentine activity of any kind until 23 April when the Bird Island group was able to report a C-130 Hercules over their area.

The parties at Lyell Glacier and at St Andrews Bay were better able to patrol across country. They wasted no time. On 5 April, Damien Sanders and Ian Barker left their hut at Lyell and approached very close to King Edward Cove. For several hours, from the slopes of Mount Hodges, they observed and photographed the Argentine activity around Grytviken and the BAS buildings. They noted the Argentine colours flying from the flagpole, the construction of two possible machine gun posts at each end of Shackleton House, and the absence of any ships in the bay. Troops could be seen moving around the buildings and along the shore. It was difficult to calculate their numbers, but they were evidently not very numerous – perhaps a few dozen at most.

Sanders and Barker returned to their hut and, that night, sent their report to Signy. Their findings were queried by the Bird Island party who asked them repeatedly to confirm and amplify their first signal, and this they did. It was assumed that the *Endurance* would be monitoring these exchanges and her operators were therefore given every opportunity to hear the full details. The ploy was successful. Both Cambridge and the Ministry of Defence had the verified report by the early hours of the following morning.

On 6 April, Tony North set off alone from St Andrews Bay to visit the BAS hut at Hound Bay. His main objective was to inspect the emergency supplies of food and fuel dumped there, but his route took him to an area, at the neck of the Barff Peninsula, where the Argentines could be expected if they were following a policy of active patrolling. North found the hut undisturbed and no sign of footprints in the first snows of the approaching winter.

On 11 April, North made another patrol, in company with Myles Plant, to visit Royal Bay, twelve miles south of St Andrews. Again, they found no trace of Argentine activity along the coast and saw no ships or aircraft. Five days later they made another visit to Hound Bay, again with negative results.

North and Plant now decided to make a prolonged reconnaissance of the terrain separating them from the Argentine force at King Edward Point. On 19 April they packed their rucksacks with spare clothing and food sufficient for a three-day patrol. Leaving St Andrews Bay early on the morning of 19 April, they skirted the 1,578-foot slopes of Mount Skittle, crossed the high pass near Luisa Bay, and headed for the Sörling Valley. Crossing the saddle of the valley, eight hundred feet above sea-level, they had a direct line of sight into Cumberland East Bay and across to King Edward Point. Having noted the absence of any ships or helicopters in the area, and having checked the thin snow for alien footprints, they stayed that night in the emergency hut on the shore of Cumberland East Bay.

The following morning brought the first severe weather of the autumn, with heavy snow showers and poor visibility. Despite the hard going, they pushed on to Ocean Harbour, further up the coast of the Barff Peninsula. Here they passed their second night on patrol, this time in the ruins of a whaling station closed in 1920. Again, they found no trace of the invader. Next morning, turning back south, they decided to return directly to Saint Andrews Bay. On the evening of 21 April, however, the wind increased in force and it was evident that a heavy gale was in the offing. Reaching Hound Bay, they made themselves comfortable in the small hut concealed behind a low hill near the sandy beach. Late that night they were surprised by a knock on the door. To their mutual consternation, the civilians found themselves talking to the first British troops to come ashore from the Paraquat force. HMS *Endurance* was anchored out there, somewhere in the windswept night, and the visitors were two Royal Marines of a Special Boat Squadron reconnaissance team. North and Plant were advised to get clear of the area at first light, and in fact they were back at St Andrews Bay by the afternoon of 22 April, having surveyed three possible anchorages for enemy shipping and having searched for any sign of military activity. Apart from the brief encounter with the SBS, they had seen nothing other than the usual wildlife.

At the Lyell Glacier hut, serious problems had emerged. One of the party was suffering from a kidney complaint and there was a shortage of food. On 6 April there was some debate as to whether or not they should surrender. They asked for advice, and Dr Laws sent a signal to the effect that they should feel free to hand themselves over to the Argentine garrison – just over the mountain at King Edward Point – if they felt that they had no choice, but warned them of the possible consequences. They decided to stay where they were for the time being, but to keep their eyes and their options open.

Ian Barker and Damien Sanders continued to reconnoitre their immediate area and made another trek up onto the high ground which separates Cumberland East and Cumberland West Bays. In theory, they could have additionally made a reconnaissance of Stromness Bay to observe any Argentine activity in that area, but this was physically almost impossible. The land route would have involved crossing three glaciers, including the mighty Neumayer, and it was a golden rule that one should never normally venture onto glaciers of that type without a great deal of careful preparation. For Barker and Sanders, the chances of success would have been unacceptably slim because they had inadequate food and equipment and no support if stranded. In any case, the negative reports from Tony North and Myles Plant, together with the equally negative reports from Bird Island and Schlieper Bay, combined with their own personal observations around Lyell, all showed a complete lack of military initiative on the part of the Argentines. It was reasonable to assume that the group of troops and civilians at Leith Harbour and the garrison at King Edward Point were both equally housebound.

These BAS patrols could not cover the whole island but, between them, Plant, North, Barker and Sanders surveyed a third of the eastern seaboard and a significant part of the hinterland. They visited all but one of the usual anchorages and inspected most of the few lower-lying areas where foot patrols would be most likely to appear. Their findings, amplified by their expert knowledge of the island, permitted the compilation of accurate assessments which are the very stuff of good intelligence work. Of equal importance was the speed with which these reports were arriving at the BAS Cambridge headquarters. Added to the databank of maps and technical documents already on file in England, they represented a very useful reference source for any British commander – planning the reoccupation of the island – who might care to make use of them.

After the initial phase, when reports were coming through hourly, Dr Laws was able to establish a prearranged schedule of radio-telephone signals. The bulk of the reports concerned the health and morale of the field parties, their reserves of food, fuel and batteries, and other topics requiring immediate decisions or advice. He later shared the workload – at night, when most of the signals came in – with John Bawden, BAS Secretary, and Dr Ray Adie, Deputy Director. At least once each day, a summary of the incoming reports was passed by telephone to Admiral Tony Whetstone, the Assistant Chief of Naval Operations in the Ministry of Defence, Whitehall. Laws was committed also to regular consultations with the Foreign Office which, since 6 April, had been trying to discover, through the good offices of the Swiss diplomats in Buenos Aires, the whereabouts of Keith Mills and Steve Martin and their men. Laws and his colleagues were engaged simultaneously in dealing with the media, with the anxious relatives of the men who had been captured or who were still on South Georgia, and with a host of other people who suddenly had developed a keen interest in the South Atlantic in general. The BAS organisation was at the same time undertaking as much as possible of its normal workload.

The morale of the field parties remained generally high until 17 April. It was on this day that the *Bransfield* was ordered to leave the area and head north. Originally it had been hoped that she might call at South Georgia on 14 April to collect the BAS personnel and the two lady film producers who were due to return to the United Kingdom. It was now clear that she risked being either sunk or captured if she came anywhere near the island, so that plan was abandoned. Her captain sent a long message to the field parties, explaining the orders which he was reluctantly obliged to obey. He closed his signal with these words: 'We ask for your understanding. Everything possible will be done in the UK to arrange evacuation and repatriation. We will be back next summer or earlier. My personal best wishes to you all. May your God be with you. Stewart Lawrence, RRS *Bransfield*.'

Nobody on the island at that time had any idea of the *Endurance*'s location, nor did they know that a Royal Navy task group was headed for the island from Ascension. Winter was

The wooden hut at St Andrews Bay, South Georgia, occupied initially by Miss Cindy Buxton and Miss Annie Price, and later shared with three BAS personnel during the brief Argentine occupation of the island. The structure is anchored firmly to the ground with steel cables to prevent it being blown away by katabatic storms.

arriving in its full savagery – with winds of up to one hundred miles an hour and temperatures well below freezing – and stocks of food and fuel were limited. Cindy Buxton and Annie Price were still flying the Union Flag at St Andrews Bay and struggling to cope with the problems of sharing their small hut and two tents with three men, but nerves were wearing thin.

The Lyell Glacier quartet were particularly dismayed by the departure of the *Bransfield*. Apart from contending with illness and a leaky rat-infested hut, they had failed to find any useful dumps of fuel or edible food at the other huts in their area. They had only two weeks' reserves left, their radio batteries were almost exhausted, and they had no winter clothing. In a candid message to Dr Laws they had the courage to admit their feelings in terms which reflected their anger and their anxiety. As they viewed it, they could hand themselves over to the Argentines at King Edward Point (and perhaps never be heard from again), or stay at Lyell Glacier with no apparent prospect of rescue (and there possibly die of exposure and starvation).

Their Director could not tell them about the Royal Navy's activities because he knew very little about them and, in any event, the radio network was totally insecure. He simply asked them to stay at Lyell for as long as possible. He guessed that King Edward Point might soon again become a battle ground, and it was best to keep his own people well away from any fighting. Furthermore, the Foreign Office was currently attempting to arrange, through the International Red Cross, an evacuation of all British civilians on the island. Those negotiations might be compromised if the Lyell party were to give themselves up at that stage.

Throughout April, the BAS headquarters in Cambridge was visited by numerous representatives of the Army's Special Air Service Regiment (SAS), the Royal Marines' Special Boat Squadron (SBS), the Royal Navy, and others having a specialist interest. The relatively small scientific and administrative staff at the Madingley Road offices was well accustomed to receiving visitors, but this sudden influx of demands from the armed forces placed a tremendous strain upon the organisation. It was not easy to explain to all these enquirers that South Georgia's almost uniquely harsh environment was in many respects quite different from those of Norway or Nepal. Only with personal experience of the sub-Antarctic, and preferably having a scientific background, could one readily comprehend the difficulties of moving men and matériel across such a violently sculpted landscape and in such unpredictable weather.

An added burden for Dr Laws was the moral dilemma into which he and his colleagues had been plunged. It was the same problem which had faced Steve Martin at King Edward Point when, on 31 March, he suddenly found himself caught up in a military operation. In short, the BAS was a purely scientific institution, entirely civilian in character, and deeply involved in a close and harmonious collaboration with similar organisations based in many other countries. The future success – even survival – of the BAS was in part dependent upon the skill of its administrators in displaying a totally apolitical attitude to those topics which usually exercise the minds only of diplomats and politicians. Indeed, the whole spirit of the Antarctic Treaty, to which the United Kingdom had become a signatory in 1959, was founded upon international consensus and collaboration. Wars may come and wars may go, but reputations – good and bad – endure for a long time. It would have been very damaging to the high international standing of the BAS if, as a result of the 1982 war, it had later been perceived by others to be, even in part, a tool of the intelligence community.

Countering the desire to remain academically aloof, there was the natural wish to help one's own country at a time of crisis, and the need to meet the requirements of government. The BAS was a component body of the National Environment Research Council which, in turn, was funded by the Department of Education and Science. Financial pressures had already, by early 1981, resulted in the decision to close or severely curtail the scope of the station at King Edward Point (the future major BAS effort on South Georgia being concentrated at the Bird Island base). Clearly, Dr Laws was obliged to associate his organisation unstintingly with the British government's current policy towards Argentina. His approach was summarised in a comment made in 1985: 'The BAS is a civilian organisation and, although we were willing to help the military recapture South Georgia, this was in conflict with the general peaceful situation in Antarctica.'

The potential perils to the BAS – of involving itself in international politics – had been well demonstrated at the beginning of the crisis when, on 30 March, the BBC World Service included the following report in its summary: 'There has been a row in Parliament in London over the latest dispute between Britain and Argentina. The dispute arose because a group of Argentines recently landed on the island of South Georgia – a dependency of the Falkland Islands. They are still there. Britain maintains that the landing is illegal. In the House of Commons, a government statement said the dispute was potentially dangerous, and efforts to solve it by diplomatic means would go on. A Labour opposition spokesman accused the government of gross dereliction of duty, of crippling the Royal Navy – a reference to recent Defence cuts, in particular a decision to scrap the ice patrol ship *Endurance*. The government statement said that the *Endurance* would remain on station in the Falklands area as long as necessary and its long-term future would be reviewed. Argentina says it will protect its workers on South Georgia, and has sent several warships to the area. In the meantime, a second British ship, the *John Biscoe*, has arrived in the area with about forty marines.'

The BBC had appeared to be telling the world that the *John Biscoe* was somewhere near South Georgia and that she was engaged in some sort of military or naval operation. The sub-editor who drafted the report was evidently unaware of the eight hundred miles which separate South Georgia from the Falkland Islands. Only in the very broadest sense could they be described as being in the same area. The decision made by Dr Laws on 23 March – to give the Ministry of Defence a helping hand – had produced an unwelcome backlash for him and his organisation. The *John Biscoe* had in fact delivered the Royal Marines party to Port Stanley and was already making her way back toward the United Kingdom when the report was broadcast. Laws himself was back in Cambridge, having disembarked from the *Bransfield* at Port Stanley on 28 March and having travelled on one of the last civil flights to leave the islands. There was nothing he could do to put the public record straight. The damage had been done and it had been another phase in the pattern of escalating pressure upon Laws and his people.

It was to their credit that they overcame the problems and produced an answer to almost every question put to them by the armed forces. Whether or not this information would be turned to good advantage – in operational terms – was something over which the BAS could have no control. But all of that lay in the future. We must return now to the scene around King Edward Cove on the afternoon of 3 April.

As darkness fell, Keith Mills and the thirteen BAS personnel were taken out to the *Bahia Paraiso* where they were accommodated in the cabins already occupied by the Royal Marines. The Argentines had not expected so many prisoners, but conditions were tolerable. At 1830 the rising rhythm of the ship's engines made it obvious that she was under way. The tired Britons settled down to sleep.

During the next three days, after a brief call at Leith Harbour, Captain Trombetta maintained his maximum speed of eighteen knots as he headed directly for Argentine mainland waters. Apart from Corporal Peters, he had a large number of wounded men on board, all needing medical attention beyond the scope of the ship's sick bay.

The prisoners quickly fell in with the ship's routine. Breakfast was brought to them in their cabins, lunch and supper were eaten on a mess deck. Their guards supervised them closely, and the Britons noted the tenseness with which the Argentines watched every move. Each guard carried a sub machine gun, his finger constantly on the trigger. Keith Mills was unhappy with the possible risks which this implied. People were afraid that the dropping of a fork at mealtime might cause a bloodbath. He mentioned his anxiety to one of the ship's officers and there was subsequently a slight easing in the atmosphere, but it was evident that the Argentines were taking no chances with their captives. Presumably they feared that the Royal Marines might attempt a hi-jack of the ship.

Most of the guards were sailors, members of the ship's company, but some were marines. Several of them recognised Keith Mills' men from a previous encounter. In February, when the *Endurance* made her extended visit to Mar del Plata, the detachment had been invited to spend a day with the Argentine marine forces at their local base. They had received excellent hospitality and watched the Argentines carrying out various tactical exercises. It had been a pleasant and relaxed meeting between the marines of the two nations. The *Endurance* detachment had not at the time anticipated that they would be returning to Argentina only a few weeks later as guests in a very different guise.

The cabins were located under the *Bahia Paraiso*'s helicopter deck and immediately above the engine room. The deck soon became unpleasantly hot and the prisoners sweated a great deal. They were supplied with fresh clothing, toiletries and soap powder, and so were able to keep themselves clean and respectable. There was no opportunity during this time for physical exercise, but the Argentines gave them magazines, printed in Spanish and English. Using the pattern of the square deck tiles, the captives organised outsize games of 'battleships', but these were abandoned when someone decided that they might provoke the guards. Too many Argentine frigates were being 'sunk'!

Time passed slowly in the crowded cabins. Conversation centred upon two main topics: the prospects for returning to the United Kingdom, and reminiscences of the recent battle. Peter Leach admitted to Keith Mills that he had been wrong in wanting to fight on when the decision to surrender was made. Steve Martin cleared the air by telling Peter Leach that he now accepted the wisdom of having risked the *Albatros* on his journey to retrieve the observation party from Jason Harbour. Despite the obvious differences between them, the civilians and the marines quickly established a bond of friendship.

On 7 April, after a fast passage, the *Bahia Paraiso* made landfall off the coast of Tierra del Fuego. By 1600 she was off the port of Rio Grande and the prisoners heard a succession of helicopters arriving and departing from the flight deck above them. Captain Trombetta was off-loading the dead and wounded to the local military hospital. These flights continued for a long time, and the prisoners concluded that the invasion force had indeed suffered a heavy casualty list. Among the evacuees was Nigel Peters. He recognised the port, having played rugby against a local team on an earlier occasion.

Departing Rio Grande, the *Bahia Paraiso* headed northward at her most economical speed of fourteen knots. She had been ordered to place herself midway between the Falklands and the mainland and to act as a refuelling point for helicopters making the long flight to Port Stanley. Several touched down briefly on her flight deck during the afternoon.

The prison regimen improved steadily. The ship's stores began to run out of soap and other minor luxuries, but the captives were permitted thirty minutes of physical training each day in the large hangar. Here they saw the much punctured Alouette. The BAS personnel had difficulty in matching the pace set by the Royal Marines, but Bob Headland used these occasions to strike up casual conversations with the watchful guards. The only information he could obtain was that the ship would reach a mainland port 'soon'.

On 10 April, the seventh day of their incarceration in the ship, the prisoners were for the first time allowed on deck. To port they saw a coastline which several recognised as Patagonia. On this day, Steve Martin and Bob Headland were also allowed access to the baggage to retrieve books and some clothing. The Argentines supplied sets of chess and yang. The greatest burdens were boredom and the lack of news.

The *Bahia Paraiso* docked in Bahia Blanca on the evening of 13 April. Early the following morning, the captives were disembarked and taken by bus to a distant sector of the naval base where they were deposited in front of a large building. This was a covered swimming pool, normally used by the Argentine forces but now drained and closed for the winter. It was to be a temporary prison for the British personnel for the next four days.

Squads of technicians were busily installing heating and lighting in the dressing rooms where the prisoners were to live. A young naval officer, Teniente de Navio Roberto Alemanno, was in charge of the group and he apologised profusely for the austere conditions. It was evident that the base authorities had received little warning of the Britons' arrival. Within a short time he produced soap, razor blades, sports equipment and a complete change of clothing for every man. In the words of Keith Mills: 'That guy really leant over backwards to look after us as best he could.'

The party was visited by a barber and a doctor. An Argentine

Big lady in a hurry. While Mills, Martin and the other British captives awaited release, dozens of ships were being prepared in British ports for the recovery of the lost islands. *Queen Elizabeth 2* was arguably the most exotic visitor to South Georgia waters during the turbulent days of the Anglo-Argentine conflict. Commencing on 3 May, she was fitted with three helicopter decks and 'replenishment at sea' gear before departing Southampton with 5 Infantry Brigade and two Sea King HAS2A helicopters of 825 NAS on 12 May. Only fifteen days later she was delivering her passengers to other ships in Cumberland East Bay. Safely back in her home port on 11 June, she was swiftly restored to her normal state of luxury and returned to commercial service.

film crew arrived and – after promising to avoid military topics – interviewed a few of the marines and filmed them playing football. The producer seemed anxious to prove to his audience that the prisoners were not only alive and fit, but also being well treated.

To the great pleasure of everyone, Corporal Nigel Peters rejoined his comrades on the second day. After initial treatment aboard the *Bahia Paraiso* and at Rio Grande, he and several Argentine casualties had been flown up to Buenos Aires where his damaged arm was examined and a bullet removed. He was lucky not only to keep the limb, but eventually to regain its full use.

Peters brought with him the exciting news that he had heard a BBC World Service report announcing the sailing of a British task force. This was a reference to Admiral Woodward's Corporate ships and it was the first the prisoners knew of their government's reaction to the Falklands and South Georgia invasions. It sent morale soaring. For Keith Mills, the news was a great relief. It signified that there never had been an official ceasefire, and that he was therefore unlikely to face a court martial.

The marines wondered whether the imminent arrival of a British fleet in the South Atlantic might unsettle their guards. Would they express their nationalistic fervour, and perhaps their resentment, by taking reprisals against the prisoners? Would the Argentine government treat the prisoners as hostages, using them as pawns while seeking a diplomatic solution? Might there be a 'show trial' for propaganda purposes? With events coming to crisis point, the sooner the Britons could get out of Argentina the safer they would be.

Despite the rising tension, the guards showed no signs of personal antagonism. They were as numerous and as observant as before, but there was an obvious wish to treat the group in a correct and civilised manner. The Britons were told repeatedly that arrangements were being made to return them to their homes as quickly as possible. They could only hope that this was so.

More official visitors arrived during the second day. The scientists were asked to compile an inventory of the valuable research equipment left behind on South Georgia, and to list the frequencies and call-signs of the remaining field parties. Keith Mills was invited to go for a stroll with a suave Argentine army major who spoke polished English and who was clearly an intelligence expert. He asked a few questions regarding the battle – which Mills declined to answer – but the main drift of his subtle questioning was aimed at the sailing of the Task Force. 'Do you think Mrs Thatcher is bluffing? Do you believe that your navy will be prepared to fight? Will your army try to reoccupy the islands? How would the British public react to battle casualties?' Mills himself had no means of replying accurately to these questions even if he had wished to do so, and he privately did not yet believe that the British government would commit itself to full-scale war over the issue, but he firmly assured the major that the Argentine navy would be blown out of the water in the very near future.

On the morning of 15 April, Mills and Martin were told to prepare themselves for interrogation. Recalling the reported atrocities of the Argentine secret police, Mills felt uneasy at the prospect of close questioning. He had ordered the destruction of a helicopter and the near sinking of a warship. He was responsible for the death and injury of an unknown number of Argentine servicemen. These acts had taken place on an island which Argentina claimed as sovereign territory. What might a prosecuting lawyer make of this? Would he now be charged with murder? Would he be facing a firing squad before the end of the day? He was not reassured when the staff car in which he and Steve Martin were collected later pulled up in front of a building labelled with a sign which included the word 'Justizia'.

Mills and Martin were interviewed separately by a small tribunal of three naval officers, apparently all senior captains. Mills was quickly put at ease by their spokesman who assured him that he and his colleagues were aware of the constraints of the Geneva Convention. Mills had already given his name, rank and number, and they did not propose to question him regarding his own actions during the battle. Instead, they were inviting him to answer certain questions concerning the conduct of the Argentine invasion force. Was he prepared to co-operate? Mills could see no objection to this, so he agreed to hear the questions.

During the next two hours, the tribunal pursued every aspect of the battle as Mills had observed it. They were particularly interested in the radio conversations between Captain Trombetta and King Edward Point on the morning of 3 April. How had he phrased his ultimatum? Had he given adequate warning of his intention to land? Had he mentioned the surrender at Port Stanley? From the British side, had Trombetta been warned of a military force on the island? Was he told that a landing would be resisted? Mills had no difficulty in responding to any of these points, indeed he was glad of the opportunity to do so. He remembered distinctly the words which he, Martin and Trombetta had used. Repeating them now to these officers, he had the chance to justify in full the battle which had ensued.

The tribunal then passed on to other topics which puzzled them. Why did the Puma pilot try to land in such a dangerous location? Why did the captain of the *Guerrico* persist in re-entering King Edward Cove when he had already seen the fate of the Puma? Why did the invasion force make a frontal approach when commonsense indicated that Mills was very vulnerable to attack from the high ground to his rear? Mills was tempted to offer his personal view that the attackers had adopted a plan of remarkable stupidity, but he simply suggested that such questions were best directed at Captain Trombetta and Teniente de Navio Astiz. He was not pressed to answer any questions which he felt might compromise his own position or that of his fellow prisoners.

As the interview proceeded, Mills realised that this was not an investigation into the conduct of the British, but was instead a court of enquiry examining the actions of the invasion force. Yet another private worry was removed and his self-confidence returned in full. When asked, he refused permission for his men to be interviewed. His decision was accepted without argument. Later it was agreed that Sergeant Leach should appear before the tribunal but, as he declined to offer anything more than his name, rank and number, the Argentines made no further progress.

Steve Martin had his opportunity to tell the tribunal that there had never been a military facility on South Georgia prior to 31 March, and that the King Edward Point base was therefore in no way comparable with the Argentine installations in the Antarctic. He emphasised the peaceful and purely scientific role of the British Antarctic Survey. He also took the opportunity to voice his anxieties regarding the field parties remaining on South Georgia, requesting again that they be regarded as non-combatants. He had no way of knowing, of course, that his people had for the past twelve days been operating a voluntary and spontaneous information gathering service. He would have been even more worried for their safety if he had known that some of them were making extended patrols of up to three days duration over desolate terrain where even a mundane accident – a sprained ankle or a broken leg – could have fatal consequences.

The prisoners busied themselves with a variety of games and physical training. A strict but sensible daily routine – drawn up by the Argentines and including morning, midday and evening roll-calls – gave ample opportunity for sport and recreation. Bob Headland started a series of Spanish lessons for his BAS colleagues and for the marines. The guards became increasingly talkative and Headland tried hard to develop a relationship with some of them. His attempts to extract information met with little success, they evidently being under orders to isolate the prisoners from outside news.

While Headland and the marines cultivated their contacts among the guards – with a view to perhaps facilitating an escape – one of the BAS men was considering his own individual arrangements for release. He was Alexandro Pinheiro-Torres, a physicist and a Portuguese citizen. He composed an appeal to the local Portuguese Consul, asking that his government should intervene on his behalf. The Argentines showed no sympathy for his predicament. They were still holding the prisoners incommunicado and, if Pinheiro-Torres' letter had been delivered, it would have been the first intimation to the outside world of their whereabouts.

Pinheiro-Torres was not the only man committing his thoughts to paper. The entire group was told that they could write to their relatives, if they so wished. In the event, none of these letters were ever delivered, but one of them was the cause of some hilarity. Marine George Thomsen wrote for his family an enthusiastic and detailed account of the battle of 3 April. He concluded with the phrase: 'It was great – just like the battle of the Alamo.'

His letter was seen by the almost too charming Argentine intelligence major. He called for Keith Mills and, grim-faced, read the letter to him. He told Mills that the letter could not be released. 'This is not true. It was not at all like the Alamo. At the Alamo there were no survivors.'

On the third day at the swimming pool, the group was visited by an Argentine admiral. The Royal Marines paraded for him and he made a formal inspection, solemnly shaking hands with each man and behaving as though these were his own troops returning from a successful action. He spoke at some length with Keith Mills, congratulated him upon his defence of King Edward Point and expressed the hope that his country's own forces would conduct themselves equally well when occasion demanded. Mills found the conversation slightly fantastic, but he began to understand the Admiral's motives when, as he was leaving, he said: 'Remember, when you get back to England, tell your people that you have been well treated here. In the war which will soon come, we shall expect the same hospitality.' Evidently the Argentine high command had finally understood that, having sown the wind, they were about to reap the whirlwind.

Mills promised that he would convey the Admiral's message to his superiors in England, and he later did so. Despite the rantings of the British tabloid press – some of whose more xenophobic members were suggesting that the cruelties of the Argentine

A very short-lived triumph. President Leopoldo Galtieri made one visit to 'Las Malvinas' during the Argentine occupation. Ominously, he found himself inspecting his army's new defensive arrangements. As the April days went by, and as the peace initiatives ran out of steam, so it became increasingly clear that his forces would need to fight to hold that which they had seized by violence.

police state were characteristic of the entire nation – the people captured on South Georgia were treated humanely at all times. Their experiences certainly influenced the attitudes of the British force charged with the task of reoccupying South Georgia.

On the afternoon of 16 April, while playing a game of football, the group was told that they were returning immediately to England. After an hour of frantic packing and documentation, they boarded two buses and were taken to a nearby airfield. Here they were joined by six more Royal Marines, men who had for a while evaded capture after the fall of Port Stanley. More confusion followed when it was discovered that the waiting aeroplane was too small but, by 2300, they were airborne in two F27 Fokker Friendships. At first disbelieving that they were really going home, the Britons watched with mounting excitement the lights of Buenos Aires and the wide blackness of the River Plate passing below. By 0130 they were walking the tarmac of the international airport at Montevideo, capital of Uruguay. After only fifteen days in Argentine hands, they were once again free.

The former captives were stunned by their reception as they walked into the airport building. Apart from dozens of journalists and cameramen, representing the world's media services, there was a crowd of cheering British ex-patriots who had fled to Uruguay from Argentina at the commencement of hostilities. The British Ambassador was there to welcome them, together with representatives of the local British community and several officers recently arrived from the Ministry of Defence. Blinded by the television lights and trying to fend off the swarm of pressmen, the marines and BAS personnel suddenly realised that their release was world news. Instead of returning as an embarrassment to their government – having been defeated and flung off a British island – they were being hailed as popular heroes.

Inevitably, having been in battle, it was the marines who attracted the greatest attention from the press. They were amazed to learn that these total strangers already knew their names, the names of their wives and girlfriends, and had already interviewed some of their families. The questions came thick and fast, but the Naval Attaché, Captain Mitchell, quickly warned them against any comments of a military or political nature. The United Kingdom was treading a very careful path in respecting Uruguay's neutrality.

For Keith Mills, the style and scale of the reception confirmed his personal conviction that he had indeed done what was required of him on 3 April at King Edward Point. He had deliberately rejected the constraints of the order not to endanger life when he gave the order to open fire on the Puma, but commonsense told him at the time that it was the only acceptable course. This unreserved welcome seemed to make it clear that his superiors shared the same view.

The thirty-five Britons were taken to one of the city's most luxurious hotels, the Casino Carrasco, where the foyer was packed with more of their fellow countrymen. Offers of drink and hospitality were pressed upon them on a hugely generous scale,

RAF Brize Norton in the early hours of 20 April, and Lieutenant Mills is greeted by the broad smile of Governor Rex Hunt. The return to the United Kingdom of the defenders of King Edward Point, accompanied by full media reporting of their exciting action against the invaders, was the first good news to emerge from the débâcle in the South Atlantic.

but the marines wanted only to escape the crowd and have their own private celebration. Sitting around one large table, they ate and drank and talked throughout the remainder of the night. There was no thought of retiring to the suites which awaited them upstairs. Keith Mills described it as 'a great surge of energy and happiness, a fantastic feeling, as though we were supermen. We knew that the detachment was good news for the Corps, and we were just about bursting with pride.'

Early on the following morning, Mills was taken to the British Embassy to be interviewed by Colonel Love and Captain Mitchell. The main requirement was his confirmation of the basic story so that details could be passed back to the authorities in London. Steve Martin and Bob Headland were similarly interviewed, answering several hours of questioning regarding conditions on South Georgia prior to their capture.

The group had not consumed any alcohol for more than two weeks but, despite being confined to the hotel, they had no difficulty in making up for lost time. Everyone wanted to give them a drink, and the hours flew by. Most of the BAS people caught a few hours sleep – their first in forty hours – but many of the marines stayed awake for their second night in succession. At 0430 on 19 April they were all rounded up and taken back to the airport where a Royal Air Force VC10 was waiting for them. The former captives were de-briefed at length during the long flight north by a team of Intelligence officers from the Ministry of Defence. Every single marine was interviewed, and the shy quiet Paddy McCallion attracted particular attention.

After re-fuelling at Ascension Island, the aircraft touched down at RAF Brize Norton at 0430 on 20 April. Waiting for them was a large crowd of dignitaries, senior officers, pressmen and relatives, all headed by Governor Rex Hunt. The marines were greeted by the Commandant-General of the Corps, Sir Steuart Pringle, and the BAS men by their Director, Dr Richard Laws, and by the head of the Polar Regions Section, Foreign & Commonwealth Office, Dr John Heap.

The Corps of Royal Marines had taken to heart the lessons learned when the men captured at Port Stanley had returned to England some days earlier. That group had been filmed and photographed in the same dirty uniforms which they had worn throughout their battle and brief captivity. Mills' group was instead ushered into a room where each man was able to change into a completely new outfit. Their personal records had been checked to ensure a correct fit, and their entitlements to medal ribbons and insignia of rank. Outside waited a fleet of staff cars, ready to take each of them directly to their homes. It was, by any standard, a superbly well organised affair and the images appearing later that day on television screens around the world gave a very positive impression of British resolve and military competence. In purely political terms, the 'public relations' aspect of the return to England of the *Endurance* detachment was another step in escalating the psychological pressure on the junta in Buenos Aires.

Mills, Leach and McCallion were told that they could not go home yet – first they must undergo more de-briefing. They were flown from Brize Norton to Northwood where they met the C-in-C Fleet, Admiral Sir John Fieldhouse. He wanted to hear every detail of the battle. After congratulating Leach and McCallion, he talked privately with Mills and stated bluntly that he had 'written off' the detachment when, on the morning of 3 April, he heard of the ultimatum then being delivered by Captain Trombetta. 'I could have tried to save you by letting Captain Barker have a go with his Wasps – attacking the Argentines with the AS12s – but the risks would have been enormous and I couldn't afford to lose the *Endurance*.' Fieldhouse was plainly delighted that Mills and his men should have not only survived, but had also given the aggressor nation a very bloody nose.

Leaving Fieldhouse's office, Mills found himself being rushed in a staff car into central London. By midday he was sitting in a BBC news studio, being interviewed on live television. Then there was another dash, this time to the ITN studios to be interviewed on the one o'clock news. He had not slept properly for four nights and, although he was still feeling euphoric, the viewers saw a young man who appeared drawn and ill, his speech slurring and indistinct. Somehow he got through the ordeal and was then taken to Whitehall where he gave a half-hour lecture to a gathering of senior Royal Marines officers. 'I was pretty shattered by now, but I wanted to go on telling the story for as long as anyone wanted to listen. And everyone seemed to want to listen.'

Finally the pressure stopped and Mills went home to Anglesey for eight days' leave. Reporting back to the Royal Marines depôt at Poole, Dorset, he was grateful for the guidance and advice of the local commanding officer, Lieutenant-Colonel David Storrie, RM. Together they composed the recommendations for the awards which the detachment so obviously merited. Then came the important question: 'Do you want to go back south?'

A signal had been received from Nick Barker, requesting the return to the *Endurance* of her marines detachment as a matter of urgency. Apart from needing a military force for any action in which he might shortly find himself engaged, Barker also needed the marines as working members of his ship's company. Each had duties distinct from those of a soldier. John Stonestreet, for example, was one of the ship's cooks. Bob Ashton was the ship's butcher, and Dave Combes served as a steward in the Senior Rates' Mess.

Keith Mills and his detachment were not the only temporary absentees. The two survey officers, Todhunter and Ball, together with one senior rate and four junior rates, had been left behind on East Falkland to complete the season's work when the *Endurance* sailed for South Georgia on 21 March. Another missing key member of Captain Barker's team was his First Lieutenant, Lieutenant-Commander Mike Green, who, in January, had been taken ill in the Antarctic and flown back from Ushuaia to the United Kingdom for treatment. Events were to prevent him from rejoining his ship until months later. The 'Jimmy' is an important man in any ship, none more so than in one cut off from hearth and home for months on end and therefore requiring an experienced link between captain and ship's company. His duties were taken over by the Navigator, Lieutenant-Commander Bill Hurst, and the Charge Surveyor, Lieutenant-Commander David Ives. With four officers and seventeen marines and ratings unavailable, 'the plum' would have been hard pressed to sustain her normal routine under the best of circumstances. Under the operational conditions which she had been experiencing since the outbreak of the war, the strain was even greater.

Perhaps the severest strain was felt in the ship's Communications Department where Chief Petty Officer David Hamilton and Petty Officer David 'Wendy' Craig worked to the point of exhaustion to cope with the flood of extra radio traffic generated by the conflict.

The survey team had been swiftly evacuated from Port Stanley by their Argentine captors and had been able to rejoin their ship, by way of Ascension and HMS *Antrim*, on 14 April. Now, one month later, Mills and his men were equally anxious to get back to their shipmates. Specifically, they wanted to avenge their enforced surrender. South Georgia had been reoccupied by British forces, on 26 April, but the Argentines still held the Falklands. Their garrison was said to exceed eight thousand men, so there was every prospect of much hard work and hard fighting before the invaders could be thrown out.

The detachment departed Poole for the long haul back down to the edge of Antarctica. The convalescent Corporal Peters stayed in England, his place being taken by Corporal Steve Gogerty. After a brief stop at Ascension Island – where they had the quiet satisfaction of guarding Alfredo Astiz for two days – they boarded the cable ship *Iris* which delivered them, on 25 May, to their own ship off Grytviken. For Mills and his men, the wheel had turned full circle. □

VI To The South Atlantic, Quick March!

In the United Kingdom, the first British troops to be told that they might be on the brink of active service were the élite Special Boat Squadron, Royal Marines, otherwise known as the SBS (or simply 'the Squadron'). On 30 March, while attending a routine conference in London, the unit's commanding officer, 'Jonathan', received an indication that trouble might be brewing. He returned to his headquarters at Hamworthy, near Poole, Dorset, and quietly alerted his staff. On the following morning a signal arrived from the Ministry of Defence, giving the first official warning of a possible move.

Trained specifically to survey hostile coastlines in advance of landings by Commando forces, the SBS was a very small unit. Divided into numbered Sections, its personnel were drawn exclusively from the Royal Marines. Each Section was divided into Patrols, with four men in each Patrol. Trained as swimmers, divers, canoeists, parachutists and small boat operators, some of them had served with the Squadron for nearly twenty years. The strength of such units is their very high standard of fitness and training, and the total faith which each man has in the others.

In the early hours of 1 April, the signal reached Poole to 'stand to' three Sections. Only two of these were to be involved in the South Georgia story, 2 SBS and 6 SBS. Both were on leave when the call came, but every man could be reached by telephone and soon they came hurrying back to the sand-coloured steel-clad buildings of their headquarters. They plunged into the task of preparing their stores for a possible move at short notice.

Fresh signals continued to pour into the operations room throughout 2 April and the following morning. It was confirmed that 6 SBS was to proceed to Faslane, in Scotland, to embark in the nuclear-powered HMS *Conqueror*, while 2 SBS was to fly south to Ascension Island. At this stage there was no conviction within the SBS that armed conflict would ensue. To quote one of their officers: 'Even after the fall of Port Stanley and Grytviken, we assumed that we were being moved down there on a contingency basis only. It seemed unlikely that the government would commit us to a full-scale counter-invasion. I think we anticipated being witnesses to a great naval action, fought out on the open sea. How wrong can you be?'

On 3 April, 6 SBS departed for Faslane and went aboard the *Conqueror*. Commanded by 'David', they settled into the crowded accommodation reserved for them, while their weapons and other stores were packed into the torpedo room. Knowing that this might be a very extended patrol, Commander Chris Wreford-Brown, RN, the boat's commanding officer, had packed every inch of space with food and other stores. The passageways were decked with cases of tinned food, two cases deep, and men walked on this false floor bent almost double. Later they were able to straighten their backs only when they gradually ate their way through the stocks.

Although accustomed to travelling in submarines of the conventional diesel-electric type – for relatively short periods – the men of 6 SBS had not previously taken passage in a nuclear-powered boat. They did not know it, but the *Conqueror* was to be their home for the next twenty-two days. If the opportunity arose, Wreford-Brown might be able to bring them close to shore, launch them from the casing in their Avon Gemini inflatable boats, and leave them to make their way to any lonely beach where a secret landing might be made. Nobody, at this stage, had any idea of where, when, or even whether, such a landing might be required.

Also on 3 April, 2 SBS travelled by road to RAF South Cerney in Wiltshire. This was the air trooping centre for personnel about to depart from RAF Lyneham. With them they had two corporals of the Special Air Service who were needed to make up numbers and to act as liaison contacts with the SAS should the need arise. Completing the group was a command party headed by 'Jonathan'. Two days later, on 5 April, they lifted off from RAF Lyneham in a C-130 Hercules, bound for Gibraltar and Ascension Island. There were only twenty passengers on board, but the belly of the Hercules was packed with tons of equipment and stores – canoes, inflatable boats, weapons, ammunition and food. The unit was travelling to a part of the world where there were no British military facilities of any kind. 'Jonathan' intended to be as autonomous as possible. The only type of kit not included in large quantities was Arctic warfare clothing and equipment. It would be nearly another week before he was told of the decision to reoccupy South Georgia. It was his initial assumption that any SBS operations would be directed at the Falkland Islands where the climate and terrain do not require skis, snow-shoes or white combat uniforms.

Arriving at Wideawake Airfield at 0330 on 6 April, 2 SBS found that they were the first British troops to reach the island of Ascension. The American ground personnel were slightly surprised by the sudden disturbance to their usually uneventful lives, but they quickly and efficiently allotted accommodation and storage space.

Prior to the move, 2 SBS had spent three months in Norway engaged in Arctic warfare training. They were confident that they could deal with any harsh weather conditions they might encounter in the South Atlantic, but the next three days, spent in Ascension's blazing sunshine, came as a shock to the system. The unit busied itself with route marches and weapon testing, while the senior officers awaited further instructions regarding the next move. At this stage there was some debate as to whether or not the

Belfast G-BEPS photographed while making a routine flight over East Anglia. The fact that Heavylift Cargo Airlines Limited had two aircraft readily available in the United Kingdom in early April was a stroke of good luck for the Ministry of Defence. It was even more fortunate that a British commercial organisation had kept the type in flying condition long after the Ministry had decided that it was no longer needed by the armed forces.

Section might be embarked in one of the first two submarines to have departed United Kingdom waters – the *Spartan* or the *Splendid* – for deployment on the Falkland Islands. This would have meant delaying the high-speed passage of one of the two boats as they raced south to commence the blockade which soon was to be announced by the British government, so that plan was abandoned. Instead, the Section was told to embark in the RFA *Fort Austin* as soon as she arrived from Gibraltar.

Only a few hours after the arrival of the SBS on Ascension, four more aircraft flew in from England, bringing with them a unit of the Special Air Service (otherwise known as 'the Regiment'). Apart from numerous attached specialists, this group consisted of the entire strength of 'D' Squadron SAS. The Squadron comprised four sub-units – Boat Troop, Air Troop, Mobility Troop and Mountain Troop – each Troop being divided into several small Patrols. Commanding the party of approximately one hundred men was Major Cedric Delves of the Devon & Dorset Regiment.

'D' Squadron had spent the early weeks of the year in Africa on a variety of training exercises. By late March they had completed a short leave and reassembled at Hereford as the current 'standby squadron'. Late on the afternoon of Friday, 3 April, they received their first alert – a 'heads up' signal – to the effect that they might be deployed in the South Atlantic. The 'go' signal arrived on the following morning and the unit was ready to move thirty-six hours later.

Major Delves later described his preparations: 'Like everyone else, we had been following developments in the Falklands but, initially, it looked like being a job for the Navy rather than for us. However, it is very much in the spirit of the regiment to get up front – to get to where the action is. Our services were offered and off we went. The CO's orders to me were drawn in very broad terms: to assist amphibious operations with surveillance, path-finding, and diversionary assaults. We had no idea what might be involved, but it was obvious that we ought to get down there quickly. We took off on the Monday, with two VC10 out of Brize Norton and two Hercules from Lyneham. We needed a big lift because we took a great deal of kit with us – everything from CS gas to Milan missiles. Operating so far from base made it necessary to have the right equipment for whatever job we might be asked to perform. We even managed to get one of our people back from 23 SAS – a Permanent Staff Instructor with the Territorials – because he spoke Spanish and that seemed useful. Africa hadn't been the best acclimatisation for operations in the South Atlantic, but we felt ready for anything.'

The SAS and the SBS were accustomed to working together and knew each other's ways. There was, of course, one fundamental difference between them. The SAS was essentially an army-oriented organisation, the SBS was at heart a branch of the naval tradition. As one commentator has said: 'The SAS look at the sea from landward, the SBS look at the land from seaward.' The nuances reflected in this phrase were to show themselves during the coming days. In the meanwhile, both groups greeted old friends and concentrated upon the task of preparing for the arrival of the *Fort Austin*.

Back in England, the Fleet Air Arm had swung into action on the day that Port Stanley fell. At the Royal Naval Air Station Yeovilton, Somerset, orders arrived on 2 April to despatch immediately a force of helicopters for deployment to the *Fort Austin*. A big ship, she could handle up to five helicopters. Captain Peter Williams, the Commanding Officer of HMS *Heron* (otherwise RNAS Yeovilton), did not have a standby unit ready for instant action because expensive helicopters cannot be held on the ground on the off-chance that they might or might not be required. Within a few hours, however, the staff had allocated three Lynx 2 of 815 Naval Air Squadron to be commanded by Lieutenant G. R. Moodie, RN, and two Wessex 5 of 845 Naval Air Commando Squadron, to be commanded by Lieutenant-Commander John Lomas, RN.

Many of the Fleet Air Arm aircrew and maintainers had gone home for the weekend. Intense activity in HMS *Heron*'s operations centre and telephone exchange resulted in a flood of calls giving them three hours in which to bid farewell to their families, pack their bags and report for duty. Navy wives are accustomed

to long separations from their menfolk, but the speed of this mobilisation called for the highest level of domestic loyalty. Many homes in the Yeovilton area – and in other parts of the United Kingdom – were entering a long period of anxiety. The naval air station itself was plunging into an equally long period of intense round-the-clock activity.

Transporting the Lynx by air to Ascension Island did not present a major problem because this relatively small helicopter could be carried within the freight hull of the C-130 Hercules. Three Hercules C1 of Royal Air Force Support Command arrived at Yeovilton, each loading a Lynx with spares and maintainers, and departed for Gibraltar, Dakar and Ascension Island on 3 April. The three helicopters were XZ730 (HMS *Minerva*'s flight), XZ242 (HMS *Newcastle*'s flight), and XZ247 (from 815 Squadron's Trials Flight).

The Wessex were a different proposition. Not only is this a heavier helicopter but, more importantly, it is larger. It would not fit inside the Hercules and the RAF possessed no other aircraft big enough for the task. The Ministry of Defence turned to the civil charter market and contacted the only British firm able to accept the challenge, Heavylift Cargo Airlines Limited. The cornerstone of their fleet was, and still is, the Belfast – an aircraft capable of lifting a payload of 80,500 pounds. The huge cargo hold is 84 feet long, 16 feet wide and 13 feet high, hence it could accommodate two stripped Wessex helicopters. Heavylift owned five Belfasts, but the only one immediately available was G-BEPE, *en route* from Sokoto in Nigeria for its home base at Stansted Airport, near London. The pilot was contacted over France and told to divert to Yeovilton. He landed at the naval air station where the two Wessex 5 of 845 Squadron's 'B' Flight were already being stripped. They were XT468 (Yankee Delta) and XT765 (Yankee Sierra). By 1500 on the next day, Sunday 4 April, Captain Keith Sissons and his crew were lifting off from Yeovilton's runway and heading south with the first Commando helicopters to be involved in the conflict.

Twenty-six hours later, a second Belfast lifted off from Yeovilton with the two Wessex 5 of 'C' Flight 845 Squadron. These were XT464 (Yankee Foxtrot) and XT473 (Yankee Alpha), both earmarked for the RFA *Tidespring*. Like the *Fort Austin*, she had a hangar and full helicopter handling facilities, but did not have any aircraft embarked at the commencement of the crisis. The usefulness of both ships would be vastly enhanced once they had a full helicopter 'lift' capability.

At 0850 on 5 April, G-BEPE touched down at Wideawake Airfield, having made a brief refuelling call at Dakar during the night, and 'B' Flight's maintainers and aircrew immediately started to reassemble their helicopters. They were joined on the following morning by 'C' Flight when Captain Keith Sissons (who had changed aircraft at Dakar) brought G-BFYU down onto the long runway carved through the volcanic ash hills of Ascension Island. By 7 April, all four helicopters had been air-tested and were fully operational. The three other pilots in command of various 'legs' of these initial freight flights were Captain Bob Reynolds, Captain Peter Jobling and Captain Colin Penny, all of whom were former RAF Belfast pilots.

There was an urgent need to send the *Fort Austin* further south without delay. The stores she was carrying were needed by the *Endurance*. Following the fall of King Edward Point on 3 April, the red ship had hidden herself for forty-eight hours in the coastal reaches of South Georgia. Reserves of food were running low and

Facing and right: The recently rebuilt Sea King HC4 serial ZA312 of 846 Squadron was taken from storage at RNAS Culdrose, flown to RNAS Yeovilton, stripped, and swiftly loaded into Belfast G-BFYU. Re-assembled at Wideawake Airfield by Naval Party 1222, this helicopter performed general delivery and replenishment duties on the island until late May when it was transferred to the MV *Elk*. Subsequently it operated from HMS *Fearless* and from forward bases on East Falkland. These two photographs were taken at RNAS Yeovilton on the afternoon of 7 April.

Wessex HU5 serial XT460 is loaded into the cavernous belly of Belfast G-BEPE on the night of 7/8 April for delivery to Ascension Island. This 845 Squadron helicopter later flew on general load-lifting and delivery duties throughout the build-up period of Operation Corporate and operated from HMS *Intrepid* and on East Falkland.

2 April 1982, and HMS *Antrim* replenishes at sea from the Type 26 Leander class frigate HMS *Ariadne*, one of the Exercise Springtrain ships ordered back to the United Kingdom. The helicopter is the *Antrim* Wessex 3, flown by Lieutenant-Commander Ian Stanley, RN, still carrying her peacetime paint scheme and fuselage lettering. This elderly machine is destined to become a central character in the drama of Operation Paraquat.

Captain Barker had imposed a strict system of rationing. Breakfasts were cancelled, lunch consisted of a bread roll or a cup of soup, and the evening meal was restricted to one course. The honour of eating the last egg in the ship went to John Tippey, the film producer. The order from Northwood, to leave the vicinity of the island and to rendezvous with the *Fort Austin* on Latitude 22, to replenish with food and other dry stores, came as very welcome news. The fact that the meeting would take place more than 2,000 miles to the north, and therefore seven days later, was a relatively minor burden to bear. After two weeks of involuntary dieting, the *Endurance*'s people were developing a lean and hungry look.

The embarkation at Ascension Island of the combined SAS/SBS force of approximately 120 officers and men was completed on the morning of 9 April and the *Fort Austin* sailed shortly afterwards. Soldiers and marines alike were impressed by the warm welcome of Commodore Sam Dunlop and his civilian crew. Their ship was the first British surface vessel to be heading into the area of conflict. The feeling in the ship was, according to one of her people, 'keen but slightly confused'. Apart from the obvious need to go the the aid of the *Endurance*, it was hard to understand why all these troops had been embarked and where they were supposed to be going.

It seems likely that there was – in London – the thought that the *Fort Austin* and the *Endurance* might join forces and proceed together to South Georgia. Depending upon the local military situation on the island, and depending upon the progress of the continuing international diplomatic initiatives, they could then launch an immediate assault on the Argentine forces at King Edward Point and Leith Harbour before they had been reinforced or consolidated. Alternatively, the SAS and SBS could be landed to form an advance base for a subsequent reoccupation, or simply make a quick raid and evacuate the civilian field parties. There was much to recommend such ideas. The *Fort Austin*, once having embarked her helicopters, would have the means to defend herself against small surface vessels. The *Endurance* had her two Wasps with their AS12 missiles. Neither was a warship in the conventional sense, but speed and boldness might reap their just reward. If successful, a short sharp blow against the Argentines would serve to show British pluck and would give due notice that London had not thrown in the towel.

In the event, the possibility was never developed. The risk was judged, understandably, to be excessive. With no support nearer than Gibraltar and Ascension, Dunlop and Barker would have been in desperate straights if their ships had encountered any significant opposition at sea, or had suffered serious storm damage. Any prospect of proceeding with such a scheme evaporated entirely when, on 12 April, the two ships met in mid ocean. During the replenishment operation, both of the *Fort Austin*'s Wessex aircraft became unserviceable. With only three Lynx and two Wasps remaining, Barker and Dunlop could not have lifted the men and material needed to make a viable insertion onto South Georgia's shores.

Although their future employment was still undecided, the SBS and SAS units aboard the *Fort Austin* started to test their equipment for whatever contingencies might arise. The SBS donned their dry-suits and, while the ship hove to for a short while, dived in the open ocean. There was a moment of uncertainty when a large whale approached the swimmers. Through the clear water, they could see its vast shape cruising slowly to within forty feet. Nobody stayed in the water long enough to identify it by species.

The Boat Troop of 'D' Squadron SAS practiced the launching of their Avon Gemini boats from the high ship's side. In common with the SBS, the SAS had no complaint with the inflatable boats, but they were unhappy with the reliability of their forty-horsepower outboard engines. Contrary to popular myth, the SAS did not enjoy unlimited funding by the Ministry of Defence. Over a period of nearly ten years they had argued that this particular design – the product of an otherwise reputable manufacturer – was ill-suited for operational use. Indeed, most of the engines were themselves ten years old and had long completed their useful working lives. The criticisms had fallen on deaf ears. Continuing Defence Budget economies had left the Boat Troop with engines in which they had little if any confidence.

In mid ocean, the Boat Troop went through the drills of loading their boats by passing men and kit down scrambling nets. The procedures were tested to the full, the swell causing the little boats to rise and fall ten feet along the big ship's hull. Having embarked, the crews tried to motor around in the vicinity, but soon paddled back as the engines failed one after another. These engines were to become one of the largest single preoccupations of the SAS as the days went by. Despite much expert help from the marine engineers serving in the *Fort Austin* (and later in the *Antrim*), and despite more test runs on the open ocean, they obstinately refused to run sweetly. Their deficiencies were to seriously jeopardise the success of the United Kingdom's response to the Argentine invasions.

Not all of the passengers enjoyed their passage southward from Ascension. The *Fort Austin* was a big ship but, even so, her gentle movement across the ocean was enough to unsettle the stomachs of some. Like all soldiers who have ever been transported by sea, they received their share of mariner's humour. 'Not to worry,' they were told, 'you'll be alright when you get aboard the *Endurance*. She's as steady as a rock.' The irony of this advice was revealed when, early on the morning of 12 April, the sufferers saw the approaching *Endurance* 'bouncing around all over the horizon like a flaming cork,' as she was later described. She had in fact been known on some occasions to roll through more than sixty degrees of arc.

Replenishment of the *Endurance*'s almost exhausted stores commenced as soon as the ships had closed to within three miles of each other, the bulk of the load-lifting work being done by the two Wessex. One of the first loads was an underslung supply of frozen lamb chops. The *Endurance*'s hungry company watched the approaching helicopter with keen anticipation. They were not

The beginning of a long and embarrassing series of incidents relating to the ill-chosen obsolescent outboard engines operated by the SAS and the SBS. Heading south from Ascension Island, HMS *Antrim* heaves to for a while so that her 'D' Squadron passengers can rehearse their Gemini embarkation procedures. Even under these ideal conditions, the engines quickly reveal their unreliability.

Cases of food, canned beer and other essential stores being transferred from the RFA *Fort Austin* to the flight deck of HMS *Antrim*. Parked on the stern of the *Fort Austin* is Lynx HAS2 serial XZ247, one of the Fleet Air Arm helicopters airlifted to Ascension Island on 2 April. This aircraft saw much service, operating from a number of Operation Corporate ships and taking part in the Sea Skua attacks on the Argentine patrol boat *Alferez Sobral*. The Wessex with the underslung load is either XT468 or XT765 of 'B' Flight, 845 Squadron.

pleased when the pilot experienced a sudden loss of engine power and was obliged to jettison his precious cargo into the ocean.

The aerial traffic between the two ships continued for several hours. Apart from stores, the helicopters also delivered to the *Endurance*'s small flight deck the men of 2 SBS, commanded by 'Roger', and of Mountain Troop, 'D' Squadron 22 SAS, commanded by Captain John Hamilton.

Passengers were moved in the opposite direction. Francis Ponsonby and John Tippey, with his film crew, said goodbye to the red ship which had been their home for so many weeks. Ideally, they would have preferred to stay with Nick Barker and his people in order to record the impending operations for the recovery of South Georgia, but accommodation in the *Endurance* was at a premium.

Cross-decking between the *Fort Austin* and the *Endurance* was completed on 13 April. 'The red plum' turned back south to make her way slowly down ocean, waiting to be joined by other ships coming south to her support. The *Fort Austin* wished her 'goodbye and good luck' before reversing her course back to Ascension Island. On board, the RFA ship still carried three Troops of the SAS – Air Troop, Mobility Troop and Boat Troop – and a small command party of the SBS. They were to be disembarked back onto Ascension Island while the *Fort Austin* was to join the main Task Force now forming for Operation Corporate – the reoccupation of the Falkland Islands.

The initial decision – in the opening hours of April – to despatch the SBS and the SAS to Ascension Island, had been an instinctive reaction by the Ministry of Defence and Northwood to the obvious need to have some sort of cutting edge to a British military response. Exactly how they would be employed, or whether they would even be needed, was not immediately apparent. After the trauma of those opening days, firm plans and ideas started to evolve.

The Prime Minister's instruction to the First Sea Lord, Admiral Sir Henry Leach, to mount a campaign for the recovery of the islands, had led in turn to a series of conferences and decisions regarding the strategy to be adopted. Meetings between the MOD in Whitehall and the Commander-in-Chief Fleet at Northwood resulted in the decision to make South Georgia the priority target.

The Falkland Islands were the big prize, but they must come second. It was first necessary to eliminate South Georgia from the scene, and for four principle reasons: to increase the diplomatic pressure on the Argentine Government, to achieve moral ascendency over the Argentine armed forces, to obtain a safe haven for cross-decking and the repair of damaged ships, and to prevent the Argentines from using the island as a forward operating base. Admiral Woodward would find it very difficult to concentrate all his efforts against the Falkland Islands if, for example, he still had the possibility of a submarine threat to his eastern flank, that threat coming from one or more boats lurking in the various inlets and anchorages of South Georgia's north-east coast. The bright red *Endurance* had succeeded in hiding herself in that wild scenery, so too could units of the Argentine navy. According to Admiral Fieldhouse, 'the decision to mount Paraquat was partly political and partly military – in the broadest sense of that word. And you cannot really separate one consideration from the other.'

The decision to reoccupy South Georgia was conveyed to Admiral Woodward on 7 April. Late that same night, Northwood signalled Captain Brian Young, commanding the County Class guided missile destroyer HMS *Antrim*, to detach from the main battle group (Task Unit 317.8), and to proceed at best speed for Ascension Island. Young was told that he was now in command of a 'forward combined force' (Task Group 317.9) with the sole objective of re-establishing a British presence on South Georgia. The target date set for the initial landings was 21 April.

One of the early problems encountered at Northwood was the lack of a staff structure able to deal with the huge additional operational planning requirements resulting from the crisis. Northwood was run as a command centre for the everyday activities of the Royal Navy, and for all NATO operations in the Eastern Atlantic and the English Channel. The international flavour of the NATO command was reflected in the staff personnel. It included officers drawn from several different navies: American, German, Dutch, British, Canadian, Danish and Norwegian. They were committed to the continuing NATO function, and none of them, not even the British officers, could be utilised for the campaign against the Argentines. Indeed, it would have been unthinkable to attempt to utilise in any way the representatives of allied navies in what was exclusively an Anglo-Argentine quarrel.

Over a period of many years – since the ending of the Suez campaign of 1956 – the British armed forces had been told by their political masters that the United Kingdom would never again 'go it alone'. Hence, when Argentina invaded the Falkland Islands and South Georgia, and when the decision was made by Mrs Thatcher and her Cabinet to recover the lost territories, the Navy found itself with an inadequate national planning staff. A much larger and purely British staff structure was needed – instantly – to deal with the crisis.

Vice-Admiral David Hallifax, Chief-of-Staff to Admiral Fieldhouse, solved the problem by transferring the entire staff of the Maritime Tactical School (MTS) from its usual home at HMS *Dryad*, Southwick Park, near Portsmouth, to 'the hole', the underground bunker at HMS *Warrior*, Northwood. The MTS was the permanent establishment attended by all sea-going commanding officers and senior operations officers from time to time during the course of their careers. Here they were trained and re-trained in the skills of operational planning and operational tactics of all types. In the early hours of 1 April, the bedside telephones of various officers currently employed at the MTS started to ring. 'Our friends in the south are on the move. Please report for duty immediately.' Although suspecting an April Fool's joke, they struggled into their clothes and drove to the shore base of HMS *Dryad*. Four days later they started to commute regularly to and from Northwood, setting in motion the wheels which would permit Woodward and his people to carry through their orders.

Accommodation at Northwood was already largely taken up by the personnel normally based there, but arrangements were made to accommodate the additional Operation Corporate staff. They consisted of twelve officers and six WRNS. Directed by Captain Gordon Walwyn, CVO, RN, and his deputy, Commander Jimmy James, RN, their first task was to assemble every available item of information regarding the events of the past few days, the current state of play in the South Atlantic, the resources available to recover the islands, and to prepare an assessment for Admiral Fieldhouse. As he told Walwyn at their first meeting: 'I want to know all the options. This time it's for real.'

The military appreciation drawn up by Walwyn and his people – with specific regard to South Georgia – was based upon a wide variety of factors, some of them certain, some of them not. The most glaringly obvious consideration was the huge distance over which any force sent to reoccupy it would need to travel and the lack of logistical support. A concomitant problem was the small number of ships available for the task. If they were intercepted by the Argentines while on passage, and if they suffered loss or damage as a result, then there was no prospect of immediate rescue. Any kind of defeat or setback – coming so soon after the loss of the islands on 2 and 3 April – might sound the death knell

In contrast with the intense activity in the Paraquat force, this was the scene only a few days earlier when the Exercise Springtrain ships had been berthed in the peaceful sunshine of Gibraltar harbour.

Guy Sheridan, commander of the Operation Paraquat land forces, shown in this photograph after his subsequent promotion to Lieutenant-Colonel.

to any hope of sustaining public and political enthusiasm for a major campaign to recover the Falkland Islands. It was vital that any force – no matter how small and no matter what the diplomatic constraints – should be entirely and swiftly successful.

An analysis of the capabilities of Argentina's naval and air forces suggested that the greatest threat to the Paraquat ships was represented by their major surface units such as the cruiser *Belgrano* and the two Vickers Type 42 destroyers *Hercules* and *Santissima Trinidad*. The heavily-gunned *Belgrano* had a range of 7,600 miles and could therefore roam the South Atlantic without needing frequent refuelling at sea from tankers. With the support of Boeing 707 reconnaissance aircraft, these Argentine warships could – if handled with boldness – lay a deadly trap for any British units sailing down the eastern half of the South Atlantic. The latest reports suggested, however, that the Argentine navy was confining its activities to home and Falkland Island waters, at least for the time being.

The air and sub-surface threats were judged initially to be far less substantial. The distances were considered too great for strike aircraft to operate from mainland or Falkland Island bases, and the Argentines would be most unlikely to hazard their one aircraft carrier, the *Veinticinco de Mayo*, so far from home. Submarines could, theoretically, try to ambush Brian Young's ships in mid ocean but, as Admiral Fieldhouse later observed: 'It would have been a very remarkable achievement if they could have managed that with the small number of conventional boats available to them. The usual problems – time, distance, speed – were against them. Once we reached South Georgia, it was a different matter altogether. It was logical to expect a submarine threat in the local waters. The island was the focal point for any approach track which we might choose, so that was where they would wait for us if they knew we were coming. It was important not to let them know that.'

It was with all these thoughts in mind that the decision was made to blanket Captain Young's future activities in total secrecy. The South Georgia operation would be treated as a quite separate project, with its own planning staff and its own channel of communications. It was assigned the codename Paraquet, soon to be corrupted by a typographical error to the more appropriate Paraquat. On the morning of Sunday, 10 April, the order was given to establish a sealed Paraquat planning cell at Northwood. The officer selected to set up and run this cell was Commander Dennis Davis, RN, a torpedo and anti-submarine specialist currently serving with the MTS as an instructor. He was given a small room in 'the hole' and an initial staff of two WRNS (later he was allocated two Lieutenant-Commanders as watch-keepers). Ships earmarked for Paraquat were assembling at Ascension Island and were due to sail in less than forty-eight hours. There was no agreement yet as to what they should do when they arrived at South Georgia, and it was Davis' job to draw up an operational plan as quickly as possible for consideration by Admiral Fieldhouse and the Chiefs of Staff. The workload was intense for everyone. Davis himself did not leave 'the hole' but worked twenty-two hours per day until he collapsed from exhaustion and was rushed to hospital on 23 April.

Commander Davis recalls his immediate problems: 'All eyes were fixed on the Corporate build-up. Captain Young would be out on a limb. It was clear that nothing should be allowed to leak out which might compromise his safety. Apart from the obvious military considerations, it was important that South Georgia should be taken quickly and bloodlessly. The overriding requirement was secrecy. One of the first directives, therefore, stated that all Paraquat signals would have a special prefix. This meant that only certain named individuals could see them. There were only sixteen names on the list, starting with the Prime Minister. Others included Admiral Fieldhouse, Vice-Admiral Hallifax, John Nott, the Flag Officer Submarines, the Assistant Chief of Staff (Operations), and a few more at that sort of level. Not even their own staffs were permitted to see them. It was strictly on a "need to know" basis. If the enemy got wind of the operation – if they had time to reinforce South Georgia without our knowledge – then South Georgia would be a disaster and the Falkland Islands would never happen.'

In the main operations rooms at Northwood and in the Ministry of Defence in Whitehall, the position of every ship afloat was carefully plotted on huge wall maps. On 10 April, the markers representing the Paraquat ships were taken down. The curious were told 'forget them, they're on a special operation, you will be told the details if and when necessary.' From now on, only Commander Davis and the fifteen other people – all cleared by the security agencies at the same high level – would have access to the new plot installed within the locked room of the Paraquat cell. Davis started to study the tactical problems and the forces available. They had been assembling over a period of several days.

In the closing days of March, high in the Pyrenees and far removed from the mounting crisis, a Royal Marines officer had been enjoying a long-awaited Easter family holiday. Major Guy Sheridan, second-in-command of 42 Commando RM, had just completed three months of tough Arctic warfare exercises in Norway. His unit was the only Commando that year to have benefitted from this essential training. The government's finan-

The 'hole' at HMS *Warrior*, Northwood, on the northern suburbs of London. Staffed by naval personnel and representatives of the other armed services, this is the busy nerve centre for the United Kingdom's campaign to recover the South Atlantic islands. Few, if any, of the people working in the main operations room are aware of the plans being prepared in the nearby top secret 'Paraquat cell'.

cial economies had bitten deep. Now, with his wife, Molly, and their young daughter, Peta, Sheridan was able to relax in the isolated mountain farmhouse which he had rented. Having no telephone or radio, and receiving no daily newspapers, he had no warning of the storm which was about to burst upon his life and those of his fellow marines.

On the evening of Friday, 1 April, after a day spent in the high peaks, the family came back to their chalet to find a telegram pinned to the door. In terse French, it told Sheridan to immediately contact the British Embassy in Paris. Although suspecting an April Fool trick, he went down to the village and called the Defence Attaché. 'You are to return to your unit as soon as possible. I'm sorry, but I cannot tell you why.' By 2030 the family had packed their bags and were on the road. Sheridan and his wife slept briefly in a field beside the highway – their daughter sleeping in the car – and they then raced on for the Channel coast. All the way across France they listened in amazement to the BBC reports of the battle at Port Stanley that morning which had resulted in the loss of the Falkland Islands.

From many other countries around the world, the officers and men of 42 Commando were similarly hurrying back to the unit's barracks at Bickleigh, near Plymouth, by whatever means they could arrange. One officer was in America to get married. The recall telegram arrived while he was at the altar. His best man called the airline office and reserved two seats. At the reception he showed the signal to the bridegroom. 'I'm very sorry, chum, but you've only got two hours to consummate.'

The Sheridans boarded a ferry at Cherbourg and landed at Weymouth on Saturday night. Guy Sheridan found a telephone and called the Adjutant's office. 'Go home, Sir, get some rest. The CO wants you here tomorrow.' Major Sheridan deposited his family at their Devon home and, a few hours later, reported to his commanding officer, Lieutenant-Colonel Nick Vaux. 'We are going south. We have a lot to do before we sail.' After the extended exercises in Norway, the unit's equipment was suffering from wear and tear. The next forty-eight hours were devoted to sorting and packing the flood of new stores arriving in abundance from various ordnance depôts. By Tuesday, 5 April, the Commando was fully assembled and ready to embark in the SS *Canberra*.

It was at this stage that Colonel Vaux was told to detach part of his command and to send it ahead by air for an operation to recover the island of South Georgia. As far as the Royal Marines were concerned, Operation Paraquat had started. Vaux called his number two into his office. 'I want you to take a team down to Ascension where you will board a ship for the second leg. I'm giving you "M" Company, with supporting elements.'

It is the traditional role of the second-in-command of an infantry battalion or a Royal Marines Commando to take over in the absence of his commanding officer. Under normal circumstances, Sheridan would have been concerned mainly with matters of administration. Now, however, he suddenly found

Major-General Moore had himself at one time commanded 42 Commando. Even earlier in his career, as a Captain, he had led 'L' Company when it made its dashing attack at Limbang, on the Limbang River in Sarawak, to rescue fourteen hostages from well-armed rebel forces. On the morning of 12 December, 1962, 5 Troop of 'L' Company had been the first to fight its way ashore and, with losses on both sides, drive the defenders inland. Moore was awarded a bar to his Military Cross for the action and, it is said, always in later years showed a special interest in the progress of 5 Troop's successors. In this photograph, taken at Bickleigh on 8 April 1982, he has a friendly word with Lieutenant J. R. J. 'Jerry' Burnell, 5 Troop's current commander. Burnell and his men were to see much action on East Falkland, taking part in the Mount Harriet attack and conducting several mine clearance and probing operations in advance of the main force.

himself leading the spearhead of Great Britain's offensive to recover one of her lost South Atlantic territories. Events were moving very fast.

He consulted Captain Chris Nunn, the officer commanding 'M' Company, and they agreed a list of the additional personnel required to make their small force as autonomous as possible. Colonel Vaux had already allocated two sections of the Reconnaissance Troop and a section of the Mortar Troop (comprising two 81 mm mortar teams). Once ashore, communications would be important, so Sheridan obtained the services of four signallers. Then, assuming the worst, medical facilities might be needed. The team was augmented by Surgeon Lieutenant Crispin Swinhoe, RN, with a Petty Officer and a Leading Medical Attendant. The force had a total strength of one hundred and thirty-two men.

On Thursday, 8 April, the entire Commando paraded at Bickleigh in full operational order for Major-General Jeremy Moore, the general officer commanding operational Commando Forces RM and himself a former commanding officer of 42 Commando. The event was given a pleasingly dramatic touch when Nick Vaux, at the end of the inspection, gave the order '42 Commando will march past in column of route. 42 Commando, to the South Atlantic, quick march!' The long column of weapon-laden troops swung confidently out of the gate and down the road where a fleet of buses was waiting to take them to Southampton. As they went, 'M' Company wheeled away and headed for the Gymnasium. Here, for the first time, Colonel Vaux announced that they would not be sailing in the *Canberra*, but would be operating independently, under Guy Sheridan. He gave no further details. Everyone was confined to barracks, no

Lieutenant Nick Vaux, 42 Commando's commanding officer, requests permission to march off the Bickleigh parade ground before giving the order which provides the title to this chapter.

outside telephone calls were permitted, and security became very tight indeed.

Sheridan's party travelled by road to RAF Brize Norton on Friday, 9 April, and departed that night aboard two well-laden VC10. The aircraft could not reach Ascension without landing to refuel. The Foreign Office had succeeded in arranging for staging calls to be made at Dakar, capital of Senegal. The marines travelled in uniform, their weapons packed in rolls and stacked in the aircraft cargo hold together with their Bergens and company stores. Sheridan and Nunn drowsed uneasily in their seats as the long hours slowly passed. Beneath their feet lay the tools of their profession, ahead were the imponderables of an undeclared war. It seemed a curious way to pass the greatest of the Christian festivals.

Arriving at Ascension at 0830 on Sunday, 10 April, Sheridan's party was welcomed by 'Jonathan's' small SBS command team which had reached the island four days earlier. After unloading the VC10, one of the first tasks was to re-zero 'M' Company's rifles. The sights had been knocked about during the transit south in the cargo hold. The problem had been foreseen by 'Jonathan' who had earlier organised a makeshift 30-metre rifle range for his own people. Testing the weapons occupied most of the morning, but soon the few available helicopters were lifting Sheridan's men out to three of the ships which constituted the Paraquat force.

There was one other military component to be integrated into Brian Young's task group – a Naval Gunfire Support (NGS) party. As the name implies, the NGS technique is one in which the Royal Navy uses its conventional guns against shore targets under the direction of specially trained personnel. In its simplest form, a ship's guns are directed against a land objective while an observer – flying overhead in a helicopter or light aircraft, or concealed ashore within sight of the objective – can call down fire at the most propitious moment and then correct the fall of shot.

Although NGS parties include a proportion of Royal Navy personnel, the bulk of their strength is drawn from the Royal Artillery. The application of shellfire against a land target is an art best understood by army gunners. Whether the shells are fired by a ship's gun, by a tank, or by a field gun, is a matter of relative unimportance. The end result is all that matters. However, the degree of skill needed to control naval gunfire is higher than any other because it is not only the target which may be moving – the ship itself is likely to be under way and therefore occupying a different position at the moment of firing each shot. Ensuring accuracy demands a very high standard of expertise by the Naval Gunfire Forward Observer (NGFO), spotting the fall of shot, and the Naval Gunfire Liaison Officer (NGLO), working alongside the ship's Surface Warfare Officer in each ship's operations room.

They have three prime responsibilities: to destroy the target with the most economical use of shot, to prevent shots falling on friendly forces already ashore, and to protect their own ships by keeping them close to shore no longer than necessary. A maximum efficient concentration of fire in the minimum period of time is the ultimate objective. It requires much intensive training to achieve the highest standards.

By 1982 the concept of NGS appeared to be – in some respects – an anachronism. It had reached its apogee in 1944 when the Allied navies blasted Hitler's Atlantic Wall along the coasts of Normandy and the Cherbourg Peninsula. After that war, the Royal Navy's new commitment within NATO was directed

Deeply alarmed at the prospect of full-scale war between two of her allies, the United States of America tried hard to act as broker between Prime Minister Thatcher and President Galtieri. In a dedicated personal search for compromise, the US Secretary of State, Alexander Haig, shuttled busily between Washington, London and Buenos Aires throughout the month of April. Here he poses with Mrs Thatcher on the evening of 8 April. At this stage he is still pursuing the 'even handed' approach to mediation which is in such sharp contrast to the 'open handed' American support on Ascension Island. Haig did not finally abandon his attempts at reconciliation until 28 April (by which time the British had displayed their resolve with the successful conclusion of Operation Paraquat).

primarily towards meeting the surface, sub-surface and aerial threats in the area of the Eastern Atlantic, North Sea and English Channel. True, there was a continuing potential 'out of area' need for NGS, and there was also a NATO responsibility for NGS deployments along the coast of Norway, but it seemed very unlikely that the Navy would ever again find itself supporting a full-scale invasion of a hostile shore. The construction of its ships reflected this political philosophy. Conventional guns had been eliminated entirely from the design of many of Great Britain's newest ships. The number of 'gunned' ships in service had been reduced – between 1960 and 1980 – from approximately one hundred and fifty to fifty. The Navy's plunge into high technology was leading to an almost total dependence upon rocket-propelled missiles. The tradition of NGS was gradually withering on the vine.

The last remaining vestige of a former glory was 148 (Meiktila) Commando Forward Observation Battery, Royal Artillery. Based at Poole, Dorset, the battery formed part of 29 Commando Regiment RA. This regiment consisted of three 105 mm gun batteries – 7 (Sphinx), 8 (Alma) and 79 (Kirkee) – each of which were permanently allocated to 40, 42 and 45 Commandos RM. The Meiktila battery possessed no guns. Its function was that of providing up to seven self-contained five-man spotting teams for allocation to the Commando batteries, to the SAS, to the SBS, or to individual ships, as required.

Changes in the Defence Budget had whittled down the strength of 148 Battery to the point at which it could raise no more than five NGFO teams. It had no regular NGLOs – these were part-time Territorial Officers (two of whom were professional politicians). Despite the long-standing threat of disbandment, all members of the fifty-five man unit were highly trained in mountain and arctic warfare techniques, the operation of small boats, insertion from submerged submarines, and descent by standard (PX) and steerable (SS2) parachutes. As one of them said: 'You never know where an NGFO team might be needed, so we were trained to get on target by any means available.'

Commanding the Battery was Major Mike Morgan. Given the worldwide scale of its activities – working in conjunction with the forces of many countries – the Battery's activities were co-ordinated by Lieutenant-Colonel Keith Eve, an officer with thirty-three years' service and the premier British exponent of NGS. He had a roving commission and rejoiced in the title of Royal Artillery Liaison Officer for Naval Gunfire Support (RALONGS). He was to play an unexpected role in the Paraquat story.

The Battery had returned at the end of March from three months of exercising in Norway. It was placing most of its equipment in store and preparing for the Easter holiday when Eve first became aware of the rapidly mounting crisis in the South Atlantic. He had one NGFO team (Captain Nigel Bedford and two other ranks) about to depart aboard HMS *Exeter* for Belize. She was due to conduct a routine NGS practice on the Seven Hills range. On his own initiative, Keith Eve arranged for a set of Falkland Island maps to be drawn from 8 Map & Stores Depôt, Royal Engineers, at Guildford. 'There was no real reason to suppose that we would be involved in any action in the South Atlantic, but it seemed prudent to think ahead, just in case the *Exeter* was diverted.'

Having closed down his office on the afternoon of 1 April, Eve said goodbye to Mike Morgan and motored north to his home at Orcheston St Mary, near Salisbury. At 0530 on the following morning his bedside telephone rang. It was Morgan. 'Have you heard the news?' 'No, not really.' 'Well, the Argentines have invaded the Falklands. Everyone is being recalled.'

Eve packed his 'ready to move' Bergen and by 0815 was back in his office. It was the beginning of five days of 'on-off' planning which was to culminate with a Lynx flight to Commando Forces HQ, Plymouth, to be briefed on 7 April by Major-General Jeremy Moore and his Chief of Staff, Colonel Richard Preston. Moore revealed the outline of the secret decision to send the Paraquat force to reoccupy South Georgia as a preamble to the major battle for the Falkland Islands. Keith Eve met Vaux, Sheridan and Nunn, and received details of the impending flight to Ascension Island.

Keith Eve takes up the story: 'I had a problem. The only NGFO team still immediately available was Captain Chris Brown's. Three others had been already earmarked for the Corporate force. The other NGFO, Captain Willie McCracken, had gone to New Zealand on leave, and was still making his way back. Furthermore, because this was not a full-scale war, there

In April and May, Wideawake Airfield, Ascension Island, became one of the busiest airports in the world. The civilian Pan American air control and ground staff, accustomed to dealing with a handful of air movements each week, suddenly found themselves handling dozens each day. Here a Fleet Air Arm Sea King prepares to deliver more stores to ships anchored offshore. In the background, a VC10 and a Hercules operated by the Royal Air Force.

Fleet Air Arm maintainers start to identify cases of parts and stores as they prepare to reassemble two Wessex just delivered at Wideawake Airfield by Belfast air freighter.

was no Queen's Proclamation to embody the Territorial force. Consequently I had no NGLOs. It was therefore a case of getting our act together as best we could. We needed two complete teams, so I decided not only to command the Paraquat party but also to act as one of the NGLOs. The other would be Battery Sergeant-Major Brum Richards. It was not really his job either, but he had years of experience and was first class in every way. Initially, before General Moore's briefing, we had no idea what we might be doing. However, we had assumed that we might have to join the Corporate force as it made its way down to Ascension. This would mean being flown down and making a parachute drop into the sea. So we practiced water drops and generally waterproofed everything in sight. It was something of an anticlimax to learn that we would be travelling by VC10.'

By 9 April the party had completed its preparations. Willie McCracken had returned from New Zealand and another man from Cyprus. Guildford had supplied their last remaining twenty sets of 1 : 100,000 scale maps of South Georgia. Full Arctic kit was packed and ready for air freighting. The other NGFO teams had departed for Portsmouth to board the Operation Corporate ships. Now it was Eve's turn.

With McCracken, Brown, Richards and eight other ranks and ratings, he took off from Brize Norton at midnight on 9 April. The other passengers were the balance of 'M' Company, 42 Commando. By the following morning they were being lifted by Wessex out to the waiting ships – Eve, McCracken and Brown with their teams to HMS *Antrim*, Brum Richards to HMS *Plymouth*.

Lieutenant-Colonel Eve, by virtue of his rank, was allocated the Admiral's cabin in the *Antrim*. He gently explained to the Navy that his army rank did not entitle him to this privilege which more properly belonged to Major Sheridan as Commander Land Force. With much good humour, the two officers exchanged accommodation. For the first time, the Navy realised that crowns and pips on a military epaulette do not always reflect the status of the wearer.

The ships assigned to Captain Young were, apart from his own, the elderly Type 12 frigate HMS *Plymouth* (Captain David Pentreath, RN), the big fleet tanker *Tidespring* (Captain Shane Redmond, RFA) and, of course, the ubiquitous *Endurance*. To some of those few people who were aware of the operation, it seemed a modest force for such an unprecedented venture.

Admiral Fieldhouse's resources were so limited at this stage, however, that he had little choice in the matter.

The Commander-in-Chief's initial plan for the reoccupation of the Falkland Islands depended upon attempting to draw the Argentine navy and air forces into battle – and defeating them – before the arrival of the main landing force. To do this, he ordered Admiral Woodward to despatch a spearhead battle group, led by Captain John Coward in the Type 22 destroyer HMS *Brilliant*, to place itself, with all despatch, in an area three days' sailing north of the Falkland Islands. Consequently, the British maritime forces were, at this stage, forming into three distinctly separate groupings: Coward's battle group headed for the Falklands, Young's small Paraquat force earmarked for South Georgia, and then the main body of the Task Force – the carriers, the assault ships, and the many Royal Fleet Auxiliary and STUFT (Ships Taken Up From Trade) ships – starting to load and make their way down from the United Kingdom to Ascension Island. Dozens of ships were strung out, in a very long and very thin line, over several thousand miles of seaway.

To counter the threat to the Paraquat force of attack by the *Belgrano* or any other long-range Argentine naval unit, the nuclear-powered submarine HMS *Conqueror* was ordered to divert towards South Georgia. Her captain, Commander Chris Wreford-Brown, was given a twofold task: to prevent any interference to Young's task group by hostile warships, and to prevent any reinforcement of the Argentine force already on the island. Travelling deep, at speeds in excess of twenty-five knots, he hoped to arrive off South Georgia on 18 or 19 April. His main preoccupation was the risk of encountering large icebergs after crossing the Antarctic Convergence. They would oblige him to reduce speed and, if met in large numbers, to frequently change course.

Brian Young's own immediate difficulties were almost entirely logistical. The *Antrim* and the *Plymouth* had been at sea for many days – taking part in Exercise Springtrain – and their on-board supplies were limited. Large quantities of stores had been taken from other Springtrain ships during the huge cross-decking operation on 2 April, but Young anticipated that he would need eight weeks' supplies to see him through the South Georgia operation and beyond. He must allow not only for the needs of his own ships' companies, but also for the military personnel who would be travelling south as passengers.

Captain Young could have wished for better circumstances under which to launch this most important enterprise of his service career. He was headed for an operation of considerable complexity, against a foe of unknown capability and resolve, on an island 4,000 miles distant, and commanding a small group of totally disparate ships. His scratch force had never previously worked as a team, none of his captains were trained in amphibious warfare and, as became apparent, not all of his military passengers would be embarked in the ships from which they subsequently were to operate. A great deal of cross-decking would be needed during the passage south. The indispensible Fleet Air Arm helicopters would be hard worked during the days ahead.

If there was one man in the Paraquat force who understood the principles and application of naval air power, it was Captain Brian Young. The oldest officer in the force, he had joined the Navy as a Cadet in 1944 and subsequently specialised in aviation. Having completed pilot training with the United States Navy at Pensacola and Corpus Christi in 1952–53, he flew Sea Hawk jet fighters with 803 and 804 Squadrons, Fleet Air Arm. In 1956 he flew in the ground attack role from HMS *Bulwark* against Egyptian military airfields around Cairo, and later still saw active service during the confrontation with Indonesia. After further service with Hunter, Scimitar and Sea Vixen squadrons, he commanded HMS *Wiston* and HMS *Danae* and held a number of senior staff appointments. The latter brought him into regular contact with Admiral Sir John Fieldhouse and the two men came to know each other well. It was possibly Fieldhouse's personal knowledge of Brian Young's experience and character which led to his selection as officer commanding the Paraquat force.

The *Antrim* and the *Plymouth* anchored off Georgetown, Ascension Island, on the morning of 10 April. The big RFA *Tidespring* arrived later the same day. The massive and highly efficient logistical facilities which were to evolve on the island during the coming weeks were not yet in evidence. The large quantities of cold weather clothing required by the ships' companies had not yet arrived by air from England, and would not do so for at least forty-eight hours. In the meantime, Young and his Commander, Angus Sandford, were involved in an intensive period of planning conferences. So many things needed to be extemporised, there and then, and with whatever resources happened to be available at the time.

A Magnavex satellite navigation installation was taken out of HMS *Sheffield* and fitted into the *Antrim*. This equipment was later to prove invaluable when persistent cloud in the South Atlantic prevented navigation by traditional means. The *Plymouth* soon completed the loading of stores and Captain David Pentreath moved to a position south-west of the island where he replenished his fuel stocks from the 40,000-ton RFA *Appleleaf*. The *Antrim* stayed close inshore, her flight deck constantly in use by a procession of helicopters.

On the island itself, 'B' and 'C' Flights, 845 Naval Air Squadron, were flying non-stop. The main task was to lift stores and tons of ammunition out to the *Tidespring* and the *Antrim* and, on the morning of 11 April, to complete lifting the bulk of Captain Chris Nunn's 'M' Company group to the *Tidespring*. Barges from Georgetown had been used initially, but with unsatisfactory results. To the *Antrim*, the helicopters delivered Major Sheridan's specialists (the mortar section and signals and medical units) and the naval gunfire support unit commanded by Keith Eve. Lastly, 'C' Flight embarked itself on the *Tidespring*, the ship from which it would be operating in the coming days. The entire loading programme on the tanker – co-ordinated by Chief Officer Alec Bilney and his Bosun, Chief Petty Officer Ron Kennedy – was completed in just twenty-four hours. It was a remarkable feat.

An account by the *Tidespring*'s Master, Captain Shane Redmond, reflects the great speed with which Operation Paraquat was being mounted: 'We had no advance planning figures whatsoever. Everything that arrived alongside or underneath a helicopter came as a surprise. It was all accepted and slotted in as it was revealed. Although the marines were an immense help in moving the stores around, they could not create storage space or magazines. Apparently even they didn't know what they were to get. By the time the magazines were full of ammunition, and after we had drained the swimming pool and turned that into a reserve magazine, we still had ammunition looking for a home. It just had to be stored wherever we could find a corner – all of this in a tanker with Class 'A' fuels aboard! There were stores and people everywhere. I soon realised that a major worry would be fresh water supplies. Our normal daily consumption was thirty-five tons so, when the troops joined, we could have easily trebled that if there had been no strict rationing. The operation depended as much on evaporators as it did on firepower.'

Late on the morning of 11 April, the *Antrim* and the *Tidespring* weighed anchor and proceeded to a position south-west of Ascension Island. The warship topped up her bunkers, but then

The Operation Corporate force assembles off Georgetown, Ascension Island. Included in this scene are the P&O liner *Canberra* and three of the Royal Fleet Auxiliary 'landing ships, logistic'. Ascension Island had been fortified and garrisoned by the British in the previous century, but never had it witnessed military activity on this scale.

returned to Georgetown to load the cold weather clothing being flown down from the United Kingdom. Unfortunately there was no such special clothing for the civilian crew of the *Tidespring* and this deficiency was to cause difficulties later. While HMS *Antrim* waited off Georgetown, the *Plymouth* and the *Tidespring* joined forces and started to move down track at an easy pace.

Captain Young was told that his command was to be reinforced by the bulk of 'D' Squadron SAS, currently coming northward towards Ascension Island aboard the *Fort Austin*. The Hercules bringing to the island the extra clothing would also deliver additional Arctic equipment, the property of the SAS. Over the following ten hours a series of flights touched down at Wideawake Airfield and their cargo was lifted swiftly to HMS *Antrim* by the two Wessex 5 of 'B' Flight. There was little sleep for anyone on board that night as more and more stores were packed into the hot and humid interior of the ship. The ship's company were much relieved when, at 0100 on 12 April, HMS *Antrim* departed Georgetown at twenty-two knots to catch up with the *Plymouth* and the *Tidespring*.

These two ships, meantime, were heading for a rendezvous with the *Fort Austin* and, in fact, all four ships met more or less simultaneously on the morning of 13 April. The *Antrim* and *Plymouth* received food and more ammunition from the *Fort Austin*'s by now almost exhausted stocks. The *Tidespring* also took on additional stores, the Royal Marines forming a human chain and handling forty tons of cases during the morning. At the same time, all helicopters were put to work lifting Major Delves and his Air Troop and Mobility Troop to the *Antrim*, and the Mountain Troop to the *Plymouth*. Accommodation in the two warships was now very crowded – not only by all the additional officers and men but also with their immense quantities of kit, weapons and ammunition. Space had been at a premium in the *Tidespring* since first embarking the 'M' Company marines. The heat of Ascension created problems with sanitation and hygiene which were reduced only when the ships moved south into cooler waters.

The transfer of men and equipment was a very large task with the limited number of helicopters and flight decks available, and was not completed until shortly after midnight on 14 April. Finally the job was done and the *Fort Austin* turned away northward to resume her journey to Ascension. She took no further part in Operation Paraquat.

The *Antrim*, *Plymouth* and *Tidespring* continued southward at their best speed and, on the morning of 14 April, they caught up with the *Endurance*. To find herself sailing in company with conventional warships was a novel experience for the *Endurance*. It was even more unusual to find that she had been elevated to the status of heroine. There was more than one damp eye as the *Antrim* and the *Plymouth* manned their sides and gave her three cheers, while the distant *Tidespring* flashed her own greetings. In the normal course of following her lonely business in southern waters, the *Endurance* very rarely had any contact with 'the grey funnel line'. As one of her ratings phrased it: 'they're in the Navy, we're in the plum.' But there was no concealing the sense of relief at this meeting with Captain Young's force. At last there was the prospect of supporting firepower and, most importantly, there was the opportunity to fill the near empty fuel bunkers.

The *Tidespring* was a highly specialised large fleet tanker, designed specifically for fuelling naval vessels at sea. With a full fuel cargo of 13,000 tons, her total displacement exceeded 27,000 tons. She could transfer fuel while under way, either by having her customer come alongside or by sailing ahead and then streaming a pipeline from her stern. This latter technique is unusual, but does not present any particular problems for ships which are equipped to handle the long snake of pipeline trailing back towards them across the surface of the sea. The *Endurance* had never had either the need or the opportunity to practice this evolution because she always operated alone, hence this particular replenishment was accompanied by some anxiety in the red ship.

Chief Officer Alec Bilney and Petty Officer Terry Simmons were flown over to the *Endurance* to explain the procedures for a replenishment at sea (RAS). They brought with them the adaptors necessary for the connection of *Tidespring*'s hoses with

In ideally calm weather, HMS *Endurance* takes aboard tons of badly needed diesel oil and aviation fuel from the RFA *Tidespring*. Insertion of the hose through an *Endurance* scupper was achieved by suspending a young seaman - held firmly by the ankles - over the ship's side.

'the plum's' fuel intake points. An initial trial with a three-inch hose - transferring aviation fuel - was completed successfully. It was followed by the six-inch diesel fuel hose. What might have been an exceptionally difficult RAS was accomplished with remarkably little fuss.

Every ten tons of fuel pumped over to 'the plum' represented another twenty-four hours of power for her big diesel engine and for her boilers, evaporators and three generators. It was fortunate that the weather on this day was kind. The two ungainly merchantmen remained coupled together for six hours, their disparate sizes causing the *Endurance* to surge ahead and astern and to yaw from side to side. Skilful work by the conning teams and the helmsmen enabled the deck parties to complete the task without a single break.

The four ships continued southward - in radio and radar silence - at the best speed of the slowest ship. This, inevitably, was the *Endurance*'s thirteen and a half knots. With her bright red hull and white painted upperworks, she was an incongruous stranger in this grey-painted company. Only the long sleek *Plymouth* - 2,800 tons - was smaller than her, and both were dwarfed by the massive form of the *Tidespring*. The *Endurance* was a vital member of the team, however, and for two reasons. Firstly, she was the only ship to be sailing towards familiar waters. Her ship's company knew the island of South Georgia and had experience of the navigational perils of the area. Secondly, she carried equipment and personnel which permitted her to listen to international radio frequencies which were not accessible or comprehensible to the other three vessels.

At the earliest opportunity, Brian Young consulted Nick Barker regarding the hazards which were likely to develop as the task group headed towards the Antarctic Convergence and the true Southern Ocean. These centred upon the unattractive mixture of fog, icebergs and severe gales which usually spring up with little warning.

The force was still 2,000 miles - seven days' sailing - from its destination, but there was an intimidating amount of work to be done in the interim. At first glance, the huge distance was an obstacle, but in some ways it was a positive advantage. On the one hand, there was the political and military desirability of recovering one of Britain's lost territories as quickly as possible. On the other, Task Group 317.9 was a scratch force - thrown together under a variety of conflicting pressures - and the component parts needed time to 'shake down' or, in the language of the day, 'get its act together'. The commander of this relatively small force had a considerable problem on his hands.

A continuing anxiety was fuel. The *Tidespring* had been assigned to Brian Young with the intention of giving him operational independence over a long period, but she carried the bulk of his military component and had therefore become, in effect, his main assault ship. She would be obliged to come close into the approaches to South Georgia. Proximity to a hostile shore might inhibit her freedom to operate in her primary refuelling role. She might be damaged or sunk. It would be advisable, therefore, to 'top up' all four ships before committing them to action.

This scenario had been anticipated at Northwood where the fuel status of every British naval and auxiliary ship around the world is monitored daily. In the Paraquat Cell, Commander Davis was conscious of the coincidental need to keep the *Tidespring* well laden, not only to meet the future fuel requirements of the Task Group but also to maintain her stability for the operation of helicopters. The lighter she became the greater would be the movement on her flight deck, and helicopters were a vital component in the plans to put troops ashore.

Coming west around the Cape of Good Hope was the large fleet tanker *Brambleleaf* (Captain Mike Farley, RFA). This unarmed ship had departed Portsmouth on 27 January with orders to take two cargoes of oil from Bahrain to Singapore. In the event, having made the first of these deliveries, she had been ordered to replace the RFA *Olna* as support tanker for three warships of the standing Gulf Patrol. All four ships were heading for Mombasa when, on 3 April, Captain Farley was told to detach and to head for the Cape of Good Hope pending further orders. On 11 April he hove to in heavy weather south of Port Elizabeth. Forty-eight hours later he was told to head for a rendezvous with a Task Group at a position of approximately 35°00′ South 30°00′ West. He was further instructed to proceed in radar and radio silence and to darken ship at night. Farley had already adopted these measures - after leaving Mombasa - and had painted over all bright colourings and identification marks.

Departing the Cape on 13 April, the *Brambleleaf* ploughed steadily westward - totally unaware of the destination or composition of the Task Group she was intended to supply but receiving numerous changes of rendezvous as Brian Young pushed ever southward. Young was told that his request for additional tanker support would be met as soon as the *Brambleleaf* joined him, the estimated date for the rendezvous being 21 April. The fuel problem, for the moment, seemed to have been solved.

Tidespring, the Royal Fleet Auxiliary tanker which unexpectedly found herself acting as troop transport and potential assault ship for Operation Paraquat. Her role, subsequent to the surrender of Argentine forces on South Georgia, on 25 April, proved to be even more surprising.

Another of Captain Young's difficulties was the fact that HMS *Antrim* lacked the radio equipment necessary for the tactical control of a combined amphibious landing. He and his operational staff would need to be able to talk with the other ships, with the helicopters, with covert shore patrols of SBS and SAS, with the naval gunfire support officers, and with the main Royal Marines assault force once it had landed.

This implied a multiplicity of circuits, far more than the relatively large number already available in a ship of HMS *Antrim*'s type. There was an accompanying need for a military operations room where the various services could work together as a team. Accordingly, the elegant furnishings were removed from the Admiral's day cabin and replaced by tables from the messdecks and map display boards fabricated by the shipwrights.

An extensive range of naval and army communications equipment was installed. A hole was cut in the deckhead to allow the various aerial feeders to be routed from the operations room via the bridge to a group of new HF aerials rigged between the bridge roof and the foremast. A VHF whip aerial was installed on the port bridge wing and a spare UHF discone aerial mounted on the deck abaft the bridge. These *ad hoc* arrangements, designed and organised by Lieutenant Kevin White and largely executed by one of the ship's UHF maintainers, Leading Radio Mechanic David Bullen, were to work very well when the time came for the landings. In the meanwhile, all ships continued to follow strict radio and radar silence. Captain Young was determined to conceal his approach until the last moment.

The helicopters carried by the four ships were continuously in the air, cross-decking ship to ship with personnel required for planning meetings and moving stores so that 'the right things would be in the right place at the right time'. The Wessex 3 operated by the *Antrim* was equipped with sonar, but this heavy equipment was removed in order to increase the load-lifting capability.

All helicopters were repainted in dark colours, the usual bright markings and insignia being obscured. In the *Antrim*, the maintainers experimented with a GPMG borrowed from the *Tidespring* and, by the judicious use of plywood and a home-made mounting bracket, succeeded in fitting the weapon so that it could be fired from the helicopter's rear cabin window. A similar 'locally manufactured' bracket was used to fit a borrowed GPMG to Wasp 445 on board HMS *Plymouth*.

It was assumed that the helicopters would be engaged not only in covert operations (inserting reconnaissance patrols) but also in searching for enemy surface units. A great deal of time was therefore invested in training the aircrew in the 'pop up' technique. Simply stated, this involved launching each helicopter at night, using only hand torches and without use of radio. The pilot flew out to a distance of fifteen nautical miles, at very low level, before climbing rapidly and then switching on his radar for a quick all-round surface search. Having made his climb, he stayed just high enough above the sea to permit the use of his UHF radio to report any sightings back to the mother ship. One of the main objectives of the technique was to conceal the location of the

Royal Marines of 'M' Company practice their abseiling techniques on the bridge superstructure of the RFA *Tidespring* while HMS *Endurance* refuels astern (the red ship's bows can be seen off *Tidespring*'s port quarter).

mother ship from any hostile units which might be in the area. It called for a high degree of skill, not only by the pilot and his observer but also by the flight deck officers and handling crews. Repeated practice flights produced a high level of competence and, despite the often bad weather, no accidents.

During the following days of the passage south, 'C' Flight, in the *Tidespring*, continued to train hard for the task ahead. Under the leadership of Lieutenant Mike Tidd, RN, the Flight was involved in round-the-clock operations. By day, they and the other helicopters of Task Group 317.9 took part in a constant round of cross-decking and stores deliveries. By night, they practiced formation flying, ship controlled approaches (SCA), and poor visibility approaches (PVA). Ships and aircraft adopted wartime 'darkened ship' routines, and the helicopters therefore flew formation on each other by having fixed to their rotor tips the radioactive Beta lights which gave pilots a guide by which to navigate.

It was found that the Wessex 5 could be fitted with up to three general purpose machine guns. The Flight had embarked with a full war scale of weaponry and ammunition. This included strap-on launchers for 2-inch rocket pods, each pod containing fourteen missiles. Live firing practice was conducted against 'splash' targets towed astern of the *Tidespring*. The ship did not carry this type of target as part of her normal stores, but her carpenter, John Hocking, obtained the correct MOD specification and built them with materials which happened to be at hand. His handiwork was well tested by the impressive firepower of the rocket salvoes.

Aircrew and maintainers practiced with rifles and 9mm Browning pistols against gash bags thrown over the ship's side. The Flight intended to prepare itself, in all respects, for a real shooting war. They had no indication at this stage of the frustrations which lay ahead.

As the four ships ploughed their way steadily down the South Atlantic, the infrastructure of an assault group gradually evolved at meeting after meeting in HMS *Antrim*'s new operations room. It is a long-established principle that all amphibious operations must fall into two distinctly separate phases. First, the senior naval officer must deliver the assaulting force to the shore where it has been ordered to operate. This officer remains in overall command of the combined force until such time as the troops are firmly established on shore. At that point, all military activities on land become the responsibility of the senior army or Royal Marines officer, while the senior naval officer provides logistical support and protection from seaward. This process was broadly understood by most of the officers engaged in Operation Paraquat and it should not have presented any problems, but Paraquat was no ordinary operation.

To comprehend the difficulties under which all parties were labouring, and to explain the decisions flowing from their planning conferences, it is appropriate to analyse the command structure.

Between 7 April and the morning of 3 May, the senior officer commanding the entire force – naval and military – was Captain Brian Young. His title was Commander Task Group (CTG). For all military advice he depended upon Major Guy Sheridan, the designated Commander Land Forces (CLF). It was Young's task to safely place on South Georgia the principal military force, 'M' Company, commanded by Captain Chris Nunn, RM. To facilitate this task, to help it achieve its objectives, he had a reconnaissance force of SAS and SBS. He had been given to understand, by Northwood, that these specialist units would consist of an SAS Troop, probably under the command of an army captain, and of 2 SBS, commanded by a lieutenant, Royal Marines. These thirty-odd men would act as the force's eyes and ears, gathering information in respect of the terrain and the enemy dispositions. Subject to their advices, Young and Sheridan would then decide upon the main landing area and 'M' Company would go ashore for the main assault.

Where the situation started to become less clear was in the sphere of deciding the precise details of the operation. Sheridan was a major, Royal Marines, a rank which equated to that of a lieutenant-colonel in the army. Lieutenant-Colonel Eve was therefore of equal rank to Sheridan, but Sheridan was the designated Commander Land Forces and, in any event, Eve would not be involved in the operation until such time as naval gunfire support might be required. Major Delves was junior to Sheridan, but he carried the aura of his own élite organisation.

Another important member of the planning group was 'Clive', a lieutenant, Royal Marines, representing the interests of the SBS and acting as liaison officer with the other services involved in the operation. 'Clive' had joined the marines as a boy, more than twenty years earlier, and had worked his way up through the ranks. Having spent almost his entire career with the SBS, he had more knowledge of clandestine amphibious reconnaissance work

Keeping their eye in. Royal Marines firing at gash bags thrown from the stern of the RFA *Tidespring*. Already tuned to a high state of readiness by their recent exercises in Norway and Africa, all embarked military personnel continue to train hard during the long voyage south.

than any other man in the Paraquat force. Initially he had believed – as had Sheridan – that the force would include only one Troop of SAS. It was slightly overwhelming to discover that they had been joined by the entire strength of 'D' Squadron. Despite the best endeavours of Delves and his second in command, 'Danny', it was inevitable that the reputation and sheer numbers of the SAS contingent should distort both the balance of the force and the weight of the opinions being voiced at the endless planning meetings. 'Clive' had often worked with the SAS and had great respect for their capabilities, but he felt that he could only offer general counsel and advice. It was for Delves to decide upon the best method of employing his Squadron which constituted nearly a half of the total number of troops available.

Captain Young was not accustomed to working with the military. As he said later: 'Apart from the Royal Marines, everyone was learning the amphibious warfare trade as we went along.' Many naval officers in his position might have been disconcerted to find themselves operating with such a wide diversity of strong characters from so many different branches. The situation might have been further complicated by the fact that each of the services involved in the operation was trained for a different role and to function in a very different style. For the Navy, command has always been an autocratic concept. There can only ever be one captain on the bridge. He will say 'do this' and it is done. He does not normally interfere with the manner of its doing. Officers of the army and Royal Marines will say 'we shall do this', and then give the opportunity for questions to be raised by junior officers and other ranks at the conclusion of the 'orders group'.

SAS and SBS 'orders groups', on the other hand, always act more in the manner of a council of war. The commanding officer will say 'this is what must be done, how are we to do it?' He calls his people together, places the problem before them and then invites each man to state his thoughts on the best way of achieving the objective. This system has much to recommend it, but is feasible only with very small units composed of highly trained troops accustomed to think for themselves.

It says a great deal for the equable temperament of Captain Young that he was able, in the space of only a few days and despite the pressures of overall command, to adjust to such a novel situation. Although initially bemused by the interplay between 'the Royals', 'the Gunners', 'the Squadron', and 'the Regiment', and although taxed by standards of superficial discipline alien to a naval officer, he left it to the military to work out the solutions to their mutual problems. Subsequently he received this accolade from someone who shared those days with him: 'Young was excellent, just the right man for the job. A rather quiet style, didn't interfere, simply trusted us to use our own training and experience to the best advantage. With such a small force, and a rather unbalanced force at that, things could have gone wrong if anyone in his position had thrown his weight around. He didn't play it that way. He simply asked what we wanted to do and then did his damnedest to make sure that it all worked.'

Basic ideas for the reoccupation were already taking shape when, on Thursday 15 April, a Nimrod aircraft of the Royal Air Force appeared overhead. It made three low passes over the ships, dropping a message container on the second, and saluting the force with a waggle of its wings before disappearing back towards Ascension Island after the third.

The recovered cannister contained Admiral Sir John Fieldhouse's formal operational order, dated 12 April, for Paraquat. The recipients were pleased to see that it was signed on behalf of the Commander-in-Chief by a Lieutenant Nelson. This was interpreted as a good omen. In essence, the bulky document told Brian Young that he must reoccupy South Georgia with minimum loss of life and with minimum damage to property. The latter consideration was based mainly upon the need to preserve all-weather accommodation for garrison troops over the impending winter. The initial landing target date of 21 April was reaffirmed.

Planning for the reoccupation of South Georgia commenced in earnest aboard HMS *Antrim* on 14 April. During the following days, numerous formal and informal meetings – 'brainstorming sessions' – were held in the makeshift operations room. It was quickly agreed that the forthcoming operation would be divided into two separate phases. Initially, the SBS would be put ashore from the *Endurance* to reconnoitre the Grytviken/King Edward Point area. At the same time, the SAS would make a reconnaissance of the Leith/Stromness/Husvik area. For communications purposes, the SBS insertion would be codenamed Floater and that of the SAS would be Foxhound. Both surveys would be conducted covertly and in such a way as to minimise the risk of alerting the Argentine garrison to the presence of a British task group in local waters. It was agreed that a period of five days would be required for the two scouting parties to make their landings, to make their approach marches to their objectives, and to then observe them for an adequate period of time.

The details of the second phase of the operation would be dictated by the intelligence gathered by the reconnaissance groups. In the event that either the SAS or SBS patrols hit trouble, elements of 'M' Company would go ashore to their assistance. They would be known as the Quick Reaction Force (QRF). If, on the other hand, all went well, the main landing would be made by the entire strength of 'M' Company with supporting elements, this to be known as the 'M' Company Group.

There were other options which could have been considered. For example, given the naval firepower at his command, Captain Young could have simply blasted the two Argentine bases into submission without first committing any troops to a landing. This possibility was never given serious consideration because it would have breached the spirit and the letter of the C-in-C's orders – 'minimum loss of life and minimum damage to property'.

Another option might have been to follow the earlier example of Captain Trombetta by using the radio to call upon the garrison to surrender without a shot being fired. This idea was not debated at any stage. Even if it had been put forward, it would have been rejected because the whole essence of the Paraquat operation was total secrecy. Any ultimatum delivered by Young would have immediately put his enemies on their guard. Further, it would have invited attack by Argentine submarines and long-range aircraft.

A third option could have been a direct assault – in the style of Iwo Jima or Normandy – frontally across the beaches at King Edward Point and at Leith Harbour. This, too, was never a viable proposition. Casualties on both sides would have been high and the Task Group ships might have suffered considerable damage.

Given the ambiguous political situation – with no declaration of war between the two countries and with both sides still engaged in various peace initiatives in New York – it was desirable that world opinion should remain basically sympathetic to the United Kingdom's cause. Any excessive use of force and spilling of blood could lead to a rapid swing in sentiment. At the same time, it was vital to demonstrate British resolve and British military capability. Failure on South Georgia would destroy credibility and would compromise the much larger operation for the recovery of the Falkland Islands.

All of those who took part in the *Antrim* meetings confirm that they were acutely aware of the importance of the operation upon which they were about to embark. One of them summarised their feelings: 'In purely military terms, it didn't look too bad. We believed that they had about a hundred soldiers or marines on the island. We had just over two hundred. Ideally one would like to have an advantage of three to one – especially for an amphibious operation – but we had to make the best use of what we had, and what we had was troops of the highest quality. The inhibiting factors were the island's very peculiar natural characteristics and the need to avoid excessive violence. We knew what we were up against when he hit our first heavy gale on the final leg of the approach.'

It was the opinion of some of the marines that the military competence of the Argentines was unlikely to be high, but that the physical obstacles to be overcome were by far the greatest problem. One of their officers has stated: 'We didn't know much about the opposition. There had never been any previous need to study the Argentine national character or the capabilities of their armed forces, but I was confident that, man for man, we were fitter and better trained.'

Delves and his people did not necessarily disagree with this broad assessment, but were prepared to give the Argentines more credit than was apparent among some of the Royal Marines and Royal Navy officers. To quote Major Delves: 'In retrospect, perhaps you could say that we gave them too much credit. At the time, we knew almost nothing of either their dispositions or their professional skills, but we had to assume that the Argentines would have used one of their better units to seize the island and to then defend it. We certainly planned on the basis that they were a force to be reckoned with.'

The Royal Marine Commandos normally spent three months each winter in Norway where they exercised in the amphibious and Mountain and Arctic Warfare roles. Guy Sheridan himself possessed as much knowledge of cold climate conditions as any other officer then serving in the British armed forces. This fact no doubt explained his selection as the Paraquat land forces commander. Forty years of age, he was extremely fit and had just completed his fourteenth winter in Norway. He had spent four seasons in the Himalayas and eight seasons in the Swiss Alps. An expert skier, he had represented Great Britain at the World Biathlon Championship in Finland in 1971, had been a member of the national team between 1969 and 1972, and, in 1981, had made the first traverse of the Indian Himalayas on skis (covering 420 miles in four weeks). Not only was he a trained Mountain Leader, he had himself commanded the Mountain and Arctic Warfare Cadre at Arbroath, teaching the same skills to other officers and marines.

It was not Sheridan's style to speak of his background and achievements. The SAS people were not initially aware of his qualifications and – until he was persuaded by Captain Chris Nunn to talk more freely on the subject – they accepted some of his views with a degree of scepticism. Thereafter his advices were weighed with much greater respect, but there was still a high degree of confidence in 'D' Squadron that they could deal with practically anything which South Georgia might throw at them. One of their corporals later described his feelings: 'Agreed, the Royal Marines were operating in their natural environment while we were not. But make no mistake – there was plenty of expertise in the Mountain Troop. Everyone was fully trained in rock-climbing and mountaineering, and we knew how to live above the snowline. Despite Ulster and so on, the skills had not gone rusty. We had only just finished three weeks of climbling in Africa – going up to 17,000 feet – so we were mountain fit. Also, besides having a Himalayan climber in the Troop, we could call on the advice of Brummie Stokes and Bronco Lane back at Hereford. They climbed Everest in 1976. All in all, we knew about mountains.'

Although Sheridan and Delves were able quickly to establish a good personal rapport, there was inevitably an element of restraint between them. Sheridan had not previously worked with the SAS. He had considerable operational experience, but that had been in Aden, Borneo, Malaya and Dhofar, ten and twenty years earlier. Delves, for his part, also had considerable operational experience, but that had been gained later in Dhofar, and in counter-terrorist operations, and never in a joint role with the Royal Marines. Each had respect for the professional skills of the other, but without fully understanding the limits within which they were able to work.

The nature of the dilemma was reflected in their later comments. According to Delves: 'Guy Sheridan was the Commander Land Force and I fully recognised him as such. At the same time, I was accustomed to working with a degree of latitude. I was the SAS adviser and I was also the SAS commander. We prefer to be given a task with stated operating parameters – we ask

HMS *Plymouth*, one of the work-horse frigates of the 1982 campaign. Following the successful conclusion of Operation Paraquat, she joined the Operation Corporate amphibious group and acted as escort during the San Carlos landings. She was seriously damaged by aerial attack on 8 June, but was once again operational by the end of the conflict. She fired a total of 909 rounds from her 4.5-inch turret during various bombardments and launched nine Seacat missiles.

then to be allowed to get on with it.' Sheridan understood Delves' independent background and deliberately refrained from imposing his own views. 'They had a job to do and I left it to them to find the best way of doing it. If they said they could do something, then I had to assume that they were competent to do it. I did feel that they were underestimating the hazards of operating in an environment such as South Georgia's, but, because so few of us ever knew what the SAS were accustomed to doing, I could not identify the point at which they might be overreaching themselves.'

The only senior officer with detailed knowledge of the island's formidable coastline was Captain Barker. He was able to involve himself in the planning process on several occasions during the voyage south. 'I was naturally keen to have a piece of the action. I had been trying to think of ways of getting the *Endurance* involved in the landings and I came up with the idea of using her to put 'M' Company ashore near Hope Point. The special forces were going to have their chaps up on the high ground above King Edward Cove and on either flank. Once they had located the defences, they would eliminate any sentries on Mount Duse and in the Bore Valley. I would then run the *Endurance*'s bows up against the beach or the cliffs where the scree comes down to Cumberland Bay, and the marines could step ashore more or less dryshod. It would have been hellish difficult for the troops because the scree there is very steep and unstable and it rises about 1,600 feet above sea-level. There would also have been difficulties with keeping the ship on station while we were disembarking. The prevailing wind along that stretch runs parallel with the shore, and this could have pushed the stern around so that she broached. Remember, she is a very underpowered merchant ship with no stern anchor. We thought of all sorts of ways to keep her steady – laying out a kedge, or using the ship's boats as tugs – and decided that it could be made to work.'

Initially this option was sanctioned by Young and Sheridan, but further discussions led to the decision to adopt it only in the event that all else failed. The proposal assumed that the vulnerable ship could be brought very close to the Argentine base without provoking a hostile response. As one officer put it: 'It might have worked, but I remembered the story of the SS *Clyde* at Cape Helles during the 1915 Gallipoli landings. 'M' Company would have been massacred if they had been caught while disembarking and trying to climb that scree cliff.'

Further discussions concentrated upon approaches which were far more cautious and which reduced to a minimum the risk of incurring casualties. They also gave greater emphasis to the need to co-ordinate the initial insertions so that King Edward Cove and Leith Harbour could be reconnoitred simultaneously. Guy Sheridan's primary requirement was information regarding Argentine strength and dispositions. This could be obtained only by direct observation from the high ground overlooking the target areas. Once he knew what the enemy were doing, he could decide the second phase of the operation – the main landing by 'M'

Two views of one of South Georgia's many glaciers, the Brögger, located in the southern part of the Allardyce Range. The surfaces of many other glaciers are similarly fissured, presenting an almost impossible barrier for heavily laden troops trying to avoid detection. The dome of the Fortuna Glacier, by contrast, is far less daunting (*vide* Appendix D).

Company. It was the first phase which was causing all the debate – how best to put the reconnaissance parties ashore without them being detected?

The island's severe terrain and jagged coastline posed difficult problems for the planners. Both of the target areas were sited deep within mountain-ringed bodies of water. There were very few beaches where small craft could safely land troops, and the defending force, if professionally competent, would be keeping constant watch on the seaward approaches. These constraints obliged the planning team to look at landing sites which were many miles from the targets.

In the case of the SBS – and their assigned task of observing Grytviken and King Edward Point – there was only one viable route. The *Endurance* would land them at Hound Bay and they would make a waterborne approach in their Gemini inflatables. In the case of the SAS – tasked to survey Leith, Stromness and Husvik – there were more options, but each was fraught with risk. After numerous 'brainstorming sessions', it was agreed that the SAS Mountain Troop would be landed by helicopter on the Fortuna Glacier. In view of what was to follow, it is important to appreciate the factors which led to this controversial decision.

Some of the influences at work in the planning process are described by Major Delves: 'The need to thoroughly reconnoitre a target before striking it is well proven. In the case of Leith Harbour, it meant getting in close and finding out as much as possible of the enemy's strength, weapons, dispositions, morale, and so forth. The purpose of this was to ensure that "M" Company could make their assault with maximum advantage – for their attack to be swift and, if possible, bloodless. We had been ordered to use minimum force for the reoccupation. "M" Company could achieve this only if the Argentines were taken totally by surprise. Hence my insertion needed to be made in such a way that it could not be detected. There were other considerations which also coloured our thinking. This would be Britain's first military act in the retaking of her lost possessions, and it had to go well. Anything less could have jeopardised the diplomatic efforts then under way. So that was the background – we were determined to obtain the required intelligence, but without incurring any risk of compromising the final outcome. That brought us to the question of how to get ashore and where. There were two choices – by Gemini or by helicopter. The idea of using the boats was not attractive for three main reasons: one, our engines were not good; two, we had been warned what the local weather could do; three, we assumed that the Argentine defences around Leith would be oriented towards the sea. Other likely approaches along the nearby coast might be covered by observation posts. Certainly one could expect some local patrol activity. From the enemy's viewpoint, a landward approach from the rear or in flank would seem the less likely and hence the least guarded. Consideration turned more and more to a helicopter insertion followed by an overland march. To be perfectly honest, we are soldiers and, while not underestimating the difficulties, we all of us felt that, once ashore and back in our own environment, we would make a go of things. The problem then distilled into one of route selection.'

A great deal of time was invested in choosing the landing site during the conferences in the *Antrim*. Many opinions were invited, and they varied widely. Given their knowledge of local conditions, the officers of HMS *Endurance* were consulted as to the availability of alternative sites along the coast stretching north-east of Stromness Bay. In particular, they were asked how far the noise of helicopter engines and rotors was likely to travel. Unanimous in advising against a landing on the Fortuna Glacier, they felt that the ambient noise of the island – wind, bird cries, barking seals and the crash of calving icebergs – would mask the helicopters and enable them to approach much closer to the target. They advised against any attempt to move laden troops

across any of the glaciers because the extreme conditions – combined with the unpredictability of South Georgia's weather – loaded the dice against success. Similar advice came from Dr Laws in Cambridge when he was asked to comment upon the feasibility of traversing the Fortuna. Although not told the reason for the enquiry, he guessed what was being planned and advised against it. Even his own glaciologists, experts in their field, were subject to tight safety rules when venturing onto some of the island's glaciers. Guy Sheridan, too, in the light of his Himalayan experiences, was unhappy about the choice of the Fortuna, but he preferred to counsel Delves rather than deny authority for the operation.

In the face of the adverse advice, Delves continued to look hard at the alternatives. He discounted Hercules Bay and Fortuna Bay because they were overlooked by high ground where the enemy could be expected to have placed one or more observation posts. They were also uncomfortably close to Leith Harbour – the proximity carried the risk of helicopter noise being heard by the garrison. Captain Young was himself an aviator and he could well understand Delves' preoccupation on this count. He guessed that the sound of a helicopter might carry for long distances over such mountainous terrain. 'On a windy day on South Georgia you can hear almost nothing but, on a still day, sound does carry a very long way.'

The officer commanding the Mountain Troop was Captain John Hamilton of the Green Howards. He had joined the SAS only recently, but already had earned a reputation as a courageous and hard-driving individual. He and his Troop were taking passage in HMS *Endurance*, so they had ample opportunity to learn about the island's notorious weather and topography. They studied the detailed maps held in the ship's Hydrographic Department and the private photograph albums owned by many of her officers and ratings. Hamilton was warned of the crevasses to be found on the Fortuna Glacier. The question of crevasses was central to the debate regarding the practicality of marching across the great high dome of this very large glacier. Where its five arms flow down to the sea, it is indeed veined with hundreds of deep fissures and seracs (pressure ridges). The top of the glacier, however, is comparatively level. Here the ice moves but slowly, its great weight pressing downwards rather than sideways. The surface is certainly crevassed, but these fissures are not very deep and are only two or three feet in width. A laden man, falling into such a crevass, is unlikely to drop more than waist-deep because his rucksack or Bergen acts as a wedge. It is then the work of only a few minutes for his companions to haul him back onto the level surface.

Delves contacted Hereford and asked for advice from the Everest climbers, Brummie Stokes and Bronco Lane. They stated that the glacier should not be an impossible obstacle. Contact was made also with John Peacock who, as a Captain, Royal Electrical & Mechanical Engineers, had been a member of the Combined Services Expedition team which, in 1964, had successfully followed Ernest Shackleton's route across the glacier. 'The Fortuna is a goer,' was his response.

Cedric Delves listened to all the divergent views and, in the customary SAS style, discussed them with Hamilton and the troopers who would be making the reconnaissance. As he approached his final decision, he again consulted Guy Sheridan and Brian Young, both of whom indicated that they would support whatever course of action he might settle upon. 'Major Sheridan's written order to me was clear and straightforward. I was told to "determine Argentine strength and dispositions in Husvik, Stromness and Leith." How I achieved that aim was down to me. At the end of the day, somebody had to make a decision, and it was my decision to go for the Fortuna. Given the same circumstances today, I would not alter my plan.'

The saga of Ernest Shackleton had great influence upon the judgement of Hamilton and Delves. 'Even though he was a sick and starving man when he made his transit, he had been able to walk right across the island – and go over the Fortuna – all in thirty-six hours. It seemed to us obvious that, if he could do the whole thing in that time and his condition, then fit trained troops ought to be able to complete the final leg without too much difficulty. Shackleton must have covered thirty or forty kilometres, but we would be only ten from our target. Allowing for the usual detours, we might need to travel twice that distance, but only the first two or three kilometres would be on the glacier. After that it would be firm ground. We had allowed five days for the recce and I was prepared to allow the entire first day for crossing the glacier if necessary. What I did not know, of course, was that Shackleton had been very lucky with his weather. As things turned out, we were not.'

The decision to land on the glacier by helicopter was made with the full agreement of the aviators concerned. They felt that there should be no difficulty in lifting the patrol up to the 2,000 feet contour provided that the weather was reasonably favourable. However, it would need to be a daylight operation. Only one of the aircraft was fitted with a radar, and the pilots did not have night vision goggles. There would be no ground party to guide them in with smoke or illumination, so a night landing would have been excessively hazardous.

In anticipation of the landing, Delves arranged for his Air Troop, which had been travelling aboard HMS *Antrim*, to be exchanged for the Mountain Troop which, until then, had been accommodated first in the *Plymouth* and then in the *Endurance*. The Air Troop, specialising in free-fall parachuting, were feeling frustrated by the lack of any apparent opportunity to exercise their special talents. Although it was never seriously contemplated that anyone should be parachuted onto the island – the local conditions making such a venture almost suicidal – they spent much time talking to the *Endurance*'s Meteorological Officer, Lieutenant-Commander Andrew Lockett, pleading for a weather forecast which might just conceivably allow them to make a covert descent onto South Georgia from a high flying helicopter. Lockett, understandably, was not to be drawn. The Air Troop's frustrations were tempered by the discovery that the *Endurance* offered better food than the *Antrim* and carried a larger stock of beer.

At Northwood there were weightier topics to occupy the mind of Commander Davis in the Paraquat Cell. His initial task, between 10 and 12 April, had been to draft a detailed operational order for discussion and approval by Admiral Fieldhouse and the Chiefs of Staff. It was this document which had been flown south to Ascension Island and then dropped to HMS *Antrim* on 15 April. It was too long and too sensitive to be sent by radio. In Dennis Davis' words: 'The initial pressures related directly to the maximum operational range of a Nimrod aircraft. We had to get the document down to Ascension before Captain Young had sailed so far south that a Nimrod couldn't reach him. If the planning and drafting process had been slow, we would have had to either recall him or keep him cooling his heels on the outer radius of the aircraft's range. There were two reasons why we did not want him to be delayed. Firstly, and most importantly, there was the political desirability of a quick result. Secondly, the longer he was in the vicinity of Ascension Island the greater was the risk that his ships would be seen and identified. He had already been overflown once by a Soviet aircraft – on 8 April, while on passage down to Ascension from Gibraltar – and we didn't want that to happen again'.

For several years there had been close ties between the government of Angola (formerly Portuguese West Africa) and the governments of Cuba and Russia. Part of the military accord between them specified that Russian long-range Bear reconnaissance aircraft could use the airport at the Angolan capital of Luanda as an operational base. A pattern had developed whereby very long-range flights were made out into the Central Atlantic to survey maritime traffic on the north–south sealanes. The Russians were interested also in American activities on Ascension Island. When the British government announced the despatch of a Task Force to back its diplomatic effort in resolving the Falkland Islands conflict, the Bear flights (deployed initially from North Russia) became much more frequent. By the time the vanguard of Royal Navy ships reached the island, sorties were being flown from Luanda every day by these enormous aircraft. They were circling the island two or three times, observing and eavesdropping, before either returning to Angola or flying north-west to Cuba.

A Soviet Bear D aircraft, intercepted by a Sea Harrier of 800 Naval Air Squadron, Fleet Air Arm, and photographed in 1983. The deceptive camera angle distorts the comparative sizes of the two aircraft. The Bear is nearly thirty times larger than the Sea Harrier.

The name Bear was a NATO identification code. More correctly, they should be described as the Tupolev Tu-95 and, in its maritime reconnaisance form, the Tupolev Tu-142. By any standard, they were impressive aeroplanes. With a crew of ten, they had a maximum all-up weight of 167 tons and could fly 10,000 miles non-stop. The version operating from Luanda was thought to be equipped specifically to search for and identify naval vessels. The Bear carried onboard facilities for recording and analysing the electronic impulses emitted by various types of search and missile guidance radar system. This and other categories of electronic intelligence would have been invaluable to any hostile forces planning an attack against the Paraquat ships.

There was no evidence in London of direct collusion between the Russians and the Argentines. The flights were thought to reflect little more than the customary Russian interest in all and any activity by the forces of NATO members. However, it was believed that the Russian Embassy in Buenos Aires was engaged in regular and friendly contact with the Galtieri regime and it was reasonable to suppose that Moscow would be happy to pass on any snippits of intelligence which they judged to be useful. Although Russian national interests were not involved in the Anglo-Argentine dispute, they would clearly prefer to see the United Kingdom – as a pillar of NATO – humiliated rather than victorious.

The Bear flights were particularly relevant because the USSR did not at that time possess a 'spy' satellite capable of maintaining a constant watch on the South Atlantic. It had not occurred to them – any more than it had occurred to the British and the Americans – that this vast stretch of ocean might become the scene of intense naval activity. By the same token, it was thought that the Russians did not have any of their submarines or naval surface ships deployed in the area. Their closest naval units – a Krivak class destroyer and an amphibious warfare group – were based at Conakry, the port of Guinea, West Africa, but that was a long way from Ascension Island. It seemed improbable that a Krivak destroyer would be sent down to below the latitude of Ascension Island when there was so much for it to observe further north. The Krivak did not in the event venture much further from Conakry than its customary patrol area. In short, if the Argentines were to receive any assistance from Moscow, it could be only as a result of visual and electronic surveillance by the Bears.

It was to counter this threat of electronic detection that Captain Young's ships had maintained strict radio and radar silence since leaving Georgetown. All radio traffic to and from Northwood was being conducted via the satellite communication system which, at that time, was considered to be relatively immune to unfriendly monitoring. The other threat – visual observation – could be countered only by removing the ships from the area of Ascension

Island as quickly as possible. According to Dennis Davis: 'On only one occasion did we have a scare. That was on the second day after Young's departure when a Bear crossed his wake. Our calculations showed that it failed to detect the Paraquat ships. At Northwood, the few people in the know were cock-a-hoop. Young was away and running, and the other side didn't know. It was a perfect start to the operation. We had done everything possible – with the resources available – to ensure that he could do his job in the allotted timescale. We were there also to back him up as best we could once it all started to happen. But after 15 April, when he got his detailed orders, it was very much his show. He was the man in charge and it was his on-the-spot judgement which counted.'

Captain Young still had the fuel problem uppermost in his mind. He wanted to rendezvous with the RFA *Brambleleaf* in order to replenish the *Antrim*, *Plymouth* and *Endurance*, and also to pump over an additional reserve to the RFA *Tidespring*. Then, if the *Brambleleaf* was ordered away to some other part of the ocean, at least he would have the *Tidespring* well gorged to again replenish his entire force. Calculating the present speeds, headings and positions of the *Brambleleaf* and the Paraquat group, Young realised that the anticipated rendezvous – one hundred miles north-east of the island – could not take place before the afternoon of 21 April. Allowing for the time needed to conduct a pumpover, plus three ship replenishments, and allowing also for possible interference by the weather or by hostile action, Young decided that his first insertions of the SAS and SBS could not occur before the morning of 23 April.

At first, Northwood seemed to be content with this plan, but Captain Young then received another call from Northwood, telling him that – because of political requirements – it was essential to stick to the 21 April deadline. The only way to do this was by changing course and, accepting the increased risk of discovery by long-range Argentine aerial reconnaissance, to head directly for South Georgia. Even with this short-cut, Young was obliged to change his plans by deferring the replenishment until 22 April. In other words, he now intended to commence the landings *before* having topped up his ships. This was the only occasion when pressure from Northwood caused the Task Group Commander to amend his own judgement.

Each ship was the scene of intense activity as the group ploughed steadily southward. By day, the Wessex, Lynx and Wasp aircrew were engaged in a busy programme of surveillance and transferring men and stores from ship to ship. By night, they continued to train hard for the impending landings. Even at this early stage, pilots were feeling the effects of inadequate relaxation and sleep. The maintainers were hard pressed to keep each aircraft fully operational and it was to their great credit that none of the helicopters became unserviceable during this period.

On 16 April, the first Wandering and Black-browed Albatrosses joined the force, tracking effortlessly back and forth across the wakes of the advancing ships. It was becoming cooler by the day. Latitude 40° South was passed on 18 April. The force was now further south than the Cape of Good Hope.

All ships were hove to, on 17 April, to permit the SAS and SBS to again practice launching and recovering their Gemini inflatable boats. Drills were carried out with the ships' boats so that seamen and soldiers would be familiar with each other's ways if and when a seaborne landing took place.

These were busy days also for Lieutenant-Colonel Eve and his

NGS party. As early as 12 April he had arranged the first 'dry' practice shoots by the gunnery officers and the 4.5-inch gun teams in the *Antrim* and *Plymouth*, using the technique of indirect control. There had also been a 'live' practice, with twelve salvoes being fired at smoke targets under direct control. In his own words: 'It was fortunate that we had only recently – in February – worked with *Antrim* on her annual practice shoot at St Alban's Head, between Poole and Portland. The ship was well worked up. Also, she had 800 rounds of 4.5 in her magazines, all fused for naval gunfire support. From my point of view, she and *Plymouth* were the right ships for the job. Both carried the twin Mark 6 turret. This was an old but well-proven system, semi-automatic, relatively free of recurring faults. The rate of fire was sixteen rounds per barrel per minute. The weight of the shell was fifty-five pounds. With two barrels, this meant that each ship could hit the target with three-quarters of a ton of high explosive every sixty seconds – the equivalent delivery rate of a complete battery of 105 mm light field guns. The maximum range was 18,000 yards and – properly handled – it was extremely accurate.'

Keith Eve spent much time with Sheridan and Delves, preparing the maps which would be common to all those involved in calling down fire on shore targets. The three most likely targets were the areas of Leith Harbour, King Edward Point and Grytviken. The fact that each place was surrounded by very high ground placed considerable restrictions upon the available lines of fire. In gunner's language, there was a severe 'crest clearance' problem.

Eve made several visits to the *Plymouth* to talk with the other NGLO, Brum Richards. There were long and productive meetings with the ship's commanding officer, Captain David Pentreath, and with his Gunnery Officer, Lieutenant-Commander David Sayer. Eve formed a very favourable impression of the *Plymouth*: 'She was a happy ship. Everyone seemed to know exactly what they were doing and what was going on. Pentreath was good at keeping his people in the picture. It helped a lot. Further, the *Plymouth* was under no illusions – she knew she was going to war. Her wardroom was stripped out, all loose gear was either stashed or ditched, her damage control and medical parties were well drilled, and she was in every respect up on her toes.'

In the *Antrim's* operations room, Eve was involved in several practice exercises designed to test the ability of the command team to deal with future operations involving the employment of naval gunfire. There were minor misunderstandings and disagreements regarding the chain of command, but these were resolved with time. The 'instant staff' created for Operation Paraquat made itself into a cohesive team remarkably quickly and with few apparent tensions.

Sustaining morale in the Paraquat force was not a significant problem for Captain Young. 'One or two of the older senior rates – "the old and the bold" – were not entirely enthused by the prospect of shot and shell whizzing in their direction. They seemed to feel that this was not exactly what they had joined the service for. I had my say on the subject, and they soon sorted themselves out. But the overwhelming feeling among the whole ship's company was that they had a job to do and they wanted to get on with it. There was a strong current of professional pride at all levels. People had trained hard for many years on endless peace-time exercises, and here was a chance, perhaps the only chance of their entire careers, to put that training to the test in a real shooting-war situation. No, morale was not a problem.'

The Royal Marines were well accustomed to the problems of living at sea and at the same time keeping themselves physically fit and mentally alert. It has been recognised, since the very earliest days of amphibious warfare, that troops transported into action by ship will deteriorate rapidly in health and morale unless active steps are taken to keep them on their toes. In *Tidespring*, Chris Nunn had drawn up a busy training schedule with Lieutenants Ian Grant, Andrew Milne, Ewen Duncan and Daniel Downes, and with Company Sergeant-Major Paul Juleff, in preparation for 'M' Company's expected commitment to action. Starting on 12 April, the marines went through all the basic drills – individual weapon training, fire control orders, fire and movement, skirmishing, abseiling, first aid, Gemini embarkation drills, and so on. Particular attention was given to methods of searching and securing prisoners of war. Everyone was reminded that the Argentines had treated their captives at Port Stanley and King Edward Point in a correct manner, and the same policy must be followed in dealing with any prisoners resulting from the impending Paraquat landings.

In the *Endurance*, 2 SBS were similarly engaged in final training and planning. The sailors watched in silent awe as they and John Hamilton's SAS troopers performed prodigious feats of physical training. The apparently casual discipline of the passengers was not the only hallmark which set them apart from the ship's company. There was a polite silence, an inner resolution, which commanded respect. As one matelot observed: 'They weren't big roughie-toughies, they didn't look different to the ordinary bloke in the street. But there was something there. You know?'

The wealth of personal experience to be found in the *Endurance* was made available to the other ships by frequent visits and briefings. The Wasp Flight Commander, Tony Ellerbeck, accompanied by the ship's senior hydrographer, David Ives, gave a series of lectures to the troops in each ship, describing the island on which they were due to land and illustrating with colour slide projections its topography and buildings. Having been captured at Port Stanley and subsequently released, having therefore seen the Argentine army at close quarters, the hydrographers commanded particular attention. Detailed large-scale maps of the various whaling stations were made available to the SAS and SBS so that they could work out lines of approach, hiding places for observation posts, and other matters of fine detail which would enable them to gather intelligence without being seen.

On Sunday, 18 April, a final formal planning meeting took place in the makeshift operations room in the *Antrim*. 'Roger', commanding 2 SBS, presented his plan for the insertion in the south, at Hound Bay. Major Delves confirmed his intention of putting his Mountain Troop onto the Fortuna Glacier. Major Sheridan approved the plans drawn up by Captain Chris Nunn for the main assault force. In essence, this would be the bulk of 'M' Company, with attached mortars, medical team, signallers and land forces headquarters. It would remain aboard the *Tidespring* until such time as the two covert reconnaissance patrols – SAS and SBS – reported that they were in contact with the enemy or had at least determined the Argentine strength and dispositions. Everyone agreed that this would not happen before 26 April but, in the main, flexibility was the overriding requirement. To quote Guy Sheridan: 'It was impossible to know exactly how the land operation would develop, but I felt that, once it all started, speed and boldness would be the keys to success.'

As the Paraquat force sailed ever closer to South Georgia, so did the military components become increasingly confident that they were ready. For Brian Young, however, the pressures and problems escalated even further. He was fairly certain that his movements had not been discovered by the Argentines and that he had preserved, thus far at least, the element of surprise. However, the closer he came to the island the greater would become the submarine risk and the probability of discovery from the air. There were the continuing worries of the sinking levels of fuel in the bunkers of his group's ships and the increasing iceberg collision risk. To crown all, there was the uncertainty of the local weather for the key period 21 to 26 April.

Despite the volume of meteorological information being reported by the BAS and other weather stations around Antarctica and the southern hemisphere, all of which was channelled through to the United Kingdom weather centre at Bracknell, Berkshire, the Paraquat force was operating in a near void for much of the time.

The *Endurance* was equipped with a receiver which permitted her to obtain a regular flow of good quality visual and infra-red pictures from the polar orbiting weather satellites. She also carried equipment for the reception of surface synoptic data, but this receiver was prone to atmospheric interference and, during the final critical stages of the operation, it became necessary to use it for other communications purposes.

The *Antrim* shared the task of meteorological forecasting. Northwood, as the Fleet Weather Centre, was able to supply a limited amount of information from the geostationary satellites, and the Argentine weather service was still broadcasting a restricted volume of data from Buenos Aires. Satellite information served to identify approaching fronts, but much depended upon the intuitive sense of the man on the spot and the traditional householder's method of 'tapping the barometer and looking out of the window'.

The climate of South Georgia is notoriously fickle – 'the only place in the world where you can have all seasons in one day' – and any interpretation of local forecasts must be based upon a knowledge of the island's peculiarities. No matter how extensive or sophisticated the equipment available, totally accurate predictions can never be possible in such a strange place as this. Given Brian Young's considerable dependence upon helicopters and small boats as the principal means of disembarking two hundred or more fully-armed troops, forecasts based upon such sparse information did nothing to reduce his anxieties.

Some of Captain Young's preoccupations were about to be resolved, at least in part, with the arrival in South Georgia waters of HMS *Conqueror*. The submarine had overhauled the Paraquat surface force two days earlier, on 17 April. Her passage between the two hemispheres had been fairly uneventful. The only incident for the submariners and their passengers had been an encounter with an undersea mountain. While travelling at high speed, the echo-sounder unexpectedly showed that the ocean bed was rising rapidly towards the *Conqueror*'s keel. For the men of 6SBS, the voyage was otherwise crushingly boring. Each morning was devoted to lectures in tactics, first aid, weaponry, and communications, and they had access to the boat's gymnasium 'bicycle', but they had little to occupy either the mind or the body. They had no news of world events and the only confirmation of having crossed the Equator was the rising temperature inside the hull. Later it became much cooler as the submarine passed through the Antarctic Convergence.

The men of 6SBS had spent the previous six months in Norway and were accustomed to the wide open spaces of Scandinavia. After only one week of leave, they had been recalled and incarcerated in this submarine. They tried hard to keep fit – using the deck-head pipework for their 'pull-ups' and helping the sailors to load and unload the torpedo tubes – but they could sense that they were becoming less fit by the day. The situation was aggravated by the fact that one of their number was suffering from influenza when he first embarked. As a consequence, almost everyone in the boat caught the virus.

Commander Wreford-Brown arrived off South Georgia on 19 April, having slowed considerably during the final hours. He, too, had been warned of the possible presence of ice in the area. To avoid detection, his sonar had been turned off for most of the journey south but, during the final twenty-four hours, his operators made intermittent searches for icebergs. None was encountered, but the sensors detected the underwater roaring and hissing of the Fortuna Glacier at a great distance.

On the morning of 19 April, Wreford-Brown commenced a cautious submerged approach to the entrance to the Cumberland Bays. The weather was poor and, even in the lee of the headlands, the 4,000-ton hull was rocked by the action of the waves. His brief scans shoreward revealed nothing – a dense fog had reduced visibility to less than a mile. The only sign of life was a family of fur seals following playfully in the wake of the periscope's exposed tip. Then, at midday, the fog suddenly lifted to reveal, five miles away, South Georgia's amazing snow-capped mountains rising almost vertically out of the sea. It was Wreford-Brown's first sight of land since departing Faslane, two weeks earlier.

The *Conqueror* cruised closer inshore and reconnoitred Cumberland East Bay, Stromness Harbour and Leith Harbour. There were no signs of activity around the buildings of Leith and the anchorages were devoid of shipping. Wreford-Brown withdrew and moved to a position fifteen nautical miles north-west of the island's northern tip. His report to Northwood, re-transmitted to Captain Young, made it clear that the Paraquat force was unlikely to find any hostile ships in the immediate vicinity of the target area.

Shortly afterwards the submarine was obliged to surface to repair her storm-damaged radio antennae. She then began patrolling the island's northern approaches to detect any attempted reinforcement of the Argentine garrison. The next three days passed uneventfully, the only moment of excitement being a sonar contact on 21 April. The boat went to action stations and approached the target, ready to react. It was difficult to identify and Wreford-Brown decided that the target was biological, probably a large whale. For the rest, the hours slipped by with nothing to disturb the boat's routine. With no specific duties to perform, the 6SBS marines tried to fight off their increasing boredom. One of them started to read the same long novel for the third time. Wherever the Argentine navy might be, it was certainly not in this area.

There was still the question of the enormous stretch of ocean further north of the island. Were there any icebergs up there, a terrible menace to darkened ships, sailing in company at night, in total radar silence? And were there any Argentine warships, waiting to intercept a reoccupation force such as the *Antrim* group? It was almost inconceivable that Galtieri and Anaya should have placed their flag on the island, boasting to the world that they had now exercised their claim to sovereignty, without at the same time taking steps to protect the small garrison.

The build-up of facilities on Ascension Island, five thousand miles to the north, had continued at a frantic pace since the *Antrim* departed Georgetown seven days earlier. Among the many aircraft now crowding the tarmac at Wideawake Airfield were several Handley Page Victor K.1A in-flight refuelling tankers of No 55 and No 57 Squadrons, Royal Air Force. These four-engined 90-ton former bombers were, in some cases, almost as old as the five-man crews who flew them, but their performance, in 1982, was still impressive. Designed to carry heavy loads over great distances at sub-sonic speeds, they were the only aircraft which could reach South Georgia and provide Brian Young with the information he needed.

A Victor K2 of 55 Squadron RAF, showing the refuelling drogues extended from the underside of each wing. Apart from the mission flown on 20 April, Victors of 55 and 57 Squadrons were involved in numerous other operations during the 1982 campaign. Some were maritime reconnaissance missions, others were flown in support of Vulcan 'Blackbuck' bombing raids on Stanley Airfield.

The Royal Air Force was aware – at very senior level – that the Royal Navy was engaged in operations relating to South Georgia. This was a frustrating time for the RAF. Like the two other services, they had for years trained hard but, now that the crisis had come, the scene of operations was thousands of miles away from their home bases. It was also too far away for the employment of all their latest and most expensive aircraft and weapons systems. Northwood was soon receiving urgent enquiries as to the possible involvement of the RAF.

According to Commander Davis: 'I could understand their problem. They wanted to get in on the act. If Paraquat succeeded in persuading the Argentines on the Falklands to pack their bags and go home quietly, then South Georgia would turn out to have been the only scene of action. The navy and the marines could show that they had done their stuff, but the RAF would be left empty-handed. So I received frequent offers of help. Initially I advised against it. The last thing we wanted was to alert the enemy that they were about to be hit, and they would certainly get the message if they learned that we were sending reconnaissance flights down in that direction. Eventually, however, when the Task Group was getting very close to the island, it made sense to look at the area to the north and north-west where we had no alternative means of surveillance. A flight was authorised with the proviso that the aircraft's track did not take it within sight of the coastline.'

In the early hours of 20 April, four Victors lifted off from Wideawake and headed south in a loose gaggle. All carried a maximum load of fuel. One of them, piloted by Squadron Leader John Elliott, RAF, was about to make the longest operational reconnaissance flight on record.

Two hours down track, a thousand miles south of Ascension Island, two of the Victors transferred all their spare fuel to the other pair and turned back north. Elliott continued south for another two hours before taking on board a further load of fuel from the third Victor. After that he was on his own. As the sun came over the horizon, he descended from his cruising height of nearly 50,000 feet to the optimum altitude for a maritime radar search, 18,000 feet. Down at sea-level, at 0920, a break in the clouds permitted several watching officers on the *Antrim*'s bridge to identify the contrail of the aircraft as it overflew the Paraquat ships.

Starting 90 miles off the east coast of South Georgia, Elliott approached the centre of the island and then turned north. After 120 miles, he turned to port, flew 90 miles due west, then 120 miles due south, again 90 miles west, and finally headed back north towards Ascension Island. Late in the afternoon, after nearly fifteen hours in the air, Elliott and his crew were back on firm ground with radar photographs covering 200,000 square miles of ocean. According to the pictures, there was not a single ship, nor a single iceberg, anywhere within 200 miles of South Georgia's northern coastline. □

VII | No Luck on Fortuna

IN PREPARATION for their landing, John Hamilton and his men moved to the *Antrim* on 17 April. Before leaving the *Endurance* they sought permission to examine the crates – still packed in her hold – which contained most of the equipment owned by Bob Veal's Joint Services Expedition. The ship had departed Port Stanley so suddenly on 21 March that there had been no time to off-load this freight. It now provided an unexpected supplement to the Mountain Troop's own supplies. One of them later explained: 'We had brought tons of gear down from the UK, but you've never got everything you need. For example, most of us were wearing the standard army DMS boot. We guessed that these would soon get torn apart if we were going to operate on a glacier. We found that the JSE crates contained their Swedish mountaineering boots – civilian manufacture – which were of much better quality. Likewise, some of their climbing kit was better than the usual issue from our own stores at Hereford. We were given authority to help ourselves, and we were very glad to do so.'

Final preparations and briefings continued throughout 18 April and the following day. The Task Group was now closing the final miles to South Georgia. Brian Young's attention was still concentrated very much upon his force's fuel requirements. He had hoped to conduct a massive pumpover from the *Brambleleaf* to the *Tidespring* on 21 April, but this was put back by twenty-four hours so that he could make the planned SAS and SBS insertions on schedule. The lonely *Brambleleaf* was ordered to a holding position 100 miles north-east of the island while the Paraquat force moved much closer to the coast. The *Antrim* and the *Tidespring* headed for a position just north of Antarctic Bay, while the *Endurance* and the *Plymouth* went down the coast to prepare for the insertion of 2SBS at Hound Bay. Their adventures will be described later.

On 20 April the helicopters were busy making final transfers, in bad weather, of men and equipment between one ship and another. It was planned that the Mountain Troop would be lifted from the flight deck of the *Antrim* by Lieutenant-Commander Ian Stanley in the ship's Wessex 3 and in the Wessex 5 of 'C' Flight 845 Squadron The two Wessex 5 would be flown by Lieutenant Mike Tidd, RN, and Flight Lieutenant Andy Pulford, RAF, both of whom were 'jungley' pilots. In other words, they were trained in Commando helicopter flying techniques. Each had operational experience in Ulster and had exercised with Commando forces in Norway. In contrast with Ian Stanley, a specialist in anti-submarine warfare, Tidd and Pulford were accustomed to inserting troops on unfamiliar terrain. They had discussed with Tony Ellerbeck the geography of the area and the rapidly changing weather patterns. Given reasonable visibility, they were confident that, with the relatively light loads they would be carrying, they should be able to place the SAS party safely on the Fortuna Glacier as requested.

HMS *Antrim*'s helicopter, best known by the affectionate nickname 'Humphrey', was the key component in the operation. As a 'pinger' (anti-submarine) aircraft, it was fitted not only with radar and sonar but also with a computerised Flight Control System (FCS). In effect a fully automatic pilot, this system permitted the human pilot to instruct his helicopter to fly to a particular point over the sea and to hover over it without changing height, heading or position. This facility is vital when the helicopter is 'dunking' a sonar transducer at night or in poor visibility over a featureless surface. The Doppler system which normally provides positional information for the FCS would be equally important in the conditions to be expected over the Fortuna Glacier. It would help the pilot to keep track of his movement over the ice if he lost contact with visual reference points. Navigational information would be calculated and given to the pilot by the observer operating the ground-stabilised radar fitted in the back of the aircraft.

The two Wessex 5 flown by Tidd and Pulford did not carry any of this equipment. As 'utility' helicopters, they carried the comparatively unsophisticated Automatic Stabilisation Equipment (ASE) which, with the aid of gyros, maintained the altitude of the aircraft but did not hold it positionally. In bad visibility, the pilots could determine their height by radio altimeter but, to discover whether they were travelling backwards, forwards or sideways, they were dependent upon magnetic compass and personal observation of reference points. As far as the Fortuna Glacier insertion was concerned, the limited instrumentation of the two Wessex 5 meant that they would need to be guided by Ian Stanley's Wessex 3 if they lost sight of the ground.

The only available route up from Antarctic Bay to the chosen landing site entailed a zig-zag between steep mountains on either side. Once over the glacier, Tidd and Pulford would be committed to flying by visual references. They would not be able to turn away from any patches of low cloud because that would take them towards the cliffs where turbulence could make an aircraft drift unpredictably. The pilots would not be able to climb over any such clouds because that would incur the risk of ice quickly forming on the metal of the fuselages and, more significantly, on the rotor blades. In Mike Tidd's phrase: 'It was vital to maintain eyeball contact either with the ground or with the Wessex 3. One or the other. But preferably both.'

During the night of 20/21 April the two ships of the northern group moved westward in worsening weather. Northwood gave final authority for the insertions – codenamed Dandelion – to

The sight which greeted Lieutenant-Commander Ian Stanley as he crossed Cape Constance and approached the confined waters of Antarctic Bay. Through the mist and light snow can be seen one of the steeply descending arms of the Fortuna Glacier.

proceed. By 0800 on 21 April the *Antrim* and the *Tidespring* were in their planned launch position. Although only fifteen miles from South Georgia's soaring peaks, they could see nothing. A half gale was blowing from the north-east, horizontal visibility was down to two miles, the solid cloudbase was at 400 feet, and the barometer was falling steadily.

Captain Young ordered a short delay until there was an improvement. Then, at 0930, Ian Stanley lifted off without passengers to inspect the weather inshore. He flew fast and low through the dawn murk and entered Possession Bay where Captain Cook had made his own landfall in 1775. This was the first time that Stanley and his crew had seen the sheer cliffs of South Georgia. They were a breathtaking sight. The contour lines marked on the maps in the *Antrim's* operations room gave no indication of the size and grandeur of these extraordinary mountains.

Turning to port, Stanley crossed the flat neck of Cape Constance and entered Antarctic Bay. He circled for a while, satisfying himself that the conditions were acceptable and that there were no Argentine ships or military emplacements in the vicinity. The saw-tooth face of the Fortuna Glacier – up which the three helicopters must soon fly – was exactly as the officers of the *Endurance* had described it. Stanley turned and flew back to the *Antrim*. Young, Sheridan and Delves heard his report and agreed that the SAS insertion could proceed. Departure was delayed by a snow squall until 1030, and he then loaded his four passengers. Hovering off the destroyer's quarter, he waited while the two Wessex 5 came across from the *Tidespring*, landed on the *Antrim*'s flight deck, and embarked their own passengers. All three aircraft then formed up in 'echelon port', with Stanley in the lead.

By the time the three aircraft reached Possession Bay the visibility had become extremely limited. In Stanley's phrase: 'We hit a solid wall of snow.' He had made the approach by dead reckoning and knew where he was, but it was foolish to persist in these conditions. He announced his decision to Tidd and Pulford by UHF and the echelon banked carefully onto a reciprocal course. Twenty minutes later the Mountain Troop was once again sheltering in the *Antrim*'s hangar.

There was further consultation in the operations room and Stanley explained the patchy nature of the inshore weather. Cedric Delves was understandably anxious that the few precious daylight hours were slipping by with alarming speed. He requested a first-hand inspection of the coast. At 1100, 'Humphrey' lifted off with Delves and Hamilton on board. The Wessex followed the same track as before, entering Possession Bay, crossing Cape Constance at low level, and approaching the foot of the awesome glacier.

Stanley describes this second reconnaissance sortie: 'The weather was much better this time. I worked my way up the slope of the glacier – hover-taxiing about thirty feet above the ice – and got up onto the top without difficulty. We cruised around for a while to get the feel of the place and to let Delves and Hamilton have a look. You could see the crevasses clearly enough, especially the big ones near the edge of the glacier. John Hamilton didn't

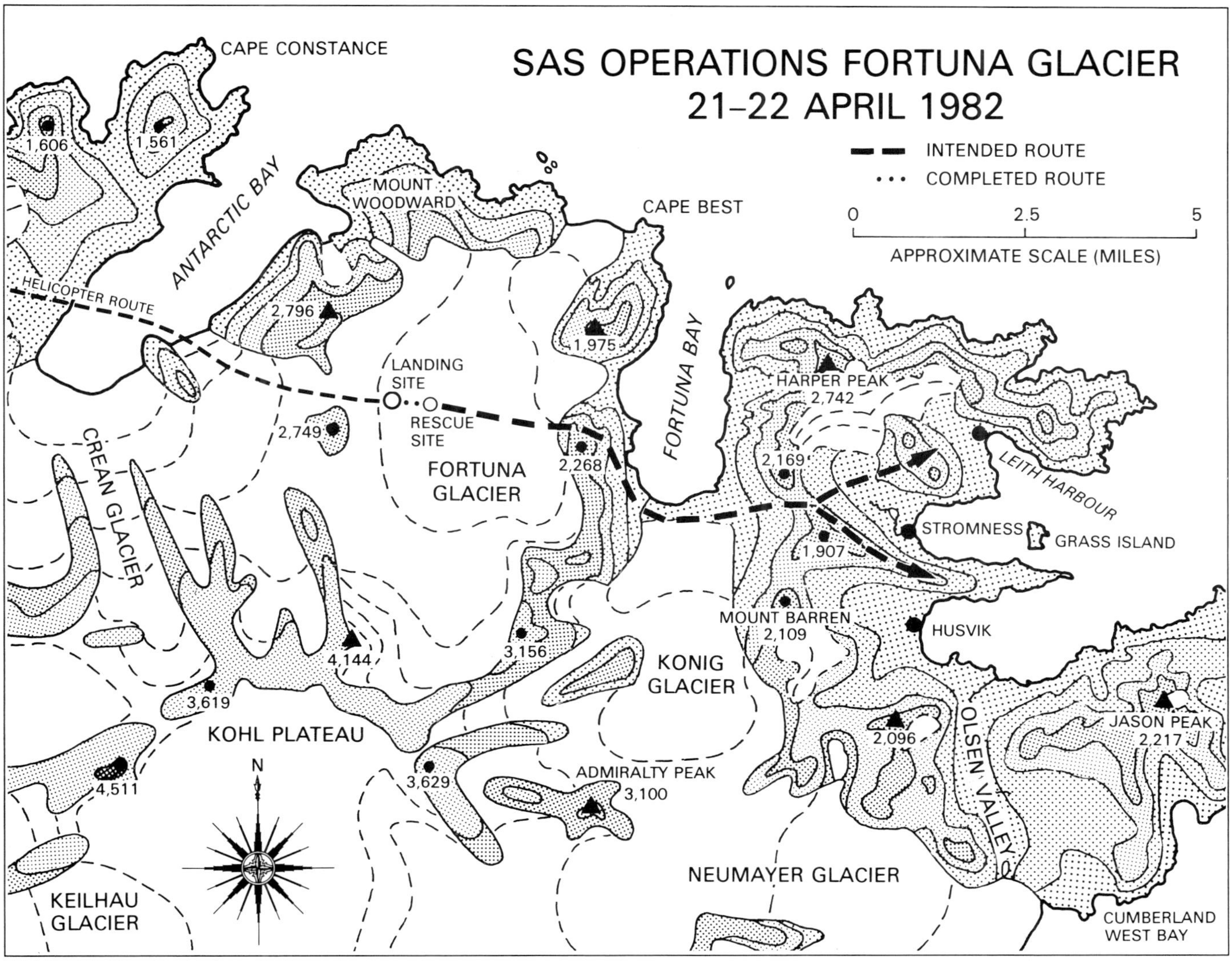

At 1140 on 21 April, the British committed themselves to the first major step in recovering their lost South Atlantic possessions. Finding a sheltered corner of HMS *Antrim*'s bridge, Lieutenant Kevin White took this photograph of the three Wessex helicopters engaged in the 'D' Squadron Mountain Troop insertion. Two of the helicopters are already loaded and airborne. The third Wessex stands on the RFA *Tidespring*'s flight deck, ready to fly to the *Antrim* to collect its passengers.

The three helicopters approach the base of the Fortuna Glacier. Ian Stanley's intended route will take them directly to the base of the ice fall and then upward, only thirty feet above the surface, until they ascend to the agreed SAS landing site on the glacier's highest slope.

seem too enthused at the prospect. Cedric Delves' comment sticks in my memory because it became a catch-phrase in the *Antrim* for days afterwards – "You've got to get on, John." They soon agreed that the top of the Fortuna was not really as bad as they had been told, so the plan could go ahead.'

'Humphrey' was back on the *Antrim* thirty minutes later. Cedric Delves disembarked and gave a situation report to Young and Sheridan. The embarkation of the Mountain Troop was again delayed by a snow squall which swept over the two ships – preventing the Wessex 5 from leaving the *Tidespring* – but all three aircraft were once more flying in echelon towards the coast by 1145.

The trio followed the route pioneered by 'Humphrey' and made their way in radio silence up to the top of the glacier. Lieutenant Chris Parry, the observer in 'Humphrey', plotted the course on his chart by referring to the passing cliffs and mountain peaks. It was just after midday when he informed his pilot that they had arrived over the planned landing site. The insertion, in Stanley's words, was 'a bit daunting. There was plenty of snow flying around, but no real problems apart from severe buffeting. I had the advantage over the other two pilots because, apart from having FCS, I also had three other pairs of eyes – Chris Parry's, Stewart Cooper's and Fitz Fitzgerald's. They were looking out of the windows and door while I concentrated on the instruments. They were giving me directions as I descended and it was slightly depressing to hear their steady chant in my headphones – "You are going down, you are going down!" Even with their help, I still nearly got into trouble. At the last moment, Fitz shouted that I was putting one of the wheels into a crevasse.'

Lieutenant Tidd remembers his first sight of the glacier: 'We had been warned to expect plenty of wind, so I was prepared for the buffeting. The main problem was visibility. There was simply no horizon. Ice, snow and cloud all blended into one. I could see Ian and the occasional crevasse and bits of rock, so I had visual reference points but, even then, I overshot at my first attempt. I got down OK, but had difficulty in keeping the cab firmly on the ground. I noticed that the airspeed indicator was registering gusts of up to sixty knots while I was stationary.'

Flight Lieutenant Andy Pulford has a similar recollection of the insertion: 'I was flying number three in the formation and therefore had the other helicopters as references. It was virtually impossible to distinguish between glacier, cloud or snow-covered cliff. I made my approach using Ian's aircraft as a reference on the ice and landed forward and right of him. As we were unloading our troops, the visibility reduced to about fifty metres in driving snow. Then, a few moments later, it cleared again just as rapidly.'

One of the SAS troopers who was there takes up the story: 'The visibility was very poor, but our pilot put us down exactly where we wanted to be. We jumped out and started to unload the gear. I was standing by the door on the leeward side, and I was catching the hot air coming out of the engine exhaust, so I didn't at that stage realise just how cold and windy it was up there, but I did notice that the helicopter was being pushed along sideways. It was only when the pilot lifted off – and I felt the full force of the wind for the first time – that I knew what we were up against.'

The first pilot to disembark his passengers, Mike Tidd, took advantage of a temporary improvement in the visibility and lifted off. Quickly he headed for the rim of the glacier and dropped

The massive form of the RFA *Tidespring* pitches through the increasingly steep-sided waves as the storm gathers strength.

down into Antarctic Bay where he reduced speed. He flew slowly north towards Possession Bay, waiting for the two other helicopters to catch up. On the glacier's desolate dome, Ian Stanley and his second pilot, Stewart Cooper, watched from the cockpit of 'Humphrey' as the SAS troopers carried their equipment to a safe distance. Their leader, John Hamilton, turned and raised his arm, giving Stanley a 'thumbs up'. The two Wessex lifted off in unison and flew back the way they had come. A few minutes later, over Possession Bay, they rejoined Tidd's aircraft and the three flew back in formation to the waiting ships. In Lieutenant-Commander Stanley's words: 'The insertion had gone well and the trip back from the coast was no problem. It was good to get back onto *Antrim*'s deck. I remember saying to the others "I'm glad we'll never need to go up there again!"'

As the helicopters dipped out of sight over the lip of the glacier and back down to sea-level, the sixteen SAS men roped themselves together into four groups of three and one group of four. They then roped themselves to their three 'pulks' – long lightweight sleds designed for man-hauling and preloaded with food and ammunition.

Preparing to move off, they checked their weapons and discovered that they had frozen solid. Even though they had borrowed a supply of special machine oil from the *Endurance*'s artificers, they now found that the weapons – the metal of which was comparatively warm while they were in the *Antrim* – had attracted, during the flight, a film of condensation which instantly froze on exposure to the Fortuna wind. Even the rifling inside the barrels was affected.

Leaning hard into the rising gale, the Troop started to walk eastward along the gentle incline of the glacier. The surface was composed mainly of hard exposed ice, in places polished smooth by the weather, in others covered with a thin layer of fine powdery snow which the wind whipped horizontally across the path of the advancing men. This snow lodged in any indentation, concealing the ice beneath. It also formed itself into a covering over the many crevasses, forming what to mountaineers are known as snow bridges. In many cases, the bridges entirely concealed the fissures. Within a few yards the SAS men started to fall through these bridges. The lead group acted as pathfinders for the 'pulk' haulers struggling along in the rear. Although initially alarming, the falls soon became part of a pattern of slow steady progress. A man would suddenly drop waist-deep into the ice, his colleagues would anchor the rope with ice-axes, carefully haul him out, and the advance then continued until the next fall. The party was encountering a significant crevasse, on average, every thirty yards.

John Hamilton and his men succeeded in covering half a mile before failing light brought the decision to halt for the night. He was not dissatisfied with his progress. The wind was strong, but not unbearably so. The crevasses were a hindrance, but no more than he had expected. Many were visible and could be avoided. He was safely ashore, nobody had been hurt and none of the kit had been lost. If the launch from the *Antrim* and the *Tidespring* had not been delayed that morning, and if the wind had been not quite so strong, he might by now have been off the ice and onto *terra firma*. Provided that the weather did not deteriorate further, an early start in the morning should bring him back almost to the planned schedule. The Troop searched around for the least exposed patches and, in a rising wind, dug themselves in for the night.

Spread over an area of two hundred yards, working in pairs, they hacked at the hard ice, digging shallow trenches large enough for two men to lay down side by side. It was impossible in those conditions to erect their small tents but, having stuck an ice-axe vertically in the ice to act as a marker, each couple pulled the nylon over themselves and settled down to sleep.

Once again the weather played havoc with SAS intentions. The wind continued to grow in strength, bringing with it a heavy fall of snow. Blowing violently across the surface of the glacier, it quickly filled the man-made trenches, burying the occupants. Several men slept on, snug in their sleeping-bags, while others stayed awake to shovel out the snow. Some trenches were better placed than others. Corporal Lofty Arthy and his companion abandoned theirs when the wind, now gusting in excess of 100 miles per hour, tore their tent from their hands and whirled it away into the darkness. Hoping for a more sheltered haven, they spent ten minutes cautiously moving 400 yards downwind. Finding only bare slope, they then needed nearly three hours to crawl back, on their hands and knees, to their point of departure.

Fifteen miles out at sea, the six thousand tons of the *Antrim* pitched and rolled through increasingly violent waves. Although the ship was fitted with stabilizers, the people on board were obliged to re-lash all loose gear and move through passageways with great care. In the wardroom, off-watch officers sat on the carpet and watched a Hollywood adventure film, the projector held firmly by its operator with hands and knees.

The wind rose to Force 11 and Captain Young invited some of the ship's company up to the darkened bridge so that they could view the waves bursting over the length of the ship and 'one day tell their grandchildren about it'. He remained on the bridge throughout the night, wondering how long his ship could endure such a battering without suffering serious damage. With no Royal Navy repair facilities closer than Gibraltar, storm damage had been a constant worry for Young since crossing the Antarctic Convergence. The violence of this storm – judged even by South Georgia's exceptional standards – was illustrated by the plunging line of ink on the ship's barograph. The angle of descent traced by the trembling nib was so acute – falling to 965 millibars – that photocopies of the barograph paper were later purchased by the ship's company, for charity, at ten pence per sheet.

It was a night of acute anxiety for the ship's Wessex flight. The hangar door on HMS *Antrim*, and other ships of her class, was not sited in the customary position, at the after end of the superstructure, but on the port side. To get the helicopter inside, it was necessary to push it along the deck on the side of the ship and then gently persuade it through a gap with only inches of tolerance. It was not an easy task at the best of times but, with waves running green down the length of the decks, it was now impossible. Ian Stanley had to consider the risks of losing not only the aircraft but also the maintainers who might be swept overboard with it. 'Humphrey' remained on the open flight deck all night, flailed by wind and spray and punished by the violent movement of the ship.

Aboard RFA *Tidespring*, Mike Tidd had identical problems with 'C' Flight. One Wessex 5 was stowed in the hangar, but the other had been secured on the flight deck, ready for an emergency launch. As the wind gained in strength, the canvas socks used to secure the tips of the rotor blades started to tear apart. Heavy manila ropes were lashed around the blades which, with the gusts increasing to 120 miles per hour, were creating sufficient 'lift' to threaten the safety of the aircraft. Mike Tidd describes the long night as 'bloody diabolical – Chief Petty Officer Bill Raines and his maintainers were out there in the open, secured with body-lines, trying to prevent the helicopter from being ripped apart. It was impossible to move it into the hangar. That would have meant folding the rotor, but the blades would have been torn from the hands of anyone attempting the task. Rotorhead and blades would have been smashed and people killed or injured. We just had to hope for the best.'

Both of the exposed Wessex survived the dark hours. As the wind started to die down and the waves abate, so did both teams of maintainers emerge from their hangars to check the airframes, engines and electrical systems. All three helicopters were ready to fly before first light.

Mike Tidd had planned that he and Andy Pulford would remain on 'standby' until the early hours of 22 April and then hand over to two other pilots, Lieutenant Ian Georgeson and Sub Lieutenant Andy Berryman. In the event, the storm prevented any of them from resting and the roster became meaningless. It did not require a vivid imagination to anticipate that the SAS might be in trouble. Further flights up to the glacier were a strong possibility and Tidd laid his plans accordingly. He decided that Berryman, at the time a relatively inexperienced pilot, should not be exposed to the abnormal conditions inshore. He would fly in Berryman's place, in Yankee Foxtrot, and accompany Georgeson who would be flying Yankee Alpha. Neither would have a second pilot with them – space and 'lift' capacity would be the prime requirements. Aircrew would be restricted to Leading Aircrewman Tug Wilson, flying with Tidd, and Leading Aircrewman Jan Lomas, in Yankee Alpha with Georgeson.

Before first light the men on the glacier emerged from their holes and checked to ensure that everyone was safe. Men and kit were buried in snow. At least one man was reluctant to leave the comparative comfort which he had created with his own body warmth under the white mantle. Hamilton called his people together and, crouching on all fours, they packed their equipment and reloaded the 'pulks'. The planned route lay along the slope of the glacier rather than directly across the top. Theoretically, this tactic should have provided some protection from the worst of the wind. In practice, the weather was showing little sign of improvement. Hamilton could neither advance nor operate in any military context. For three hours he and his party sheltered wherever they could, hoping for some lessening of the wind. Gradually it eased, but by now the physical condition of his men was declining.

By mid-morning he accepted the inevitable. Not only was the party getting cold but, more dangerously, their clothing was very wet. If they stayed another night under these conditions, they would suffer casualties with hypothermia and frostbite. Bitterly disappointed, Hamilton drafted a signal to Major Delves, advising him of his situation and asking whether evacuation was possible. Even this simple procedure was complicated by the fact that the signaller was obliged to remove his gloves in order to operate his set. Within seconds his hands became completely numb.

Hamilton's signal was received in the *Antrim* at 1000 on 22 April. Delves and Sheridan agreed that they must accept Hamilton's on-the-spot judgement, that he should be lifted off as quickly as possible. The *Antrim*'s sick bay was alerted to stand by for exposure casualties and the helicopter crews were briefed for a rescue mission.

Heavy squalls sweeping across the sea prevented the three helicopters from taking to the air before 1050. Eventually, between snowstorms, they lifted off and flew in through Possession Bay. Flying conditions were, in Stanley's phrase, 'quite taxing'. He told Georgeson and Tidd to land on the flat neck of Cape Constance while he went on alone to ascend the glacier. Thick cloud commenced at one thousand feet – halfway up the face. He spent thirty minutes trying to find a way over the nightmare of shattered ice and soaring black rock. 'There was just no way through to the place where the SAS were thought to be. The wind was very unpredictable and on several occasions I lost tail-rotor control. Ice was building up on the fuselage and the aircraft was difficult to handle. Watching the altimeter was an amazing business – it was flickering madly from thirty feet to two hundred feet and then back to thirty feet as I passed over each crevasse. I decided that it was too dangerous to bring the others up here.' Stanley turned back, searched for Tidd and Georgeson on Cape Constance, and, having found them, led them back to the ships. At 1330, having refuelled, they made another attempt. This time they found clearer weather and Stanley led them straight up the ice face.

Arriving over the glacier, the three helicopters searched at low level. The weather was, for the moment, much improved. On the ground, the soldiers could hear the engines. They ignited three smoke grenades and Stewart Cooper quickly spotted the streaks of orange dye discolouring the blue-white of the ice. All three aircraft touched down, their blunt noses pointed into the wind, the rotor discs tipped forward and turning hard to keep the aircraft from being blown backwards. Equipment was thrown quickly into the back of each helicopter and the SAS men climbed in behind. As they did so, cloud and snow swept over the area once again.

In Yankee Foxtrot, the aircrewman Tug Wilson told his pilot,

Yankee Alpha, piloted by Ian Georgeson, flies in past the soaring 2,000-foot cliffs of Possession Bay to rescue the SAS reconnaissance party. Weather conditions at this stage are acceptable. Turbulent winds cause problems with buffeting and down draughts, but a bleary sun provides better than average visibility. Yankee Alpha is destined to be destroyed just twenty minutes later.

Mike Tidd, that loading was complete. Having been the first to touch down, they were the first to be ready to leave. Looking to his left, Tidd saw a snow squall sweeping towards him. Ahead, he could still see – 800 yards away – a high crest of the glacier and clear air beyond. Urgently he called Ian Stanley, requesting permission to lift off. The answer was affirmative and, ahead of the other two pilots, Tidd headed north towards safety. At this point the snow squall overtook him and he immediately suffered a phenomena known to pilots as a 'whiteout'. He could not see the ground and there was no horizon. As another pilot describes it: 'When that happens, you have no idea where you are or what is going on. Its like trying to see through milk. You become disoriented. The only thing is to fly on your instruments – assuming you've got the right ones, of course.'

Tidd knew that there was a mountain about 700 feet to his left and another about half a mile away to the right. 'All aviators try to steer clear of cumulo-granite, and I was keen not to wander too far off track. However, I knew that I had passed over a cluster of rocks shortly before. If I could find it, it would give me a reference. I was flying with one eye on the altimeter and the other looking ahead. I had about forty feet of height. As the snow swept over me and I lost sight of the ground, I shouted "Tug! We've got a problem." He was in the back, serving out a thermos of coffee to the SAS guys. He jumped to the door, looking down for sight of the ground. I was banking hard to the left, looking for the rocks. Then I saw the altimeter unwinding very fast and I knew we must be close to the surface. I put on full power and that softened the impact, but we hit at about thirty knots. The left undercarriage was torn off and then she thumped down on her port side. I was hanging there in my straps, thinking "Mrs Tidd isn't going to like being a widow," while seven tons of helicopter went bashing across the glacier for about eighty yards with bits flying off in all directions and the cockpit rapidly filling with soft snow and broken Perspex. It was lucky we went down on the port side – the left-hand side of the cockpit was smashed. If I had had a second pilot, he would have been killed. We stopped, and I went to turn off the fuel cocks and electrical circuits, but the controls were buried under the snow. I clambered out and there was Tug, getting the passengers out of the door. The helicopter was a wreck, but for some peculiar reason the windscreen wiper was still working. When Ian landed on I walked over and plugged in my headset so that we could talk. The first thing he said was, "You messy bastard, you've left your wiper running!"'

Ian Stanley and Ian Georgeson had watched helplessly the fate overtaking Yankee Foxtrot. Stanley did not even have time to give advice by radio. 'Mike was going well when the squall hit him. He was keeping very low to stay under the clouds. He was fine as long as he had visual references. Once he lost those he was in trouble. I saw him banking to the left and the tail and rotor tips hit the surface of the glacier. The aircraft slewed around and went sliding through a slight depression full of fresh snow which must have acted as a cushion. He came to rest on the far crest of the dip.'

Yankee Foxtrot has hit the ice and crashed. Ian Georgeson, in Yankee Alpha, lands nearby and some of his SAS passengers hurry across to the wreck to give aid to the survivors.

Stanley called Ian Georgeson and they lifted off together to hover-taxi to the crash site half a mile away. The snow squall had passed and visibility was again 800 yards. Landing beside the wreck, they were amazed to find that nobody had been killed or seriously hurt. The only visible injury was a small cut over the right eye of Staff Sergeant Phil Curass. The survivors were divided into two groups and loaded with their kit into Yankee Alpha and 'Humphrey'. Tug Wilson and four SAS soldiers went with Georgeson, Mike Tidd and two soldiers with Ian Stanley. Georgeson dumped part of his fuel to compensate for the additional weight and the two pilots lifted off in unison ten minutes later. The passengers retained their side arms, but much of the SAS equipment was left behind at the crash site.

The two Wessex headed towards the rim of the glacier which, at this stage, was again visible. Yankee Alpha was flying three lengths behind the leader, in port echelon, and slightly higher. For the first three minutes all went well, the leading helicopter maintaining a height of thirty feet and the Wessex 5 flying at seventy feet. Then, at precisely the wrong moment, the next snow squall swept across their track. It was another 'whiteout'. For Stanley, this did not present a problem because he could depend upon his altimeter and the navigational information being fed to him by Chris Parry to maintain his height and heading. The flight control system permitted him to partly override the computer and he was therefore flying in a part-manual and part-automatic mode. Behind him, Ian Georgeson could still navigate because he could simultaneously watch his altimeter and use 'Humphrey' as his reference point. Both pilots continued to follow the intended route to safety.

What they could not know was that they were approaching a ridge of ice which runs parallel with the rim of the glacier but some hundreds of yards inland from it. Stanley lifted over this ridge and down the other side without having any way of knowing that he had done so. He was flying in total whiteness. The first sign of impending trouble for Ian Georgeson was the sudden disappearance of the leading Wessex when it dipped down the far side of the ridge. It vanished below the nose of Yankee Alpha and the pilot had lost his one and only reference point. Glancing down again at the altimeter, Georgeson saw that he was losing height. 'I had no way of knowing why the other helicopter had disappeared so rapidly. It was just engulfed somewhere in the whiteness. I continued to fly in the same direction, trying to maintain level flight. Then my altimeter indicated that I was about to hit rising ground ahead. There was a sort of fateful inevitability to it. I reduced speed and prepared for an involuntary landing. The wheels touched and the whole aircraft shuddered for several seconds. I almost got away with it. What I didn't know was that the wind had slewed me around so that I was broadside on. It was gusting to forty knots or more. It pushed the aircraft broadside across the ice until a wheel caught in a small crevasse and over she went onto her starboard side.'

Because helicopters have much of their machinery mounted high on the fuselage and, because the airframe is designed for lightness, they are top heavy and prone to roll if landed sideways. Yankee Alpha came to rest in a tangle of smashed rotor blades and broken Perspex.

Aircrewman Tug Wilson, who had slightly injured his back in the first crash, describes the scene inside the fuselage: 'I found myself lying on top of Jan Lomas with an SAS soldier across my back. At first neither Jan nor I could move because our legs were

'Humphrey's' aircrewman, 'Fitz' Fitzgerald, ploughs through freshly fallen snow to consult with one of the SAS survivors from the Yankee Foxtrot crash. Other SAS troopers stare glumly at the debris. The piece of wreckage in the top right-hand corner of the picture is Yankee Foxtrot's tail rotor gearbox cover.

trapped under ammunition boxes and weapons. The troops seemed stunned – they were making no move to evacuate the aircraft. Jan shouted to them to jettison the port cabin windows and get out. They quickly did so and then Jan said that he could see the pilot's legs jerking around in the cockpit. I disentangled myself and went around to the front.' He saw that Georgeson was trapped in the harness of his personal life-raft. Several SAS passengers had noticed that he was not with them and, although they believed that the wreck might be about to catch fire, ran back to help. Lomas reached into the cockpit and freed the grateful pilot who then crawled out through the smashed top cockpit window.

In the Wessex 3, a sickened Ian Stanley had been listening to a running commentary by his second pilot, Stewart Cooper, keeping watch astern through the port cockpit side window. 'Fine, OK, OK, following us at fifty yards, steady, steady, steady, all OK. Oh God! He's gone in.' There was nothing that Stanley could do other than head back to the *Antrim* to offload his passengers. Standing behind the pilot's seat – listening to Cooper's report on his headset – was Major Delves. Slowly he turned and faced the other passengers. Drawing a finger across his throat, he indicated to them what had happened.

As they left Possession Bay, escaping from the blanketing effect of the mountains, Chris Parry had his first chance to make a radio report to the people waiting in the ship's operations room. 'Departing coast now. ETA fifteen minutes. Regret we have lost our two chicks.' There was a moment of silence before he heard a laconic, 'Roger.' The *Antrim*'s controller was struggling to find an adequate response. He gave up. There were no words to express the impact of Parry's bad news.

'Humphrey' is safely back on HMS *Antrim*'s heaving flight deck. White-uniformed SAS soldiers disembark while the *Antrim*'s maintainers and medical team check the condition of the helicopter's passengers.

Weather conditions had become much worse by the time 'Humphrey' touched down on the destroyer's flight deck. Taking on a half load of fuel and a quantity of blankets and medical supplies, Stanley returned to the coast with the intention of rescuing the Yankee Alpha survivors. Again an anxious Cedric Delves flew with him. 'I hover-taxied up the side of the glacier and managed to get up to the top. The conditions were very violent. It was impossible to attempt a landing – I couldn't even find the wreck – but I raised Ian Georgeson on the 243.0 distress

Aircrewman 'Tug' Wilson and two members of the Mountain Troop photographed beside one of the liferafts inflated and utilised as temporary shelters while the Fortuna Glacier party awaits rescue.

frequency and he told me that everyone was safe. I went back to the *Antrim*, was debriefed, and tried to decide what to do next. The weather was showing no sign of improvement. All we could do was wait and hope.'

In HMS *Antrim*'s operations room, it was accepted that it would be almost suicidal to attempt another rescue mission until the gale eased and visibility improved. This might not happen for hours and, in the meanwhile, the men stranded on the glacier would be suffering cold and exposure injuries. It seemed likely that they would be spending a second night on the ice, so their condition might be even worse than the medical staff had first feared. The wardroom was converted into an additional emergency medical reception centre and every available stretcher assembled near the flight deck.

The first reports of the two crashes had caused deep concern in the operations room. Captain Chris Nunn was monitoring the radio traffic and he realised that his own specialist mountaineers might be needed to rescue the survivors. He sent a signal to the *Tidespring*, alerting the 'M' Company Reconnaissance Troop. A two-man team must prepare for a helicopter transfer up onto the glacier. The order was passed to Sergeant Mac McLeman, a Mountain Leader First Class (ML1) with great experience in mountain and Arctic warfare survival techniques. From the radio message, he understood that one of the Wessex had tipped over and had fallen bodily down a deep crevasse. McLeman assumed that not only would there be injured men down there, but also uninjured men who would be unable to climb out. It was his belief that they could not do so without specialist assistance. He explains his assessment: 'A few days earlier I had spent some time with the SAS guys and had checked through their kit with them. In my view, it wasn't very good. It was standard army issue – not as good as our stuff. There are recognised techniques for getting yourself out of a crevasse, and these involve the use of special crampons and jumars and so forth. It was my opinion that these guys just didn't have that sort of gear, that they would not be able to haul themselves up the insides of a crevasse maybe a hundred feet deep. I got hold of an ML2 – Corporal Pat Mitchell – and we made up a supply of equipment for a full-scale rescue. Then we heard that they weren't down a crevasse after all, and the launch was cancelled. But they were lucky not to have lost any of their people.'

High on the Fortuna Glacier, the survivors gathered in the lee of Yankee Alpha. The aircraft's MS9 life-raft was broken out and inflated. Lomas and Wilson climbed inside with the injured Phil Curass so that his lacerations could be cleaned and dressed. They were joined by an SAS signaller whose radio had survived the crash. While he and Ian Georgeson tried to string up an aerial, the rest of the party set off through the thick driving snow to recover their Bergens and other equipment from the wreck of Yankee Foxtrot. When they returned, Wilson and three troopers made a second trip, bringing back with them another life-raft. Everyone climbed inside the two shelters and huddled together for mutual warmth. Only their massed weight prevented the inflated boats from being blown away. Hamilton and Georgeson discussed the situation and agreed that rescue was unlikely before the following morning. There was little they could do pending the arrival of the last operational Wessex in the Paraquat task group.

Miles out at sea, aboard the heaving HMS *Antrim*, Chief Petty Officer 'Fritz' Heritier and his maintainers fussed over the venerable 'Humphrey'. These repeated flights in turbulent con-

The loneliness of command. Major Cedric Delves waits on HMS *Antrim*'s flight deck as the maintainers secure 'Humphrey' after the helicopter's return from the Fortuna Glacier.

ditions might have unwelcome effects upon the helicopter's airframe and control systems. Everything was thoroughly checked to ensure that the crew would not be prevented by mechanical failure from rescuing the men on the glacier.

Ian Stanley had a meal and withdrew to the relative quiet of his cabin. 'I must admit to feeling depressed and bombed out. We had lost two helicopters. I was in charge and the whole thing had gone wrong. I asked myself if it was my fault, if there was anything I could have done to avoid the disaster. Was I right in authorising Mike Tidd to take off when he did? Should I have told him to stay on the ice until that particular squall had passed? Should I have followed a different route off the glacier? And so on and so forth. I suppose it was natural to have a period of doubt but, the more I thought about it, the more I realised that everyone had tried their hardest and it was simply the weather which had defeated us. The main thing was to concentrate on the next phase – getting the survivors off. I certainly didn't relish the prospect. The "jungley" Wessex – the Mark 5 – has two engines, but the Mark 3 – like "Humphrey" – had only one. The Mark 3 did not have a good reputation for engine reliability and, with only one of them, I would have no margin of safety when we went back up there. Earlier in my flying career I had twice experienced sudden and total engine failures with the Wessex 3. On one occasion I had to make an emergency landing in Dorset, and another time I ditched in the sea and two of the aircrew were injured. With all this mind, I was somewhat nervous about having a third engine failure in the conditions in which we were now operating. "Humphrey" had been up there five times in the past twenty-four hours and now we were faced with trip number six. You tend to wonder how long your luck will hold out.'

By 1600 the gale was diminishing and the cloudbase rising slightly. Stanley spoke with his crew and decided that there was time to make two more rescue flights before nightfall. There were sixteen men waiting to be brought off the glacier, plus a quantity of important equipment. If all went well, they could be lifted off in two parties of eight.

Some days earlier, several of the aircrew had followed the wartime practice of drafting and signing their Wills. When he heard that they were going back to the Fortuna Glacier, Chris Parry slipped away to his cabin and wrote a short letter to his wife, Alison. He took off his wedding ring and sealed it in the envelope. He had no doubt that the impending mission would be hazardous.

By 1630 the dark grey Wessex was once again flying in over the neck of Cape Constance and approaching the Fortuna ice cliff. Ian Stanley recalls the scene: 'Conditions inshore were worse than they had been when we left *Antrim*. The wind was stronger and there was a lot of cloud. We decided on a completely different tactic for this approach. Instead of going up the glacier at low level, I climbed to 3,000 feet and, accepting the icing risk, went up through the top of the clouds. This gave me about 600 feet clearance over the glacier itself, but nothing over the mountains. I stooged around for a few minutes and then, in a sudden break, we were looking straight down at the crashed helicopter and the bright dayglo of the life-rafts. I dived down through the gap and landed alongside. The wind was not only strong, it was shifting direction every few seconds. I had a quick talk with Ian Georgeson and assessed the situation. It wasn't good.'

The weather was again fast deteriorating. The break in the clouds closed over. Stanley realised that there was no prospect of being able to return for a second load before nightfall, nor would he be able to climb back through the cloud. This, like the earlier flights, would be a low-level exit. He decided there and then to take everyone in a single lift. Bergens and other equipment must be abandoned – lives were more important than kit. Within a few minutes the fuselage was tightly packed with grateful bodies. The SAS troopers insisted on keeping their side arms but, somehow, they were all fitted in. As one of them remembers: 'When we took off, there were arms and legs sticking out of the door and windows.' This is not an exaggeration but a plain statement of fact. To anyone who has ever squeezed into the cramped interior of the famous 'Humphrey' – which now resides on permanent display at the Fleet Air Arm Museum, RNAS Yeovilton – it is astounding to consider how Petty Officer Fitzgerald managed to fit sixteen armed men in among the four-man crew and the mass of radio and radar equipment. In order to watch his radar screen, Chris Parry was obliged to sit on the chest of a SAS soldier who was laid across the seat with his head jammed in the side blister window. Three more soldiers were lodged in the sonar well and Fitzgerald himself hung halfway out of the open door.

Normally, the maximum all-up weight of the Wessex 3 was 13,600 pounds. When Ian Stanley departed the Fortuna Glacier, he did so with a weight in excess of 15,000 pounds. This in itself was not as alarming as it might seem, as the pilot explains: 'For lift off, my problem was how many bodies we could jam into the hull rather than the amount of power available. The engine could produce all the drive I needed and the high wind helped to create lift. The only limitation is the amount of torque you can safely apply to the rotor-head gearbox. I didn't need to dramatically over-torque for the lift-off, but I knew I would have problems

Wessex 406, 'Humphrey', reunited with its mother ship, HMS *Antrim*. During an action-packed period of twenty-four hours, this elderly machine has made six flights into one of the most dangerous environments on earth. Two other helicopters have been lost but, almost miraculously, nobody has been killed or seriously injured.

landing on. When I got back to *Antrim*, the air would be slightly warmer, there would be less wind, and I would be contending with the movement of the ship. Consequently, when she came in sight, I didn't go through the usual manoeuvre – hovering alongside and gradually moving in sideways – I simply came down in a long descending slide over the port quarter. The landing was a bit heavier than usual, but nothing broke.'

On board the *Antrim*, the helicopter's return was watched first with anxiety and then, as 'Humphrey' thumped safely onto the steel deck, with inexpressible relief. An admiring soldier described it as 'a bloody marvellous controlled crash landing'. Willing hands helped the passengers to disentangle themselves and they were hurried away for hot drinks and a medical examination. One man was found to be showing the first signs of frostbite and some of those who had been through both crashes were slightly shocked but, apart from being very cold, all the survivors were surprisingly well.

By a most extraordinary display of skilled and courageous aviation, Lieutenant-Commander Stanley and his crew had not only saved the two *Tidespring* aircrews and the sixteen SAS soldiers, they had done so without damage to their own aircraft. In the broader context, they had prevented the damage to morale which would have been inflicted on the Paraquat force if men had died with nothing achieved.

That night John Hamilton's Mountain Troop sought out 'Humphrey's' crew and presented them with a bottle of whiskey. It was opened with appropriate ceremony and aviators and soldiers drank each other's health. It was a moment to be savoured. The adventure might so easily have had a less happy ending. It was a tired but contented aircrew who soon after crawled into their bunks.

And now we must turn to the adventures of 2 SBS who, since 12 April, had been living on board the *Endurance*. Following several briefings by the ship's officers, and having attended two full planning conferences on board HMS *Antrim*, the Section's commanding officer, 'Roger', knew exactly what he wanted his men to do. He would insert the three Patrols on the beach at Hound Bay, on the north coast of the Barff Peninsula. Assuming that the coast was clear – a particularly apposite phrase in the circumstances – they would make their way on foot over the slope of the Sörling Valley and down to the edge of Cumberland East Bay. The *Endurance*'s Wasps would then deliver to them two Geminis. All three Patrols would embark in the two boats, make their way along the west side of the Bay and, if all went well, go around the tip of Dartmouth Point to make a landing at Hestesletten. Leaving the boats there, they would climb the slopes of Brown Mountain to obtain a perfect view straight down on to Grytviken and King Edward Point. 'Roger' was prepared to devote two full days to the approach phase. This would allow three further days for careful observation of the main Argentine garrison before 'M' Company made its final assault.

All of this assumed that nothing went wrong. In the event, the SBS were to encounter no better luck than had the SAS on the Fortuna Glacier. The island's extraordinary climate and fearsome topography were to prove overwhelming.

On 20 April, HMS *Endurance* and HMS *Plymouth* parted from the *Antrim* and the *Tidespring* to proceed further down the coast. While the frigate remained off shore – providing cover from seaward – the *Endurance* made her way into Hound Bay and anchored three miles from the beach. On the following afternoon, a Wasp piloted by Tony Ellerbeck flew south to St Andrews Bay to deliver Chief Petty Officer Tommy Scott to the small hut currently occupied by Cindy Buxton, Annie Price and Peter Stark. This was the first contact with the civilians since the departure of 'the plum' for her rendezvous with the *Fort Austin* on 5 April.

Ellerbeck stayed on the ground long enough to describe the current situation and to emphasise the need for secrecy. He explained that he was leaving Scott at the hut for a few days so that he could exercise some control over the way in which the BAS field parties made use of their Squadcall radios. Now that the Paraquat force was on station, operating its helicopters along the coast and landing reconnaissance patrols, it was vital that their presence should not be revealed to the Argentines. If the ladies saw a helicopter overhead, or a patrol on the ground, or a ship on the horizon, they might unwittingly mention it during one of their conversations with other parties on the island. In the event, these fears were groundless. The ladies had already seen the distinctive red profile of the *Endurance* several miles off shore and had used their own commonsense in not telling the others, but Scott did arrange for future transmissions to be kept to a minimum.

HMS *Endurance*'s Wasp 434, piloted by Lieutenant-Commander Tony Ellerbeck, prepares to touch down at St Andrews Bay in the murk of late afternoon. Peter Stark raises the Union Flag to indicate the wind direction and to confirm that the area is still unvisited by the Argentines.

The other reason for leaving Tommy Scott with the lady film producers was to provide them with some armed and uniformed protection against any Argentines who might appear on the scene and who might decide to take advantage of the women's obvious vulnerability. With Plant and North away on their extended walk up the Barff Peninsula, the only remaining male at St Andrews Bay would have been Peter Stark. Captain Barker was asking him to fly out to the *Endurance* and to stay aboard for several days. He had lived on the island for two years and knew it intimately. He was needed by the military for his detailed knowledge of the buildings at Grytviken and King Edward Point. Tommy Scott would act as guardian for the women while he was away. Cindy Buxton and Annie Price elatedly questioned him about the impending operation while Peter Stark quickly gathered together his personal effects and boarded the Wasp. Tony Ellerbeck returned to the *Endurance* at low level, hugging the sides of the high cliffs.

There was at this time a potentially dangerous breakdown in communications. The ladies had told Ellerbeck and Scott that Myles Plant and Tony North were making a reconnaissance of the Barff Peninsula. The third BAS man, Peter Stark, was aware of their intended route and he was now on board the *Endurance*, being questioned by various people regarding the latest situation in the locality. Somehow, the fact that Plant and North were very probably in the area of Hound Bay, and that previous visits to the Barff Peninsula had revealed no signs of Argentine activity, did not find its way to the SBS personnel who, within hours, were due to land there.

The three Patrol leaders, named 'Kiwi', 'Flip' and 'Willie', had finished preparing their kit for the evening's insertion. In anticipation of several days' intense activity on land, they were resting or asleep. Consequently, they were unaware of Stark's arrival or the value of his information. Their leader, 'Roger', spoke briefly with Stark but the visitor was then taken away for a long discussion with an SAS operative who was building an exquisitely detailed model of the King Edward Cove area. The modeller needed to know the position of every window and door, and the internal layout of each building, so that observing or attacking troops could find their way into the heart of the Argentine positions on the darkest night.

The difficulties of the situation are well reflected in 'Roger's' own recollections: 'I remember Stark. I had a talk with him. I don't think he made me aware of the two men at Hound Bay but, if he had, I would have briefed my team. You must understand that, by 22 April, we had spoken with many people with intimate knowledge of the area. We were swamped with knowledge. But what we really needed to know – the state of the ice in Cumberland Bay and enemy dispositions – we had to get into the weeds to find out. I felt that Stark had nothing new to offer and decided not to disturb my busy or sleeping team leaders.'

As dusk fell, the SBS made their final preparations. A stiff and strengthening wind was blowing dark scudding clouds and heavy rain from landward, but the weather was not so severe as to prevent flying operations. Ellerbeck and his observer, David Wells, prepared to fly the first of the Patrols ashore at the planned hour of 1730.

Although they would not be taking their Geminis with them, the SBS did have a large quantity of kit. The two rear passenger seats and the observer's seat were stripped out to create maximum space. 'Kiwi' and one of his men were seated on the floor in the rear while David Wells lay stretched along the edge of the cabin with his head hanging out in the open air. Holding a night vision aid (NVA), it would be his job to guide the pilot in over the coast and to identify the chosen landing point. Tony Ellerbeck takes up the story: 'We were determined not to show ourselves to anyone ashore. The ship was maybe three miles out and, in the normal run of things, Dick Buckland would have guided us in under the control of the ship's radar. She, however, was in total radar and radio silence. The Wasp itself was not equipped with radar, so it was a case of flying visually. The value of the NVA was that it enabled David to tell me when the tube passed from dark – which meant that we were over the sea – to light – which indicated land. In particular, the contrast would show up a distinctive group of rocks which I knew from past flights in the area would take me to the setting down point. I flew in at low level – about fifty feet – and made a routine landing on a stretch of tussock just behind the beach on the south shore of the bay. The passengers unloaded their stuff which included a "Bardic T". This is a piece of kit which you can set up quickly to act as a flight path indicator for night landings. With night coming on, I would need the lights for the other planned seven flights.'

Ellerbeck returned to the *Endurance* and collected two more men – the other half of 'Kiwi's' Patrol – and delivered them ashore. While returning from this task, he noted that the weather was fast deteriorating. By the time he had landed, he knew that no further flights could be made until there was an improvement. He and Wells shut down and went in search of an evening meal. To everyone's intense frustration, the wind increased in force and showed no sign of abating.

On shore, 'Kiwi' soon guessed that nobody would be joining him for some time. Leaving two men at the agreed landing site, he set off in the dark with 'Scobes' to reconnoitre the surrounding area. Moving cautiously through a protesting colony of penguins, they circled the low-lying stretch of tussock grass behind the beach. 'Kiwi' suddenly shouted in alarm as something very large and very black reared up beside him. He had trodden on the tail of a sleeping elephant seal.

The adrenalin flowed again a few minutes later when he saw a light. He and 'Scobes' advanced cautiously towards it and discovered that it was the curtained window of a small hut. After a whispered conference, it was agreed that 'Kiwi' would knock on the door while 'Scobes' adopted a covering firing position. When the door opened it was to reveal two heavily-bearded and very startled civilians. Myles Plant and Tony North had been about to settle for the night. After brief explanations on both sides, the two SBS men were invited inside for a cup of tea. It seemed an unlikely way to start a war.

'Kiwi' advised the two civilians to stay in the hut overnight but to return to St Andrews Bay at first light. Hound Bay might conceivably become a shooting gallery if the landing was detected and the Argentines contested it. 'Kiwi' then learned that Plant and North had been walking around in this area for the past two days and had seen no sign of human life. The proposed SBS route over the Sörling Valley should be clear. Pleased by this good news, 'Kiwi' and 'Scobes' rejoined the other two men at the beach.

The gale continued to blow at Force 9 into the early hours of the morning. By midnight the situation was reaching crisis point. One-third of the SBS were ashore, two-thirds were in the ship, and the weather was still very bad. Ellerbeck offered to make another attempt. At 0300 he lifted off with 'Flip' and another man crammed into the stripped rear half of the cabin.

For the next twenty minutes, Ellerbeck tried to fight his way through the roaring darkness. The gale was veering northerly, bringing with it fierce squalls of sleet and snow. Hanging on to whatever handholds they could find, the two marines were thrown about violently as the small lightweight Wasp struggled against the wall of turbulence. The night was very dark and they could see nothing.

On two occasions the helicopter almost struck the surface of the sea and 'Flip' wondered glumly what their prospects of survival would be if Ellerbeck was forced down. The same thought was on the mind of the pilot: 'We were flying without burning our navigation lights, but this time it had been decided to go in under the ship's radar control with Dick Buckland giving me directions by radio. It was a truly awful night and pitch black. Dick was doing his best, but the wind kept changing direction so unpredictably that his orders for course alterations became almost meaningless. At one point I was fifty degrees off course. This was a classic katabatic wind, sweeping down off the 9,000 feet of Mount Paget and then bouncing around the lower mountains so that it could change direction by ninety degrees in just a few seconds. It was gusting to seventy knots and we were taking a real hammering. David Wells and the two marines could not be strapped in because they had no seats. I was worried that they might be thrown out. I was trying to fly at two hundred feet, but I had only limited control. At one stage we were hit by a down-draft which must have pushed us very close to the surface of the sea – not that I could see it, of course. I remember having full right stick on, with the aircraft rolling eighty degrees to the left. I ruddered the nose straight, but the airspeed dropped and she started to fall. David was watching the altimeter and calling out the heights. I had recovered from this drop, with full power on, when David called "forty feet and climbing", so we must have come pretty close to contact. This happened twice and my fundamental orifice was biting holes in the seat cushion.'

Eventually the pilot decided to abandon the insertion. Snow had accumulated to a thickness of four inches on the windscreen and he had lost forward visibility. Large pieces were breaking off and flying back over the canopy. There was a real possibility that these slabs of frozen snow might enter the engine intake, with inevitable results. Turning eastward, Ellerbeck was swept back to the *Endurance* in much shorter time than he had needed to leave her, and succeeded in setting down on her heaving flight deck.

With insertion by air now impossible, 'Roger' was obliged to adopt the only other alternative form of transport – Gemini inflatables. The boats themselves were reliable, but they were powered by elderly 40-horsepower outboard engines. Like the SAS, the SBS did not have any great faith in this particular design. They were liable to fail in adverse conditions. In this severe offshore wind, a dead engine could be disastrous. A Gemini would be swept out to sea before the *Endurance* could even know that it had gone.

Despite the reservations, it was agreed that 'Roger' and his Section Sergeant-Major, 'Phil', would act as coxswains and would take the two remaining Patrols ashore by boat. Captain Barker agreed to move his ship much closer to shore, dropping anchor only 800 yards from the beach. Quickly the engines were warmed up, the boats dropped over the side, and the voyage commenced.

The events of the next two hours are described by CSM 'Phil': 'The weather had changed again by the time we got the boats into the water. The wind had reduced to a stiff breeze and some of the cloud had blown away. Intermittent moonlight was bouncing off the ice caps on the mountains and giving us reasonably good visibility. We had two problems. Firstly, the gale had kicked up a heavy swell and – with six men in each boat, plus a lot of gear – we didn't have much freeboard. Secondly, we could not approach the south shore – where "Kiwi" had been landed – because there were thick beds of kelp in that area. We were obliged to go for the north shore of the bay where there was a belt of heavy breaking surf about three hundred metres wide leading in to the beach. Everyone was well soaked by the time we had landed and unloaded. The next problem was to make contact with "Kiwi" around on the other side of the bay.'

Instructing the new arrivals to remain where they were, 'Roger' and 'Phil' set off along the dunes behind the beach with a single marine as escort. It was assumed that Argentine troops might be in the area, so the search party could not use lights or shout to attract 'Kiwi's' attention. During the next forty minutes they advanced stealthily around the curve of the bay. 'Phil' resumes his account: 'The area was thick with seals and penguins. They were making plenty of noise before we arrived, but the giveaway to anyone keeping watch was the fact that they shuffled off into the darkness as we walked along, and such a lot of movement made it very obvious that there were humans in the area. We kept going, and then suddenly there was a soft voice behind us. "Kiwi" and "Scobes" had seen us coming and gone to ground. They let us walk through their position and then, when they recognised our silhouettes, identified themselves. After that it was easy. We moved back to our landing place, briefed the lads and sent them on their way. "Roger" and I relaunched the boats and – because they were now unladen – we motored back through the surf to the *Endurance* without too much difficulty.'

Nick Barker withdrew from Hound Bay with the intention of once again fading into the scenery further down the coast and conducting another search for Argentine surface ships in the numerous bays and inlets. During the next twenty-four hours, while the SBS were commencing their reconnaissance, he was to launch his two Wasps on a series of searches as far down as Cooper Island. Before leaving Hound Bay, however, he had the irritation of a windlass motor drive failure, and was obliged to slip his starboard anchor and three shackles of cable. For ships employed in coastal duties, an anchor is a very important item of equipment. The loss could not have come at a worse time.

An Avon Gemini inflatable boat motors across a South Georgia bay in 1983. Conditions such as these would have been ideal for 2 SBS when they attempted their night-time crossing of Cumberland East Bay. In the event, fortune did not favour the brave.

On shore, the three Patrols started their march at first light and ascended the slopes of the Sörling Valley. In good weather this is a relatively easy walk. Cindy Buxton and Annie Price had covered the same route in February and, although well-laden with rucksacks and camera equipment, had traversed the neck of the peninsula in two hours. The marines were much more heavily laden and there was a covering of several inches of snow on the higher slopes, but they reached the shore of Cumberland East Bay by mid-morning.

The first tasks were to search for signs of Argentine patrols and to find a suitable launching point for the Geminis which were to be airlifted to them by one of the *Endurance*'s Wasps later in the day. The party moved along the shore towards the foot of the Nordenskjöld Glacier with the thought that there might be a suitable launching place in this area. To the contrary, they found the surface of the water littered with ice. The deeply fissured face of the glacier – two miles broad and a hundred feet in height – was constantly shedding more huge chunks of ice. The great frozen mass creaked and groaned under the weight pressing down from the higher slopes. Even though the wind had died down, this was no place for rubber boats.

It had been intended that the first day and night would be spent reconnoitring the area, choosing a launch site, and, having reassembled the air-dropped Geminis, setting off across Cumberland East Bay under cover of darkness. Depending upon progress, the party would then land either at Dartmouth Point (giving direct but long-distance observation of King Edward Point), or they would go around Dartmouth Point, across Moraine Fiord, and set up an observation point on Brown Mountain. It had been calculated that all three Patrols could be carried in the two boats. All of these assumptions proved to be impractical.

The scene which first greeted the party is described by 'Willie': 'The glacier was extremely active. It was calving growlers and icebergs every few minutes. The set of the current was carrying this stuff around to the north-east corner of the Bay. It was packed really tight – bits of ice of all shapes and sizes – growlers, icebergs and brash. However, the foreshore where we were established was still fairly clear at this stage. Consequently, when we saw the Wasp dropping our Geminis at the agreed spot – late in the afternoon – we reckoned that we could still carry through with the plan.'

Ellerbeck had flown in over the coast with the two boats and the outboard engines slung under Wasp 434 in a net. He could not fly in daylight over the saddle of the Sörling Valley because – in clear weather – he might be seen from King Edward Point. He therefore chose a more southerly route which brought him over the higher slopes of the Nordenskjöld Glacier. His knowledge of the area reminded him that there is a narrow gorge, about forty feet deep and forty feet wide, running down the length of the glacier to sea-level. This chasm is the result of a warming in the world's climate, the glacier having shrunk back from the neighbouring mountainside. It provided an ideal concealed line of entry into the Bay. Unfortunately, in flying so low, Ellerbeck may have unknowingly brought the freight net into contact with a jagged rock. One of the Geminis received a gash, either at this stage or during the descent to the agreed delivery point on the foreshore.

Wasp 434 departed for the *Endurance* and, two hours later, the SBS party reached the boats and reassembled them. 'Willie' again takes up the story: 'We realised that the initial plan – to lift everyone in two boats – was not going to work under these conditions. We couldn't safely get that number of men – and all their kit – into only two boats. It was agreed, therefore, that "Flip" would stay behind while "Kiwi" and myself made the crossing. Then, after we had inflated them and about an hour before launch time, we found that "Kiwi's" boat had a bad leak in one of the front tubes. This meant that he and his Patrol also would have to stay behind. I would make the crossing on my own, with just three others. My Patrol was the obvious choice because my coxswain, "Thommo", had just completed two years with the Boat Team and was very experienced in handling small craft.'

By nightfall, the surviving boat had been loaded and made ready for launching. During the past hour, however, the wind had shifted to the north and was gaining rapidly in strength. The effect of this half gale was to push the mass of ice from the far side of the Bay down to the south-east corner where the SBS were located. It piled up along the shoreline and extended out over several hundred yards. Only one gap remained – a channel one hundred yards wide – and this was gradually diminishing.

'Willie' decided to accept the obvious risks. 'We got away fairly well and started to motor through the gap in the ice towards open water. It was a pitch-black night and we couldn't see any of the surrounding shores. "Thommo" was steering by compass bearing, but the boat was skittering around in increasingly heavy conditions. Sea and wind state were by now Force 5 to 6. The boat was digging her bow into the waves and we were shipping a lot of water. The kit was well waterproofed and all four of us were wearing dry suits. The cold didn't trouble us too much, but there was a lot of movement. We managed to cover about eight hundred yards when it became obvious that we were in trouble. The motor was working very hard and we were using far too much fuel. We still had five miles to go. "Thommo" was having problems because the electrics were so wet and one of the spark plugs was firing badly. Soon he was losing power and – with the wind getting stronger all the time – we weren't getting anywhere. The main fear was that the engine might die completely. Then we would be swept back onto the ice and smashed to bits. There was no way you could climb onto that stuff, or try to walk across it. We agreed it was far too dangerous to carry on. I told "Thommo" to turn back and he put us onto a reciprocal bearing.'

In the violent black turmoil of breaking waves, 'Willie' and 'Thommo' could only guess whether the compass would take them back to their original point of departure. One of the Patrol was placed in the bow with a shaded hand torch and, every minute, he flashed a signal towards the hidden shore. The brief flashes were seen by 'Flip' who, guessing that the boat might be in trouble, was keeping careful watch. He flashed his own torch in reply and so was able to guide 'Thommo' back to the rocky beach. With a spluttering engine, the Gemini surged in through the breakers and was quickly man-hauled up onto higher ground. The landing was made only just in time. The gap in the ice had shrunk to twenty-five yards and, minutes later, it closed entirely. Disaster had been avoided by the finest of margins.

The three Patrol leaders discussed the situation and decided to wait until the morning before making any further move. Perhaps the weather would improve during the coming twelve hours. If the wind and sea subsided, if the ice dispersed, and if visibility was poor, there was a reasonable prospect of crossing the Bay in daylight without being observed. The party settled down in the shelter of the rocks and slept through until dawn.

The morning of 23 April brought no cheer. The Bay was still blocked by a jagged clutter of ice. A stiff onshore breeze made it plain that the pack was unlikely to clear in the foreseeable future. The three sergeants held another discussion, reviewing their various limited options. It was suggested that the Gemini could be pushed out through the ice and the motor used once the boat was in clear water. This was rejected on the grounds of excessive risk. The probability of being punctured, combined with the poor prospects of survival for anyone struggling to swim in the ice-choked Bay, made it plain that the party should be withdrawn and a fresh line of approach adopted. They decided to send a radio report to 'Clive' in the *Antrim*, requesting evacuation.

Atmospheric conditions were poorer than usual on this morning and the party failed to make contact. The radio was carried up onto higher ground and further attempts made. They too brought no response. What the SBS party could not know was that – during the morning – Captain Young had received a BAS report of an Argentine C-130 Hercules reconnaissance aircraft off the north-east coast of South Georgia. In addition, Captain Young could not rule out the fact that an Argentine submarine of the 'Santa Fe' type could be in the area, even within a hundred miles of the Task Group. These reports led to the decision to make a speedy withdrawal of all ships from the island's inshore waters. In the excitement of the moment, nobody thought to tell the shore party of these developments prior to leaving normal radio range. In the words of 'Clive', the SBS liaison officer: 'It all happened very fast and, down in the ops room, we didn't appreciate just how fast the *Antrim* was moving when she took off.'

Ironically, rescue was available not many miles away. If he had known of their predicament, Captain Barker could have sent in his Wasps, but the SBS appeals did not reach him. All of the SBS communications people, together with their code books, were with 'Clive' aboard the *Antrim*. The *Endurance*'s own communications people were fully stretched to cope with the flood of radio traffic from other sources and had no reason to monitor the SBS frequency. Even if they had, lacking the appropriate codes, the operators could have made nothing of the signals. The first intimation that their men were in difficulty came to 'Roger' and 'Clive' when, on the night of 23/24 April and after three fruitless nights ashore, they finally made contact with HMS *Antrim* and asked to be evacuated.

There was some dissent in the *Antrim*'s operations room as to whether or not this request should be granted. Surely, if they tried again, they could still reach a good lying-up position? The radio report had been very brief, but 'Clive' had no doubts. 'I knew all three Patrol leaders as very experienced and very capable marines. If they said they had problems, then they had problems. They would never have given up if there was the least chance of getting through and remaining undetected. We had to pull them out.'

Over the previous forty-eight hours, the *Endurance*'s two Wasps had been busy flying reconnaissance missions up to the tip of the Barff Peninsula and south to St Andrews Bay and Cold Harbour. Captain Barker was now ordered to return to Hound Bay to rescue the SBS team. This was effected during the afternoon when both helicopters again flew in very low and picked up the twelve men. The Geminis were left behind on the beach at Cumberland East Bay, deflated and hidden under rocks and weed. Each boat weighed 525 pounds and each engine 150 pounds. To lift this equipment would have required yet more flights. It was simply not worthwhile to keep the conspicuous *Endurance* loitering in Hound Bay any longer than absolutely necessary. Captain Barker got under way again as soon as the two aircraft had brought back the last of the party.

While the SBS were suffering their three nights of frustration on the Barff Peninsula, a great deal was happening elsewhere. Much of this activity concerned Captain Young and his attempts to carry though the complicated refuelling of his various ships. On 22 April he still had the big unaccompanied RFA *Brambleleaf* in a holding position one hundred miles to the north-east of the island. He himself, in the *Antrim*, was a few miles off Stromness Bay, attempting to insert the SAS. To the east was the RFA

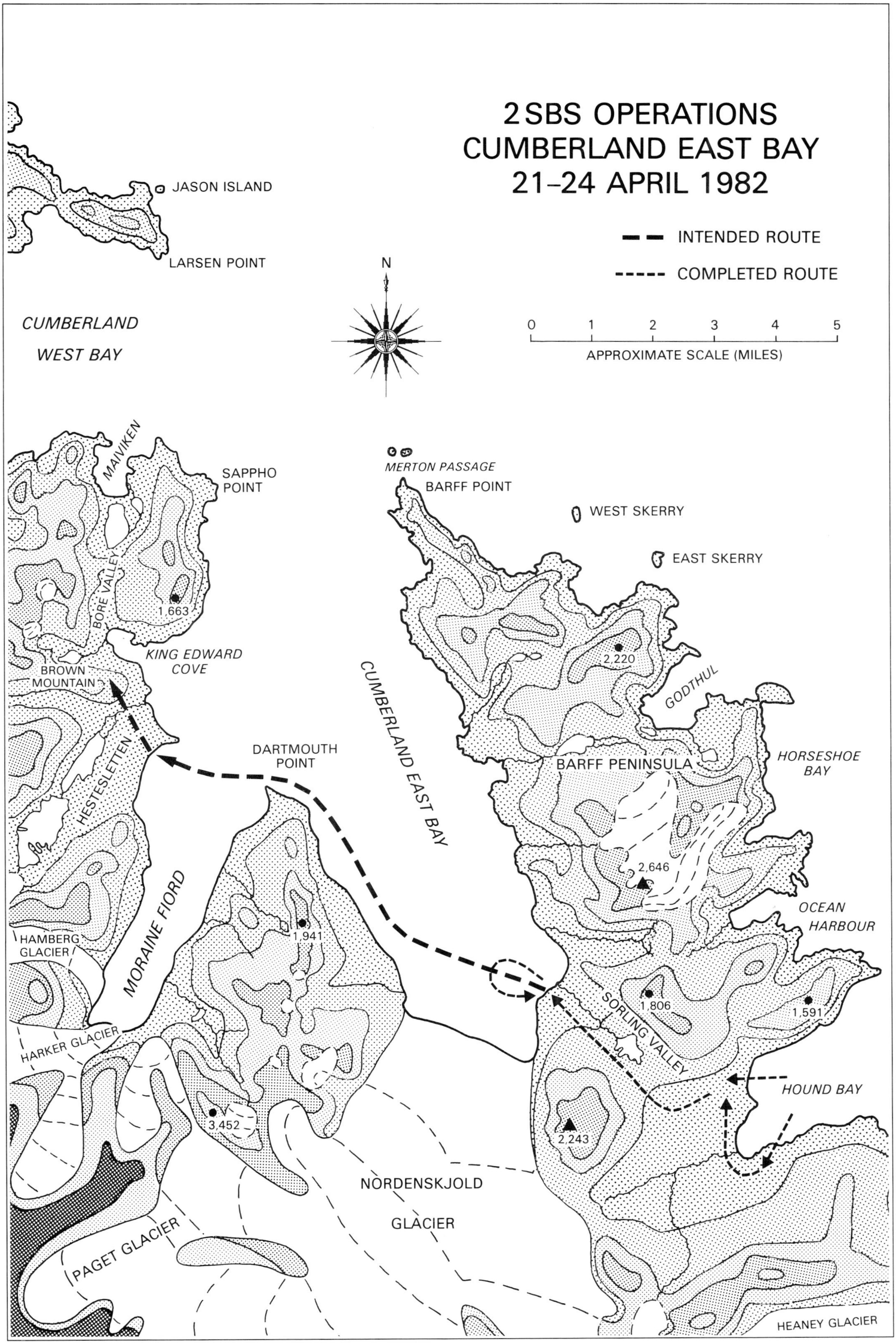

2 SBS OPERATIONS
CUMBERLAND EAST BAY
21–24 APRIL 1982
INTENDED ROUTE
COMPLETED ROUTE
0 1 2 3 4 5
APPROXIMATE SCALE (MILES)
N
JASON ISLAND
LARSEN POINT
CUMBERLAND WEST BAY
MAIVIKEN
SAPPHO POINT
BORE VALLEY
1,663
KING EDWARD COVE
BROWN MOUNTAIN
HESTESLETTEN
DARTMOUTH POINT
CUMBERLAND EAST BAY
MERTON PASSAGE
BARFF POINT
WEST SKERRY
EAST SKERRY
2,220
GODTHUL
HORSESHOE BAY
BARFF PENINSULA
2,646
OCEAN HARBOUR
MORAINE FIORD
HAMBERG GLACIER
1,941
SORLING VALLEY
1,806
1,591
HARKER GLACIER
HOUND BAY
3,452
2,243
NORDENSKJOLD GLACIER
PAGET GLACIER
HEANEY GLACIER

Tidespring, keeping clear of danger but ready to close the coast if any opportunity arose to disembark the 'M' Company group. To the south were the *Endurance* – in the vicinity of Hound Bay and St Andrews Bay – with the *Plymouth* covering her further out to sea.

Young had intended to send the *Tidespring* to join the *Brambleleaf*, for the massive pumpover of 6,700 tons of oil, on 22 April. This would take place in a position sixty miles due east of the island. It was unlikely that the *Tidespring* would be needed for military operations on that or the following day because, if the SAS and SBS reconnaissances went according to plan, they would last five full days (21 April to 26 April). In the event, the weather once again intervened, the foul conditions at sea causing the pumpover to be put back to the following day. It had been intended that the delicate pumping operation – with two ships displacing a combined total of more than 60,000 tons sailing side by side through heavy seas – would occupy twenty hours. They would be extremely vulnerable during that period. Captain Young had intended to join and protect them during the operation. However, following the failure of the Fortuna Glacier insertion, he was obliged to remain close to shore so that a fresh landing attempt by the SAS might be made. The need for intelligence was becoming extremely pressing. He therefore signalled David Pentreath, in the *Plymouth*, and ordered him to take his place as guardian for the tankers. The *Endurance* would continue to remain by herself in her favourite hiding areas, close to Hound Bay by day and further down the coast each night. It took some time for Young and his people to make these dispositions. In the meanwhile, his military commanders were trying to decide their next step.

While the last of the Fortuna Glacier survivors were being taken below, Cedric Delves was starting another 'brainstorming' meeting with Guy Sheridan in HMS *Antrim*'s makeshift operations room. The loss of the two precious helicopters drastically reduced the options available to them for inserting the SAS reconnaissance party, but it did not prevent them from trying other methods.

It had always been assumed that a landing by Gemini might be necessary at some stage, and it was evident that the time had come to put this option into practice. Major Delves takes up the story: 'Fortuna had gone wrong, and it was up to me to press on with an alternative. The entire philosophy of Fortuna had been based upon a desire to approach Leith Harbour from the most unlikely direction – through the back door. Now I had to accept the fact that we would have to go in through the front door. We did consider a landing in Fortuna Bay, but the Boat Troop had serious reservations about the thick beds of kelp in that area. I was not particularly happy about the frontal approach, but the lack of suitable landing sites for inflatables meant that we must go straight into Stromness Bay. This was a bit like going through the front door and then halfway down the hall! It also meant asking Captain Young to place his ship in a position where no ship's captain would really want to be. But he and his Navigating Officer were completely cool and co-operative. They, and Major Sheridan, agreed that there was no time to be lost on half measures. The *Endurance* had put the SBS people ashore further down the coast and we were already twenty-four hours behind schedule. It was important to synchronise our own activities with theirs. It didn't take long to decide that Stromness Bay was the only route in. The next step was to talk with the Boat Troop – who would be doing the job – and get their ideas on how to tackle it. The consensus view was they should land on Grass Island and lay up there. It would give them a line of sight straight into Stromness and along the waterfront area of Leith. At that stage we had no reliable information concerning enemy dispositions ashore, so we felt that Grass Island would give us a useful observation platform before committing ourselves too fully.'

By 1930, on 22 April, the decision had been made and the *Antrim* started her stealthy approach to the coast. Fifteen men of the Boat Troop prepared themselves for a task the success of which would depend almost entirely upon the outboard engines. As the ship slipped quietly through a dark calm night, the SAS men tested the motors yet again. Below decks, aided by the ship's engineers, they clamped the motors to sawn-off forty gallon oil drums, filled with water, and gave them another thorough test run. They ran perfectly. Even so, distrustful of the temperamental machinery, and of one engine in particular, they asked for the loan of the *Antrim*'s own 35-horsepower outboard. It was relatively new and much more reliable.

The plan agreed with Brian Young was that he would bring his ship into the mouth of Stromness Bay and launch the five Geminis at a point approximately 3,000 yards from Grass Island, and therefore 5,500 yards east of Leith Harbour. The Bay is well charted and plunges down to eighty fathoms at its centre. The southern half of the Bay is littered with moraine reefs which represent a considerable navigational hazard for a ship as large as the *Antrim*, but Young felt that the greatest risk was the possibility that the Argentines might have installed radar and stand-off weapons on one of the nearby headlands. If they had, then they would catch him in a very vulnerable position and at very short range. The experiences of the *Guerrico*, as reported by the recently debriefed Lieutenant Keith Mills, were fresh in Captain Young's mind at this stage.

The launch of the Geminis had been planned for the darkest period of the night, about 0330. The reluctance of one outboard engine to run sweetly delayed by half an hour the lifting of the small rubber boats over the destroyer's side. The engines were run up but kept below until the last possible moment – to keep them warm – and were fitted to the boats shortly before they were launched. The SAS crews restarted each motor so that they were running as their propellers entered the black smooth sea. This involved the risk of over-heating the water-cooled machinery and, despite the silencers, also created a great deal of noise, but the Boat Troop was determined to minimise the risk of mechanical failure.

The weather conditions at this stage were ideal. A cold offshore breeze was gradually pushing the *Antrim* stern first towards shallow water and her ability to manoeuvre was limited by the presence of the small boats alongside and in the vicinity. Young and his Navigating Officer, Lieutenant-Commander Chris Morrison, were quietly anxious to be away, but the business of getting the SAS men and their equipment down into the Geminis from the two launch points on the destroyer's starboard side was soon completed. After a few quiet calls of 'good luck' the five boats circled the big darkened ship and, assured that all five motors were running correctly, they slipped away in formation towards the shore. The officers on the *Antrim*'s bridge breathed a sigh of relief as she turned towards the open sea and away from this dangerous place.

For communications purposes, each Gemini had been assigned a codename – Alpha, Bravo, Charlie, Delta and Echo. There were three men in each – a coxswain at the stern, a gunner in the bow, and a third man huddled with a heap of weapons, ammunition, food and survival gear in the centre section.

Although they were only two or three feet above the surface of the sea, they quickly found their bearings. Grass Island's dark outline could be seen in the distance, vaguely silhouetted against the mass of the mainland. Initially it seemed that all was well.

A photograph which illustrates the type of clothing and equipment used by 'D' Squadron's Boat Troop during Operation Paraquat. Trooper 'E' is pictured shortly after his return to HMS *Antrim* from the Stromness Bay insertion. Having formerly served with 29 Commando Regiment, RA, he has retained his Greenlander boots (a Naval stores item). The gaiters are privately purchased, made of canvas for the civilian market. On his back is the standard SAS rucksack, with an SAS 'ops' waistcoat draped over it. His weapon is the AR15 Armalite with a 30-round magazine.

Trouble struck within the first ten minutes. The 40-horsepower engines fitted to Alpha, Bravo and Charlie each failed almost simultaneously and refused to respond to treatment. Echo turned back and took Alpha and Bravo in tow, while Charlie was passed a line by Delta. Again the boats settled into ragged formation and headed towards their objective. The coxswains in the two leading boats could only hope that their engines would keep turning while, behind them, every man dug hard into the icy water with the emergency paddles.

South Georgia's fickle climate now once again stepped in to frustrate Cedric Delves' best laid plans. With almost unbelievable speed, the offshore breeze grew suddenly in strength and, within five minutes, had become a roaring gale. Gusting to Force 8, and then Force 9, it kicked up a short steep sea with waves rising to six feet in height. Despite energetic bailing, the Geminis started to fill with water.

On board the *Antrim*, by now two miles out to sea, Delves felt the buffeting wind and knew instinctively that his Boat Troop must be having problems. 'The launch had gone very smoothly. It was a gorgeous night, almost no wind and a smooth sea. Seeing the lads over the side had been like something out of a World War Two movie. There had been no chance of a rehearsal but, with those conditions and with all the work that had gone into getting the engines properly set up, I was pretty confident. Then that gale came out of nowhere and my stomach sank into my feet. I remember thinking – here we go again, another bag of worms!'

In Stromness Bay the Gemini crews were making little progress. Only Delta was able to push forward through the increasingly heavy seas. Still towing Charlie, she reached the small sheltered cove on the seaward side of Grass Island shortly before first light. Her commander, 'Tom', stayed long enough to ensure that the men in Charlie were making a safe and unopposed landing before turning back to search for the others. In the circumstances, it was a bold decision and it had serious repercussions for 'Tom' and his two crew members.

Alpha, Bravo and Echo were facing severe difficulties. Echo had been able to tow the other two Geminis for a while, but then her own engine also died. The three remained tethered to each other for as long as they could to avoid being separated among the dark and roaring waves, all nine men trying to paddle towards Grass Island, but it was an impossible task. The heaving seas threw the small boats against each other, entangling ropes, paddles, gun mountings and swearing men in an unmanageable muddle. They decided to cast off the lines – 'it was a case of every boat for itself'.

In Alpha, the officer commanding the operation and known as 'Tim', tried to stay in visual contact with the two other boats, but the weather gradually drove them apart. He could see the outline of Grass Island about 3,000 yards away, but his boat was being pushed quickly away from it. His crewman, 'Tony', struggled for twenty minutes with the outboard engine, spraying petrol into the carburettor and tugging repeatedly at the starter line. Eventually he succeeded in getting under way and motored straight through to the landing site.

In Bravo, the tall slender 'Chippy' tried time and again to restart his engine, but it would not respond. His two fellow voyagers worked hard with their paddles as the Gemini was swept along the shoreline of the Busen Peninsula. As he said later: 'It was only about two hundred metres away, tantalisingly close, but the wind was pushing us out to sea all the time. We saw Harrison Point go by, and then Busen Point, but after that there wasn't much between us and Australia. We made a last big effort to head in to Jason Island, but by then we were pretty well knackered. The boat was full of water, I had one man bailing with two of us paddling, and we really drove hard to get ashore, but the wind was too strong and the waves too big. We just kept on drifting out towards all that ocean.'

The commander of Echo, known as 'Geordie', had initially counted himself lucky by virtue of having the modern 35-horsepower motor on loan from HMS *Antrim*. When it, too, failed he had found himself in a similar predicament to that experienced by 'Chippy' in Bravo. The wind was blowing from the east, veering northerly, and this drove his waterlogged Gemini along the shore of the Busen Peninsula at disconcerting speed. The crew paddled hard across the track of the waves which continuously broke over them, but they could not approach closer than 200 yards to the swiftly passing cliffs. Halfway between

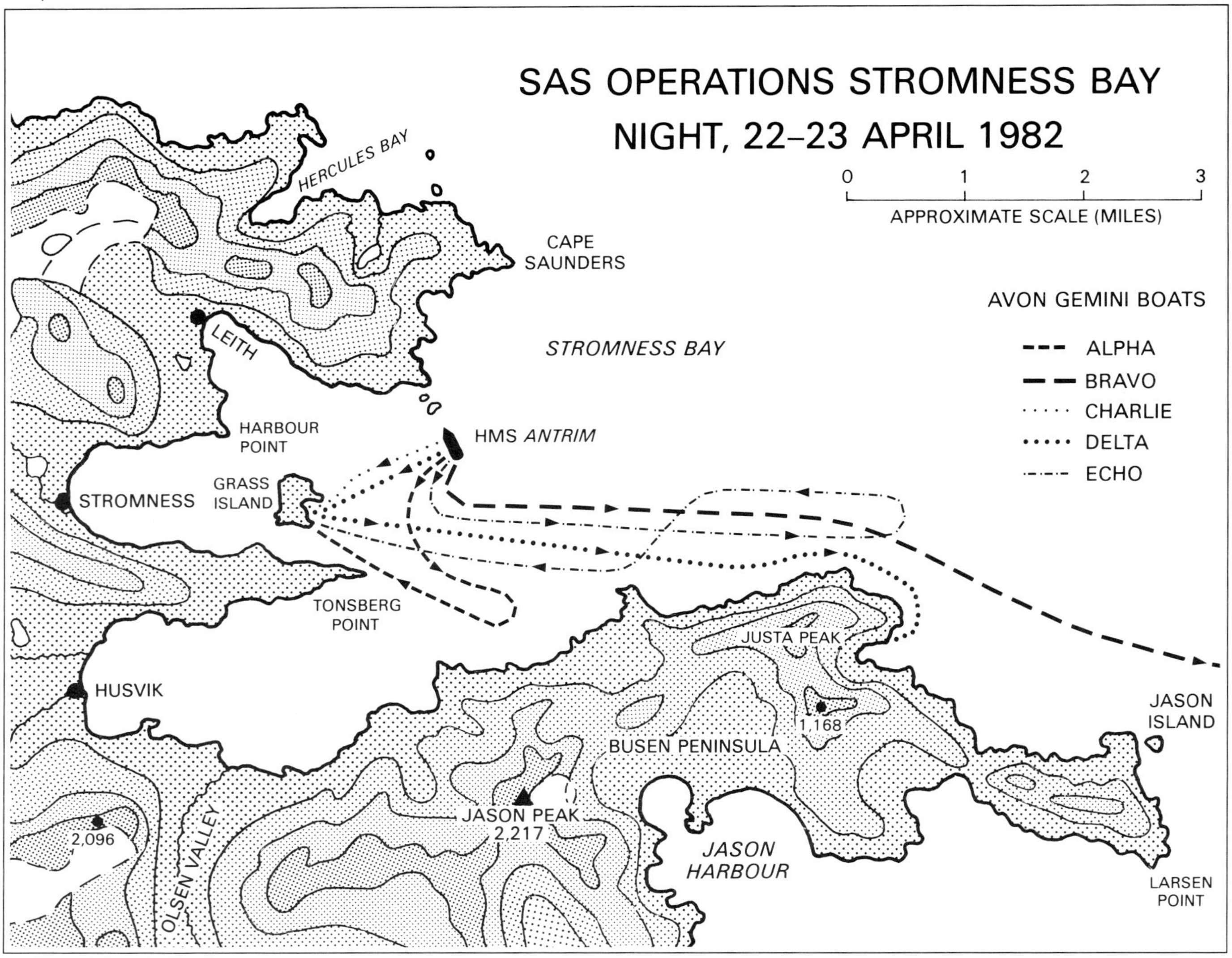

Harrison Point and Busen Point, 'Geordie' decided to break radio silence and put out a distress call on his VHF set. This was heard by the Alpha party on Grass Island who, in turn, passed it to Brian Young aboard the *Antrim*. There was nothing he could do to launch a rescue at that stage.

The feeling in Echo was by now one of deep concern. One of its crew, 'Dave', takes up the story: 'We were getting very tired. I tend to get seasick, and the movement in that Gemini was horrendous. The waves were chucking us all over the place. Then, when I realised that we were not going to get anywhere with the paddles, I got very angry with the engine and I literally attacked the bloody thing. I pulled and pulled the starter rope and eventually it fired. By then we were somewhere off the western tip of Busen Point. I handed the engine over to "Geordie" who skilfully manoeuvred us through and over the waves. The gunner, "Ginge", kept on bailing while I knelt in the bows and navigated. The wind gradually abated and we reached Grass Island at first light. We ran into the little cove there, and got a lot of leg-pulling from the others. It was not something that I would really want to go through again!'

Having earlier delivered Charlie to the island, and having decided to come back for the others, the men in Delta now found themselves in trouble. After an hour of vain search, the Gemini was swamped by an exceptionally large wave which drowned the engine and split the metal bracket of the GPMG mounted on the bow. Just as had happened with Echo and Bravo, the combination of wind and waves started to drive Delta past the cliffs of the Busen Peninsula. Her crew paddled furiously towards the shore and, surprisingly, they succeeded in reaching a small sheltered beach. Even better, they found a cave near the beach where they could hide themselves and the deflated boat. It was 'Tom's' intention to lay up here throughout the daylight hours and then relaunch after dark on the following night for a second attempt at reaching Grass Island. After a quick meal, he began to dismantle the engine and dry out the flooded components. He could not report his location to HMS *Antrim* – even if he had felt it necessary to do so – because his was one of the two Gemini parties not equipped with a radio.

By 0730 on the morning of 23 April, the people in HMS *Antrim*'s operations room had a fairly clear picture of the night's events. It was not very encouraging. Three boats – Alpha, Charlie and Echo – were known to have reached Grass Island. Their arrival had been seemingly undetected by the Argentines. The Geminis had been deflated and concealed in the long tussock grass which grows in abundance on the island and which justifies its name. 'Tim' had established his observation point and was already transmitting, by his secure radio equipment, his first reports of the scene around Stromness and Leith. However, the two other Geminis – Bravo and Delta – had disappeared without trace. Given the violence of the short-lived gale, there was a real possibility that six men had drowned during the night.

Still drifting uncontrollably out to sea, the men in Bravo decided that their situation was becoming desperate and that they must report it on their 349 radio. They were heard by 'Tim' on Grass Island who passed their report to HMS *Antrim* shortly after 0800 hours. Brian Young reversed course and headed back towards the coast. Now that 'Tim' and the others were established ashore, he was prepared to accept risks which would have been unacceptable when the first distress call was received from 'Geordie' – in Echo – some hours earlier.

0810, 23 April, and 'Humphrey' is yet again deployed to undertake an unusual and potentially hazardous task. As the hard-worked Wessex lifts off from HMS *Antrim*'s flight deck, Ian Stanley can be seen at the aircraft's controls while David Fitzgerald sits at the door. Below the helicopter dangles the strop which, ninety minutes later, is to save the lives of three SAS troopers.

Ian Stanley lifted off at 0810 with orders to search for the tiny craft 'somewhere off Busen Point'. Chris Parry studied his charts and the latest weather reports as the helicopter flew in towards the mouth of the Cumberland Bays. Basing his calculations upon the strength and direction of the gale, he plotted the likely track of the Gemini and its probable position. Ian Stanley flew to the area of windswept sea identified by Parry and commenced a box search. Flying increasingly long 'legs' to the pattern, Stanley and his crew spent the next one and a half hours vainly searching for the small black dot. Success depended entirely upon eyesight – their vicinity to the Argentine base precluded the use of radar. On the very last leg, just as shortage of fuel was forcing Stanley to decide that he must return to the *Antrim*, Parry heard the faint tone of the SAS homing beacon at extreme range. The boat was already three miles east of Busen Point.

'Humphrey's' aircrewman, Fitz Fitzgerald, recalls the problems of the mission: 'The met officers had not given us much hope of success at the pre-flight briefing. In fact, the cloudbase was even lower than we had been warned. At between one hundred and fifty and two hundred feet, visibility was very poor indeed. We were getting towards the end of our endurance when I suddenly saw something at four o'clock and about one and a half miles away. It was the Gemini. Our track then was taking us towards the coast and there was real concern that we might have been seen and reported by Argentine observation posts.'

Down on the heaving surface of the swell, 'Chippy' and his two companions could hear the sound of a helicopter in the distance. 'We could see it flying around a mile or so away, but the weather was murky and we weren't certain that it might not be an Argentine aircraft. Then we recognised the shape of a Wessex and we knew it had to be one of ours. We switched on the beacon and set off an orange smoke flare. The pilot came hurtling over and winched us up. The last man out slashed the rubber and the boat went down like a rock. We were very glad to get back to *Antrim*. We were more tired than cold, even after seven hours in the Gemini. The navy had loaned us some divers' dry suits before we launched from *Antrim*, and these kept us dry and reasonably warm. Without those suits, and plenty of underwear and gloves, we might well have been in a very bad way. They made all the difference. Our only concern after that was to rejoin the Troop and to find out what had happened to Delta.'

On Grass Island, meanwhile, the SAS observation team spent the day watching the entire coastline of Stromness Bay. The island itself, at its widest extremities, is about a thousand yards square. On the landward side are two hills, the highest being 245 feet above sea-level. These high points were ideal for the purposes of the SAS. They could see, but not be seen. The principle disadvantage was that there was no direct line of sight into the hinterland behind Leith Harbour, and the old Husvik whaling station was entirely hidden behind Tonsberg Point. The only movement noted by the watchers on 23 March was a lone Argentine workman who wandered out on to one of the Leith Harbour jetties for a few minutes before disappearing back

Leith Harbour, the old whaling station which was the target for all 'Tim's' endeavours during the period 21/26 April. His opponent, Teniente de Navio Alfredo Astiz, had fortified the small hill inland from the base buildings, but lacked the military strength needed to make the place a significant defensive position. Given enough troops, and given the requisite determination, Astiz could have turned the complex of sturdy buildings into a minor fortress.

among the buildings. It was decided that the next step was to get ashore and to observe at much closer range.

At 2300 on the evening of 23 April, the nine men attempted to cross the few hundred yards of water separating them from their planned landing site near Stromness. In rough seas and poor visibility, they set off from the Grass Island cove but, within minutes, the outboard engines failed. All three Geminis were paddled back to the launching point. The decision was made by 'Tim' to wait until the next night in the hope that the weather might be kinder. Once again, those SAS men not actively on watch assisted 'Geordie' in his familiar task of stripping and reassembling the three engines.

At midnight on 24 April the SAS party on Grass Island made their second attempt at getting ashore. True to form, two of the 40-horsepower motors failed after the first few minutes and Echo was obliged to take Alpha and Charlie in tow. Paddles and muscle power were the prime movers in permitting the party to reach the beach which curves north of the whaling station. The group came ashore north of the stream which flows down into Stromness Harbour from the high ground to the rear.

One of the soldiers who was there described his first impressions: 'It was very dark and you could barely see your hand in front of your face. The beach area was covered in boulders and scrap iron, and everyone was slipping and falling around. We pulled the boats up beyond the highwater mark and the Boss told one man to hide them while the rest of us went into Stromness. It was a very eerie place, a ghost town. The wind was stirring the old corrugated iron sheets on the cladding of the sheds, making weird screeching noises. Stromness is a big place – lots of buildings – so it took a while to search it, but we found no signs of occupation. After that we split into two groups – five men staying behind to secure the area while the Boss went north with three others to have a look at Leith.'

Following almost in the footsteps of Lieutenant Mills and Sergeant Leach – who had made their own reconnaissance of Leith Harbour a month earlier – 'Tim' and his three companions

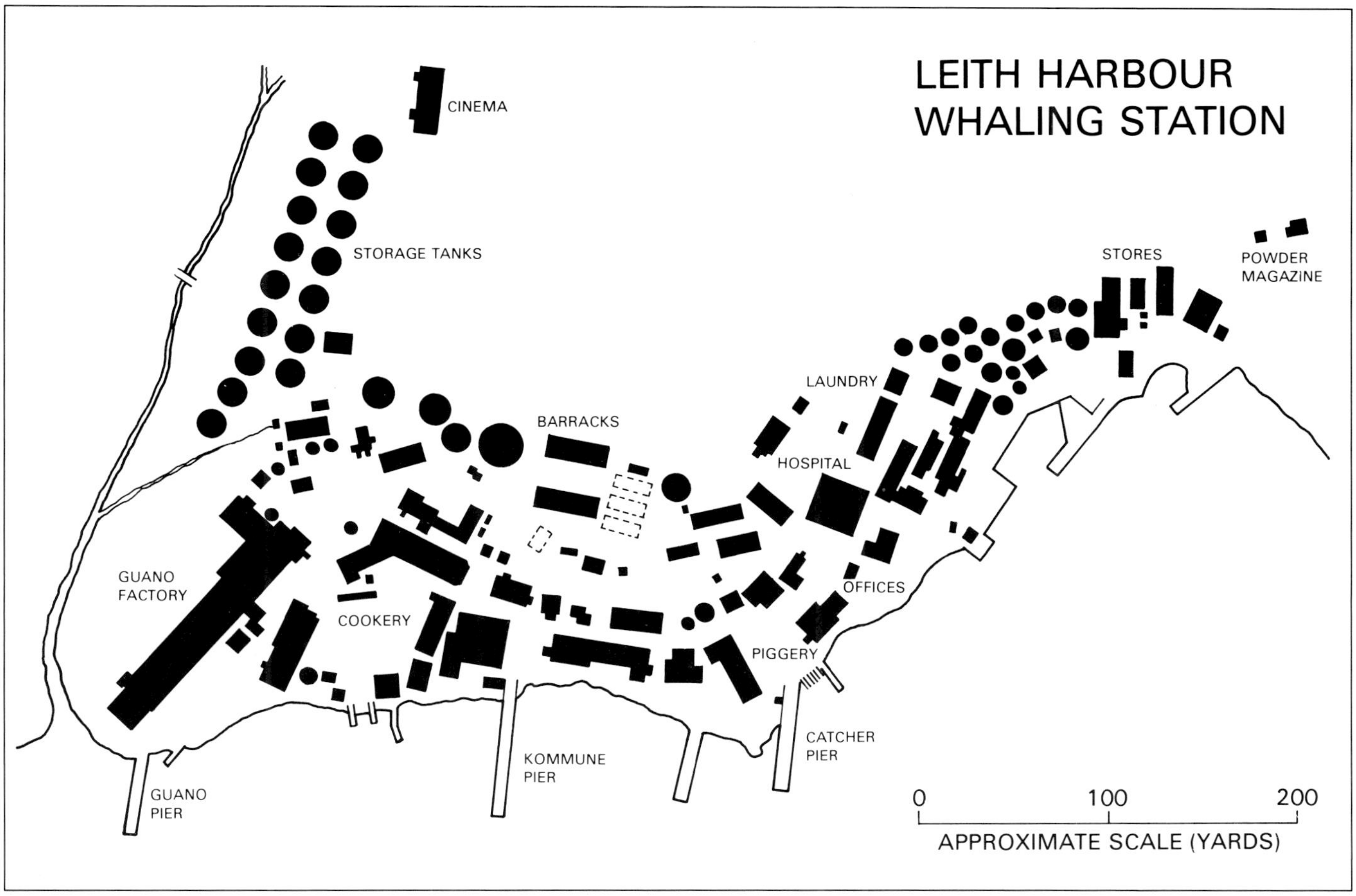

climbed to the crest of Harbour Point and looked down on the lights of Leith. They heard a diesel generator and saw occasional movement around the dormitory building, but the place appeared to be quiet. At 0200 the lights went out as the Argentines settled for the night.

Not all of the garrison were asleep. In the early hours of the morning a heavy rockfall on the northern cliffs of Leith Harbour created a disturbance which alarmed an Argentine sentry. He fired a long burst with his machine gun in the general direction of the fall, but thereafter fell silent. The area remained peaceful until dawn.

As early morning sunlight flooded into the Bay, 'Tim' watched in quiet satisfaction as the sixteen Argentine marines guarding the whaling station performed their first 'stand to' of the day. An officer could be seen making his rounds of the trenches which his men had dug along the rising ground behind the station. His route, interrupted by short conversations at each trench, enabled the SAS watchers to plot the precise location of each defensive position. One of these was barely two hundred yards down the slope in front of the British spectators. The sole occupant of this trench was facing towards the sea, oblivious to the danger concealed behind him.

A mortar tube could be seen pointing up from a large trench on the forward slope of the low round hill to the rear of the station. Nearby was a strongly constructed headquarters dugout. The positions were well laid out, but the force was too small to resist a determined assault.

'Tim' reported his findings back to Cedric Delves in HMS *Antrim* via his own radio operator at Stromness. He added his assessment that the beach running along the northern shore of Leith Harbour offered a good prospect for a landing force. His advice was acknowledged, but he was told that Leith was now unlikely to be the first target. The Navy had attacked a submarine off Cumberland East Bay and preparations were in hand for an immediate assault on King Edward Point. After several days of frustration, events suddenly were moving very quickly indeed. □

VIII | Search to Destruction

FOR CAPTAIN BRIAN YOUNG, the pressures of command increased dramatically on the morning of 23 April. The SBS reconnaissance party had been inserted at Hound Bay by the *Endurance* and, although there had been no subsequent radio contact, appeared to be following its original plan. The SAS landing on the Fortuna Glacier had failed, but part of 'Tim's' Boat Troop was established on Grass Island and starting to send back detailed reports. The *Tidespring* and the *Brambleleaf* had made their rendezvous and, accompanied by the frigate *Plymouth*, had commenced the long pumpover. Losses so far consisted of two Wessex helicopters and possibly one Gemini and its SAS crew, but Young was still only one day behind schedule. Then, at 1130, a signal was received from the *Endurance* to the effect that a C-130 Hercules was in the area. It had been sighted in the past half hour by Bill Doidge and Paul Martin – the two BAS biologists working at Schlieper Bay on the island's north-western tip. They reported it on their Squadcall radio and the transmission had been monitored by 'the plum's' operators. It was obvious that this aircraft must be Argentine.

The Hercules flew down the north-east coast of South Georgia, its altitude varying between two and five thousand feet as it flitted through the banks of cloud and overcast. Evidently it was searching for any British ships which might be anchored in the numerous bays or operating off shore. Its radar emissions were detected by HMS *Antrim* as it came further south toward Stromness Bay, but apparently it failed to find the ship.

Banking away from the coast, the Hercules started to search out toward the area where the *Brambleleaf*, *Tidespring* and *Plymouth* were engaged in their pumpover. Despite the poor visibility, the Argentine pilot found them and closed to within visual distance. He soon turned away, presumably to avoid any missiles which might be launched by HMS *Plymouth*. It seemed unlikely that he could have identified them with any degree of accuracy, but Young had to assume that the enemy was now fully aware of the Paraquat force and had a fair idea of its strength.

Following the departure of the first Hercules, another took its place, at 1330, and it too began a search at low level off the island's north-east coast. Then, at 1456, the *Endurance* reported to Captain Young that her electronic warfare operators had intercepted a high-frequency transmission which they believed to have come from an Argentine submarine. The strength of the signal suggested that the boat was within one hundred miles of the *Endurance* and therefore, possibly, even closer to the other Paraquat ships. Later on, after being told of this detection, Young received further information which told him that there was indeed an Argentine submarine approaching South Georgia and all the evidence suggested that she was the *Santa Fe*.

For the Task Group Commander, the situation had changed entirely. Apart from the initial reports by 'Tim', regarding the situation at Leith Harbour and Stromness, no useful intelligence had yet been received in respect of Argentine positions ashore. At least part of the Paraquat force had been seen from the air. There was the threat of an imminent submarine attack. The whole emphasis of Brian Young's thinking now shifted toward the need to safeguard his ships and their embarked troops. The military reoccupation of the island must, for a while, take second place.

At about this time, Captain Young received a call from Vice-Admiral David Hallifax, Chief of Staff at Northwood, on the secure voice satellite communications system. Hallifax wanted to amplify an earlier signal which had stressed the need to avoid shipping losses. Young was able, during this conversation, to request a change in his operating parameters. He had been under orders, so far, not to open fire on any Argentine aircraft. Now, with an enemy submarine in the vicinity and the probability that it would be guided towards his ships by the Argentine air force, he was anxious to have a free hand. Hallifax replied that his request would be considered but, until advised otherwise, no action could be taken against the surveillance flights.

Young decided to order David Pentreath to break off the *Brambleleaf/Tidespring* pumpover. He was told to move all three ships to the south-east at the tankers' best speed of seventeen knots. Having distanced themselves from the approaching submarine, they were then to regroup forty miles north-east of the island. The pumpover was far from complete, but the safety of the tankers was paramount.

The two tanker captains were acutely aware of their vulnerability to torpedo attack. Captain Mike Farley describes the *Brambleleaf*'s first day with Task Group 317.9: 'When we arrived at our final rendezvous, late on 22 April, we had no idea why we were there or who we would meet, but I guessed that we were due to refuel another RFA. During the early hours I was called to the bridge and was informed that a darkened ship had flashed at us with a Morse light. We flashed back at the indistinguishable shape – using a challenge code – but saw no reply. The wait was nail-biting, to say the least. As dawn lit the sky it was a relief to find the *Tidespring* there. The pumpover was quickly organised and we started to transfer fuel in foul sea conditions. We had managed to transfer seven hundred tons of fuel oil and two thousand four hundred tons of diesel when the submarine report came through. We were under the tactical command of Captain Pentreath in the *Plymouth* and he ordered an emergency breakaway. This is an evolution which the RFA practice regularly. It is important if a ship suffers an engine or steering failure in mid-pumpover, so it was quickly accomplished. We turned south with

Brambleleaf, the Royal Fleet Auxiliary tanker which came from the east coast of Africa to support Captain Young's task group. Here she is photographed fresh from dockyard hands, travelling in ballast. She lost much of her smart paintwork during the great storms off the Cape of Good Hope and off South Georgia.

orders to, "Make best speed". It takes time to build up momentum with a tanker and I was soon being asked to go faster. When I said that I was already making my best speed I was told to, "Make better than best speed!" Soon it got dark and we were in among the icebergs. Until then we had been in radar silence, but Captain Redmond and I both decided that the ice was now a greater risk than Argentine torpedoes, so we switched the sets on for the first time in more than a week. The screen showed that we were surrounded by bergs and growlers. I had a few grey hairs before Operation Paraquat, but I had a lot more after that night.'

The severe cold was causing difficulty for the civilian tanker crews. 'At Ascension,' recalls Captain Redmond of the *Tidespring*, 'we had been given just twenty sets of army camouflaged jackets and trousers, and twenty pairs of ordinary army woollen gloves. This stuff was not waterproof and gave little protection against the cold. The longest a man could bear to stay on lookout was twenty minutes. After that he became too chilled to do his job efficiently. Similarly, while we were pumping over, the hands working on the deck parties had to be brought inside every fifteen or twenty minutes because they became so wet and cold that we would have had accidents if they had stayed out in the open any longer. It was very impressive the way the lads stuck to the task under those conditions.'

Late in the afternoon of 23 April a conference to review the options was held in HMS *Antrim*. Both Young and Sheridan had received telephone calls from their superiors querying the lack of military progress. For the Task Group Commander, there were still the two conflicting requirements – to get the troops ashore, but not to hazard the ships which were carrying them. Unless resolved soon the air and submarine threats could cause the entire operation to become very protracted, thereby defeating its original purpose. The pressure was now on – quick action was needed to maintain momentum.

Captain Young recalls his feelings during that period: 'The Argentine air activity was a factor I could do nothing about until such time as the Rules of Engagement were amended. The submarine, on the other hand, posed an immediate danger and our capability to counter it was very limited. My main anti-submarine weapon – *Antrim*'s Wessex 3 – had been stripped to enhance its troop-lifting capacity, and it would take time to reinstall the sonar equipment. After some discussion it was agreed that steps had to be taken to secure a military foothold ashore while at the same time safeguarding the two tankers and the main embarked force, 'M' Company. Accordingly I gave orders for David Pentreath in *Plymouth* to escort *Tidespring* and *Brambleleaf* to a position two hundred miles north-east, clear of the Maritime Exclusion Zone, overnight on 23 April, and to complete the pumpover on 24 April. Meanwhile, Nick Barker in *Endurance* was to return to the vicinity of Hound Bay to support the SBS insertion, while *Antrim* would proceed to put the SAS Mountain and Mobility Troops ashore in Stromness Bay at first light on 24 April. The intention was that they should seize Leith Harbour. This would eliminate at least one of the Argentine bases and would be a concrete result. Having achieved that objective, I would rejoin the tanker group and replenish. At the same time, the Wessex 3 would be reconfigured to the ASW role and we would await the arrival of another frigate which was on her way to join us. The additional firepower, helicopter lift and ASW capability brought by this ship would enable us to mount an offensive against the submarine and generally push forward with a reoccupation of the island.'

Reasonably content with the new plan, Captain Young gave his orders and the ships moved towards their new positions. HMS *Antrim* started to move in towards Stromness Bay with the intention of making a direct frontal assault against Leith Harbour in the morning. Young was prepared, if necessary, to use naval gunfire – firing over open sights – to encourage a speedy capitulation. Then, late in the evening, yet another Hercules arrived on the scene. It made a radar search down the coastline, and Young had reason to believe that this time the *Antrim* had been discovered. This meant that his position would be reported to the approaching *Santa Fe*. To go ahead with the Leith Harbour operation would involve placing the big destroyer in a very dangerous position. She might find herself with no searoom, trapped by the submarine inside Stromness Bay. The risk was unacceptable and the operation was cancelled. The *Antrim* turned back and headed out towards the tanker group.

Yet another complication surfaced at the end of what had been a difficult and eventful day. Whenever conditions permitted, the men of the Task Group listened to the BBC World Service news

Captain John Coward, RN, photographed on the bridge of HMS *Brilliant*. His ship (facing page) was the second vessel (after the RFA *Brambleleaf*) to be assigned to Operation Paraquat when it became evident that reinforcements were needed. Like her five sister ships of the Type 22 *Broadsword* class, HMS *Brilliant* was designed primarily for anti-submarine operations. In the tradition of former generations of destroyer captains, John Coward pressed his ship hard to reach the scene of imminent action.

reports. Not only did they provide a link with home, they also provided a useful political background against which their own activities could be judged. At 2150 on 23 April they heard this report from London: 'Mr Costa Mendez has said in Argentina that he would not be making his trip to Washington unless he had some hope of a settlement. In an interview with a BBC correspondent in Buenos Aires, Mr Costa Mendez said that he put the chances of a war between Argentina and Britain at half and half. Some of the British warships in the main Naval Task Force heading for the Falkland Islands are reported to be within striking distance of the island of South Georgia, which is eight hundred miles from the Falklands. South Georgia is said to be defended by less than two hundred Argentine troops. The Ministry of Defence in London has denied that British troops have already landed on South Georgia.'

This was the first time that any BBC news report had made specific reference to South Georgia since 4 April, nineteen days earlier. The men of Captain Young's force had been following the accounts of the highly publicised departures from United Kingdom ports of the Operation Corporate ships, but they had assumed that the ultra-secret Operation Paraquat would not be compromised – as they interpreted it – by the British Broadcasting Corporation. They heard this announcement with a marked lack of enthusiasm. They were even less enchanted when, on the following day, the BBC reported that, 'two British warships and a troop carrier have been sighted about fifty miles off South Georgia.' Similar reports appeared on 24 April in British newspapers, repeating and giving credence to the reports carried by the Argentine news agency TELAM and published in the Argentine press. The pilots of the Hercules reconnaissance aircraft had evidently done their job all too well.

The sightings prompted General Galtieri to proclaim that his troops on South Georgia, 'will fight on to the last drop of their blood.' He also promised that, if the British attacked his garrison, his government would ask the Organisation of American States (OAS) to invoke the Rio de Janeiro mutual defence treaty of 1947.

In the event, the OAS did not directly involve itself in the conflict between the two countries, but the open discussion regarding the presence of the Task Group and its probable next move meant that the element of surprise integral to the Paraquat plan had largely evaporated. Brian Young weighed the implications and decided that he had no choice other than to pursue the plan agreed earlier in the evening. He must ensure that the pumpover could be safely resumed in the morning, and meanwhile await the arrival of his reinforcement.

The unit ordered by Northwood to reinforce Young's group was the Type 22 destroyer HMS *Brilliant*, commanded by Captain John Coward. For the past two weeks, Coward had been leading the spearhead battle group of Admiral Woodward's Operation Corporate force down towards the Falkland Islands. Coward, a softly spoken but fiercely determined officer, was revelling in the novelty and challenge of his role. It had been defined by the Commander-in-Chief, Fieldhouse, when he flew down to Ascension Island for a meeting with Admiral Woodward. The general thrust of the instructions given to Woodward, subsequently passed to Coward, was that the spearhead force should 'go fast and go far' to establish a British naval presence close to the Falkland Islands. If the continuing negotiations between Argentina, the United States, Great Britain and the United Nations were to result in a suspension of further hostilities, the British government would at least have control of a stretch of ocean which could be exploited at the peace table. In Captain Coward's phrase: 'It suited the Navy to be frozen well down track rather than at Ascension.' At the same time, and assuming a failure of the diplomatic initiatives, the presence of the forward group might tempt the Argentine navy to come out and fight. Its defeat would ensure the safety of the troop transports destined for the Falkland Islands.

At this stage the British government was still handling the crisis with kid gloves. It was important that the international community should not view the British response as a vengeful act of retaliation. Within the Navy itself, attitudes varied, according to one senior officer, between two extremes – 'ask them over for lunch, on the one hand, or roast 'em for breakfast, on the other.' The orders given to Coward were that he could not open fire at will, but only in defence of his own ships. He interpreted these instructions as implying that any Argentine submarine, warship or aircraft closing the task group could be doing so for one reason only, and that would be with hostile intent, and could be attacked and destroyed. In the event, the long-range reconnaissance flights by military Boeing 707 aircraft – which soon were to cause much concern to Admiral Woodward with the main task force – failed to detect Captain Coward's small group of ships as it raced southward.

Coward's force consisted of the *Brilliant* (4,000 tons), the Type 42 destroyers *Coventry*, *Glasgow* and *Sheffield* (4,100 tons), and the Type 21 frigate *Arrow* (3,250 tons). Accompanying them was the support tanker RFA *Appleleaf* (40,200 tons with full load). Whenever the weather permitted, Coward increased speed to 25

knots. The maximum speed of the *Appleleaf* was 15.5 knots so, from time to time, she fell behind the warships. Captain Coward's view was that, 'she could always catch up if we ran out of fuel. The main thing was to get south. Initially I assumed that we were headed for a battle with surface ships. This changed as the days went by. Then I became more concerned by the air threat, and attack by Exocet missile in particular.'

The high-speed passage was clouded not only by lack of knowledge of Argentine intentions and capabilities, but also by the condition of the British ships. Neither the *Sheffield* nor the *Coventry* were in first class order because they were overdue for dockyard attention. The *Sheffield* had engine room problems and the *Coventry* had engine and steering problems. The little *Arrow* had difficulty in keeping up with her bigger sisters and, in heavy weather, there were worries that she might not stand up to the repeated hammering.

By 22 April, the spearhead group was one thousand miles north-east of the Falkland Islands. Having trained hard all the way south, Coward and his fellow commanding officers felt ready for whatever orders Admiral Woodward might give them. Coward himself would have preferred to sail straight into Port Stanley and start bombarding the Argentine positions. The spirit was admirable, but the lack of air cover and the political restraints made such a project invalid. Then came the order from Admiral Woodward for HMS *Brilliant* to detach from the spearhead force and to proceed immediately for South Georgia. Having lost the two helicopters on the Fortuna Glacier, Brian Young's Paraquat force was desperately short of troop-lifting capability. Further, it was under threat from submarine attack. The *Brilliant*'s hangar accommodated two Lynx helicopters and she was equipped specifically for anti-submarine warfare. She was a natural choice as reinforcement for Paraquat.

In total radar silence, Captain Coward headed for South Georgia at the maximum speed he could safely manage. It had become much colder, and he ran through a series of gales but, with South Georgia now emerging as the setting for a real shooting war, he was determined to place himself under Young's orders without delay. Nelson's injunction – 'lose not an hour' – fitted the mood precisely. 'I knew from past experience that the *Brilliant* could go fast into bad weather. *Sheffield* and *Coventry* tended to bury themselves, but *Brilliant* would go through a full gale at twenty-five knots if you kept your nerve. It got very noisy on board, and a few things got broken, but there was no structural damage. Perhaps I was lucky, but I didn't slow up until I neared South Georgia and entered the iceberg area.'

Coward continued to press his ship hard, but he knew that he could not join Brian Young before the afternoon of 24 April at the earliest. If the weather was bad, he could not join before the morning of 25 April which was when Young expected to see him. In the meanwhile, much could happen. Coward's main worry – apart from icebergs and the presence of HMS *Conqueror* somewhere along his track – was that he might arrive too late to join in whatever was about to happen around South Georgia. So secret had been the conduct of the Paraquat operation that even Coward knew very little about it, and what little he did know was based upon personal deduction rather than official advices.

Having decided to abort his attack on Leith Harbour, Young pondered his next move as, during the night of 23/24 April, the *Antrim* headed north-east to join HMS *Plymouth* and the two tankers. Two signals received from Northwood during the early hours of the morning caused him to make a fresh assessment.

Twenty-four hours earlier, the nuclear powered submarine HMS *Conqueror* had been ordered by Northwood to leave South Georgia waters and to proceed north-westward at high submerged speed. Wreford-Brown was to act as forward screen for Task Group 317.8 – the spearhead battle group from which HMS *Brilliant* had recently detached and which was now approaching the Falkland Islands. Detection of the *Santa Fe* had caused Northwood to change that plan and Wreford-Brown had new orders to return to South Georgia at maximum speed. The

Conqueror's sophisticated underwater listening devices might be an invaluable aid in finding the Argentine submarine. Captain Young was told that the *Conqueror*'s new designated patrol area was the stretch of sea off South Georgia's northern tip.

Censored by the Ministry of Defence.

For the ship's company of HMS *Endurance*, the departure of HMS *Antrim* was a depressing development. The red ship was now once again entirely alone and in an increasingly hostile location. To quote one of her senior ratings: 'The Navy had gone off two hundred miles, and there we were, on our own, with our diesel engine booming away like Big Ben, acting as a magnet for the *Santa Fe*. All we could do was cross our fingers and hope for the best.'

In the event, 'the plum' found plenty to do on the morning of 24 April. Returning to her old haunt off the Barff Peninsula, she launched Tim Finding in Wasp 435 who delivered Peter Stark back to the company of his BAS colleagues at St Andrews Bay. Finding stayed long enough to give the civilians a brief account of the latest developments before embarking Chief Petty Officer Scott and taking him back to the ship. Wasp 434, meanwhile, was making a series of flights to the far side of the Sörling Valley where Lieutenant-Commander Ellerbeck collected the stranded SBS party. The marines were in good physical condition but deeply disappointed that they had failed to get through to their target area. Even on this third day of waiting, their intended route across Cumberland East Bay was still choked with a thick mass of growlers and brash ice.

Over the horizon, miles to the north, the RFAs *Tidespring* and *Brambleleaf* had resumed their pumpover at first light. It was anticipated that this operation would last most of the day. Captain Young planned to replenish his own ship and HMS *Plymouth* as soon as the big ships had finished, but a fresh flurry of signals received during the afternoon caused him to accelerate his timetable. The *Endurance* reported that she had intercepted a message from the BAS station at Bird Island. Paul Goodall-Copestake and Julian Hector were announcing that they had seen two warships, two small jets and a Boeing 707 on the horizon. This report was, in fact, one of the summaries which Bird Island had been sending for onward transmission to Cambridge. Some of the details referred to previous events when the scientists may have spotted two of the Paraquat ships on the horizon. The 'small jets' may have been the ships' helicopters observed at great range and in murky weather. Such a mistake would be easily made. The big Boeing, however, was correctly identified and this aircraft was already approaching the Cumberland Bays area.

The new arrival was one of the three 707s operated by Escuadron II of Grupo I de Transporte Aereo, based at El Palomar on mainland Argentina. The Argentine air force had first taken these aircraft into service as long-range military freighters. As recently as 25 March, one of them, TC-93, had visited Standsted Airport (London) to collect stores and equipment purchased from the British Ministry of Defence. Their flight endurance made them suitable for protracted maritime surveillance tasks and one of them, TC-91, was the first aircraft to locate and identify Admiral Woodward's Corporate force on 21 April. It may have been the same aircraft and crew which flew to South Georgia three days later.

Like the three Hercules of the previous day, the 707 flew down the north-east coast of the island at approximately 3,000 feet but then, instead of turning away when it reached Stromness Bay, it continued southward to Hound Bay and passed near HMS *Endurance*. This happened at a moment when Captain Nick Barker was talking on the satellite telephone with Vice-Admiral Hallifax at Northwood and reporting the evacuation of his SBS party. The Chief of Staff repeated, in passing, the new instructions he had given to Captain Young – authority had been granted to open fire on enemy aircraft but only if they offered a direct threat.

Barker felt that his treasured red ship was being risked unnecessarily. 'One moment, Sir.' He held the telephone receiver toward the door of the radio cabin for a few seconds and then spoke again. 'Hear that, Sir? That racket is my ship going to action stations. We have an Argentine aircraft overhead. The pilot is talking in plain language to a submarine telling him where I am and what I am doing. I really do think my position is becoming rather marginal.' The 'do not fire' order was repeated and Tony Ellerbeck was despatched at speed from the bridge to pass the instruction to the Oerlikon gunners and to an eager group of SAS marksmen gathered on the hangar deck.

The Bird Island report had been received by Captain Young with considerable scepticism. If two Argentine surface ships were in the area, surely they would have been heard by HMS *Conqueror*? In theory, nothing could get past her. On the other hand, in tactical planning terms, Young could not entirely discount the possibility that one or more enemy frigates were even now slipping quietly down the island's north-east coast toward the Cumberland Bays. They could land substantial reinforcements for the garrisons at Leith Harbour and King Edward Point, and could attack and sink the lonely *Endurance*.

There were similar uncertainties regarding the 'jets' allegedly seen from Bird Island. Little was known, at this early stage of the campaign, of Argentine capabilities with regard to in-flight refuelling. It was not even known whether the airfield at Port Stanley was being exploited as a forward operating base for combat jet aircraft. In broad terms, it seemed unlikely that any of the combat or naval strike aircraft types operated by the Argentine air forces could venture this far east, but the possibility could not be ignored. As a former naval strike pilot, Brian Young was fully aware of the damage which could be inflicted on his ships by a skilfully handled pair of Skyhawks.

Captain Young studied the latest signals from Northwood and the current assessments prepared by his warfare team. He decided to stop the tanker pumpover forthwith. It was not yet complete, but the *Tidespring* had taken on enough fuel to sustain the Paraquat operation for several days. The *Brambleleaf* could be released for other tasks to the north and her departure would signify that he had one less vulnerable tanker to worry about. Young now proposed to replenish his own ship and HMS *Plymouth* and to then turn back for South Georgia immediately. The *Brilliant* had not yet joined and was not in fact expected to do so until the morning of 25 April. Young gave orders to Captain Redmond that he should turn north-west overnight and meet the *Brilliant* when she reached local waters. Having refuelled herself from the *Tidespring*, the *Brilliant* could then leave the tanker loitering outside the Maritime Exclusion Zone while she herself hurried south-west to catch up with the *Antrim* and the *Plymouth*. In the event, John Coward's ship made better time than expected and she was able to start refuelling from the *Tidespring* that evening.

Captain Brian Young, RN, pictured outside Buckingham Palace when subsequently he was honoured by Her Majesty the Queen for his services off South Georgia. The citation accompanying his award of the Distinguished Service Order, published in *The London Gazette* on 4 June 1982, makes specific reference to his 'leadership, bold initiative and cool appraisal of the situation'. Significantly, the citation concludes: 'The importance of this operation (Paraquat) to the overall strategy of re-establishing British administration in the Falkland Islands and their dependencies cannot be over-stated, nor can Captain Young's personal contribution to its success'.

As the *Antrim* and the *Plymouth* headed back toward the island during the closing hours of the day, Young and his team formulated their plan for their operations against the *Santa Fe*. The submarine presented a deadly threat. In common with all such vessels, she could not be boarded, arrested or shadowed. She could only be either ignored or sunk. The Standard Operating Procedure (SOP) following detection of a hostile submarine in these circumstances was 'search to destruction', a phrase which, when he heard it, appealed immensely to Cedric Delves and the SAS philosophy.

Yet another signal from Northwood was received at this time. It stated that if the *Santa Fe* were to enter South Georgia inshore waters within the next few hours, it was reasonable to assume that she had a specific reason for doing so, and a prime possibility was that her captain intended to land reinforcements for the garrison. That achieved, he would wish to distance himself from the immediate coastline. Once back in deep water, he would be free to strike wherever and whenever he chose. The *Santa Fe* might be old, but she still carried a powerful punch in the form of her ten 21-inch torpedo tubes.

It now seemed clear that the Argentine submarine would be closing the coastline that night, 24/25 April. She had two possible destinations – either Leith Harbour or King Edward Point. If she went to the former she would be seen and reported by 'Tim' and the SAS Boat Troop on Grass Island. If she berthed at King Edward Point, on the other hand, her arrival would be unobserved by the British. In theory, the BAS scientists at Lyell Glacier could have been asked to make observations from Mount Hodges, but Captain Young was unaware of their existence. The only means of detection clearly available to Young were his helicopters and he issued his orders accordingly.

The plan for the Paraquat helicopter force was simple. The principal role of finding the submarine – either on the surface or submerged – belonged to the only helicopter equipped with both radar and sonar, HMS *Antrim*'s Wessex 3 (the ubiquitous 'Humphrey', known more formally by the call-sign 406). It would be Lieutenant-Commander Stanley's task to go into the lion's den at first light and locate the *Santa Fe*. Assuming that the submarine made a brief overnight visit to King Edward Point to offload her passengers and to take on fresh stores and, possibly, fuel, she would presumably start to make her way seawards sometime around dawn. With any luck, Stanley would find her off Cumberland East Bay.

The orders sent to HMS *Brilliant* were that her two Lynx should survey the offshore area north of the entrance to the Cumberland Bays. Their primary role was to search for the two 'warships' reported by Bird Island, but they could also support Ian Stanley if and when the *Santa Fe* was located.

The Wasps operated by HMS *Endurance* and HMS *Plymouth* were to wait at alert on the flight decks, armed with AS12 air-to-surface missiles and ready to launch. It would be their role to attack any surface ships located by the *Brilliant*'s Lynx or support the *Antrim*'s Wessex 3 as necessary.

While the *Antrim* and *Plymouth* headed toward the island, HMS *Brilliant* made her rendezvous with the RFA *Tidespring*. The latter had taken on 3,500 tons of fuel oil and a large quantity of fresh water during the day from the *Brambleleaf*, and now she pumped part of her reserves into the thirsty *Brilliant*. A Royal Marines 'M' Company officer embarked in the *Tidespring*, Lieutenant Ewen Duncan, watched the operation: 'It was a pitch black night. Apart from a few dim red lights, both ships were in total darkness. It seemed incredible that they could do the job under those conditions.' It was in fact completed without incident and the *Brambleleaf* turned away north for a rendezvous with the chartered BP tanker *British Esk*. Subsequently she continued northward for Ascension Island to make good the structural damage suffered during the great storm of 21 April and to deal with a microbiological contamination problem in her tanks. Captain Farley and his crew played no further part in Operation Paraquat.

Captain Coward was anxious to have a personal briefing from the Task Group Commander. He was lifted off in Lynx 341, flown by Lieutenant Nick Butler as pilot and Lieutenant Barry Bryant as observer, to fly down the wake of the rapidly departing HMS *Antrim*. Despite the use of intermittent radar, Butler was forced to turn back after an hour of vain search.

Coward was uncertain regarding the latest situation and what it was that was required of him. Matters were clarified during the night when he was told to close the coast, aiming for the area of Possession Bay where his two Lynx were to conduct an anti-surface ship and anti-submarine patrol at first light.

HMS *Antrim* and HMS *Plymouth* continued to head for the entrance to the Cumberland Bays through the uncertainties of a moonless windy night. It was at this stage that Captain Young received two further items of intelligence. First, Northwood had extended the *Conqueror*'s operational area to cover the entire northern and north-eastern coast of South Georgia. This news was received by Young with mixed feelings. It inhibited his movements to the extent that his ships could attack only surfaced submarines which were positively identified as being Argentine. He calculated, however, that the *Conqueror* could not reach local waters before the afternoon of 25 April. In the interim, provided that Wreford-Brown was informed of his dispositions, he could operate freely along the eastern seaboard. This was agreed, and Young concentrated his mind upon the second piece of intelligence.

The Argentine submarine had departed Mar del Plata on 9 April. At Northwood, in the Paraquat Cell, Commander Davis had made full use of his specialist background in underwater warfare to estimate the boat's likely track and speed. It had been his view that Captain Bicain would arrive off the Cumberland Bays some time during the period 24/25 April. The electronic detection by the *Endurance*'s operators had validated Dennis Davis's predictions in full. Northwood was now able to inform Brian Young that the *Santa Fe* would probably enter Cumberland East Bay. It seemed likely that she would waste little time alongside the jetty at King Edward Point and would quickly put to sea again with the intention of attacking the British ships. Any previous doubts could be put to one side – the *Santa Fe* would indeed be emerging from Cumberland East Bay at dawn.

Captain Young made his final dispositions as the *Antrim* and the *Plymouth* raced through the dark early morning hours toward the coast while, further north and increasing speed to twenty-five knots, HMS *Brilliant* headed in toward Possession Bay. Having started two hundred miles out, the three warships could not expect to reach the coast before approximately 1030. That would be too late to catch the submarine on the surface. Only the Fleet Air Arm helicopters had the speed and the agility to be in the right place at the right time. Each ship's flight prepared itself accordingly. For the maintainers, the next few hours would put to the ultimate test of combat the concentrated effort which they had devoted to the six helicopters in their care. If only one engine or component were to fail, the enemy submarine would have a hugely enhanced prospect of escape.

At 0730 the radio office of HMS *Plymouth* intercepted a live voice transmission in Spanish on the VHF Channel 16 civilian maritime frequency. The operator did not understand what was being said, but the transmission was reported immediately to Captain Pentreath who informed the operations room in HMS *Antrim*. The submarine was undoubtedly on the move and must have been experiencing a communications problem. Channel 16 is an open frequency used by merchant ships for distress calls and normal commercial traffic. It was extremely unlikely that there were any Spanish or Latin American-owned merchantmen in the area, so, given the limited range of VHF radio, the British could assume that this was Captain Bicain talking to the radio shack at King Edward Point. It could be further assumed that Bicain had departed King Edward Cove and had tried to announce his departure by HF radio to his mainland base, but had been frustrated by the high intervening land mass of South Georgia's mountains. Presumably he was now asking the garrison operators to do the job on his behalf. It was a remarkably ill-considered action on his part. In the freakish propagation conditions often encountered in the area, it was almost inevitable that the transmission would be heard by one of the British ships. There was now very little doubt that the Fleet Air Arm was about to engage its first major naval target since Lieutenant Robert Gray, RCNVR, won his Victoria Cross on 9 August 1945.

The first helicopter launches were made shortly after 0800. Lynx 341 departed from HMS *Brilliant*, armed with a single Mark 46 aerial torpedo. The pilot had orders to patrol the north-east coast. The flight deck of HMS *Antrim* saw the departure of 'Humphrey', armed with two Mark 11 depth charges. The Wessex 3 had been refitted overnight with its active sonar system and was therefore ready to revert to the full submarine hunting role. The normal time required for installing and testing the 'pinger' equipment was two to three days. Chief Radio Technician Jim McKee and his team completed the task in just fourteen hours and, to their immense credit, the system worked perfectly at the first attempt.

Further south, off Hound Bay, Wasp 434 stood on HMS *Endurance*'s small flight deck, waiting for the order to launch. The pilot was Tony Ellerbeck and the observer David Wells. Fitted to the sides of the helicopter were two AS12 air-to-surface missiles. A French-designed weapon with a maximum range of 7,000 yards, this missile was wire-guided. When the time came, it would be Wells' job to steer it in flight by sending command signals through a joystick control and along the wire. He had made many previous practice 'firings' on the ship's simulator, but had never before fired a missile in earnest. After the past weeks of frustration, when they had not been permitted to incur the risks of direct action against the Argentines, the two aviators and the rest of the ship's company were tense and expectant.

Lieutenant-Commander Stanley entered Cumberland East Bay at 0845. Since leaving the *Antrim*'s flight deck he had maintained radio and radar silence. Now, flying at 200 feet, he continued to rely solely upon the 'mark one eyeball' to find his opponent. A fresh offshore breeze was carrying a thick mist down from the mountains and horizontal visibility was no more than half a mile. A layer of cloud at 400 feet was capped by another at 1,000 feet, the two combining to shut out most of the early morning sun's feeble light.

Stanley flew down the east side of the Bay before altering course to starboard towards Sappho Point. Then, turning north, he followed what he hoped would be the track of the departing submarine. Stanley deliberately stayed away from King Edward Cove in order to preserve the secrecy of the operation. Beside him, Stewart Cooper stared out through the port cockpit window. 'Fitz' Fitzgerald crouched behind the GPMG mounting, at the door, keeping watch to starboard. In the observer's seat, Chris Parry continued to calculate the helicopter's speed, heading and drift, and pass navigational directions to the pilot.

Parry describes the function of an ASW helicopter observer: 'Essentially you are there to navigate the aircraft, to plan tactics, and to operate the weapons systems. There are other things, of course, such as communications with ships and other helicopters, but on that particular morning it was a question of assisting Ian to get into the Bay without being detected and then helping him to find the submarine. It had been a longish flight in from *Antrim* and visibility was very poor indeed. I couldn't use the radar because the enemy would have detected us approaching the coast, so everything was done by dead reckoning. We came in at exactly the right point – on the east side of the Barff Peninsula. Once we had visual contact with the ground, Ian took us around the tip and then down the inland side of the Peninsula. It was very grey and gloomy inside the Bay and all we could see was the vague shape of occasional cliffs and a pretty rough sea underneath us. Then we made our turn and Ian headed back out towards the mouth of the Bay. I was starting to feel worried because there didn't seem much prospect of sighting anything under those conditions. I suggested to Ian that I should switch on the radar for a single one-second sweep. One was all I could make – any more would have told an enemy skipper that he was under surveillance from

HMS *Conqueror* returns to her home base of Faslane on 3 July. Following her brief involvement in Operation Paraquat and the destruction of the powerful cruiser *General Belgrano*, she maintained a close watch on the Argentine coastline. Following the custom of the service, she flies the unoffical piratical Jolly Roger. Apart from the emblem of an enemy ship sunk, the flag carries also a dagger symbol. This records at least one clandestine operation involving special forces.

the air and he could plot our speed and course. I illuminated and immediately picked up five or six traces in the area ahead and out to sea. I was expecting this because I had marked on my chart several icebergs which we had been plotting for several days and which all had their individual names – Fred, Charlie, and so forth. But there was just one blip – about five miles due north of Barff Point – which showed up where it shouldn't have been. It was faint, but I felt instinctively that it was our submarine. I reported it to Ian and gave him the range and bearing. Shortly afterwards, in a very cool tone, he said, "Submarine ahead."'

Stanley, Cooper and Parry conferred rapidly, trying to decide whether this was indeed the *Santa Fe*. There was the slight possibility that it might be the *Conqueror*. Parry was the Flight's recognition officer. He studied the boat through his blister window and confirmed that it was a Guppy class submarine. He flicked the switches arming the two depth charges and quickly calculated the relative speeds and bearings of the hunter and the hunted. 'There is a rule of thumb which states that a surfaced submarine will travel twenty yards per knot. The forward throw of the Mark 11 depth charge is one hundred and sixty yards from a helicopter travelling at one hundred knots and an altitude of one hundred feet. I made another one-second sweep with the radar and this gave me the submarine's speed – eight knots. The arithmetic was therefore easy – eight knots at twenty yards per knot meant that we would score a direct hit if we released when directly overhead. The charges would throw forward one hundred and sixty yards and the target would go smack over the top of them as they sank.'

Ian Stanley increased speed to precisely 100 knots and approached the submarine from astern. The fin (conning tower) was unmanned and the periscope was in the raised position. It seemed likely that she was about to dive. In Ian Stanley's words: 'It was a fairly unsophisticated form of attack. The Mark 11 depth charge is a very old-fashioned weapon – just a can of explosive and a big bang. Normally it would be dropped on a shallow target under the control of a second helicopter operating a sonar. In this case the target was on the surface and I was on my own, so it was simply a case of flying over the top and dropping by eye. I made my approach in a shallow dive, calling out the decreasing range to Chris in the back. As the submarine disappeared under my nose, I said "Now!" and Chris completed the attack.'

Chris Parry had been waiting for this moment for the past thirty seconds. He had lifted the small red cap covering the weapons release button and his finger was poised. There was a small gap in the floor where the sonar well joined the fuselage and he could see the white-crested waves flicking past below the aircraft. Suddenly he saw the stern of the target and jabbed the button at the same moment that he heard Stanley's shout of 'Now!' in his earphones. The two depth charges fell away, arching into the sea ahead of the fin and on the port side.

Ian Stanley hauled the Wessex into a sharp turn to port. Two splashes were clearly visible among the tumbling waves. Several seconds passed as the charges sank thirty feet below the surface, the submarine meanwhile passing overhead. At thirty feet the hydrostatic pistols detonated and two tall columns of foam and spray burst into the air. Both weapons had exploded little more than twenty feet from the submarine's port quarter.

According to Chris Parry the effect of the underwater blast was dramatic. 'I was watching through the blister and could see everything very clearly. The stern lifted half out of the water, showing the rudder, part of the propeller and what I would call the barnacle line. The effect on the people inside must have been very frightening. The stern whipped about twenty feet to one side and then she started careering all over the place, constantly changing course.' From the pilot's seat, Ian Stanley could see the target 'turning her helm violently every few seconds, careering around but still more or less in the same area. Then "Fitz" opened up with the GPMG at four hundred metres and started to spray the fin to keep their heads down. I circled around the boat, which had apparently now decided to head back towards Grytviken, while Chris got busy on the radio.'

Lieutenant Parry announced the attack to HMS *Antrim* on a UHF frequency, but received no reply. Atmospheric conditions were poor and the ship was still forty-five miles away. He switched to HF and repeated his report. 'This time I made contact with a *Brilliant* Lynx pilot and he headed down from the north to join us. Coincidentally, our HF transmissions were picked up by several ships in the Corporate force and they had a first-hand running account of the battle. I gather that everyone up there got pretty excited. Some of the aircrew in the *Invincible* even recognised our voices and, because the Fleet Air Arm is such a small family, that gave a very personal touch to the proceedings.'

A subsequent analysis of Ian Stanley's attack suggests that one of the 300-pound charges may have fallen so accurately that it grazed the forward casing before bouncing into the sea. The 'killing range' of the weapon was nineteen feet, so the *Santa Fe* must have come very close to being sunk. As it was, the port after ballast tank was ruptured and several electrical circuits damaged. Oil started to leak from a split fuel tank, staining the wake of the submarine as she reversed her course and headed for King Edward Cove. She was no longer able to dive with safety, but her wounds were far from mortal.

Stanley and Parry debated how best they could further cripple the submarine. Fitzgerald was told to aim his fire at the two sonar housings – one on the bow casing and another at the foot of the fin – and at the vulnerable periscopes and radio antennae. He stopped firing only when he had used his supply of 500 rounds of 7.62 mm ammunition. Stanley then went into a hover while Fitzgerald 'dunked' the sonar transducer. With signals already being received in the helicopter it would be much easier to keep track of his movements if the submarine commander did decide to submerge. The Wessex continued the sonar watch for several minutes until Parry heard that the second helicopter had arrived over the target.

Lieutenant Chris Parry, RN, Observer in Wessex 406 and the officer whose navigational and tactical planning skills brought the helicopter to the right place at the right time.

Thirty miles to the north, Lynx 341 was still flying in towards the north-east coast when the crew, Nick Butler and Barry Bryant, heard Parry's first sighting report. Bryant was plotting his course on a photostat copy of the only air map of South Georgia to be found in the *Brilliant*. The helicopter was armed with a single GPMG mounted outside the starboard cabin door. Having only just entered service, and despite being a very stable gun platform, the Lynx was not fitted with the standard gun mounting. The ship's Flight simply did not possess one. The maintainers had instead constructed one to their own design by sawing through an office swivel chair and fixing to it the universal joint of a valve operating rod. An initial test firing had shown this contraption to be more serviceable than the official design. It was now about to be tested again, this time under battle conditions.

The other weapon carried by Lynx 341 was a single Mark 46 air-launched torpedo. This potent weapon – weighing only 230 kilograms but capable of seeking a submerged submarine down to a depth of a thousand feet – had been in service since 1971. It was still regarded as the best anti-submarine weapon in the Fleet Air Arm's armoury. A 'drop and forget' weapon, it carried self-contained homing sensors which permitted it to find its quarry and then explode against the hull. These onboard depth sensors were designed to obviate any risk that it might start hunting a friendly surface ship. It was believed at the time, however, that there was a chance of it attacking deep-draught surface targets if dropped with shallow settings applied.

Barry Bryant describes the tactical plan: 'We had a good supply of Mark 46 in the ship, but we carried only one on this patrol because of the weight factor. The weight of an additional torpedo would have reduced our flight time significantly. So this one was important – we couldn't afford to waste it. From what Chris Parry was telling us, the submarine had turned back, but was still under control and had apparently increased speed from eight knots to ten or twelve. This suggested that it might still be capable of diving. Everything hinged on what we found when we arrived on the scene.'

Most of the damage inflicted upon the Argentine submarine was caused by AS12 missiles. The example shown here is mounted on the side of Wasp 445, photographed just after the action as HMS *Plymouth* approaches the South Georgia coast. The *Santa Fe* has been turned back, but the *Plymouth*'s Wasp has been rearmed and is ready to launch again if necessary.

Nick Butler takes up the commentary: 'The *Antrim*'s Wessex guided us in towards the target and then told us he was going back for more depth charges. A moment later I saw the target. It was apparently not seriously damaged, but showed no sign of preparing to dive. Barry and I therefore agreed on the "shallow setting" option, the idea being to discourage the enemy from submerging for at least another six minutes – an optimum running time for the Mark 46.'

Lynx 341 ran down the wake of the *Santa Fe* at 130 knots and released the torpedo into the sea off the boat's port beam. It immediately went into a shallow dive and, like a gun dog sniffing for a scent, began to track backwards and forwards under the target's keel. The sound of its motor was heard by her sonar operators. Captain Bicain had watched through his periscope as the helicopter released the weapon and, recognising it as a Mark 46, knew that he must continue to run on the surface.

Following the launch, Butler pulled into a sharp turn and prepared to make another pass along the target's port beam. Bryant clambered into the rear cabin to fire the GPMG. Aiming at the vulnerable periscopes, he fired a series of long bursts, spraying 200 rounds over the side of the fin at a range of only fifty yards. Nick Butler banked away sharply to avoid not only the machine gun fire coming at them from a gunner positioned at an aperture at the front of the fin, but also to avoid Barry Bryant's ricochets, some of which were bouncing back from the submarine's hull.

Butler climbed and circled the target. Ian Stanley had gone back to the *Antrim* to rearm and it would be some time before he could return. What could be done to harass the Argentine commander in the meantime? 'You have to understand,' explains Barry Bryant, 'that we were in a very exhilarated frame of mind. This was it – the culmination to years of training. We had dropped our torpedo and, after about two hundred rounds, the GPMG had jammed. We were almost desperate to throw something – anything – at that submarine, so Nick made another low pass while I fired off a magazine with a Sterling SMG. He circled while I changed magazines. This time he went down the other beam so I crawled across to open the port cabin door. As I did so, the one hundred and twenty knot slipstream whistling through the two open doors ripped open a plastic envelope containing several skillets of "chaff". The interior of the aircraft was instantly filled with a blinding cloud of millions of tiny aluminium dipoles. Nick headed for a more dignified altitude and it took a few minutes for us to regain our composure and get the stuff out of our mouths and nostrils. I unjammed the GPMG, firing off another hundred rounds or so, and then had a go at the target with four magazines of SLR. I also fired off the other magazine of SMG. After that there was not a lot more we could do. I saw Nick starting to pull his 9mm Browning out of its holster, but then he decided that it might not be much good against a submarine and put it back again. We sent a report to Chris Parry and also told *Brilliant* that we were on our way back to rearm. I asked the Controller to have ready another torpedo and a Hoover. He didn't query it – even though he didn't know about our "chaff" problem – but when we touched down, there it was waiting for us, one Hoover!'

In the beleaguered interior of the *Santa Fe*, the communications team was still having difficulties trying to talk to any Argentine mainland base. Subsequent conversations between the submarine's officers and their British captors revealed that they tried a number of different frequencies, but failed to make contact. Throwing caution to the wind, they resorted to an insecure frequency and Captain Bicain transmitted his desperate report: *'Me atacaron helicopteros enimigos, me atacaron helicopteros enimigos.'* This open channel report was heard by the British and was of considerable assistance in confirming Chris Parry's first reports.

A third helicopter now arrived on the scene; it was Wasp 434 from HMS *Endurance*. Tony Ellerbeck recounts the events of the following fifteen minutes: 'We had been told the night before to stay at alert on *Endurance*'s flight deck pending a sighting report. I was not too happy with this idea and nor was Captain Barker. It was clear that the success or otherwise of the plan to sink the *Santa Fe* would quite probably hinge upon communications. That was all very well, but the *Endurance*'s Wasps were fitted with radio kit which was entirely different to the rest of the Royal Navy. Because of the ship's special role, both helicopters carried maritime VHF and maritime UHF, but no HF. The other helicopters in the Paraquat force were fitted with aeronautical UHF and aeronautical HF. The Lynx also had VHF, but it was AM

Lieutenant-Commander Tony Ellerbeck, RN, HMS *Endurance*'s Flight Commander and the pilot who flew Wasp 434 on a wide variety of hazardous missions during the months of March and April. This photograph was taken in 1986, at which time he was Second Lieutenant of Her Majesty's Yacht *Britannia*.

and not compatible with our FM. The result of all this was that Tim and I could not talk to the others once we were airborne. We could have passed messages back and forth through the mother ships, but that would have been a painfully slow process. We in *Endurance* decided, therefore, to get up to the likely scene of action and, to a large extent, do our own thing. To this end, I flew slowly up the seaward coast of the Barff Peninsula, looking for any signs of activity. I had reached the tip of the Peninsula when mother called us to say that she had heard a mayday call from the *Santa Fe*. The skipper was reporting his position and stating that he had just been attacked by a helicopter. I made a quick ninety-degree turn to port and flew in over the Right Whale Rocks. Almost immediately I saw a submarine travelling quite fast on the surface from right to left. I told my observer, David Wells, what I could see, but he said that he thought he could see instead a periscope feather travelling from left to right. This worried me because I thought the *Conqueror* might be in the Bay and that something was going wrong. Then David picked up the target at two miles and we went through the pre-release checks. He tried to fire the port missile but nothing happened. He repeated the checks and tried again. This time the AS12 went away and David steered it down through the turbulence and obtained a hit on the forward part of the fin. This missile injured a machine-gunner who had been firing at Nick Butler from an aperture in the fin. The gunner was knocked back through a door and he fell down the hatch into the submarine's control room. His sudden arrival, with a mangled leg and covered in blood, caused a lot of anguish to the people there.'

Ellerbeck's first round did not explode on impact because the fin did not form part of the pressure hull and was constructed only of glass reinforced plastic (GRP). Offering little resistance, this material did not initiate an immediate explosion in the missile's warhead. Instead, it passed through to explode near the surface of the sea twenty yards beyond.

The firing of the AS12 caused some consternation in Lynx 341. Circling at three hundred feet, Butler and Bryant had been unaware of Ellerbeck's approach. The bright streak of the missile flashing past below them provoked Nick Butler to ask, 'Bloody hell! What was *that*?'

Ellerbeck lined up for another attack and David Wells launched the starboard AS12 at a height of one hundred feet and a range of three miles. His point of aim was the junction of the fin with the hull. This time he was not so successful – the missile dived into the sea thirty yards short of the submarine's port quarter. Wasp 434 turned away and flew back to the *Endurance* for a reload.

During the next twenty minutes, the *Santa Fe* continued to head due south into Cumberland East Bay under spasmodic rifle and machine-gun attack from Lynx 341. The fire caused little damage other than to add to the number of holes in the GRP fin, but it obliged Captain Bicain to remain in his control room, conning the submarine through the periscope. The diesel propulsion system was still intact and he was able to maintain a speed of ten knots. His sole aim now was to get back to King Edward Cove and either tie up alongside the jetty or, *in extremis*, run his boat up on a beach. In theory, he could have made a dash for the open sea and, accepting the risks resulting from his damaged ballasting system, tried to escape by diving. In practice, however, he had to assume that his attackers would drop more Mark 46 torpedoes. Bicain further reasoned that the streamlined shape of the fin had been spoiled by the missile strike. The damaged GRP panels would cause cavitation under water, and the resultant noise would be quickly identified by British sonars. The game was up as far as offensive action was concerned – it was now a case of simply trying to save the lives of his crew.

While a deeply worried Captain Bicain consulted with his officers in the noisy fume-laden control room of the *Santa Fe*, a very different scene was being enacted in the operations room of HMS *Antrim*. Chris Parry's initial reports had been received with a profound sense of relief. Broad smiles greeted the news that Captain Young had ordered battle ensigns to be broken out. The *Antrim* and the *Plymouth* increased speed to twenty-eight knots, the *Brilliant* hurrying south to join them on a converging course.

During the following fifteen minutes – between 0945 and 1000 – the attacks against the *Santa Fe* reached their peak of intensity. While Lynx 341 circled overhead, four more helicopters appeared on the scene.

Lynx 342, piloted by Lieutenant-Commander John Clark with Lieutenant Paul McKay in the observer's seat, had launched from HMS *Brilliant* twenty minutes earlier. John Clark describes their experiences: 'Paul and I were on board *Brilliant* when we heard that the submarine had been found and attacked on the surface. My helicopter was fitted with a Magnetic Anomaly Detector (MAD), a device for detecting submerged submarines by measuring the disturbance to the earth's magnetic field caused by the target's metal hull. I therefore decided to get airborne to assist "Humphrey" in locating the submarine should she dive. I closed the Bay at one hundred and thirty knots. Visibility was

In rapidly clearing weather, HMS *Antrim*'s Wessex 406 speeds back to 'mother' to collect another Mark 11 depth-charge. By the time the helicopter has been rearmed and has returned to Cumberland East Bay, the Argentine submarine is no longer seaworthy and further depth-charge attacks are unnecessary.

poor and the cloudbase very low. It was one of those days when the sun is gradually burning off the cloud and mist, but leaving unpredictable patches here and there. We emerged from the offshore gloom and suddenly I spotted the *Santa Fe* about five miles ahead, on the surface. She was heading for the shore, illuminated by brilliant sunshine, and I closed on her port quarter. I did not at first see any other helicopters. I decided to attack with the machine gun mounted on a swivel chair in the cabin door and manned by Paul McKay. I circled the submarine and Paul engaged it. We could see the tracer rounds bouncing off the hull and passing through the fin. Having spent years training to find and attack submarines to destruction, it was intensely frustrating to find that I was now powerless to do any real harm.'

Next it was the turn of Wasp 343. Tony Ellerbeck had made a high-speed return to the *Endurance* where the flight deck team quickly rearmed his helicopter with another pair of AS12. Coming back into the Bay around the tip of the Peninsula at one hundred feet, he flew through the small swarm of other helicopters and approached the submarine from her port quarter. David Wells fired the port missile, but it was a 'rogue'. Curving sharply away to the left, it disappeared from view and was never seen again. The Wasp then closed to two miles on the target's port beam and Wells successfully launched the starboard missile. His aim was true and the AS12 plunged through the fin. This time there was no explosion. John Clark was circling nearby, watching the action, and his exultant shout of 'Bullseye !' was heard by HF radio listeners in the Corporate Task Force. The Argentine boat continued to run for King Edward Cove, gradually reducing speed. Ellerbeck orbited the submarine for several minutes and then flew back to the *Endurance* to rearm again.

Two of the aircraft circling the area, assessing the situation and looking for a clear line of fire, were Wasp 435 from the *Endurance* and Wasp 445 from HMS *Plymouth*. Flown by Lieutenant-Commander John Dransfield, with LACMN Joe Harper as aircrewman, the second of these two Wasps had launched from its parent frigate thirty minutes earlier when the warships were still forty-five miles from the coast. The constraints of distance and fuel meant that Dransfield could carry only one AS12. Flying in towards the coast, he and Harper had seen the snow-capped mountains bathed in brilliant sunshine, but visibility inside Cumberland East Bay was still limited. At first they saw nothing, but then 'it all happened'. There was the stricken *Santa Fe* – turning in to a final approach to King Edward Point.

John Dransfield watched Tony Ellerbeck's second attack and then closed in – 'ducking and weaving to avoid the friendly fire' – so that Joe Harper could fire his single AS12 at 5,000 yards from a height of one hundred feet. The target was so close to King Edward Point that some of the attacking helicopters were coming under inaccurate but heavy fire from rifles, machine guns and anti-tank missiles. Dransfield carefully stayed beyond the range of the groundfire and steadied his Wasp for the moment of firing. Despite the difficulty of judging the distance accurately, Harper succeeded in guiding his weapon to a point very close to the submarine's starboard quarter. Evidence obtained subsequently suggests that this missile did not make direct contact with the pressure hull but exploded in the water very close to a port after ballast tank, rupturing it. The detonation may also have caused a serious leak into the interior of the boat, either through a split in the pressure hull or by breaking the seal on the stern gland.

Dransfield banked away from the threatening groundfire and

hurried back towards the fast approaching trio of warships – now less than thirty miles from the coast – so that he could rearm and hopefully make a second attack. His place was taken by Wasp 435, piloted by Lieutenant Tim Finding, from the *Endurance*. Beside him he had his aircrewman, POACMN Bob Nadin, hero of the Stanley Shield football match. Attached to the fuselage were two AS12 missiles.

Tim Finding now found a free line of sight at the submarine. He recalls his feelings: 'Bob Nadin and I had had a desperate urge to join in the fray ever since we first heard that *Antrim*'s Wessex had made an attack, but, although the heart was thumping, when it came to our turn, we did everything by the book. Missile checks, aircraft checks, the right height and speed, it was all in accordance with our normal training on the simulator. It wasn't until the enemy started firing back – something which never happened on the simulator – that we were brought back to reality.'

The *Santa Fe* was very close to the BAS jetty. A thin streak of oil trailed from her stern, marking the course which she had followed down the length of the Bay. With the Wasp at a range of four miles, Nadin held the target firmly in his rangefinder and launched the port missile. Thirty seconds later it struck the surface of the sea, just fifty yards short of the submarine. Finding continued to close at sixty knots while his aircrewman reselected for a launch of the starboard missile. At a range of two miles it flew straight ahead of the helicopter, Bob Nadin watching it through his optical sight and controlling its course with the joystick. This time his aim was true. The missile went through the tall fin, inflicting further damage to the panelling but failing to explode.

Wasp 435 was by this time less than a mile from King Edward Point. The pilot turned away to starboard, long arcs of tracer coming up from the ground as the Argentine garrison tried to hit the small agile target. He hurried back towards the *Endurance* with the intention of rearming with two more missiles. Crossing the Barff Peninsula at Godthul, he passed Wasp 434 coming in the opposite direction. Tony Ellerbeck had obtained a third pair of AS12 from 'the plum' and was returning to make yet another attack. Tim Finding warned his Flight Commander of the hostile reception waiting for him in the target area.

It was at this stage, shortly after 1000, that Ian Stanley returned from the *Antrim*, rearmed with another Mark 11 depth charge. He and Parry assessed the condition of the submarine and decided that there was no purpose in depth-charging it again. All six helicopters had attacked and three of them were still buzzing like angry bees over the increasingly sunlit waters of the Bay. It was an astonishing scene – streams of tracer and missile trails arching up from the ground and more bursts of tracer coning down towards the tattered *Santa Fe* as she made her way in towards the BAS jetty. Chris Parry had not been able to talk by radio with the two *Endurance* Wasps, but he had been co-ordinating the movements of the two Lynx from HMS *Brilliant* and the Wasp from HMS *Plymouth*. Studying the condition of the target, assessing the volume of groundfire, and calculating the reserves of fuel remaining in each aircraft, Parry felt that the attacks should cease. He passed his view to the two Lynx pilots and all three aircraft turned away, the Wessex heading out to the *Antrim* and the Lynx climbing to a greater altitude. It was during this phase of the action that Wasp 434, flown by Tony Ellerbeck, once again reappeared on the scene.

Flying across Cumberland East Bay, Ellerbeck was startled by an urgent warning from David Wells, seated beside him and maintaining watch: 'Two bogeys, eleven o'clock high, menacing.' Both men were aware of the Bird Island report of two jet fighters having been seen in that area, and Ellerbeck experienced a moment of acute anxiety. It quickly passed when the aircraft were identified as the two Lynx circling at higher altitude.

Ellerbeck had earlier watched John Dransfield make his missile attack. 'It was a copy-book attack and it was a pity that John was carrying only one AS12. A second shot might have been decisive. It was a pity also that, owing to the lack of radio communications, he went all the way back to the *Plymouth* to rearm instead of popping over the hill to the *Endurance*. Then he could have had a second crack at the sub. We couldn't talk to the other pilots, but we saw the Lynx and the Wessex starting to withdraw. I decided to have one final attempt on my own. I went in and hovered over Dartmouth Point, about two miles from the submarine. David selected the port launcher and, having acquired the target in his rangefinder, tried to fire. It was a misfire. He made a second attempt while I circled around, trying to keep away from the groundfire. After a second misfire we decided to forget the port missile and to make an attempt with the starboard one. By this time I was over Hobart Rock and had lost track of the submarine. I turned in towards the Point at sixty knots and two hundred feet, about a mile out, perhaps less. Tracer was coming up at us from the ground, apparently quite slowly in the distance and then flashing past the cockpit at speed. Neither of us had ever previously been under fire, but we were too busy to feel really frightened. David had had a difficult time with the selector – later we found it was a control panel wiring fault and we ditched the missile in the sea – and now he was trying to find the target again in the rangefinder. Neither of us could see it – it seemed to have vanished. Rather than hang around any longer, I decided to have a go at the radio shack. I reasoned that there would be various officers or other important personnel in there, so it was a useful target. David fixed it in the viewer and pressed the starboard tit. The missile went away and David flew it along towards the building. I was counting down the final seconds – the time when the operator lowers the flight path just before impact – when David suddenly shouted, "I see it! I see it!" I thought he was talking about the radio shack and said, "Don't worry, keep going", but in fact he had caught sight of the *Santa Fe* tied up alongside the BAS jetty. I saw the missile go into a climb as David brought it around over the jetty and steered it into the back of the submarine's fin. It must have struck something solid inside the fin because there was a big explosion and bits of metal and panelling flew into the air. I discovered later that Captain Bicain had already landed on the jetty and was standing there, shouting for a doctor to come and look after his casualty, when our AS12 screamed over his head and hit his boat only twenty yards away. Poor man, it just wasn't his day.'

This final strike wrecked the periscope, radio aerials and snorkel. Clark and McKay, in Lynx 342, were still in the area. Their mother ship, HMS *Brilliant*, would need some minutes to recover and stow Lynx 341, so there was no need for 342 to hurry back. John Clark decided to inspect the target at close range. He dived and raced low past King Edward Point – enemy machine gun fire churning the sea fifty feet to starboard – and saw oil spreading across the Cove and men running away from the BAS jetty. The *Santa Fe* was clearly in a distressed condition and had been effectively destroyed as a direct threat to the Paraquat ships. He and Ellerbeck followed the example of Stanley and Butler by heading out to their mother ships.

When Wasp 445 came hurrying back to renew the attack – having taken on more fuel and another AS12 from HMS *Plymouth* – John Dransfield was advised that the submarine was no longer a worthwhile target. He was turned back as he started to enter Cumberland East Bay. A pioneering action – the first ever

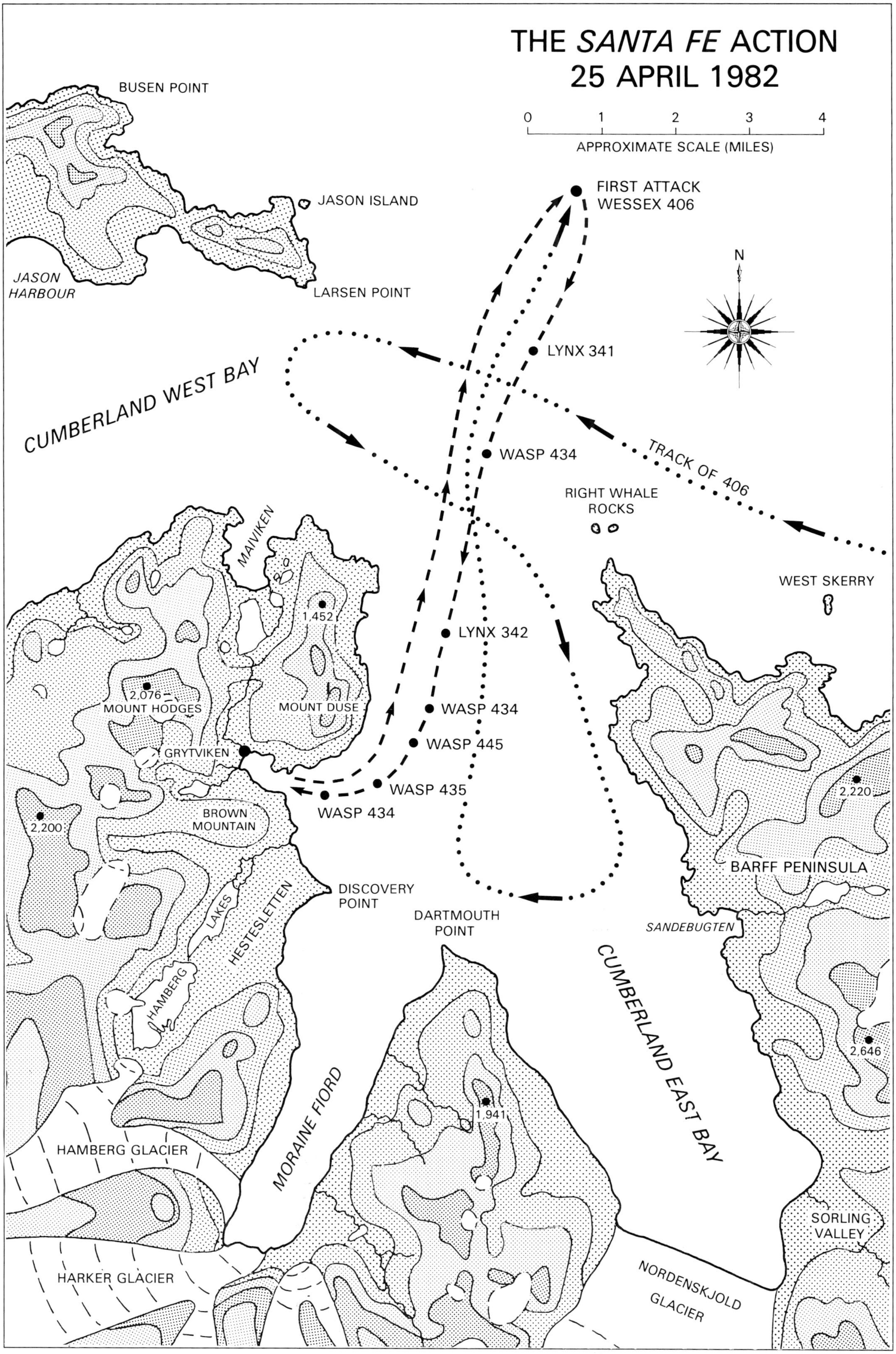
THE *SANTA FE* ACTION
25 APRIL 1982
0 1 2 3 4
APPROXIMATE SCALE (MILES)
N
FIRST ATTACK WESSEX 406
BUSEN POINT
JASON ISLAND
JASON HARBOUR
LARSEN POINT
CUMBERLAND WEST BAY
LYNX 341
WASP 434
TRACK OF 406
RIGHT WHALE ROCKS
WEST SKERRY
MAIVIKEN
1,452
LYNX 342
2,076
MOUNT HODGES
MOUNT DUSE
WASP 434
WASP 445
GRYTVIKEN
WASP 435
WASP 434
2,200
BROWN MOUNTAIN
2,220
BARFF PENINSULA
DISCOVERY POINT
DARTMOUTH POINT
SANDEBUGTEN
LAKES
HESTESLETTEN
HAMBERG
CUMBERLAND EAST BAY
2,646
MORAINE FIORD
1,941
HAMBERG GLACIER
SORLING VALLEY
HARKER GLACIER
NORDENSKJOLD GLACIER

anti-submarine operation carried through exclusively by helicopters – was at an end.

By 1045 the helicopters had all returned to their parent ships. Despite the groundfire, none had been damaged and the aircrew were all safe. There was great jubilation in the *Antrim*, *Brilliant*, *Plymouth* and *Endurance*, while those aboard the *Tidespring* – miles to the north-east – could only send their envious congratulations. Events were coming to a climax, and the frustrated 'M' Company marines began to suspect that they would not be 'in at the kill'.

There now began a period of conferences and de-briefings. Captain Young had intended to sink the *Santa Fe*. He had not anticipated that she might only be damaged and able to return to King Edward Point. How badly crippled was she? Might she make another attempt to put to sea? Exactly what had the pilots seen as they left the area? It soon became clear that the submarine was not seaworthy and no longer presented a threat. Her crew had managed to get alongside the BAS jetty at King Edward Point and secure her. Some of them had been seen streaming up towards Shackleton House. She was down by the stern, listing to port, and leaking diesel oil over the surface of the Cove.

There were several suggestions as to what should be done next. Keith Eve wanted the *Antrim* and *Plymouth* to sail into Cumberland East Bay and bombard the garrison positions. This was firmly turned down by Young on the grounds that it would cause casualties and damage property. Guy Sheridan wanted to make an immediate helicopter-borne landing, close to the Argentine positions, and to rush them while the defenders were off balance. 'It was obvious to me that the Argentines must be feeling very demoralised. Until that morning they had not even known for certain that we were in the area. Then they suddenly saw their submarine being hammered, forced back to base, and abandoned by its crew. To them it must have seemed that the air was full of our helicopters and that we had overwhelming resources. I was very anxious to go straight in before they had time to catch their breath.'

This suggestion was accepted in principle by Captain Young and, at 1040, the troops embarked in HMS *Antrim* were placed on one hour's notice to move. There was a great deal of work to be done, however, before anyone could go ashore, as Brian Young explains: 'The decision to go straight in was not as simple as it looked. I was still concerned about the possibility of stand-off weapons on the headlands surrounding Cumberland East Bay. I doubted very much whether the Argentines could have had either the time or the facilities to install land-based Exocets, nor did I really believe that they had laid sea mines, but they could have had other heavy weapons waiting for us. I was particularly concerned at the possibility of long-range anti-tank missiles having been deployed. Remember, we still had no detailed intelligence regarding Argentine dispositions on shore and, although the helicopter pilots had not seen any prepared positions during their pursuit of the *Santa Fe* into King Edward Cove, I could not discount the possibility that the Argentines had learnt the lesson of the *Guerrico* incident. Again, just because the *Brilliant*'s Lynx had not seen any surface ships further up the coast two hours before, it didn't mean that there was nothing up there. Everyone was very keen to get ashore, but it would have been foolhardy to rush in without proper preparations.'

As soon as the aircrew de-briefings were completed, the *Antrim*'s operations room filled with the numerous officers involved in making the decision as to what should next be done. Events had taken such an unexpected turn that there was no previously agreed plan instantly available and ready to be put into effect. The atmosphere became heavily charged with impatience as, over the next ninety minutes, the military and naval components struggled to hammer out a plan which met the requirements of all concerned. It was with a sense of relief that Young, Sheridan and Delves, and their various specialist advisors, were able at midday to co-ordinate their strategy.

Four important decisions emerged from the morning's discussions. First, and despite the action with the *Santa Fe*, they would continue to pursue the political directive regarding 'minimum loss of life and minimum damage to property'. Secondly, the 'M' Company Group in the RFA *Tidespring* – still two hundred miles away – must be excluded from their immediate plans. Too much time would be lost waiting for her to arrive. Thirdly, the 4.5-inch guns carried by HMS *Antrim* and HMS *Plymouth* would engage targets ashore, but only in the intimidatory role (otherwise known as 'frightening fire'). Fourthly, an assault force would be put ashore by helicopter at the earliest opportunity.

In tactical terms, the first step was to decide the composition of the landing force and the order in which it would be lifted ashore. The only aircraft with significant capacity were HMS *Brilliant*'s two Lynx (each able to carry eight fully armed troops) and HMS *Antrim*'s Wessex 3, 'Humphrey'. Ian Stanley's aircraft had been re-fitted with the bulky sonar equipment, so he, too, could take no more than eight passengers. The total 'single lift' capability of the helicopters for the initial wave was therefore only twenty-four men – a very modest number for a daylight assault in full view of an alert enemy thought to be numbering more than one hundred.

Major Sheridan's plan for the assault fell into two main phases. The first wave would be composed of men drawn exclusively from the resources available within HMS *Antrim*. They were a heterogeneous cross-section of the entire Task Group. He had his own people – a small command group, two sections of 42 Commando's Reconnaissance Troop, a section of assault engineers, and an 81 mm mortar detachment. Then he had the *Antrim*'s integral Royal Marines Detachment of nine men commanded by Sergeant Gordon Kendell. More marines were available from the small SBS command team and signals element, commanded by 'Clive'. Lieutenant-Colonel Eve would remain aboard as the NGLO, but he could spare six of his gunner and matelot signallers – the Naval Gunfire Assistants (NGAs) – led by Captain Willie McCracken, RA. Finally, there was 'D' Squadron's Mountain Troop – now fully recovered from its ordeal on the Fortuna Glacier – commanded by Captain John Hamilton. Few of these people had ever before worked together, but all were trained in the basic infantry skills as well as being specialists in their own right. For communications and organisational purposes, these elements were entitled Composite Company Group.

The intention was to put this small scratch force ashore on an area known as the Hestesletten Flats. Situated at the entrance to Moraine Fiord, the flats were sheltered by the intervening mass of Brown Mountain from the Argentines concentrated at King Edward Point. Once safely ashore, the assaulting force would begin to advance around the curve of the mountain's lower slopes and approach Grytviken and King Edward Point from the south. Simultaneously there would be a second phase of helicopter landings, with more troops being lifted ashore from HMS *Plymouth* and HMS *Endurance*. In the former was the 'D' Squadron Mobility Troop and in the latter were Air Troop and 2 SBS. It was Sheridan's intention to deploy them on the high ground overlooking King Edward Cove as soon as the helicopters had finished delivering his Composite Company Group. He describes the details: 'I sent radio orders to the two other ships, instructing the military commanders to stand by for a helicopter lift in the next ninety minutes. The Navy were initially not too happy with the idea of putting the Lynx or the Wessex down on

Some of HMS *Antrim*'s small integral Royal Marines detachment who took part in the Hestesletten landings. Their diversity of clothing and equipment possibly reflects the speed with which the final assault was mounted.

Plymouth's or *Endurance*'s flight decks. Neither deck had been stressed or officially cleared to accept the weight of anything heavier than the Wasp. However, these doubts were soon put to one side. My orders to Mobility Troop were that they should deploy in a dominating position above King Edward Point, and the SBS on similar ground above Grytviken. It would be their mission to bring down covering fire on call, if required, while the Composite Company Group advanced around the Cove. It was also in my mind that these two groups would act as stops if any Argentines tried to escape from the Cove area and make their way up over the Bore Valley towards Maiviken. Both parties were ready to start embarking in the two Lynx when subsequently I was able to cancel that part of the plan.'

During the morning, while Sheridan and the combined operations team aboard the *Antrim* were still trying to finalise their plans, Captain Barker had been making plans of his own. Guessing that his passengers might be left out of any impending helicopter landings, he quickly thought of a way of getting them – and his ship – into action. It was basically a variation of the scheme which he had proposed some days earlier for the disembarkation of 'M' Company. He proposed to leave Hound Bay, to sail around the Barff Peninsula and proceed straight into Maiviken Cove, running the *Endurance* up against the steeply shelving beach. Her heavily reinforced bows should protect the ship from underwater damage, and the troops could get ashore and proceed up the Bore Valley to take the Argentines in the rear. This suggestion was turned down later in the morning when Sheridan and Young agreed to make the main assault at Hestesletten and to lift the SAS and SBS up to the top of the Bore Valley by helicopter. Barker's idea was rejected for the same reason as on the previous occasion – there was no purpose in risking the red ship unnecessarily.

Major Sheridan felt increasingly impatient as the morning went by. Later he described his feelings: 'There had been a fair amount of confusion and frustration after the *Santa Fe* was hit, but I was able to give hasty orders at 1345 and I went around the troops waiting in the main passageways of the ship to talk to them. They were all keen to go, faces blackened and ammunition bristling from every pocket. I had seventy-four men in all – about half the opposition I expected. So it was a gamble. I had split them into three *ad hoc* Troops, each of twenty-odd men. First in would be Cedric Delves, with John Hamilton's Mountain Troop. Then the second wave would be Chris Nunn's group and the medical team led by Surgeon Lieutenant Swinhoe. The third Troop was 'Clive' and the SBS and some others. We were taking with us an 81 mm mortar, and every man carried two HE mortar bombs in addition to his own weapon and ammunition. I intended to travel with the second wave. Command would pass to me from Major Delves when I set foot ashore.'

Extemporising his scratch force meant that Sheridan was obliged to strip the *Antrim*'s operations room of many of the men who until now had been staffing it. It was agreed that 'Danny', the second in command of 'D' Squadron, would remain aboard as senior military officer, but someone was needed to co-ordinate the helicopter operations. This role fell to Lieutenant Mike Tidd.

The bows of HMS *Plymouth* emerge from the smoke of her own incessant gunfire. The framework on the left of the picture is the Seacat missile launcher mounted on the stern of HMS *Antrim*.

He and the rest of 'C' Flight were unhappy that, having travelled all the way from Yeovilton for the express purpose of making a Commando assault, they could not participate in the impending landings. The two Wessex 5 were tailor-made for the job. Instead, their remains were gradually disappearing under the winter snows of the Fortuna Glacier.

Chris Parry, 'Humphrey's' observer, was similarly at a loose end. For this trip – in broad daylight and good visibility – there would be no call for radar or complicated navigation. Ian Stanley and 'Fitz' Fitzgerald alone would take their military passengers ashore. Parry and Cooper would stay behind in the ship. Parry was given the duty of organising the serials of troops ready for despatch from the flight deck.

A key aspect of the proposed landing was the initial NGS bombardment. Keith Eve had, at long last, an opportunity to draw up a precise fire plan for a specific purpose. 'Until then it had been difficult to know exactly how and where my two NGFO teams would be deployed. I went to Captain Young and asked him what sort of fire plan he wanted. He was tremendously busy with other things and he just said, "I leave it entirely to you – I'm sure it will be alright," which was exactly what I wanted to hear. I sat down with Guy Sheridan, Willie McCracken and Chris Brown and, because we had been studying the possibilities for the past two weeks, it didn't take us long to select targets and mark the sets of map overlays.'

Captain Pentreath was ordered to take up a position three miles to the east of the Barff Peninsula. His ship's guns would fire over the top of the long neck of land, the shells rising to 6,000 feet before plunging down onto the selected landing area of Tussock Flats, Hestesletten. Extending the gunline to the north was HMS *Antrim*. She also would be firing over the Peninsula, but only when specific targets were identified. The NGFO nominated to spot for the two ships was Chris Brown. It was planned that he would observe the fall of shot from Wasp 434, flown by Tony Ellerbeck, from the west side of Cumberland East Bay and far below the trajectory of the shells passing overhead.

The Wasp was airborne at 1400 and flying along the intended track. In the front seats were Tony Ellerbeck and David Wells, in the back Chris Brown with his HF portable radio. Visibility had been improving all morning but, at this stage, isolated clouds still hung low over the Bay. They prevented the spotter from having an uninterrupted view. Captain Brown takes up the story: 'Initially I tried to observe the target area from the east side of the Bay. The guns were not required to fire for quite some while, and I asked Tony how long he could fly around like this before needing to refuel. He said thirty minutes, so I suggested that we land on the Barff Peninsula. Sitting on the ground with rotors turning would extend the fuel to an hour. Tony found a good spot and we sat there for a few minutes, but the intermittent clouds and the distance – about six kilometres – made for difficult spotting. I asked if they could get me closer to Grytviken. Tony asked what I had in mind and I suggested Dartmouth Point. He and David looked at each other. Tony said, "OK!" and flew us across to the top of the Point where he landed. This gave me a much better view of the area I was interested in. He told me that he now had forty-five minutes of fuel – assuming that he just kept the rotors turning. This was too short for me to conduct the fire plan, so I tentatively suggested that he might shut down. He and David looked at each other again. After all, asking an aviator to deny himself the ability to crank up and depart in a hurry is asking a great deal. Again he said, "OK!" I made a quick recce of the area to make sure it was unoccupied and gave him the thumbs-up. He shut down the engine and everything became very still and quiet. I remember thinking, "I hope I've got it right." We moved up to the crest, I set up my radio, spoke with *Plymouth* and *Antrim*, and we were in business. It was all a bit like being on the gunnery range at Portland but, when *Plymouth* reported "Ready", it dawned on me that the Navy was waiting for me to order the first shells of the Falklands war.'

Out at sea, the meteorological conditions had been translated into ballistic information and fed into the ships' analogue computers. Corrections had been made for tidal conditions, the ships' heading and speed, the strength and direction of the wind, and all the other variables which determine the accuracy of naval gunfire. Thirty-five seconds after leaving the barrels of the *Plymouth*'s gun turret, the first shells impacted on the grassy peat beds of Hestesletten. Brown was able to assure Brum Richards that the shooting was accurate.

'Humphrey' launches from the flight deck of HMS *Antrim* to lead the first wave into Cumberland East Bay for the Hestesletten landings. By extraordinary good fortune, and for the first time during Operation Paraquat, the weather is ideal for British purposes.

Shortly after the firing of the opening salvoes, Captain Brown saw a man run to the second of the two flag masts on King Edward Point. The Argentine blue and white were already flying from one, and the man now hoisted an unidentifiable object up the second. The wind was blowing directly towards Dartmouth Point and Brown could not see what it was. He studied it only briefly before returning his attention to the more pressing task of observing the fall of shot.

At 1415, the first troops climbed into 'Humphrey' on HMS *Antrim*'s flight deck and Ian Stanley lifted off to hover in a holding position two hundred yards from the ship's quarter. In turn, Lynx 341 and 342 arrived from HMS *Brilliant* and touched down to collect their military passengers from the *Antrim*. Cedric Delves and John Hamilton climbed into 341, piloted by Nick Butler. He lifted off and all three aircraft formed up in a tight 'vic' of three, Ian Stanley leading in the Wessex. It was while all this was happening that Keith Eve, sitting at the plotting table in the *Antrim*'s control room, became alarmed by the radio traffic which he could hear in his headset.

During the hasty briefing sessions between 1030 and 1330, it had been agreed that all communications would be conducted by UHF, one frequency being dedicated to NGS traffic and another for helicopter control purposes. This plan assumed that Chris Brown would remain airborne throughout the bombardment. Eve had been talking to Brown by UHF while Wasp 434 was flying over Cumberland East Bay. Now, as the three assault aircraft loaded and became airborne, Eve started to hear brief bursts of conversation between their pilots and the flight controller. The transmissions – typical of all aviators' announcements – were short and staccato. However, it disturbed Eve to consider the implications of gunfire and aircraft being controlled simultaneously on a unified radio net. He had not been told of a late and unusual decision to channel all radio traffic through a single frequency. This decision seems to have been influenced by the atmospheric conditions and a desire to simplify the arrangements. In the event, the problem resolved itself when Brown arrived on Dartmouth Point and switched to his Clansman HF set. Thereafter, even though the high ground of the Barff Peninsula should in theory have limited the quality of UHF transmissions, flight control was able to communicate surprisingly well with the assault helicopters and there were no further conflicts with the gunnery net.

Leaving the *Antrim*, the three aircraft swiftly covered the three miles separating them from the tip of the Barff Peninsula. Ian Stanley held the formation at low level. 'I decided to stay right down on the deck – about ten or fifteen feet above the sea – in order to make us more difficult to locate and identify. We had agreed a flight plan, so any orders I gave were very brief. I think there was only one – "turn" – and that was when we reached the entrance to Cumberland Bay. We flew in through the narrow gap of the Merton Passage – the same route, I understand, that the *Endurance* used when she made her escape on 31 March. It was lumpy and bumpy going through between the cliffs and I was wary of down-drafts, but all went well. The track took us down the length of the Bay towards a little cove named Sandbugten before turning due west to pass in front of Dartmouth Point, then across Moraine Fiord and on to the Hestesletten plain. The Argentines must have had difficulty in seeing us at such a low altitude and against the background of the cliffs. Nobody fired at us on the way in but, even if they had, we were out of effective range for anything smaller than anti-aircraft artillery. Everything was going according to plan when suddenly, just as we were approaching the touch-down point, we were told to turn away. The message was a bit broken, but they seemed to be telling us that there were enemy troops on the ground and that we were flying into the gun target line (GTL). In other words, and although I personally did not see any shellbursts, there was some danger that we might be hit by *Plymouth*'s shells. The whole thing was running on a timetable. The gunfire was supposed to shift to the crest of Brown Mountain two minutes before we landed. Either we were slightly ahead of schedule, or the gunfire was slow in changing target, but either way we needed to remove ourselves – fast. For me, that was easier said than done. The Wessex was fat and heavy, but the Lynx can turn on a sixpence. By the time I heaved my old lady around, the other two had disappeared back the way we had come. I trundled off behind them and we orbited the northern end of Cumberland Bay while it was all sorted out.'

High on Dartmouth Point, Ellerbeck, Wells and Brown

watched the three helicopters pass below them, from right to left, and then saw them leave at speed in the opposite direction. In the back of Lynx 341, Major Delves had been unable to see out of the cabin window and was relying on Nick Butler to keep him advised of progress. 'Like the rest, I was keyed up and ready to land when suddenly the pilot turned the helicopter on its ear and we shot out of there a lot faster than we had come in. It was all a bit disconcerting.'

Five minutes later, with the communications and timing problems resolved, the three aircraft returned to Moraine Fiord and landed together on the flat expanse of Hestesletten. In Ian Stanley's words: 'It was very unconventional. This was a Commando-style assault force but we had no Commando helicopters. Our three aircraft consisted of one geriatric ASW Wessex 3 – full of 'pinger' gear – and two of the latest ASW Lynx. None of the three pilots were trained for amphibious warfare. Even the business of disembarking the troops was slightly ludicrous because the sonar equipment in "Humphrey" partially blocked the fuselage door. Instead of leaving the aircraft in a rush and fanning out to take up firing positions in the best approved Commando manner, my passengers were obliged to squeeze their way past the sonar, one at a time, while my crewman, "Fitz" Fitzgerald, handed down their kit to them through the window. It was fortunate that the landing was not opposed.'

The two Lynx – having doors on both sides of the fuselage – were better able to unload their passengers quickly. Each aircraft carried eight men and assorted impedimenta, a load which involved grossly uncomfortable travelling conditions for the tightly packed soldiers. Acting as aircrewman in Lynx 341 was Lieutenant-Commander Roger Edwardes, an aviator who had succeeded in making his way south in HMS *Brilliant*. Some years before he had married a Falkland Islands girl and subsequently had served a tour aboard HMS *Endurance*. Now he was determined to be at the forefront in the first operation to eventually liberate the islands. The helicopter was so densely packed with men that he placed himself across the door to act as a human barrier and prevent them from falling out.

The three aircraft continued to ferry troops ashore in a non-stop shuttle. There had been no opportunity to rehearse such an operation, and the *Antrim*'s interior layout was not designed for the despatch of large numbers of laden troops, but the business of landing on, loading and launching, proceeded smoothly. It was a fine performance by the flight deck team.

Fifteen miles to the south, at Saint Andrews Bay, Cindy Buxton and Annie Price and the three BAS men could hear the muffled sound of distant explosions. Guessing that the British had commenced direct action against the occupation force, Tony North and Myles Plant climbed the ridge – 1,200 feet high – to the rear of their hut and from which, on a clear day, King Edward Point is just visible. They saw nothing to explain the rolling thunder of gunfire. The shells were falling on the shoreline of Moraine Fiord and the hidden slopes of Brown Mountain.

Guy Sheridan's plan for the advance towards King Edward Point was dictated by the terrain which separated him from his objective. This consisted of the floor of the Hestesletten glaciated valley, then the sloping shoulder of Brown Mountain, and then a steep descent to Grytviken and the waters of King Edward Cove. Brown Mountain is 1,500 feet high at its peak, but the anticipated line of advance curved around the 500-foot contour, crossing a skyline ridge. The final descent to the Cove would be down a rough slope broken by areas of tussock, peat and scree. Sheridan calculated that the four-mile route from the landing site to King Edward Point would be covered in approximately two hours – subject to the opposition encountered.

Lacking any intelligence with regard to Argentine defensive arrangements, he had to assume that the landing might be contested, as he explains: 'The Hestesletten flats are big – about two kilometres by five – and I didn't believe the enemy would try to defend such a large area. Nor did I expect them to put troops right on top of Brown Mountain because they would be too high to be effective. However, the Brown Mountain ridgeline, which runs down to the shore of Cumberland East Bay, dominates the entire Hestesletten area and is a natural defensive position for anyone trying to block access to Grytviken. If we were going to hit opposition, it would be as we crossed the flats and started to ascend that ridge. Hence my orders to Major Delves – who would be landing first with Captain Hamilton and the Mountain Troop – to secure the landing site first, then "advance to contact" and secure the ridge. The NGS programme arranged with Lieutenant-Colonel Eve was designed to neutralise the ridge and its approaches as the SAS moved forward, and to then shift onto the reverse slope as they arrived on the high ground. I would land with the second wave and move up and through the Mountain Troop positions, with them giving covering fire on call as we went down the exposed slope into Grytviken.'

Having disembarked, Mountain Troop spread out and thoroughly searched the helicopter landing site. Despite its flatness, the area was well covered with thick clumps of grass and there were numerous hillocks and hollows where enemy troops might be concealed. It soon became apparent that the site – codenamed Eagle – was undefended. The small British force then formed up – facing Brown Mountain – to screen the landing site as the helicopters flew in with the second wave led by Captain Chris Nunn.

The new arrivals brought with them an 'M' Company 81 mm mortar team. Quickly the tube was set up on its base and the first rounds were fired to register the Brown Mountain ridge. Two major problems became immediately apparent. The baseplate sank into the soft peat as each round was fired, and there was confusion regarding the fireplan. Major Delves explains: 'Chris Nunn was back with the mortar, I was up front with John Hamilton and the Mountain Troop. I wanted to register certain points along our planned line of advance so that I could call down instantaneous fire if we hit trouble. It is a sound principle under those circumstances that you don't move without being certain that covering fire is on tap, and will fall on the correct co-

Lieutenant-Commander Ian Stanley, RN, HMS *Antrim*'s Flight Commander and the pilot who gave Wessex 406, the venerable 'Humphrey', a permanent place in aviation history.

One of HMS *Brilliant*'s Lynx helicopters sidles in towards HMS *Antrim*'s flight deck to embark a serial of troops for the Hestesletten landings. Amphibious operations were certainly not on the minds of Westland's designers when this type was first conceived, but it performed well enough as a troop carrier when the need arose.

ordinates if and when required. Captain Nunn was telling his mortar team to range on certain features, I was concerned with different features. We got it sorted out eventually, but some valuable time was lost and we all got a bit irritated with each other.'

Guy Sheridan arrived on the Hestesletten Flats at 1520. He was startled to find the first arrivals still in the immediate landing area. The SAS Mountain Troop had started to move forward, but Sheridan had expected to see them on the high ground by this stage. The second wave were forming up, but the business of registering the mortars was still unfinished. In a cold fury, he gave orders that, 'everyone must sort themselves out and get on with it.' The *Plymouth* was already shifting her fire to the northern slope of Brown Mountain and Sheridan was anxious to follow up behind it. His troops, who were still arriving in penny packets, formed up with the leaders to whom they had been allocated and the main advance commenced.

Taking the point were Cedric Delves and John Hamilton. Spread out in front of them was a thin screen of SAS troopers. One of them observed movement in the tussock grass near the shoreline and GPMG fire was brought down in that area. It ceased when closer examination showed the target to be a small colony of elephant seals. The fire was accurate and several of the animals were killed or wounded.

The leading troops started to climb the ridge. The advance stopped when a suspected Argentine observation post was reported higher up the slope, to the left. This object had already been seen by Chris Brown, observing from Dartmouth Point, and he too believed that the enemy had established a strongpoint on this dominant feature. Being on the crest of a steeply-sided hill, it was difficult to hit. Shells were either falling short on the northern slope or flying straight over the top into the middle distance. Several landed on the inland reaches of the Hestesletten plain, but fortunately well away from the helicopter landing site.

Major Delves studied the object. 'It looked very much like an entrenched position. The grey colour was different from the surrounding area, and there was a tall pole sticking out of the top which appeared to be a radio antenna. The approach was steep and open, so I decided to take it out before going any further. We had a Milan launcher with us. This was not standard SAS kit at that time but, luckily, John Hamilton had been Anti-tank Platoon Commander with the Green Howards before joining us, and he knew how to use it. So a missile was sent on its way, and the "sangar" went up with a bloody great bang.' Later investigation revealed the 'sangar' to have been a curiously shaped natural mound of rocks. The 'radio antenna' was a wooden spar erected years earlier by the whalers as a navigation mark.

Chris Brown watched the distant British troops moving forward onto the high shoulder of Brown Mountain. Following the pre-arranged fireplan – codenamed Peasoup – he brought down more salvoes of 4.5-inch air-burst and direct action gunfire onto the area between them and Grytviken. Both *Plymouth* and *Antrim* were now firing almost without pause. 'The noise was incredible. The 4.5 makes a pretty loud bang anyway, but the echo of the surrounding mountains magnified the bombardment so that it must have seemed to the Argentines that we had a complete battle fleet out there. The effect – on them – must have been terrifying, especially as I crept the bursts down to within a thousand metres of the old whaling station. Some of the shells buried themselves in soft ground and didn't make much of a show, but most fell on rock and scree, sending up clouds of grey dust and shattered debris.'

The gun teams in *Antrim* and *Plymouth* continued to feed the Mark 6 turrets without interruption, but the pace was beginning to tell. The working areas reeked of cordite and the men sweated heavily. Under peacetime economies, they were not required to sustain a protracted bombardment. Now they were handling dozens of heavy shells and charges – a physical effort to which they were not accustomed. By the end of the afternoon the *Antrim* had fired sixty-nine rounds and the *Plymouth* one hundred and sixty-six. In less than two hours, they fired a weight of explosive equal to several months of their normal training allowance.

As the fireplan proceeded, so it became necessary to change the position of the two gunline ships. The high crest of the intervening Barff Peninsula prevented Chris Brown from calling down fire on all of the Brown Mountain area targets. He requested that the *Antrim* should come around the tip of the Peninsula and open direct fire, if necessary, from the entrance to Cumberland East Bay. With some reluctance – given the risk that the Argentines might have deployed long-range anti-tank weapons on the headlands – Brian Young entered the Bay at 1500. HMS *Plymouth* also moved up from her earlier position, her shells passing over the *Antrim* and over the roofs of King Edward Point to impact on the far side of the Cove. For the first time in his

Partly obscured by her four Exocet launchers, HMS *Antrim*'s twin 4.5-inch gun turret fires another salvo at the hidden slopes of Brown Mountain. Already the heat generated by heavy firing has caused the paint to burn and flake off the two barrels.

career, Keith Eve was witnessing the application of NGS under operational conditions. 'It was really very impressive. The noise was fantastic. None of our shells landed too near the Argentines, but the effect of the shots tearing over their heads must have been very demoralising.'

Captain John Coward could not join in the display of British firepower. The *Brilliant* was armed only with anti-aircraft and anti-ship missiles. 'I didn't have a gun, so I just followed along at a respectful distance.'

Brown wondered what the effect of the gunfire would be on the buildings of Grytviken. Suspecting that snipers might be hidden there – waiting to ambush the British forward scouts – he asked Lieutenant-Colonel Eve to grant permission for a few salvoes to be dropped among the old iron sheds. Brian Young happened to be standing at Eve's shoulder at the time, watching the plot. 'Tell him the answer is no. The orders are perfectly clear – minimum damage to property.' Chris Brown acknowledged the reply but, a moment later, ordered fire onto grid 623827. 'Where is that?' asked Young. Eve was embarrassed. 'Well, Sir, it rather looks like the whaling station.' Captain Young's usually quiet manner gave way to irritation. 'For God's sake, no! What's the matter with your spotters – don't they understand English?' Chris Brown decided not to push his luck any further and halted the fall of shot 800 yards from the buildings.

As the shellbursts crept down towards Grytviken, so did the SAS leading troops reach the crest of the Brown Mountain ridge and gain their first sight of King Edward Cove. The intervening hillside was scarred by clusters of craters. One group was concentrated near the wrecked Puma shot down by Keith Mills and his men on 3 April. To Chris Brown, it had looked at first like an entrenched position and he had given it forty salvoes from HMS *Plymouth* before recognising it for what it was. But the shots were not wasted – the Argentines could see for themselves that a small reduction in range would eliminate King Edward Point from the map.

The SAS Mountain Troop consolidated on the crest, their weapons pointing down towards the Cove, waiting for the bulk of the landing force to move up behind them. Delves and Hamilton could see the sinister grey shape of the *Santa Fe* berthed on the other side of the Cove and they discussed the chances of scoring a hit at that distance with a Milan missile. They had decided to make the attempt when a white flag was seen to be flying on King Edward Point. Men could be seen running down to the foreshore from Shackleton House, displaying bed sheets and items of clothing. In the words of one marine: 'If it was white, they were waving it.'

Guy Sheridan arrived on the crest at this stage. It was 1645 and the light was starting to fade. Quickly he assessed the situation and ordered the advance to stop. Calling for his signaller, he made a radio report to the operations room in HMS *Antrim*. He announced the apparent decision by the Argentines to surrender and stated that the bombardment should cease immediately. It required a few minutes for Sheridan to pass his information to the ship. During this time, the *ad hoc* Troops commanded by Chris Nunn and 'Clive' consolidated on the nearby high ground with John Hamilton and his Mountain Troop. Major Delves, however, continued to move forward with three other men. Their radio had apparently, and mysteriously, become inoperative at this stage, and Major Sheridan's order that they must halt met with no response.

Cedric Delves describes his movements during the following hour as he continued to stride out in front of Major Sheridan's landing force: 'I went on with a Warrant Officer from the Royal Signals known as "Geordie", and two other men. It was obvious that it was all over bar the shouting. You had to be there to have the feeling. I could see no point in stopping, so I kept on walking. We passed through the old whaling station and went along the narrow track which curls around the Cove to King Edward Point. We passed a machine gun post – the crew just standing there with their arms in the air and looking nervous – and I said, "Good afternoon. How do you do?" to everyone we met until we reached the buildings. It was obvious to me that there would be no final shoot-out. The only thing remaining was to see who could replace the Argentine flag with ours.'

Up on Dartmouth Point, Captain Brown realised that the object hoisted on the second flagpole – alongside that from which the Argentine colours were flying – was a white flag. It had been there since shortly after the commencement of the bombard-

A post-battle photograph of the track leading from Grytviken (left middle distance) towards King Edward Point. It was along this track that Major Delves and his companions made their final approach to the Argentine defensive positions. The total lack of cover would have made their venture unusually hazardous if, at the last moment, the enemy garrison had decided to make a stand and open fire.

ment. At sea, in the *Antrim*, the radio operators heard a transmission on Channel 16 from Captain Bicain. As senior officer, he was announcing his wish to surrender. He stated that his men and the garrison troops were going down to the beach to pile their arms and that he had a casualty in need of attention. There would be no resistance. He warned that the helicopter pad and the track running along the edge of King Edward Point – where Cedric Delves and his small party was approaching – were heavily mined. This information was passed to Delves but, seeing no disturbance to the surface of the track, he continued forward. It was discovered later that the mines were controlled by wires running back to a command point near Shackleton House and could only have been set off on command.

Guy Sheridan was deeply unhappy with the way in which the advance had developed. As ordered, Hamilton's Mountain Troop was holding Brown Mountain ridge, and Nunn and 'Clive' were moving forward with their Troops in arrowhead formation, but the leading Delves group was a long way in front. Determined to catch up with them, he called for a helicopter to collect him from the shore opposite King Edward Point. Tony Ellerbeck lifted off from Dartmouth Point and delivered him a few minutes later to King Edward Point, shortly after the arrival of Major Delves. The Argentine garrison and the crew of *Santa Fe* were 'milling around the Point like a small football crowd', in the words of one SAS soldier, but most of them had gathered on the foreshore in front of Shackleton House.

Captain Bicain was standing on the mined helicopter pad, waving off the approaching British helicopters and directing them to a safe stretch of level ground. The Wasp carrying Guy Sheridan set down near the flagstaffs. The rest of the assault force made its way around the Cove in single file and gradully converged upon the BAS base where they found the apathetic Argentines waiting to be told what to do. Darkness had already fallen by the time the last of the British force – Mountain Troop – finally arrived on the scene.

Squadron Sergeant-Major Lawrence Gallagher – tragically to be killed in a helicopter accident a few days later – had arrived with a large Union Flag wrapped around his body under his combat jacket. Ignoring the bewildered garrison, he raised it in place of the Argentine colours. He was very proud of the Jack which he had brought all the way from Hereford for a moment such as this. Cedric Delves remembers: 'It was a bloody great thing, like something out of Queen Victoria's wars.'

Delves had briefly acknowledged a declaration of surrender from Captain Bicain at 1705 and the same verbal declaration was made to Major Sheridan when he arrived. The prisoners were searched, but it was clear that they offered no immediate threat. They were simply glad to be alive. The single casualty – the seaman injured during the rocket attacks on the *Santa Fe* – was flown out by Ellerbeck to the *Antrim* and later treated in the sick bay of the RFA *Tidespring*. His mangled right leg had been amputated at the knee by the Argentine medical team in Shackleton House.

A search was made for weapons, explosives and booby-traps. The mines consisted of locally assembled devices, semi-buried in the track and beach, and were soon dismantled. The *Antrim* approached the Point and anchored close to shore. The *Plymouth* and *Endurance* moved into the seaward end of Cumberland East Bay while the *Brilliant* kept watch, like a goalkeeper, between the

King Edward Point has just been captured. The two Lynx from HMS *Brilliant* start to deliver stores to the helicopter pad where the Argentine Puma made its disastrous attempt to land on 3 April.

wide jaws of the twin Bays. The Task Group's helicopters ferried personnel to and from the shore in the gathering darkness. With the 121 cold and hungry prisoners captured at King Edward Point to be secured and cared for, there was a great deal of work to be done. The wind was rising and the temperature falling fast to below freezing.

By 1730 it was clear that the Argentine will to resist – which had never been strong from the outset – had collapsed completely. There were no 'suicide squad' snipers hidden on the steep slopes above the base, no 'forlorn hope' counter-attack units concealed in the mountains. The small garrison at Leith Harbour was still at liberty – under continuing surveillance by 'Tim' and the SAS Boat Troop – but the time had come to break the good news to the world. Operation Paraquat had been carried out under the orders of the Commander-in-Chief Fleet. This was reflected in the fact that a White Ensign had been run up the second flagpole to join the nearby flag hoisted by Lawrence Gallagher. It was a delicate nuance which was echoed in the wording of Captain Young's signal to his immediate superior, Admiral Sandy Woodward: 'Be pleased to inform Her Majesty that the White Ensign flies alongside the Union Flag in Grytviken, South Georgia. God save the Queen.'

The importance of Young's signal – couched as it was in the language of a glorious British naval tradition stretching back to Nelson and beyond – far outweighed the basic military event which it described. Here at last, after twenty-three anxious days marked only by failure and national humiliation, was a success to be proclaimed to the world. That the island of South Georgia had been recovered without loss of life gave added thrust to the British government's assertion that it wished to resolve the crisis within the framework of the United Nations Charter. As the Prime Minister told the House of Commons on the following day: 'I should like to emphasise that the repossession of South Georgia, including the attack on the Argentine submarine, in no way alters the government's determination to do everything possible to achieve a negotiated settlement to the present crisis.'

This statement was heard and absorbed by the leaders and diplomats of nations around the world and they weighed the implications. For the British man in the street, however, the announcement by the Ministry of Defence that Royal Marines were established on South Georgia was highlighted by the television pictures which, that evening, showed Mrs Thatcher and John Nott on the steps of 10 Downing Street. Her heart-felt call to the waiting journalists and public – 'Rejoice! Rejoice!' – was one which stayed longer in the memory than many others provoked by the events of 1982.

There was no rejoicing in Buenos Aires. General Galtieri and his colleagues had staked their reputations on the seizure of the South Atlantic islands. The loss of South Georgia, little more than three weeks after the triumph of Operation Rosario, was the beginning of a nightmare which was to culminate with the total collapse of Argentine resistance at Port Stanley on 14 June.

It is evident that the garrison commander at King Edward Point succeeded in transmitting a situation report to the mainland shortly before surrendering. His account was quickly edited for public consumption and flashed to Ambassador Eduardo Roca, his country's permanent representative at the United Nations in New York. By late afternoon, Roca was delivering a formal protest to the President of the Security Council: 'The Government of the United Kingdom of Great Britain and Northern Ireland, having begun its aggressive escalation with the despatch of a fleet of warships that includes nuclear submarines, and having continued it with the naval and air blockade of part of Argentina's territory, today, 25 April, perpetrated an act of armed aggression against the South Georgia islands, which form part of Argentine territory, using naval units and helicopters. This aggression began during the morning hours with an attack on an Argentine submarine, which was anchored at the surface and was unloading provisions, medicines and mail in the port of Grytviken, and continued with a steady bombardment of that port carried out by British naval units. For more than four hours the defence force was subjected to constant cannon fire from the destroyer *Exeter* [*sic*], in addition to raids carried out by heavily-armed helicopters, which machine-gunned the defensive positions. After a short tactical withdrawal, the *Exeter* took up a position off Port Leith with the support of the remainder of the units of the task force (one tanker ship, one troop carrier and one logistical support ship), presumably waiting for suitable conditions for a new assault on the Argentine position. According to the latest information on the events in Port Leith, which lies north-west of Grytviken, the small naval force stationed there managed to hold out until its defence capability was exhausted.

Argentine prisoners - marines and submariners - wait placidly outside Shackleton House while their captors decide what to do with them. The coming of night and a rapid deterioration in the weather prevents an immediate evacuation. Some are taken out to the RFA *Tidespring* on the following morning and carefully searched in her hangar before being placed in their new prison quarters.

The Argentine government wishes to point out that this attack was carried out at a time when the negotiations undertaken with the participation of the United States Secretary of State, Mr Alexander Haig, were still open. These negotiations involved the simultaneous consideration of all aspects of Security Council resolution 502, dated 3 April 1982.'

The meaning of this final sentence is obscure, but the Argentines seemed to be suggesting that they were the innocent victims of a nasty and unprovoked attack just at a time when they were prepared to be conciliatory. For a regime which only a short while before had been so bombastic in its pronouncements, the submission of Roca's letter reflected a swift change in sentiment. It aroused little sympathy in New York, nor did it have any effect upon the intense activity on the island to which it referred.

The remaining hours of 25 April were busy ones for Guy Sheridan. He had nearly eighty of his own people and all the Argentine prisoners to care for, and the only normal accommodation available was Shackleton House. The sleeping and cooking facilities of that building were designed to accommodate thirty or forty men at most. The prisoners needed to be secured, and the Geneva Convention required that they be protected from the elements. Sheridan told Captain Chris Nunn to see what could be done while he himself turned his thoughts to the reoccupation of Leith Harbour.

Sheridan had another conversation with Captain Bicain and confirmed that he had no wish to use unnecessary force against Teniente de Navio Astiz and his party. Would he, Bicain, exercise his authority and superior rank to convince Astiz that there was no purpose in trying to resist and no dishonour in unconditionally surrendering his command? The submariner agreed, and was taken to the radio shack.

Guy Sheridan stood at Bicain's shoulder as the Argentine naval officer spoke with his compatriot at high speed in Spanish. None of the British personnel in the King Edward Point radio shack spoke Spanish, so they could not know exactly what was being said. However, the general gist was clear enough - Bicain was telling Astiz that it was all over. The reaction from Leith Harbour was truculent. Summarising the response to his captors, Bicain explained: 'Astiz says that he will not surrender. He says he will fight to the death.' This melodramatic reply was in some ways admirable but, given the overwhelming forces now ranged against him, totally unrealistic. Brian Young and Guy Sheridan quickly agreed a plan to eliminate the last remaining Argentine toehold on South Georgia.

Captain Young ordered the *Plymouth* and the *Endurance* to move north and to enter Stromness Bay. The Plymouth's commanding officer, Captain Pentreath, was given overall command of this operation. Apart from 2 SBS, his small assault force comprised the SAS Mountain Troop (which had re-embarked after the successful seizure of King Edward Point) and the SAS Air and Mobility Troops (which so far had been kept in the *Endurance* and *Plymouth* respectively, and had seen no action).

The initial plan was to land the force by boat on Tonsberg Point - which divides Husvik Harbour from Stromness Harbour - and send them forward to join 'Tim's' Boat Troop on Harbour Point before making a final assault. Then 'Tim' repeated his earlier recommendation that there was a suitable beach on the north side of Leith Harbour where small boats could put men ashore under cover of darkness. This would mean a direct approach, but the Argentine defenders could be obliged to keep their heads down by a combination of illuminated naval gunfire from the *Plymouth* and small-arms fire from 'Tim's' Troop on Harbour Point. This was the plan adopted.

Although David Pentreath was the senior captain, he authorised Nick Barker - as the 'local resident' - to make another appeal to the small garrison to surrender peacefully. Speaking on Channel 16, Barker had three abortive conversations with Astiz during the course of the evening. The Argentine did agree that the civilian workers should be allowed to give themselves up, but he and his troops would fight. An aerial reconnaissance, by Lieutenant-Commander John Dransfield in the Plymouth's Wasp 445, confirmed that there were no white flags flying over Leith Harbour.

At 2100, Captain Barker ordered a ship's whaler to take him to the *Plymouth* for a council of war with Captain Pentreath. With him he had Lieutenant-Commander Ellerbeck, David Ives the Charge Surveyor, and Bill Hurst the Navigating Officer. It was a very black night and the coxswain had some difficulty in finding the fully darkened HMS *Plymouth*, anchored off Husvik. It was while they were away that Astiz called the *Endurance* on Channel 16 and announced that he had changed his mind. He no longer intended to fight.

His call was received by the officer of the day, Surgeon Lieutenant Neil Munro. A non-combatant, Munro was startled to find himself accepting a declaration of unconditional surrender. Shortly afterwards, when his Captain returned to the ship, he reported the details and Nick Barker had his fourth and final radio conversation with Astiz at 2245. The Argentine confirmed that he was prepared to surrender his entire command, civilian and military.

The Argentine spoke English fluently and it did not take long to agree the procedures. The scrap metal workers would emerge first from the Leith Harbour buildings, walking along the shore toward 'Tim's' position. As soon as they had been searched and secured, Barker and Pentreath would fly ashore by Wasp to accept Astiz's surrender in person. The location for this encounter would be the helicopter landing site on the old football pitch. All of this would happen in the morning, at dawn.

On board HMS *Antrim*, Brian Young wondered how best to set a formal seal upon the Argentine surrender of Grytviken and King Edward Point. He spoke with Major Sheridan and asked him to draft a suitably worded document. He laughed when Sheridan told him: 'I'm sorry, I don't have any personal experience in surrendering. But I'll do my best.'

Another concern for Captain Young was the security of his ships. With all eyes fixed on events ashore, it would have been all too easy to forget the ever-present risk of attack by Argentine frigates or a second submarine, or attack from the air by weapon-carrying Hercules or Boeing 707. John Coward was ordered to launch one of his helicopters to patrol offshore, to survey the seaward approaches to the Cumberland Bays.

Butler and Bryant were launched in Lynx 342 and, making occasional use of their radar, surveyed several thousand square miles of ocean to the north-east. It was a foul night with high wind, heavy rain and a very low cloudbase. In some places the clouds were trailing down to sea-level. The RFA *Tidespring* was known to be approaching the island and the first radar probe found her. Nick Butler flew over her flight deck at 200 feet in poor visibility, identifying himself to the ship and receiving the correct response to his challenge. Barry Bryant concentrated upon his radar screen as the Lynx pushed further out to sea. '*Tidespring* showed up very clearly and was obviously a ship, but I was getting quite a few other contacts. Most gave a slightly fuzzy image, suggesting that they were icebergs. One in particular, however, was impossible to classify. Flying conditions were appalling and Nick gradually reduced speed as we approached this thing – whatever it was. At four hundred yards I lost contact in the ground clutter and we switched on the landing light to illuminate. It was the most enormous iceberg I have ever seen. We were at an altitude of one hundred feet, but the cliff of ice was much higher than us – maybe three hundred feet or more. It simply disappeared up into the cloud and darkness. There was a considerable rush of adrenalin in the aircraft and we decided that discretion was the better part of valour. Nick turned away and flew us very carefully indeed back to mother.'

The only other type of risk which could conceivably threaten the anchored *Brilliant* and *Antrim* was a limpet mine attack by Argentine special forces divers. Another message was sent to Major Sheridan at Shackleton House. 'I was told that there was a worry about swimmers and a Chariot-type craft. Could I invite the skipper of the submarine, Captain Bicain, and the marines officer, Captain Lagos, to dinner in the *Antrim* in order to gauge their reaction. If they declined, then the navy would assume that there were indeed swimmers in the water.' In the event, the two Argentines were delighted to accept Brian Young's hospitality. Sheridan escorted them aboard and the four officers chatted over the dinner table until nearly midnight when he took them back to the confines of Shackleton House. 'They enjoyed several glasses of wine and beer. We were under orders not to interrogate them, but they parted with some useful bits of information. I expect they thought that, if life as a prisoner was going to be like this, it wouldn't be too bad after all. Much more comfortable than a cramped smelly submarine.'

In the early hours of 26 April, the balance of 'D' Squadron (less the three men still missing from the Grass Island insertion) was put ashore on Harbour Point by HMS *Plymouth*'s Avon Searider and by the sole remaining serviceable SAS Gemini. Captain Pentreath had brought his ship close in to the shore before launching the boats. 'It was pitch dark and I wanted to give the boats some protection and cut down the distance they would need to travel. The coxswains had considerable difficulty negotiating the tenacious beds of kelp and reaching the agreed landing point on the beach. I'm glad it was so dark for I was unable to see how very close *Plymouth* was to the rocks.'

'Tim' now had sixty heavily armed men concealed along the crest of Harbour Point in positions which dominated the Leith Harbour buildings only six hundred yards away. There was no sign of movement from the Argentine positions and the next four hours passed in chilled silence.

At 0600 the *Endurance* launched her two Wasps to put 2 SBS onto the foreshore between Leith Harbour on the right and Harbour Point on the left. Shortly afterwards, forty-eight Argentine civilians came out from the old whaling station, finding their way with hand-held torches. 'Roger' and his SBS marines herded them together and searched them for concealed weapons before passing them through to the SAS positions on Harbour Point.

It was at this stage that Captain Barker had second thoughts about the wisdom of flying ashore to land in an area still controlled by armed enemy troops. He had arranged to do this while talking with Astiz on the radio but, upon reflection, he decided that the plan would place himself and Captain Pentreath in a position of unnecessary risk. He consulted with Pentreath and the plan was abandoned. Astiz was now told to leave his weaponry in the buildings and to walk out to the positions where the SBS and the SAS were waiting.

The Argentine commander had no choice other than to agree. He confirmed that he would follow the instructions, but gave warning that the entire area of the buildings was heavily mined. As day broke, he emerged with his men in single file and walked along the beach toward the British shore party. Captain Barker and Captain Pentreath watched through binoculars from the bridges of their respective ships as fourteen Argentine marines were stripped to their underwear and searched. Having proved that they were not carrying weapons, they were glad to scramble back into their clothes. The dawn air was very cold.

'Tim' gave orders for his prisoners to be escorted back to Stromness where they were held pending evacuation to the *Endurance* and the *Plymouth*. At the same time, he sent a small patrol forward to check the Leith Harbour buildings, to search for any concealed die-hard snipers and to investigate the mines

The surrender document signed by Astiz and counter-signed by Barker and Pentreath aboard HMS *Plymouth* on 26 April. The hand-written postscript added by the Argentine shows that his spirit was still resilient, but it caused some irritation to his captors. The comparative size of the lettering on the document's crest suggests that Astiz was not alone in seeking to make a point.

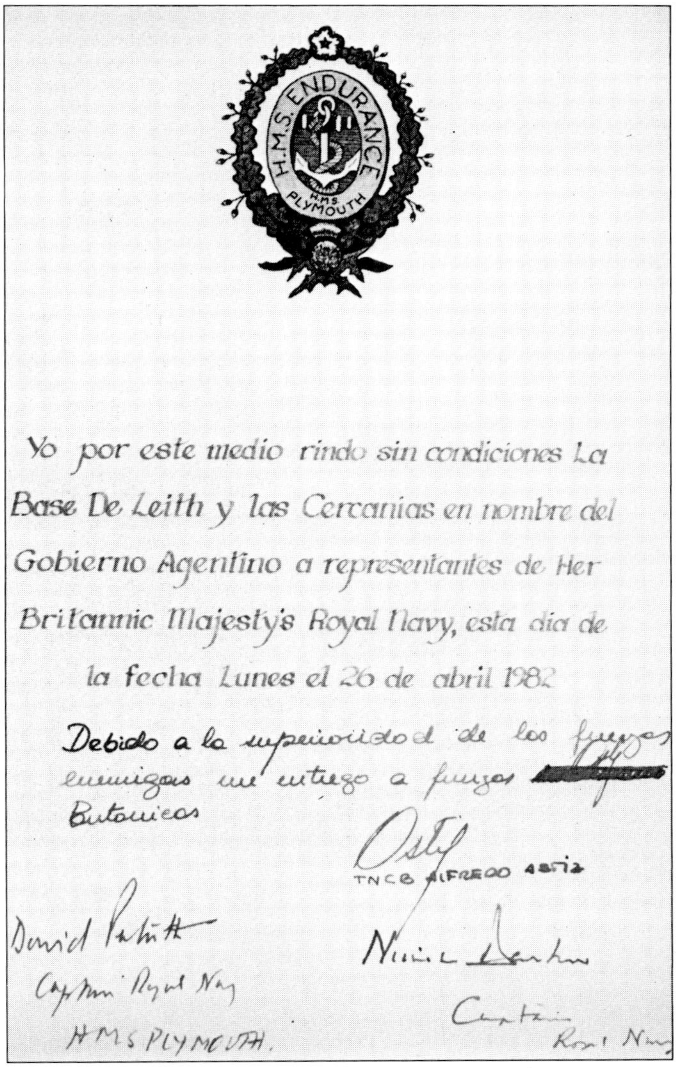

Yo por este medio rindo sin condiciones La Base De Leith y las Cercanias en nombre del Gobierno Agentino a representantes de Her Britannic Majesty's Royal Navy, esta dia de la fecha Lunes el 26 de abril 1982

Debido a la superioridad de las fuerzas enemigas me entrego a fuerzas Britanicas

TNCB Alfredo Astiz

HMS PLYMOUTH

reported by Astiz. It was found that his engineer had constructed makeshift explosive devices in much the same way as Marine Daniels had done at King Edward Point on 2 April. Approximately 500 kilograms of plastic explosive were semi-buried along the beach and among the buildings. One mine in particular gave cause for thought. Consisting of twenty kilograms of explosive packed in a steel drum with a quantity of nails, pebbles and scrap iron, it was concealed beneath a piece of canvas under the 'H' of the helicopter landing site. This was the spot where Barker and Pentreath had intended to land. Barker felt that 'this was not quite cricket', and he questioned Astiz about his intentions. The Argentine was unrepentant: 'It was the only way I could have got you.'

The charges evidently had been laid in haste. They were wired to be set off electrically from a central 'ripple switch' control point. The firing mechanism itself was of a very modern and sophisticated design – which caught the interest of the SBS and SAS demolition experts – but the wires running out to the mines were laid on open ground. The Argentines had not had time to bury them – they were crudely concealed under rubbish and sheets of corrugated iron.

Although the invaders had not given much thought to their military defences, they had evidently done some forward planning in another context. A thorough search of their living quarters revealed a startlingly ambitious supply of condoms. Nick Barker laughingly chided Astiz when this discovery was made. 'My dear Alfredo, what on earth did you expect to find here on South Georgia?' The Argentine had the grace to join in the laughter. 'Well, you never know. One must be prepared for anything.'

It was not long after the surrender that the British realised that two members of the garrison – the doctor and his medical attendant – were not accounted for. It required nearly three hours to carefully search all the semi-derelict buildings – any of which might have been booby-trapped – before they were run to ground. When they eventually appeared, they did so in an area where a bonfire had recently died down. It seems probable that they had been destroying documents and other material which Astiz wished to conceal from the British.

As a magnificent dawn brought brilliant sunshine into the deep valleys of the hinterland, boats and helicopters brought more and more visitors to the shore from the *Plymouth* and the *Endurance*. Apart from Captain Barker, many of his officers and ratings were anxious to have the chance to stretch their legs and to see their opponents at first hand. It was the first time any of them had set foot on dry land since the end of March. They were not permitted to remain for long – the area was littered with explosives, ammunition and weapons. The sooner the station could be emptied the better for all concerned.

Astiz was flown out to the *Plymouth* in Wasp 445. Captain Barker had drafted a surrender document for him to sign. During the night, the *Endurance*'s communications department had translated it into Spanish and Chief Petty Officer Ginger Woodhouse – one of the hydrographers and therefore a skilled draughtsman – had penned an impressive folio for the ceremony. In the *Plymouth*'s wardroom, facing the two captains and surrounded by representative members of the ships' various departments, Alfredo Astiz added his name to those of his captors.

Even this simple procedure was not without incident. Captain David Pentreath describes the ceremony: 'My Petty Officer Steward, Townsend, had laid everything out most tastefully on the wardroom table, including a blotter and inkwell. *Endurance* had produced overnight a splendidly decorated surrender instrument, and, at 1130, Lieutenant-Commander Astiz was brought in. He was invited by me to surrender his forces unconditionally and to sign accordingly. At this, he began to write a sentence on the parchment before him, so I asked my interpreter what he was writing. There was a pause while the interpreter studied the document and Astiz then stated, in excellent English, that he wished to record his reason for surrendering – "the superiority of invading enemy forces". "No deal," I replied, "British forces are not invading South Georgia, we are repossessing it." He eventually crossed out the offending word "invading". Frankly, I did not care for this man's arrogance and I was pleased to see him transferred to HMS *Endurance*.'

The rest of the Argentines were ferried out to the ships by boat and by helicopter. One of the SBS eyewitnesses described the civilian captives as 'a very pathetic lot. They were stinking and dirty. At first they were terrified because they thought we were going to shoot them. When they realised that we weren't, they cheered up and were just pleased that it was all over and that none of them had been hurt. Basically they were simple folk, not very intelligent.'

The Argentine marines were men of much higher calibre, but they too seemed slovenly by normal military standards. Their captors were not impressed. 'They had been living in a fair degree

Having signed the surrender document, Alfredo Astiz (centre) is flown from HMS *Plymouth* to HMS *Endurance* in the red ship's Wasp 435 piloted by Lieutenant Tim Finding. Seated at the door of the small helicopter is Astiz's personal escort, Lieutenant Dick Buckland. Astiz is one of those captured Argentine officers who now face a bleak future.

of squalor in the only building which was still habitable. They all lived in there together, and didn't seem to know much about sanitation or hygiene. They didn't smell too good. Several of their weapons were inoperative because they hadn't been cleaned properly. They wouldn't have stood a chance if they had tried to put up a fight. The SAS observation party on Harbour Point could have polished them off without any additional support, if that had been necessary.'

By midday the Leith Harbour area had been completely evacuated. It was left empty until the following morning when a small party of marines returned briefly to dismantle the explosive devices. After that, all Argentine resistance on South Georgia – active and passive – could be regarded as having come to an end. □

IX Stepping-stone to Stanley

The only men on South Georgia not accounted for were the three SAS soldiers of 'Tim's' Boat Troop. On the night of 22/23 April, the Gemini codenamed Delta and commanded by 'Tom' had come ashore near Jason Point where the crew took shelter in a cave. Throughout their first day in hiding, they worked on the outboard engine with the intention of making another attempt to reach Grass Island. Shortly after nightfall they dragged the rubber boat to the water's edge, remounted the engine and loaded their equipment.

Last light had been at 1830 and an hour later the small craft was motoring away from the beach on a lumpy sea. After 200 yards the engine failed and the party paddled back to shore. The carburettor was stripped and cleaned and they tried again. A second attempt failed for the same reason and this procedure continued until the early hours of 24 April. The party settled down to another day in their damp cold refuge. They abandoned all hope of curing the sick machinery and decided instead to make their way on foot to the distant whaling station of Stromness.

On the evening of 24 April, having deflated the boat and concealed it inside the cave, 'Tom' and his two companions started to climb the cliffs of Justa Peak. The ice-shattered rocks and loose scree rise steeply from sea-level to 1,600 feet in less than a mile. It was a hard and dangerous ascent in the dark of a windy night. One of the party recalls: 'We were hoping to get up onto the high ground of the Jason Peninsula and then walk to Husvik before dawn, but that climb really was hairy. At one stage we found ourselves inching along a ledge – about a foot wide – with cliffs shooting up to the left and a black hole down below us to the right. There was no solid footing and, with eighty pounds on my back and an Armalite to carry, I was not at all happy.'

By first light the Patrol was somewhere on the high slopes of Jason Peak. Further movement was inadvisable – they might have been seen by any observers keeping watch from Leith Harbour or the Tonsberg headland. They settled for yet another day of cold inactivity. Not having a radio, they had no idea of the fate of the rest of the Troop or of the rapidly developing situation in Cumberland East Bay. It was a surprise to them, therefore, when, early in the afternoon of 25 April, they heard the first booming explosions of naval gunfire. Uncertain as to its significance, they remained under cover.

That evening the Delta party saw HMS *Plymouth* entering Stromness Bay. 'Tom' used a hand-torch to signal 'SOS SAS'. His flashing light was seen by the warship but, in the deepening gloom, the Wasp helicopters aboard the *Plymouth* and the *Endurance* could not safely attempt an evacuation. The SAS troopers were obliged to spend a second night on the high exposed ridge.

The morning of 26 April was bright and sunny. A stiff breeze blew occasional clouds over the top of the Peninsula, but Tony Ellerbeck had often flown in conditions far worse than these. Approaching the mountain, he saw the bright orange of a smoke grenade and flew in towards the three small figures waving to him. The scene is described by 'Tom': 'We were cold and we were hungry. It was great to see this helicopter coming in towards us. The pilot obviously knew what he was doing and we could already taste the bacon and eggs that would be waiting for us in the ship. Then, when he was right in front of us and we could almost reach out and touch the Perspex, the cloud shut down like a trap door and we heard him zoom away back to the ship.' Later in the day Ellerbeck returned and lifted the Delta patrol out to the anchored *Endurance*. Later still they rejoined – after a separation of four days – the bulk of the Boat Troop at Leith Harbour.

In the *Endurance*, at a ceremony in the small wardroom, Alfredo Astiz signed the surrender document prepared by Ginger Woodhouse. Twelve miles to the south, off King Edward Point, the formal 'instrument of surrender' drawn up by Major Sheridan was signed by Captain Bicain and Captain Lagos in the wardroom of HMS *Antrim*. In Buenos Aires, the junta's information services were denouncing the British reoccupation of South Georgia as 'a treacherous act of war'. They claimed that Argentine troops were 'still fighting fiercely' and that the British had 'suffered dozens of casualties', but it was nothing more than wishful thinking.

In New York, the propaganda issuing from Buenos Aires did not impress the delegates at the United Nations. The majority preferred to accept the British account of events on South Georgia. There were quiet and unofficial words of congratulation to the British representatives from delegates sympathetic to the United Kingdom's cause, but few seemed to understand the main motive behind the reoccupation of the island. As Sir Anthony Parsons said later: 'Most of them seemed to be assuming that we had retaken the place solely to gain a bargaining counter at the peace table. In other words, they believed that this would be our one and only military operation in the South Atlantic, that it was done to save face, and that we would subsequently agree to the idea of Argentine forces remaining on the Falklands. It was assumed also, of course, that there would be a protracted period of diplomatic exchanges, with accommodations on both sides, but that our main fleet then approaching the Falklands would be utilised by the British government as a psychological rather than a military weapon.' This had never been the British intention. If the current spate of peace initiatives were to fail, or to become bogged down for such a long time that the Antarctic winter threatened the safety of the fleet, then South Georgia must be

HMS *Conqueror* prepares to disembark her 6 SBS passengers on 26 April. Nearby are HMS *Antrim* (left) and HMS *Plymouth* (right), with the tiny shape of Wasp 445, piloted by Lieutenant-Commander John Dransfield and carrying an underslung load, flitting between them. This photograph was taken by Sub-Lieutenant Stewart Cooper, from Wessex 406, fifteen minutes before the accident. During that time the wind became much stronger and the waves increased rapidly in size.

able to play its part as a forward operating base. It was against this background that work on the island went forward at a rapid pace.

The scene on and around King Edward Point was one of intense activity throughout that Monday, 26 April. At 1030 the *Tidespring* came into Cumberland East Bay and anchored nearby. She began the business of ferrying ashore the men of 'M' Company and large quantities of military stores. To the north, in Stromness Bay, HMS *Endurance* was busy clearing the Argentine positions at Leith Harbour. Off shore, seven miles from the entrance to the Cumberland Bays, HMS *Antrim* patrolled quietly in thirty-mile legs, keeping watch for any hostile aircraft or warships which might be approaching the coast.

At midday, not far from the *Antrim*, HMS *Conqueror* came to the surface. The sole purpose of her brief visit was to offload 'David' and the men of 6 SBS. It had been decided to transfer them to the *Plymouth* for eventual delivery to one of the surface vessels of the main Operation Corporate force. Lieutenant Stewart Cooper, flying 'Humphrey', was tasked to winch them up from the submarine's forward deck casing.

It was during this time, with the wind gusting to forty miles per hour, that an exceptionally large wave broke over the casing and washed an SBS marine and one of the submarine's sailors into the sea. Both of their life-lines broke as they were swept away. With them went a waiting pile of stores, several machine guns and 'David's' briefcase containing all his personal papers. Tide and wind started to carry the men away. Neither was wearing a 'once only' protective suit and the stabbing cold of the freezing sea quickly sapped their strength – the life expectancy of a man immersed in those waters is counted in minutes. The marine – a man named 'Jem' – was trained to deal with sudden stress, but the sailor was soon in difficulty.

The ubiquitous 'Humphrey' was hovering nearby, her load-lifting net ready to pick up the SBS stores. POACMN Fitzgerald was leaning out of the door and saw what was happening. 'I alerted the pilot and conned him back over the two survivors. I could see that the marine was swimming strongly, but the sailor was not moving or attempting to swim. I got a strop and lowered it down to them. It actually hit the sailor, but he made no attempt to grab it. By now I was very concerned, so I guided us over to the marine who inserted himself into the loop. I then conned Lieutenant Cooper back to the sailor, dragging the marine across the surface of the water. I gestured for him to lasso the other guy and, once he caught on to what I was trying to tell him, I managed to winch them both up to the door in a single lift. I had a hell of a job getting them both inside the aircraft because the sonar was in the way. Eventually I got them into the back of the fuselage and made them beat their arms around their bodies to combat the intense cold. The submariner didn't look too good.'

Stewart Cooper quickly delivered the casualties to HMS *Antrim* where they were rushed below to the wardroom, stripped and placed in warm baths. They had been in the sea less than three minutes, and both soon recovered, but the episode was a reminder of what would have happened if the *Santa Fe* had succeeded in torpedoing one or more of the British ships. For 'Humphrey', it was the third successful search and rescue (SAR) mission in four days.

Commander Wreford-Brown departed the island's waters as soon as he had disembarked the last of his SBS passengers. He sailed westward to help enforce the new Total Exclusion Zone imposed by the British around the Falkland Islands with effect from 1100 on 30 April. The name of his command was to gain a place in history when, on 2 May, she torpedoed and sank the 13,600-ton cruiser, *General Belgrano*.

The most pressing problem for Captain Young was the disposal of his 185 prisoners. The number far exceeded the normal accommodation capacity of the Task Group ships and, even if it had not, the gales which blew intermittently during the next two days and nights precluded any major movement between shore and ship. The counting of Argentine heads revealed that there had been forty-eight civilian scrap workers and sixteen special forces personnel under Teniente de Navio Astiz's command at Leith Harbour. At King Edward Point, there were seventy-six officers and men from the *Santa Fe*, and twenty marines under Capitano de Corbeta Lagos who had been brought in the submarine from Argentina as reinforcement for the garrison. The balance of twenty-five marines and sailors was the force defend-

Cases of Argentine ammunition are stacked on King Edward Point on 26 April, ready for disposal.

ing King Edward Point and Grytviken prior to the *Santa Fe*'s arrival.

For the first thirty-six hours, the bulk of the captives were held under guard by the Royal Marines in the overcrowded Shackleton House while the Leith Harbour prisoners were accommodated in HMS *Endurance* and HMS *Plymouth*. Plans were drawn up to transfer them all to the RFA *Tidespring* for transportation to Ascension Island and eventual repatriation, under International Red Cross supervision, by chartered civil airliner. In the meantime, there was urgent need to remove from the island some of the British personnel and move them on to new destinations.

Orders arrived at 1530 on the afternoon of 26 April for the transfer of Cedric Delves' entire 'D' Squadron to Task Force 317.8, now only 350 miles to the north of the Falkland Islands. With them would be going 2 SBS and 6 SBS and Keith Eve's NGS party. All of these specialists were needed for the initial reconnaissance and intimidation of the Argentine forces on the Falkland Islands. They were to be embarked in the *Brilliant* and the *Plymouth*.

The recovery of the SAS and SBS from Leith Harbour was not without incident. The day had been dull and overcast, with every sign of a big storm to come, and it hit the entire east coast of the island in the afternoon. The evacuation of men and equipment was nearly complete when the offshore wind became almost frighteningly violent. 'Tim' had been using his last serviceable Gemini, Echo, as a ferry from the beach to the anchored HMS *Plymouth*. The heavy rubber boat was moored on the north foreshore, loaded with crates, when suddenly a powerful gust lifted it into the air and hurled it a hundred yards along the beach.

One of the helicopters working in the area was Lynx 342 flown by Nick Butler and Barry Bryant. They were determined not to leave anyone stranded overnight. As darkness closed in, they made three more flights to lift the SBS from a small stretch of beach under the soaring cliffs. The usual precautions were set aside as passengers and equipment were packed in and flown out to the waiting ships. The final trip involved the lifting of the Gemini and the last six men on the beach. The marines struggled in the rotor downdraft to attach cables to the heavy boat as Butler hovered over the breaking surf. He then descended and sidled along the line of the waves so that his passengers could wade out and hurl themselves in through the cabin door. The SBS could not be regarded as men of nervous disposition but, on this occasion, they felt some qualms at boarding a helicopter which showed every sign of being blown away or swept against the cliff.

Lieutenant Butler recalls this as his most alarming flight of the whole campaign: 'It was pitch dark and very turbulent by the time I got away. My eyes were glued on the instruments while Barry hung over the back of my seat. He had one eye on the radar screen – watching the cliffs go by – and the other on the Gemini oscillating around underneath us. Lift-off was OK because it was into wind, but then I had to make a U-turn downwind. This is something you try not to do, especially if you are fully loaded. The Lynx lost airspeed and, at forty knots, started to lose height. Fortunately I was able to get clear without jettisoning the boat, but we were both shattered when we eventually shut down on *Brilliant*'s flight deck. We had been airborne – more or less non-stop – for seven hours. We had flown a total of nineteen hours in the previous thirty-six, much of it in darkness. Bear in mind that the peacetime book of rules states that aircrew must not undertake more than six hours' flying in any twenty-four hour period, and must have a break of at least eight hours rest in every twenty-four.'

There were many exceptional feats of aviation during that period, and this sortie by Lynx 342 epitomised the spirit and skill with which they were conducted. One of the SAS men who was there paid this tribute: 'All the aircrew involved in Paraquat were fantastic. They just never gave up.'

Captain Coward shared the same view and he made this comment in a report on the *Brilliant* flight: 'Both aircraft and aircrew are more or less dead-beat after a week's flat out flying. They have done absolute wonders and without them the invasion of South Georgia could not have happened. I greatly hope that this will not be overlooked in the accolades that follow this operation.'

There were indeed many accolades and messages of congratulation to the Task Group for the manner in which Operation Paraquat had been carried through but, for some of those who were there, it was marred by a tragedy which started to unfold on the morning of 26 April.

The early morning was calm and brilliantly sunny. It was one of those rare autumn dawns when South Georgia shows itself at its spectacular best. The few Royal Marines remaining at King Edward Point had had a short and anxious night, guarding their 121 prisoners under difficult circumstances. Those who had managed to snatch a few hours sleep were up at first light. It was going to be a very busy day.

One of the most pressing problems was the question of what should be done with the stricken *Santa Fe*. Guy Sheridan was anxious to see her moved away from the King Edward Point jetty. From his point of view, as Commander Land Forces, she was 'a total bloody nuisance'. Apart from containing several tons of explosive of unknown stability, her leaking hull was blocking access to the urgently needed jetty. It was his opinion that she should be towed away to deep water and there scuttled.

In the meanwhile, he placed a ban on all further visits to the submarine by anyone not having an authorised reason for entering her. Already several marines and SAS troopers had gone aboard in search of souvenirs and, when he himself boarded her on that Sunday morning, he discovered that there had been considerable disturbance of the boat's contents.

Early in the morning, while Sheridan was still below, he was joined by the commanding officer of HMS *Brilliant*. Captain John Coward had served twenty years in the submarine service, his last appointment having been in command of the nuclear-powered HMS *Valiant*. He had been sent aboard the *Santa Fe* by Captain Brian Young to assess the condition of the captured vessel and to determine what might next be done with her.

Coward worked his way methodically through the dismal interior of what was, in effect, a dying boat. He checked all the main controls and turned off some of the valves which had been left in the wrong positions by the fleeing Argentine crew. He also looked at the battery banks, ensuring that they had been properly disconnected and searching for signs of gas formation. Old batteries can produce highly flammable hydrogen gas. His main findings were that the boat was not seaworthy, but was probably safe enough to be moved over short distances. The batteries still carried the remnants of a charge and, once reconnected, could be used to propel the vessel on the surface at slow speed. The main diesel engines and drive shafts were intact and operative, but would not be needed for a short passage.

An examination of the electrical systems showed that they had been badly shaken by the depth-charge explosions. Many circuits were inoperative, and the internal telephone system was almost entirely dead. Only the handsets on the bridge, in the motor room and in the main control room were working, and even then the reception was poor. The bridge telephone functioned only if the damaged connecting plug was held in place while in use.

The most serious deficiencies found by John Coward were in the ballasting system. There are two distinctly separate air systems in a submarine. The high pressure (HP) system generates air, at 3,000 pounds per square inch, which is used to blast tons of sea water out of the saddle tanks in an emergency or when the boat is required to surface from depth. Reserves of air for this system are held in banks of steel bottles, available for instant use. The low pressure (LP) air system, on the other hand, delivers air at 60 pounds per square inch and this is sufficient to vent the ballast and saddle tanks when the boat is on or very near the surface (and sea water pressure is correspondingly less). In boats of the *Santa Fe*'s vintage, low pressure air was generated by a large fan-type compressor, housed under the control room deck. The controls for starting this compressor were situated against a bulkhead in the control room, and the valves for distributing the air to different parts of the ballasting system were located overhead, on the deckhead. Given the poor condition of all of the *Santa Fe*'s controls, particularly the non-return valves, it was important that there should be no delay between turning on the LP compressor and opening the overhead valves to distribute the air to whichever tank was selected. The operator had to work very quickly to avoid a blow-back effect in the compressor. He could not, of course, anticipate events by opening the valves *before* starting the compressor, otherwise the boat would flood. These requirements and limitations – which were partly the result of battle damage and partly the result of inadequate maintenance – were important in the light of what was to follow.

The submarine's commanding officer, Captain Bicain, had been allowed back aboard to collect some personal effects from his cabin. John Coward found him standing on the casing, looking thoroughly dejected. There was a brief conversation, but Bicain was in no mood to talk. His mission had been a failure, he had lost his command, and his crew were all in captivity. He tried to conceal his misery by displaying only a limited knowledge of English. In fact, he spoke and understood the language quite well.

Captain Coward was lifted out to the *Antrim* by Lynx helicopter and gave his findings to Captain Young. He reported that the submarine was, in practical terms, a wreck. There were untraceable leaks in the after part of her hull, she was down by the stern and had a twenty-degree list to port. She had not yet grounded but appeared to be slowly settling in the water. Unless something was done, she would eventually touch bottom and become a major salvage task. There was a self-evident and urgent requirement to move her. The need for swift action was underlined by the fact that the British had only just reoccupied the island. There was no guarantee that the Argentines might not launch a counter-attack, or take some other military initiative against the Paraquat force. 'A lot could happen in the next few days.' Coward recommended that the submarine be towed out to sea and scuttled.

It was soon established that such an operation was out of the question. There were neither the resources, the personnel nor the time – at that stage – to safely remove her to any distant scuttling site. One or two officers voiced the possibility that a technical survey might reveal information of interest to naval intelligence. She had been searched for cryptographic materials and documents, and some useful operational orders and charts recovered. In Coward's own view: 'there was no way we were going to learn anything more from a conventional submarine built nearly forty years earlier. But clearly we needed to park her somewhere else on a temporary basis.'

The first problem was to know where best to move the submarine. There were two alternatives – either run her up on a beach, or moor her to the old whalers' jetty across the Cove at Grytviken. John Coward discounted the beach because, in her listing condition, she would eventually ground, topple over and flood. This would defeat the aim of trying to preserve her for later examination. Grytviken jetty – also known as Tijuca Jetty – was therefore the obvious choice, particularly as Captain Coward believed it to be semi-derelict and inoperative for other purposes. This mistaken appreciation was based upon the generally deteriorated appearance of the distant whaling station and the sight of the two semi-submerged hulks (the *Albatros* and the *Dias*) alongside the jetty there. Given more time, Coward would have discovered that Grytviken jetty was not the ships' boneyard which it appeared to be, but was in fact in regular use.

Plans were laid to move the *Santa Fe* to her new berth on the following morning. Captain Bicain was asked if he would assist in the operation and he agreed. He was also asked to select the key personnel needed to man the controls for the short surface journey across the Cove. He stated that nine would be the minimum necessary. Coward disagreed, arguing that three would suffice. Eventually they compromised on six: an engine room artificer in the central control room, two electricians to operate the main switchboard in the motor room aft, a coxswain to man the helm halfway up inside the much battered fin, and two men on the outer casing, fore and aft, to handle the mooring lines. These men were asked to assist their captain, and they volunteered without argument.

Coward sought out Chris Nunn and explained his intentions. He asked for the release of Bicain and the six ratings. Nunn agreed, but with the proviso that they should be escorted at every stage by Royal Marine guards. The marines had come a long way to reoccupy South Georgia and they did not intend at this stage to lose control of their captives. Nunn went aboard the submarine and Coward showed him where each Argentine rating would be stationed.

It was decided that four Royal Marines, armed with 9mm Browning pistols, would suffice to guard the *Santa Fe*'s skeleton crew. Two would stand on the casing and supervise the two line handlers. Two more, a lance-corporal and a marine, would be stationed inside the hull, in the control room, where they would watch the artificer and simultaneously keep an eye on the helmsman. The two electricians in the motor room would be under the supervision of Lieutenant Colin Tozer, RN, one of the *Brilliant*'s officers and himself an experienced submariner. Finally, high on the conning tower, would be Captain Bicain, Captain Coward and, as guard to Bicain, Lieutenant Chris Sherman, RN, the *Brilliant*'s diving officer.

Early on the morning of 27 April the boat was prepared for the move. The hatches were opened to ventilate the interior spaces, to remove any noxious or explosive gases. The batteries were reconnected and checked. A shore party was despatched across the Cove to handle mooring lines when the submarine arrived at Grytviken jetty, and another party prepared to cast off the lines securing her to King Edward Point jetty. Coward and Bicain agreed that the submarine had settled further during the night. Before casting off, therefore, Bicain was instructed to blow low pressure air into the saddle tanks and the after ballast tanks in order to improve both buoyancy and trim.

Bicain lifted the bridge telephone and, with some difficulty, spoke with the artificer in the control room. This man was Subofficial (Chief Petty Officer) Primero Felix Artuso. He had two duties on this morning: one, to operate the LP compressor and valves prior to departure and, secondly, to operate the emergency steering controls in the event that the helmsman in the conning tower experienced any problems during the move. It was assumed that the LP controls would not need to be touched again after casting off. Consequently, he and his guards were ordered by Lieutenant Tozer to leave these and other valves alone once the preliminary blow had been completed. There were various valves, levers and switches in the control room which, in theory, could be manipulated to vent the saddle tanks and thereby scuttle the boat.

Blowing the tanks reduced the generally sluggish feel of the submarine and reduced the list to five degrees. Bicain and Coward agreed that the time had come to cast off. Bicain shouted his helm orders down the conning tower hatch to the coxswain, and called the motor room by telephone to start the motors. Slowly the boat eased away from the jetty. Everything was proceeding as planned.

Trouble struck very suddenly within two minutes. Only one hundred yards out from the jetty, the *Santa Fe* quivered and lurched further to port. The men on board grabbed for hand-holds, and several shouted in alarm. Captain Bicain turned to Coward: 'We must blow the tanks again.' 'I agree. Do it.' Given the urgency of the crisis, John Coward assumed that it was the high pressure system which would be used, but Bicain picked up the telephone and, speaking rapidly in Spanish, told Artuso to restart the LP compressor and blow the tanks.

Until this moment, Felix Artuso had given the impression to his two guards of being an amiable and slow-moving individual. When he picked up the telephone and spoke with his captain, however, his demeanour changed entirely. Looking very serious, he slammed down the receiver and sprang to the switchboard which controlled the LP compressor. The motor started immediately, generating large volumes of air, and Artuso then leaped across the control room. He reached up, grabbed the blow valve levers and pulled them full open. These were the very controls which the two marines had been told, only a short time before, must not be touched. The guards could not physically grab hold of the Argentine because there was a table between him and them. The lance-corporal was convinced that Artuso was deliberately scuttling the submarine and he therefore opened fire with his pistol. Artuso died instantly, a bullet through his elbow, two in his abdomen and a fourth through his throat.

The two marines hastily left the control room and ran to the forward hatch. Suddenly they appeared on deck where, to Captain Coward's dismay, the lance-corporal fired more shots into the air to attract the attention of the shore party. His shouts of, 'We're sinking! We're sinking!' did not help the situation.

On the bridge, Captain Bicain asked Coward, 'What is happening? Why is he shooting?' Coward had no idea what had gone wrong, but he knew that he must keep a firm grip on the situation. 'Nothing, it's alright. Carry on.' He suspected that something appalling had happened below, but it was impossible at that stage to investigate. The submarine was still moving slowly forward. The air released into the tanks by Artuso had corrected the worst of the list, and the submarine was riding better on the water, but she was still in an unpredictable condition. He and Bicain continued to nurse her towards Grytviken jetty where, shortly afterwards, waiting hands threw down mooring lines and she was secured alongside. Coward and the others quickly jumped ashore.

There was now a hurried debate as to what should be done next. Everyone was shaken by the shooting and concerned by the need to keep the vessel upright. Captain Bicain pointed out that the motor of the blower tank had been running flat out for ten minutes or more, and was now liable to catch fire. The thought of an uncontrollable fire inside a submarine loaded with several tons of explosive was enough to send John Coward back inside the hull, alone, to turn off the motor. The first lieutenant of the *Santa Fe*, who had been waiting on the jetty, ran along the casing and slammed down the fore and aft hatches. The boat was visibly settling in the water, and he was attempting to stop her being flooded.

Captain Coward questioned the Royal Marines guards as to what had happened in the control room. He gave orders that nobody must enter the hull and placed a guard on the pierhead. Apart from the shooting, he was concerned also by the possibility that the boat might sink at her berth, trapping men inside.

John Coward was lifted by his Lynx out to the *Antrim*, lying in Cumberland East Bay, and reported the facts of the episode to Captain Young. Both officers were equally upset by the tragic turn of events and it was evident that a board of enquiry must be convened. Not only would it be needed by Northwood, it was also a specific requirement under Article 121 of the Geneva Convention. A Royal Marines officer interviewed the guards and started to take the formal statements which would be required later. Surgeon Lieutenant Crispin Swinhoe, RN, Guy Sheridan's medical officer, was sent for and he descended into the control room to certify death.

Earlier in the day, Coward had instructed Chris Sherman, his diving officer, to collect his equipment and some explosive from the *Brilliant*. There was no guarantee that the British would be able to remain on the island. 'While I had specific orders to keep the submarine afloat, I was also under strict instructions to leave her unusable against the possibility of recapture.' Later Sherman placed small underwater charges around the stern and blew the submarine's rudder off its pintles.

No further action was taken that day as the *Santa Fe* gradually sank lower into the icy water and her list increased to thirty degrees. The after compartments were slowly flooding, but the control room remained accessible. Early on 28 April, a party of Royal Marines, under the command of Guy Sheridan, climbed down into the dark and eerie control room to recover Artuso's body. He was a big man, weighing fifteen stone or more, and the marines had great difficulty in lifting his remains up through the

The end of a sad incident.
The body of Felix Artuso is laid to rest in the whalers' cemetery, Grytviken.

only remaining entry, the conning tower hatch. Eventually the unpleasant task was completed and the body taken to a nearby hut for an autopsy by Crispin Swinhoe and by Surgeon Lieutenant Mike Symonds, RN, medical officer of the *Brilliant*.

Two days later, on 30 April, Felix Artuso was buried with full military honours in the old whalers' cemetery at Grytviken. The guard, bugler and pall-bearers were provided from the ship's company of HMS *Antrim*. The funeral was attended by Captain Bicain and twelve of Artuso's former shipmates. Silently they watched the turf of South Georgia placed over the grave of the only man to die in Operation Paraquat.

The Board of Enquiry held aboard HMS *Endurance* devoted two days to its hearings. Evidence was taken from all personnel – British and Argentine – involved in the affair. It was a very thorough investigation. The findings were forwarded to London where they were examined at the highest level. The decision was made that there were no grounds for charges to be brought or for any words of censure to be recorded. The bulky file was forwarded to Geneva where the International Red Cross made its own impartial examination of the evidence. The IRC made no public pronouncement before submitting the report, via the Swiss Embassy in Buenos Aires, to the Argentine government. It was accepted as it stood, and there is no record of any protest having been registered by the Argentine authorities.

The shooting was regarded, by most of the interested parties, as having been unavoidable. Even so, for the officers involved, it was a traumatic experience which they regretted deeply. Captain Bicain himself broke into tears when first told of Artuso's death. A decent and liberal man, Bicain seems to have had little sympathy with the regime which he served and, although an ardent patriot, was at heart an anglophile. He had enjoyed his visits to the *Endurance* in Mar del Plata, only six weeks earlier, and he and his wife had established a friendship with Tony Ellerbeck and his own wife, Janie. Both men were equally distressed when they met again at King Edward Point.

The *Santa Fe* shooting made it even more important that the area should be cleared of prisoners as quickly as possible. Captain Bicain and the handful of men who had been present in the submarine at the time of the incident were isolated from the other captives, but there was always the risk that anger and resentment might lead to violence. British forces were thin on the ground and they would have difficulty in suppressing an uprising without resort to gunfire. A major incident at this stage would not only result in needless deaths, it would also be ruinous to the British diplomatic effort. Work went ahead to convert the fleet tanker *Tidespring* into a temporary prison ship, a project bristling with difficulties.

The *Tidespring* was not a Royal Navy ship. Royal Fleet Auxiliary crews are civilians and the officers do not hold the Queen's Commission. The Geneva Conventions expressly forbid the guarding of military prisoners by civilians. She was the only ship large enough to transport the Argentines to Ascension Island, but there were no spare troops at South Georgia to accompany them as guards. The dilemma was resolved by Lieutenant Tidd, commanding the flightless 'C' Flight, who stepped into the breach with the suggestion that he and his people could fulfil that role. He volunteered this suggestion at a time when he believed there to be no more than eighty prisoners. It came as a shock to discover that there were more than twice that number. 'I only had twenty-five guys in the Flight, so we would obviously be very stretched. However, *Antrim* let us have their detachment of nine marines under Sergeant Kendell and we all got stuck in with the *Tidespring*'s crew to fix up some accommodation. The forehold was cleaned out and bunks built for all the Argentine marines. One of the magazines was cleared to make space for Astiz's lads, and the other earmarked for the *Santa Fe* crew. The scrap workers were mainly placid middle-aged men and didn't present much threat, so we decided to keep them in the hangar. Everyone worked hard and the *Tidespring*'s crew were very helpful – they handed over all their spare mattresses, blankets, sleeping-bags and so forth. They also let us have a lot of clothing. This was particularly important because the *Santa Fe*

The *Santa Fe* settles by the stern at her berth off Grytviken. Overhead hovers one of HMS *Endurance*'s Wasps. Although declared 'out of bounds' to non-authorised personnel, the submarine was a magnet to parties visiting the shore. Seamen from the RFA *Tidespring* pose for a snapshot which clearly shows that many of the lightweight bullets fired at the submarine on 25 April failed to penetrate the glass reinforced plastic cladding on the fin.

boys had only the gear they stood up in and it was going to be pretty cold in the magazines until we reached warmer waters. By 28 April we were ready and our guests started coming aboard that morning.'

Tidd sent a list of requirements to the other ships and obtained everything he thought he might need. This was the first time that the Fleet Air Arm had ever acted as gaolers. His only guideline was commonsense. He collected handcuffs, more blankets, packs of cards, chess sets, eating utensils, whatever was available and seemed useful. The sanitation problem was solved by cutting in half a number of forty-gallon oil drums and fixing a wooden seat on top. He decided that SLRs and GPMGs were too bulky for use in the cramped confines of the ship, and it was in any case unwise to incur the risk of high velocity rounds being let off in a ship loaded with various volatile fuels. He managed instead to borrow a number of sub-machine guns with which to arm his men. Few of them were familiar with this weapon, so they were given some high-speed lessons in handling and safety. It was a very busy time.

Mike Tidd recalls his problems: 'We were very aware of the

Members of the crew of the *Santa Fe* wait on the RFA *Tidespring*'s flight deck before being processed and quartered. To the left stands the lonely and disconsolate figure of Capitan de Fregato Horacio Bicain. He has lost his command, he and his men are prisoners, and it is evident that his naval career is ruined.

requirements of the Geneva Convention. At the same time, we knew we had some tough lads on our hands and we had to consider the possibility that they might try to take over the ship once we got under way. It was a question of trying to strike the right balance. Captain Redmond was marvellous – it could have been a tricky situation. As Master, he was responsible for the lives of everyone in his ship. You can only have one skipper on the bridge. At the same time, I was responsible under military law and the obligations of the Geneva Convention for the security and welfare of the prisoners. Fortunately we got on extremely well. Captain Bicain was senior Argentine officer and he and I had a regular conference with Captain Redmond every day to sort out problems and to hear complaints. And there were plenty of those. After the first day, when they had recovered from the shock of their defeat, the Argentines started to grumble about the uncomfortable conditions, the fact that officers and men were not separated, that they didn't like British food, and so forth. Most of the other problems were overcome with time. My thirty-four were managing to guard, in shifts, their one hundred and eighty-five, in four separate compartments around the ship. We ensured that they all got three meals per day, two deck exercise periods per day, and a shower every two days. That was more than we had time for ourselves!'

Chief Officer Alec Bilney of the *Tidespring* recalls the first turbulent days following the ship's arrival in Cumberland East Bay: 'When Captain Redmond accepted the role of prison ship, the forehold needed to be emptied of all the ship's stores which had gathered there over two decades, then cleaned out and bunks built. Initially we dismantled the awning spars from the accommodation decks – timber of beautiful quality – but we still needed quantities of plywood and planking to create sufficient bunks for a hundred plus men in a space sixty feet square. *Endurance* and *Plymouth* were very generous, sending us everything they had. Mike and I explained to the "scrappies" what was needed and, once they had grasped the concept, they worked with a will. They were a simple, easy-going bunch, physically hard and well able to handle a seven-pound lump hammer in each hand. They created their own little nests in the hangar and were no bother. The military prisoners were more difficult because conditions in the forehold were certainly not pleasant. Some of the ship's access and sounding pipes ran through the hold and could have been damaged by the prisoners if they had wanted to sabotage the ship. There were important controls in the bunker tank under the hold. To stop the Argentines from interfering with those controls, we flooded the tank with sea water. The temperature of the sea was near freezing, hence making the deck of the hold very cold. Given the lack of forced ventilation down there, they had a condensation problem. Things got easier as we sailed north. By far the worst time was the first forty-eight hours when we were concurrently embarking prisoners, disembarking marines and stores, and trying to create instant accommodation out of thin air. This meant leaving prisoners in bedraggled groups out on the exposed flight deck in bitter winds and snowstorms. But they were fairly philosophical about it all – at least they were alive.'

Adverse flying conditions slowed the business of ship-to-ship transfers over the two days following the reoccupation but, on 28 April, the *Brilliant* and the *Plymouth* departed to join Task Force 317.8. With them went Major Sheridan, his immediate task completed and his next duty being that of giving a personal report to Admiral Woodward in HMS *Hermes*. The *Antrim* and *Tidespring* remained off King Edward Point, disembarking the balance of 'M' Company with rations for sixty days. Chris Nunn's marines were told that they now constituted the British garrison on South Georgia, tasked to defend it in the event of a counter-attack by the Argentines. Prevented by circumstances from taking part in the Hestesletten landings, they learned that they were very unlikely to take part in operations on the Falkland Islands. For such highly trained troops, eager to rejoin their Commando, this was a frustrating time.

There was no lack of work to be done. The area of King Edward Point was 'a bloody awful mess' as one witness described it. 'The Argentines had not exactly vandalised the place, but they were so ill-disciplined that the outcome was the same. Electric power, for example, came from two big BAS Volvo generators. Normally you would plan to run them alternately, giving equal running time and the opportunity for routine servicing and maintenance.

Although none of the Operation Paraquat ships suffered battle damage off South Georgia, two of them were less fortunate when later they were deployed off the Falkland Islands. Both HMS *Plymouth* and HMS *Antrim* were hit by Argentine bombs. Here the *Plymouth* is seen in San Carlos Water on 8 June following a strike by Dagger aircraft.

The Argies had simply run one gennie flat out for three weeks and left the other one standing idle. The result was that we had to ask the engineers from *Antrim* to come ashore and spend a lot of time putting the machinery back into serviceable condition. Then there was the living accommodation – it was a shambles. They had gone through the BAS personal effects, pulling out drawers and leaving stuff scattered on the floor. They didn't deliberately steal or break anything, but they never picked anything up, so it was ruined by sheer neglect. The Post Office building was used as accommodation by two of their officers. They had ransacked all the filing cabinets, chucked the papers on the deck and then slept on top of it all. There were empty food tins and cigarette packets all over the place. Nobody had bothered to service the water supply or clean the latrines or any of the other routine jobs which are so important when you have a large number of men living in a confined space. You could almost feel sorry for them – they had never been taught the basics of personal hygiene and field sanitation.'

Captain Barker returned to Stromness Bay on 28 April. The accommodation at Leith Harbour was in a state of disorder similar to that found at King Edward Point and a thorough search of the extensive buildings was required before the area could be declared safe. Apart from weapons and explosives, huge quantities of food were found. A very welcome discovery was the stock of Argentine cigars, beer and wine which was distributed among the ships.

The *Endurance* was involved during these days in sustaining the morale of the BAS field parties. As soon as they knew of the Argentine surrender, each party expressed an understandable desire to be evacuated. Apart from wishing to rejoin their own, they were anxious to recover their personal possessions from Shackleton House. They were told to stay where they were for the time being – the marines had quite enough on their hands already. Two days were needed to process the prisoners and dismantle the mines. Three of the parties were in good heart, but the Lyell Glacier men needed immediate help. On 26 April, Guy Sheridan delivered to them a supply of food and a crate of beer in a Lynx from HMS *Brilliant*. Three days later, the four bearded civilians walked over Mount Hodges and presented themselves to Captain Young at King Edward Point.

On 30 April, the *Endurance* sailed south to Saint Andrews Bay to collect the five people there. Ellerbeck and Wells made a series of flights, evacuating not only the three men and two women but also a ton of film equipment and stores. Cindy Buxton and Annie Price had continued filming the penguin colony right through to the end of their stay, including the afternoon of 26 April when, in blizzard conditions, they had seen the fourteen-stone Peter Stark blown off his feet by the 100 miles per hour wind.

On 1 May, the *Endurance* went north to Bird Island and Schlieper Bay where Ellerbeck evacuated the scientists, much of their equipment, and a crate containing two South Georgia Pintail Ducks destined for Peter Scott's Wildfowl Trust at Slimbridge. All of the BAS men congregated briefly at King Edward Point to prepare a status report for Dr Laws in Cambridge and to discover what personal possessions still remained in Shackleton House. Peter Stark was depressed to find that his irreplaceable collection of colour transparencies of the island – accrued over a period of two years – had been stolen.

Some of the BAS scientific equipment had been damaged by careless Argentine curio hunters. Everything was photographed by Guy Sheridan – to show that the damage had been caused by the invaders and not by British forces – but a subsequent amnesty revealed that a certain number of BAS possessions had been 'liberated' by the victors. Chris Nunn ordered a manifest of all recovered property to be drawn up and arranged for these items to be crated and returned to the BAS headquarters in Cambridge.

Divers from HMS *Antrim* searched the sea-bed off the tip of King Edward Point in the hope of finding additional cryptographic material from the *Santa Fe*. The submarine had been watched throughout her return to the jetty following Ian Stanley's depth charge attack, and nothing had been observed to be thrown overboard during that time. If Bicain had dumped any papers, he would have done so while tied up alongside the jetty. The divers did not in fact find any Argentine material, but they did recover the sack containing the BAS base records ditched by Bob Headland on 2 April.

An unusually tranquil evening in Cumberland East Bay. Two Royal Fleet Auxiliary ships (thought to be *Regent* and *Resource*) find safe haven in company with a requisitioned tanker 'taken up from trade'. HMS *Endurance* is berthed in the deep shadow at King Edward Point.

The civilians were welcomed aboard HMS *Antrim*, the women being accommodated in the Admiral's day cabin. No longer required as an operations room, it was stripped of its maps, tables and radio sets on 30 April. Young's orders were to escort the *Tidespring* back to Ascension Island whence the prisoners could be repatriated by air under International Red Cross supervision. Young and Redmond were equally anxious to unburden themselves of the prisoners as soon as possible. The *Tidespring* was overcrowded and it was difficult to heat the prison quarters adequately. During the first forty-eight hours there had been some incidents of unrest and defiance, particularly among those men who were in contact with Alfredo Astiz. He was transferred from the tanker to HMS *Antrim* and accommodated in the comfort but isolation of an officer's cabin. Those who met him were impressed by his tall fair-haired good looks and his excellent command of the English language. Few people at this stage knew that he was wanted by the governments of France and Sweden in connection with the torture and murder of their nationals who had been seized by the Argentine secret police while resident in that country. The *Tidespring* sailed on 1 May without him.

On the following morning, HMS *Antrim* departed Cumberland East Bay with all the British civilians and hurried north to catch up with the *Tidespring*. She joined her and HMS *Plymouth* on 3 May, about 200 miles north of South Georgia. The three ships were soon found by a Boeing 707 of the Argentine air force. Captain Young reported the circling aircraft to Northwood – it was well within range of his missiles – and requested permission to destroy it. Authority to open fire was denied, his ship now being outside the immediate area of conflict. The British government was still pursuing a diplomatic solution. Brian Young watched the big jet with feelings of frustration. Eventually it flew off and HMS *Plymouth* refuelled from the *Tidespring* before going her own separate way.

With the *Brilliant* and *Plymouth* having departed on 28 April, and with the *Tidespring* and the *Antrim* having sailed for Ascension Island on 1 and 2 May respectively, Nick Barker and his ship's company once again found themselves sole custodians of South Georgian waters. There was an element of gamble in leaving the *Endurance* and 'M' Company without the protection of at least one warship. However, Brian Young and Northwood agreed that it was an acceptable risk. It had been important to move the *Tidespring* and her passengers up to Ascension Island without delay. She needed an escort and this was the *Antrim*. HMS *Brilliant* and HMS *Plymouth* were needed for the Falkland Island operations. Two large tankers, *Blue Rover* and *British Esk*, were approaching from the north and they needed protection while they carried out a major pumpover at sea. It was a matter of trying to cover a varied number of operational requirements with a limited number of ships.

In the event, Northwood was obliged to authorise numerous changes of plan during the following week as events over and around the Falkland Islands showed that the entire Anglo-Argentine conflict was coming quickly to the boil. On the evening of 2 May the *General Belgrano* was torpedoed and sunk by HMS *Conqueror*. Armed with fifteen 6-inch guns, eight 5-inch guns and seventy Sea Cat missiles, the Argentine cruiser was less than three days' steaming from South Georgia when found and destroyed. The only armoured ship in either navy – with up to eight inches of armourplate protecting her vital parts – she had posed a continuing threat to the Paraquat force. The anxiety was removed at a single stroke. Then, on 4 May, the élite Super Étendards succeeded in launching their new Exocet missiles at Sandy Woodward's force. After several days of frustration caused by bad weather and inter-service rivalry, they finally made a successful launch which hit and sank the unlucky HMS *Sheffield*.

The loss of the 'Shiney Sheff' – as the Type 42 was affectionately known – led to the first change of plan. HMS *Antrim* was ordered to detach from *Tidespring* and to act as escort

Royal Marines of 'M' Company, 42 Commando RM, patrol the adjacent shoreline and the semi-derelict buildings of Grytviken. There is time to pause and contemplate the nature of a man such as Ernest Shackleton. Crowding the plinth of his grave are the tributary plaques placed there over the years by the officers and ratings of visiting ships. To those who worked with him, Shackleton was simply 'The Boss'.

to Task Unit 317.0.4, a group of RFA logistical ships heading south for the Falkland Islands. She exchanged roles with the much smaller HMS *Antelope*, a Type 21 frigate commanded by Commander Nick Tobin.

The civilians and their kit, together with Alfredo Astiz, were transferred to the *Antelope* before she and the *Tidespring* continued the passage north to Ascension Island. Mike Tidd said goodbye to the *Antrim*'s marines as they returned to their own ship and welcomed aboard the Internal Security Detachment from HMS *Antelope*. HMS *Antrim* turned south-west and, for the next three weeks, disappeared from the story of South Georgia. Before doing so, on 6 May, she shielded the chartered BP tanker *British Esk* (Captain G. Barber) and the *Blue Rover* (Captain D. A. Reynolds, RFA) while they made their pumpover on Latitude 50° South. The *Blue Rover* then continued southward towards South Georgia, her sole military passenger being Major Sheridan. After several days of wandering from ship to ship, and having been frustrated in his attempt to rejoin 42 Commando, he found himself returning to the island for which he had already developed some affection. As a skier and mountaineer, he would have time to see more of South Georgia's spectacular scenery pending another opportunity to get ashore on East Falkland.

The *Blue Rover* entered Cumberland East Bay on 9 May. Loaded with aviation fuel, diesel oil, and ammunition, she remained there until the end of the month, acting as a filling station for other visiting ships.

The first ship to arrive off King Edward Point following the departure of the original Paraquat force was the United Towing's ocean-going tug *Salvageman* (Captain Alan Stockwell). Anchoring on 7 May, she represented the first step in converting South Georgia into a safe haven for ships in need of assistance or supplies. Her appearance heralded a procession of other vessels, large and small, naval and mercantile, which were to make full use of these relatively sheltered waters during the following weeks.

The arrival of the *Salvageman* provided the first test for 'M' Company's new defence arrangements. Having got rid of the Argentine captives and generally made the area once again safe and habitable, the marines had been working hard to establish observation posts on Brown Mountain, at Echo Pass and at Maiviken. Manned day and night, they gave a commanding view of all sea approaches to the Grytviken area. On King Edward Point itself, a network of defensive emplacements was constructed among the buildings and around Shackleton House. These were substantial structures, designed to withstand the

A picture which encapsulates the entire Paraquat story. In the centre, symbolising the heyday of the whaling industry, is the old Norwegian whale catcher *Petrel*. In the foreground, some of the scrap metal which helped to trigger the campaign of 1982. Almost hidden in the mists of Cumberland East Bay, the two luxury liners, *Canberra* and *Queen Elizabeth 2*, which played a major role in the recovery of the lost South Atlantic islands.

impact of a light shell or mortar bomb. The task was aided by the wealth of building materials to be found in the abandoned sheds. Some of the marines were skilled carpenters, others were fitters and electricians. Sergeant McLeman was a qualified welder. He took great pride in the pillbox which he constructed with high quality inch-thick sheet steel scavenged from the old whaling station.

After several days of studying the fields of fire and lines of approach, Chris Nunn was confident that no Argentine could hope to attack his position without being spotted at an early stage. He was therefore less than pleased when, at 2030 on 7 May, Captain Stockwell sailed into Cumberland East Bay and prepared to anchor without being challenged. It was only when a sentry outside Shackleton House gave the alarm that the whole company 'stood to' and an illuminating round from an 81 mm mortar revealed the 1,600-ton tug to be what she was. The effect of the parachute flare and a loud-hailer order, 'Stop or we open fire on you,' caused the *Salvageman* to execute an admirably swift emergency stop.

Captain Stockwell had entered the Bay with his ship completely darkened. His small Royal Navy communications team advised him that the garrison had been warned of his arrival and that he need feel no concern. It was when the blinding light of the flare illuminated the surrounding sea and an unfamiliar voice was heard speaking on the radio – 'Stand by to open fire,' – that he decided that all was not as it should be. A moment later the spent flare drifted down and gave him a mild blow on the head.

An officer from the tug was paddled ashore in a rubber dinghy to explain her presence and to clarify the position. The appropriate apologies were offered and accepted. Chris Nunn was unhappy with the performance of his sentries and words were spoken on the following morning. It was the one and only time that anything stirred in the Bay without being immediately reported.

There was activity in abundance as the days passed and the full impact of Great Britain's maritime mobilisation started to manifest itself in the South Atlantic. On 8 May, another tug arrived and went alongside the King Edward Point jetty. She was the *Typhoon* (Captain J. N. Morris), a 1,300-ton vessel operated by the Royal Maritime Auxiliary Service and equipped for fire-fighting, salvage and ocean rescue. On 16 May, the *Stena Seaspread* (Captain N. Williams), a diesel-electric offshore support vessel of 6,000 tons which normally worked in the North Sea Thistle oilfield, arrived from Ascension Island. Equipped with a heavy machine shop, she collected quantities of steel plate from the old whaling stations with the thought that it would be useful in repairing ships damaged off the Falkland Islands. In the event, she carried out battle damage repairs to eleven Royal Navy ships and made routine repairs to twenty-eight other vessels between 20 May and 25 July. It seemed incongruous that the scrap metal which had, in part, caused the conflict with Argentina was now being used to repair British ships damaged by the Argentine air forces.

On 23 May, the Cunard refrigerated cargo ship *Saxonia* (Captain H. Evans) arrived with thousands of tons of food for transfer to Royal Fleet Auxiliary ships. Two days later the *Stromness* (Captain J. B. Dickinson, RFA) came south from the Falkland Islands. Having departed Portsmouth on 7 April with 2,750 tons of stores, and elements of 45 Commando RM embarked, she had taken part in the San Carlos landings. Her latest passengers were the survivors of HMS *Coventry*, awaiting passage home to the United Kingdom.

Late on the afternoon of 25 May the requisitioned British

Canberra, seen dimly under the protection of a very low cloud base, while embarking 5 Infantry Brigade from *Queen Elizabeth 2*. Alongside the big white liner is one of the Hull trawlers. Ahead of her waits HMS *Leeds Castle*.

Telecom cable ship *Iris* (Captain A. Fulton), arrived in Stromness Bay. On board she had Lieutenant Keith Mills and his detachment of Royal Marines who had joined the ship at Ascension Island. As the *Iris* passed the anchored HMS *Endurance*, Mills and his men solemnly fell in on the *Iris*'s upperworks and 'presented arms' with an assortment of paddles and broomsticks. After a separation of nearly two months, 'the plum' was about to have her ship's company restored almost to full strength.

The *Iris* moored alongside the jetty at Stromness. Over the next four days she and several specialists from the *Endurance* scoured the abandoned whalers' workshops, selecting and loading stocks of steel and other metals for subsequent use in repairing the ships of the main Task Force.

On 26 May, the small offshore patrol vessel HMS *Leeds Castle* (Lieutenant-Commander C. F. B. Hamilton, RN) came to act as a fuel station for cross-decking helicopters. Her arrival, like that of the *Stromness*, was a precursor to the strangest three days ever seen in South Georgia's usually unfrequented waters.

At 1500 on 27 May, the SS *Canberra* (Captain D. J. Scott-Masson) entered Cumberland East Bay. 'The Great White Whale' had been requisitioned from Peninsular & Oriental in the earliest days of April as a troop carrier and had already been in the thick of the action in San Carlos Water. Having disembarked her first load of military passengers there, she was directed to South Georgia to collect another consignment. That night, 27/28 May, one of the largest cruise liners in the world and the pride of the British mercantile fleet, the *Queen Elizabeth II* (Captain P. Jackson), was due to arrive with the entire strength of 5 Infantry Brigade and a number of ancillary troops and marines. They were to be trans-shipped to the *Canberra* for delivery to San Carlos Water where Commando forces were already established on shore. The Brigade would provide the additional firepower of 2nd Bn Scots Guards, 1st Bn Welsh Guards and 1/7th Duke of Edinburgh's Own Gurkha Rifles.

Moving men and stores from one liner to the other required the services of smaller vessels to act as ferries. It was partly with this purpose in mind that five Hull-based motor freezer trawlers had been requisitioned and brought under temporary Royal Navy command. They arrived that morning – the *Pict* (Lieutenant-Commander D. G. Garwood, RN) owned by British United Trawlers, and the *Cordella* (Lieutenant-Commander D. C. G. Holloway, RN), the *Farnella* (Lieutenant R. J. Bishop, RN), the *Junella* (Lieutenant-Commander M. Rowledge, RN) and the *Northella* (Lieutenant-Commander J. P. S. Greenop, RN), all owned by J. Marr & Son.

Yet another new arrival was the 13,000-ton roll-on roll-off ferry *Norland* (Captain D. A. Ellerby), requisitioned from P&O North Sea Ferries in mid-April to carry stores and elements of 2nd Bn The Parachute Regiment to the Falkland Islands. She had been one of the first ships to enter San Carlos Water. Having discharged her passengers, she had been sent down to South Georgia, together with the *Canberra*, to embark 5 Infantry Brigade. Both ships anchored in Cumberland East Bay. They were joined during the day by the familiar shape of HMS *Antrim* – here to collect Major General Jeremy Moore and the rear HQ staff of 3 Commando Brigade. Moore had been designated Commander Land Forces for the main battles on East Falkland. The *Antrim* was to convey him with minimum delay to San Carlos Water where Commodore Michael Clapp, RN, and Brigadier Julian Thompson, RM, were commanding the landings from the amphibious warfare ship HMS *Fearless*.

The 67,000-ton bulk of the *Queen Elizabeth II* slid into Cumberland East Bay during the night and, at first light, HMS *Leeds Castle*, RMAS *Typhoon* and the five trawlers started the massive movement of men and material to the *Canberra* and the *Norland*. The helicopters of the *Endurance* and *Antrim* assisted with the cross-decking. Accustomed to working from the tiny flight deck of the *Endurance*, Lieutenant-Commander Ellerbeck was impressed by the very large temporary flight deck installed on the *Canberra*. 'It felt very strange to be landing the little Wasp

Above left: *Queen Elizabeth 2* **disembarks her passengers by trawler and Sea King helicopter. Weather conditions this day are ideal, but it is important to quickly remove the prestigious ship from the range of Argentine aircraft and submarines.**

Above right: The worn gun barrels of HMS ***Antrim*** **are replaced in the comparatively sheltered waters of Cumberland East Bay.**

To the victor the laurels. The news has just been received in HMS ***Antrim*** **that the success of Operation Paraquat has not passed unrecognised. The following awards have been announced in London (left to right): Sub-Lieutenant Stewart Cooper, RN, Mentioned in Despatches; Captain Brian Young, RN, the Distinguished Service Order; Lieutenant-Commander Ian Stanley, RN, the Distinguished Service Order; and Lieutenant Chris Parry, RN, Mentioned in Despatches.**

on what looked like a football pitch. It was eighty-two feet above sea-level – enough to give you vertigo! During the previous weeks I had got into the habit of always flying very much lower than that.'

The weather on this day was calm and comparatively mild, circumstances which created a thick sea mist and low cloud. Horizontal visibility was limited to one mile and the superstructure of the *Queen Elizabeth II* was hidden in low cloud. The presence of a luxury trans-Atlantic liner in such surroundings was itself strange and unreal, but the conditions on this day 'made her look like a ghost ship' as one witness remembers.

The transfer of troops, stores and vehicles was completed by the end of the day and the *Canberra* and the *Norland* departed immediately for Falkland Sound. The *Queen Elizabeth II* and HMS *Leeds Castle* sailed on the following morning, the former bound for Southampton with 629 survivors from the lost warships *Ardent*, *Antelope* and *Coventry*, the latter to take up guardship duties at Ascension Island. The trawlers sailed north to a holding area north of the Falkland Islands and later were employed in their main duty of minesweeping.

Despite the departure of all these ships, Nick Barker still had several other ships under the doubtful protection of his two Wasps and his ship's two 20mm Oerlikon guns. On 28 May, the China Mutual Steamship's motor cargo ship *Lycaon* (Captain H. R. Lawton) entered Stromness Bay with a full load of ammunition and Rapier missiles for transfer to the RFA *Stromness*.

On 2 June, the 23,000-ton fleet replenishment ship *Resource* (Captain B. A. Seymour, RFA) came in for the first of her two visits to the island. Having delivered 4,000 pallets of stores to the battle group, and having survived the worst of the Argentine air attacks against San Carlos Water, she needed a few days to reorganise her cargo. This same day witnessed the return of HMS *Antrim*, sent back by Admiral Woodward to act as guardship and to replace her worn 4.5-inch gun barrels. Their normal working life was 1,100 rounds. Having taken part in numerous bombardments, the *Antrim*'s barrels had fired their full allowance. Removing the old barrels and carefully inserting into the turret the two new ones was a procedure calling for delicate handling and a steady ship. It would have been very difficult in the rough open seas of the South Atlantic but, on a quiet day in Cumberland East Bay, Ian Stanley was able to lift each heavy tube and, hovering precisely over the turret, guide it into the apertures where waiting hands made it secure. The availability of the South Georgia anchorage and the lift capability of 'Humphrey' both, once again, proved their value.

Three days later yet another Royal Fleet Auxiliary, the support tanker *Pearleaf* (Captain J. McCulloch, RFA) appeared on the scene. She replaced the RFA *Blue Rover* as South Georgia's base

By late May, one of the few relics still accessible to shoreside visitors is the Puma helicopter shot down by HMS *Endurance*'s Royal Marines detachment on 3 April. Posing with the wreck is part of the battlefield tour programme. One man who has every right to pose for the camera is Sergeant Peter Leach, RM, (right) one of the marines who helped to shoot it down. Interestingly, the helicopter is painted in a dark combat colour scheme. The Argentines claimed that this machine was embarked in the *Bahia Paraiso* in February 1982 for peaceful survey and general delivery duties in Antarctica. The 'ski' attachments on the undercarriage are certainly consistent with snowfield operations, but the lack of dayglo panels and prominent fuselage markings suggests a premeditated hostile intent.

tanker. Two more vessels arrived on 7 June, the United Towing's tugs *Yorkshireman* (Captain P. Rimmer) and *Irishman* (Captain W. Allen), the first from Tristan da Cunha and the second from Falkland Island waters.

The great depth of Cumberland East Bay and parts of Stromness Bay was causing problems for some of the ships taking anchor in those places. On several occasions, during violent gales, ships dragged their anchors and were forced to put to sea rather than risk being driven ashore. Heavy duty mooring buoys were needed to counter the hazard. To this end, the *Wimpey Seahorse* (Captain M. J. Slack), a 1,600-ton offshore support vessel operated by Wimpey Marine, arrived on 8 June. She spent the next two weeks laying moorings off Grytviken, Stromness and Leith Harbour before sailing north to fulfil similar tasks off Port Stanley following the Argentine surrender.

Another big Royal Fleet Auxiliary, the fleet replenishment ship *Regent* (Captain J. Logan, RFA), entered Cumberland East Bay on 11 June to take on a huge volume of stores – most particularly food – from the refrigerated cargo ship *Geestport* (Captain G. de Ferry Foster). Owned by the Geest Line, the latter was usually employed in the banana trade with a primarily West Indian crew. With one exception, the latter were replaced by United Kingdom nationals when the ship was requisitioned and sent south.

One week later, on 18 June, the 33,000-ton *Scottish Eagle* (Captain A. Terras) moored in Stromness Bay where she replaced the RFA *Pearleaf* as base tanker for the following three weeks. Ships in need of fuel went alongside her to replenish their own stocks.

Throughout these busy weeks, the marines of 'M' Company continued to consolidate their positions around King Edward Cove and to maintain careful watch from their observation posts on the surrounding mountains. Winter had arrived with full force and weather conditions often made movement difficult. Whenever flying conditions permitted, the *Endurance*'s Wasps flew men and supplies up to the outlying posts. Before he finally left the island, Major Sheridan was flown to various places in search of potential sites for anti-aircraft guns and missiles. With so many important ships coming and going, it was reasonable to assume that the Argentines might be tempted to make an aerial strike at the anchorages. Bombs could be rolled from the tailgate of a C-130 – a technique said to have been used with surprising accuracy elsewhere. The British Petroleum tanker *British Wye* and the Liberian tanker *Hercules* were both bombed at sea, and on each occasion the crew identified their attacker as a Hercules.

First reports from the Operation Corporate task force suggested that Argentine pilots tended to veer away from their targets whenever they saw the flash of a Rapier or Blowpipe missile being launched from the ground. With this in mind, the two *Endurance* pilots, Ellerbeck and Finding, maintained a watch from various headlands around the anchorages, their aircraft parked on high ground and armed with AS12 missiles. These weapons could not themselves be aimed at approaching enemy aircraft, but the bright flash of ignition could, it was hoped, dissuade an aggressor from pressing home his attack. The theory, in the event, was never put to the test.

All of the *Endurance*'s people were fully committed to a number of extraneous tasks during this period. Her engineers helped 'M' Company to service the installations at King Edward Point and Grytviken. Diving parties tried to find and seal the holes in the *Santa Fe*. The jetties at Husvik and Stromness were repaired. Argentine diving gear and mountaineering equipment, of excellent quality, was retrieved from Leith Harbour. The Wasps flew some of the ship's maintainers up to the top of the Fortuna Glacier to salvage flight instruments, radios and parts from the crashed Wessex 5 helicopters. They also retrieved some of the abandoned SAS equipment and a quantity of ammunition and hand grenades. Given his intimate knowledge of the local waters, the Navigating Officer, Bill Hurst, was flown out to meet most of the incoming ships in order to pilot them to the berths assigned by Captain Barker. There were few opportunities for relaxation.

The relationship between ship and shore was excellent. The role of Commander Task Group 317.3 – equating in some ways to that of Queen's Harbourmaster – kept Nick Barker and his people fully occupied for days at a stretch, but they ensured that 'M' Company lacked neither support nor diversion. Some of the marines were depressed that they were condemned to remain here instead of taking part in the final battles for Port Stanley. Chris Nunn and his Troop Commanders kept them hard at work – patrolling, building sangars, and manning the observation posts – but parties of marines were taken out to visit ships anchored in the Bay and groups of naval personnel came ashore for exercise and to photograph the wildlife. The ambivalence of the situation prompted Lieutenant Ewen Duncan, commanding 9 Troop, 'M' Company, to ask half-jokingly: 'Is this still an operational area, or are we just catering for the tourist trade?'

The deadly seriousness of the conflict was underlined on 2 June when Chris Nunn was informed that his brother had been killed in the battle for Goose Green. Lieutenant Richard Nunn, RM, was a helicopter pilot. On 28 May, while 2nd Bn The Parachute Regiment was fighting the action which has since become legendary, he was flying a shuttle in and out of the battle area, lifting forward ammunition and evacuating casualties. Tasked to evacuate Lieutenant-Colonel H. Jones, the battalion's dying commanding officer, Nunn's Scout helicopter was attacked by a Pucara and shot down.

The loss of Richard Nunn and many other good men was partly redeemed on 14 June by the surrender of all Argentine forces on the Falkland Islands. Operation Corporate had been brought to a successful conclusion. The marines of 'M' Company and the sailors aboard the nearby ships celebrated the victory in time-honoured style, but there was no reduction in the alertness of the island's defences. General Menendez had surrendered his own command, but there had been no declared suspension of hostilities between the two nations. Unlikely though it might seem, another Argentine attack against South Georgia was still a possibility which could not be ignored. On the other hand, the liberation of Port Stanley signified that South Georgia's importance as a fuelling and transfer point would now swiftly diminish. Ships coming south from the United Kingdom, or in need of replenishment, would be able to proceed directly to Stanley Water. The time had come for a change of plan.

The only British territory still in Argentine hands was Thule Island, one of three islands forming the Southern Thule group. The southernmost of the eleven small islands forming the South Sandwich chain, it is 650 miles nearer the South Pole than the island of South Georgia. It was discovered in 1775 by Captain Cook and, at the time of the 1982 conflict, formed part of the Falkland Islands Dependencies.

The South Sandwich islands are of volcanic origin – several of them being still active – and they form part of the Scotia chain of surface and sub-surface mountains scattered between the Antarctic continent and the Andes of South America. The Argentine navy had first established a toehold at the end of 1955. One of their ships made a visit on 25 January of that year and, in the following December, returned with a construction crew. A refuge hut measuring ten feet by twelve was erected on Hewison Point, a flat promontory on the south shore of Thule Island. Several such buildings were erected by the Argentines on the South Shetland Islands and on the Antarctic Peninsula at that time. Juan Perón had just fallen from power, there was great political instability within the country and, as often happened during such periods, the Argentine ruling class distracted popular attention with renewed territorial claims and a spate of hut building.

The Thule hut was given the name Teniente Esquivel and a small party of naval personnel took up residence just before Christmas. They were hastily evacuated three weeks later, on 14 January 1956, when Bristol Island, twenty-seven miles away, erupted and released clouds of noxious fumes which drifted over the neighbouring islands.

Argentina showed no further interest in the place for the next twenty years. The Commonwealth Trans-Antarctic Expedition's supply ship *Theron* called there later in 1956, but found no sign of Argentine activity. There were other visits by HMS *Protector* and HMS *Endurance*, and by the Royal Research Ships *Shackleton*, *John Biscoe* and *Bransfield*. In March 1964, a BAS party, supported by HMS *Protector*, made an extensive topographical and scientific survey of the South Sandwich Islands and found the Thule hut to be in poor condition.

It was on 7 November 1976 that the Argentines returned to establish a far more ambitious base. Fifty technicians required four months to erect the prefabricated buildings and it was this development which caused the diplomatic incident described in Chapter II. The British government made its formal protest, but did not pursue the matter because open dispute would have jeopardised the continuing negotiations regarding the Falkland Islands. The new base, named Corbeta Uruguay, was declared officially open on 18 March 1977 and 'inaugurated' in the following November when the first meteorological observations were fed into the world weather information system. Apart from acting as a communications centre and sea and ice observatory, the base was equipped with sophisticated devices for measuring and recording conditions in the troposphere.

The station could accommodate as many as eighty men during the summer season – the majority being naval personnel – but the

Olmeda, the Royal Fleet Auxiliary fleet tanker assigned to Captain Barker's Operation Keyhole task group.

usual number was forty and this reduced to ten or eleven during the winter. It is known that the *Bahia Paraiso* called there in March 1982 to collect additional men and supplies before attacking King Edward Point on 3 April. This particular visit is frequently offered as proof of the premeditation of the Argentine invasion of South Georgia.

Following the Argentine surrender on the Falkland Islands, the British government decided that the people manning Corbeta Uruguay could not be permitted to remain there. It was most improbable – at least in the short term – that the Argentines would have either the will or the capability to reinforce the base and turn it into a significant military stronghold but, on political grounds alone, they could not be allowed to claim that they still held British territory. The order to evict them was issued to Admiral Woodward on 15 June.

The task was easier described than carried through. The southern hemisphere was approaching the shortest day of mid-winter. Pictures received from the polar orbiting weather satellites showed that the seasonal pack-ice was advancing northward at the rate of six miles per day. The vanguard was less than thirty miles from Thule. Within a fortnight the entire South Sandwich group would be encircled by an icefield so dense that even the *Endurance* might have difficulty in penetrating it.

Command of the operation was given to Captain Barker. He and Admiral Woodward discussed by radio telephone the number and types of ship needed to overcome the garrison 'with minimum force'. They agreed that four ships would suffice – the *Endurance* (for her local knowledge and helicopter capability), the *Salvageman* (as rescue tug), one frigate (to provide naval gunfire if needed), and one RFA tanker (to provide fuel for the frigate, to act as troop carrier, and to provide additional helicopter support). The military component, commanded by Captain Nunn, would consist of the *Endurance*'s Royal Marines detachment, a large proportion of 'M' Company, and, to give the *Endurance* an air defence capability, two Blowpipe detachments, Royal Artillery. With the experiences of Operation Paraquat fresh in mind, it was also agreed that Barker should be provided with sufficient helicopters for an overwhelming 'single lift' assault.

Two ships were detached from Admiral Woodward's carrier group and ordered to proceed immediately to South Georgia. They were the modified Type 12 frigate HMS *Yarmouth* (Commander A. Morton, RN) and the large fleet tanker *Olmeda* (Captain G. P. Overbury, RFA). The former carried 4.5-inch guns mounted in a twin Mark 6 turret and could therefore provide naval gunfire support. The latter was big enough to accommodate the bulk of 'M' Company, would supply the small task group's fuel requirements, had an ice-strengthened hull, and was equipped with a satellite communications system. She also possessed a good helicopter operating capability.

The *Olmeda*'s big hangar usually housed two Sea King HAS2A helicopters of 824 Naval Air Squadron. They could not both be spared for this operation. One was flown off to HMS *Invincible*, but Lieutenant-Commander I. S. McKenzie, RN, and XV649 were retained on board. Both ships started to head south on 16 June with the intention of reaching South Georgia on the 18th. There they would embark the bulk of 'M' Company.

Nick Barker did not wait. The rapid advance of the Antarctic pack-ice made Operation Keyhole a race against time. He received authority to embark a small element from 'M' Company – to augment Keith Mills' detachment – and to depart at the earliest opportunity. He was also authorised to borrow one of the helicopters currently operated by the RFA *Regent*, the fleet replenishment ship anchored in Stromness Bay. This was the Wessex HU5 serial XS486 of 'A' Flight, 848 Naval Air Squadron.

Until 1982, no helicopter larger than a Whirlwind or Wasp was officially cleared to land on the *Endurance*'s small flight deck. There were doubts regarding both the weight factor and rotor tip clearance. However, a brief touchdown by Ian Stanley in 'Humphrey' on 27 April, plus further landing and lift off trials in June, showed that the Wessex could use her deck and, equally importantly, that the aircraft could be squeezed into one half of the hangar. The two Wasps were crammed into the other half. Whiskey Whiskey's pilot, Lieutenant-Commander Chris Blight, second pilot Lieutenant Tom Mason, and Petty Officer Aircrewman M. P. Hogan, came aboard, and the ship sailed, with the *Salvageman*, at 1600 on 17 June. With her she took Captain Chris Nunn, two sections from the Reconaissance Troop, and an 81 mm mortar crew.

Only three days after departing South Georgia, HMS *Yarmouth* finds herself cruising past the dramatic ice-capped volcanic cliffs of Thule Island. Operation Matchstick provides a sharp contrast to her recent adventures around East Falkland.

On Friday 18 June, at 1030, the *Yarmouth* and the *Olmeda* met up with the *Regent* twenty miles east of Cape Constance. The *Olmeda* took aboard a large volume of food and other stores by jackstay, the intention being that she would be able to supply the Task Group for up to four months in the event that all four ships became trapped in the Antarctic ice.

It had been intended at this stage to exchange Sea King XV649 with the remaining Wessex 5 operated by the *Regent*. The Wessex was fitted for cold weather operations and the aircrew were Arctic-trained. It was now discovered, however, that the Wessex was unserviceable and in need of spare parts. Ian McKenzie, Flight Commander of 824 Squadron's 'A' Flight, made an on-the-spot decision to remain aboard the *Olmeda*. 'My reasoning was that a radar-equipped helicopter might prove useful in the same way that *Antrim*'s Wessex 3 had been so important during the Fortuna Glacier operation. Further, the Sea King would be the only helicopter with sufficient range to reach the outside world if the four ships were trapped in the pack-ice.'

The *Yarmouth* and the *Olmeda* parted company from the *Regent* and entered Cumberland East Bay at 1500 in brilliant sunshine. Swiftly they embarked the three waiting Rifle Troops of 'M' Company, together with the company headquarters commanded by the Company Sergeant-Major, Warrant Officer Paul Juleff. The marines were taken out to the *Olmeda* by helicopter and by the tug *Yorkshireman*, a ship with which they had established a particularly happy relationship. Left behind was a token garrison of thirty marines, plus the Royal Marines Detachment from HMS *Antrim*, under the command of Lieutenant Ian Grant, RM. The two ships sailed that evening. By a strange coincidence, 'M' Company had been passengers in the *Olmeda* earlier in the year during an exercise in northern Norway. Then they had been on the Arctic Circle, now they were headed for the Antarctic Circle.

The marines were accommodated in the forward end of the ship's hangar. Discussions between aviators and marines regarding the forthcoming assault led to the decision to carefully strip the interior of XV649 so that twenty-four lightly armed troops could be carried in a single lift. Apart from the observer's seat and the radar set, all other internal equipment was removed by Lieutenant Dave Smith, the Flight's engineer officer. All passenger seating was taken out and the sonar well boarded over to reduce the effect of cold air blowing into the unheated interior. Trials in the hangar showed that twenty-four men, plus the four-man crew, could indeed be packed into the helicopter. Of equal importance, the 'dry run' disembarkation drills demonstrated that the aircraft could be emptied of passengers in six seconds. With the firm prospect of armed action within the next three days, everyone worked hard to perfect these unorthodox arrangements.

A fundamental difference between Operation Keyhole and Operation Paraquat was the security aspect. No effort had been spared, in April, to conceal from the Argentines the fact that a British force was making its way towards South Georgia. In the case of Thule, it was obvious that they must be fully aware of the presence of Royal Navy units in the area. On humanitarian grounds, therefore, it was decided to announce – on an international distress frequency – the sailing of a Task Group and to invite the Argentines to indicate their willingness to surrender. The message was broadcast as the *Endurance* and *Salvageman* departed South Georgia, but the Argentines remained silent.

Captain Barker requested an aerial photographic surveillance of the maritime area between South Georgia and Thule Island to establish the location of icebergs and the possible presence of Argentine shipping. No RAF aircraft could be made available for this purpose, so he was obliged to rely upon luck and local knowledge.

The two ships reached Thule in the early hours of Saturday, 19 June. The island is small – being approximately four miles in width and three miles long at its extremities. The eastern half is dominated by a dormant volcano, Mount Larsen. The mountain rises 2,700 feet above the southern half of the island, part of which is comparatively low lying. The Argentine base was sited on the flat southern promontory known as Hewison Point.

Captain Barker takes up the story: 'We had received various estimates as to the strength of the Argentine force ashore – the most popular figure was "about eighty". This was entirely feasible. I had visited the island in January of 1981 in company with David Attenborough, the wildlife film producer. I knew that the Argentines had constructed an extensive range of buildings,

including a large hangar which was in the process of being converted into a barracks block. I had to assume that there might be a strong garrison and that they might be prepared to fight. Accordingly, I placed my two ships in a holding position about five miles north-west of the island, sheltering in the lee of a tabular iceberg three miles long and fifty metres high. This protected us from the 25-knot wind and would make it safer for helicopter operations when it got light. First light was due at 1050, last light at 1650, so time was an important factor.'

The tactical plan for Operation Keyhole was simple and direct. It fell into three main phases: initial reconnaissance, intimidatory gunfire, then a main assault. Given the continuing lack of response to his earlier radio appeals, Nick Barker called a series of operational planning meetings and agreed with Chris Nunn that they should exercise proper caution. The *Yarmouth* and the *Olmeda* were still making their way down from South Georgia. They were making their best speed but, delayed by bad weather, were unlikely to arrive before first light on the following day, Sunday 20 June. In the meanwhile, it was decided to put ashore two sections of the Reconnaissance Troop. The marines would make a covert survey of the base to discover whether or not there were any defensive positions waiting to oppose a main landing by the 'M' Company Rifle Troops. The second element in the plan – a display of naval gunfire – would be conducted by the *Yarmouth* when she and the *Olmeda* arrived.

By a happy accident of geography, Thule is located only three miles due west of another volcanic island of similar size and shape, Cook Island. Midway between the southern extremities of the two is a small steep-sided outcrop known as Twitcher Rock. It was agreed that this would provide an ideal demonstration target

Sergeant Napier's reconnaissance party embarks in Whiskey Whiskey on the afternoon of 19 June. Nearby hovers Tony Ellerbeck in Wasp 434, ready to 'ride shot-gun' protection for the landing. The British have learned from the Argentine experience at King Edward Cove: large troop-carrying helicopters engaged in amphibious operations need instant support if their landings are contested.

for the *Yarmouth*'s gunfire. The tactics are described by Battery Sergeant-Major Brum Richards of 148 (Meiktila) Commando Battery RA: 'The idea was to pound Twitcher Rock to bits so that the Argentines on Southern Thule could see for themselves what would happen if we switched the fire onto their buildings. For me personally, this sort of operation was becoming a familiar habit. I had worked as NGLO in *Plymouth* when we bombarded South Georgia on 25 April. Then I had transferred to *Avenger*, *Ambuscade* and *Yarmouth* for Operation Typhoon – the NGS programme around the Falklands in May and early June. Now here I was back in *Yarmouth* for Operation Keyhole. My modest claim to fame is that I was the only man in 1982 to take part in all three operations. It was planned to put me onto the south-west corner of Cook Island – at a low headland called Reef Point – where I could set up an OP and call the fall of shot for *Yarmouth*. I was happy to be working with her again – she knew what she was doing and had already fired one thousand four hundred and forty-one rounds of 4.5-inch shells during the campaign, more than any other ship.'

Tony Ellerbeck and David Wells launched in Wasp 434 at first light on 19 June. Their task was to seek a suitable landing site for the insertion by air of the reconaissance patrol. No obvious sites presented themselves, but Ellerbeck found one which was

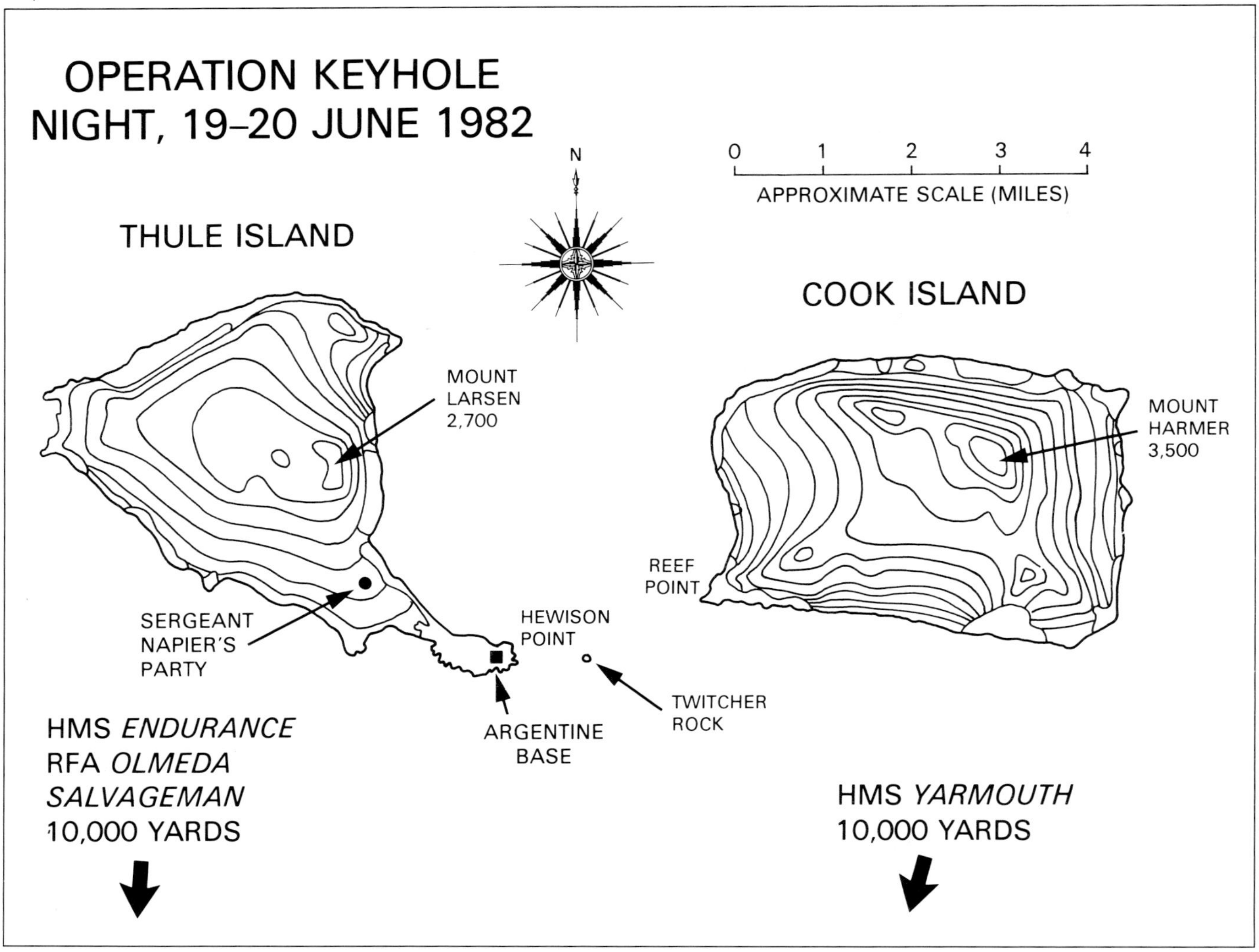

Safely ashore, the Royal Marines reconnaissance party advances across the slope of Mount Larsen in the last light of the short Antarctic day.

slightly less tenuous than the others – on the reverse slope of Mount Larsen – and he flew back to the *Endurance* to make his report. An hour later, laden with ten marines led by Sergeant John Napier, Whiskey Whiskey approached the coast at low level to deposit the patrol on the packed deep snow of the mountainside. Flying alongside the Wessex was Wasp 434, armed with two AS12 missiles, acting as close escort.

The insertion was typical of hundreds of such landings practiced every winter on exercises in Norway. Chris Blight hovered above the swirling snow, his wheels barely touching the surface, while his passengers quickly unloaded their kit and fanned out to take up firing positions. Blight and Ellerbeck turned away and flew back to the *Endurance*.

That the Argentines were expecting visitors was soon made evident. The base's commanding officer, Teniente de Corbeta Enrique Peralta Martínez, had sent out one of his technicians to keep watch. This man saw the two helicopters and raced back to announce the arrival of the British. No doubt Martinez immediately reported this to Buenos Aires.

The *Yarmouth* and *Olmeda* could not arrive before the following morning, so direct action was for the moment ruled out. It was decided to leave the patrol ashore – keeping watch – while Blight and Ellerbeck flew a number of 'spoof' missions over and around the Argentine base to give the impression of a large force over the horizon.

The west wind increased in force during the brief grey afternoon. The air temperature was minus twenty degrees Celsius. With the gale gusting to sixty miles per hour, the chill factor caused an effective drop to minus fifty-two. Over very many years, the Royal Marines have learned the techniques of not only surviving but also working operationally in conditions of

Mid-morning, 20 June. Sergeant Napier keeps watch on the Argentine base. A short distance from the tip of Hewison Point is Twitcher Rock, the demonstration target selected for HMS *Yarmouth*'s 4.5-inch guns. Beyond, almost lost in the mist, is Cook Island.

extreme cold. In the words of Mac McLeman: 'We reckon to get on with the job at temperatures down to minus thirty. If it gets much colder than that, it is a good idea to lay up for a while. Any kind of physical exertion makes you sweat. Even if you have several layers of clothing, the sweat will freeze on your body when it gets really cold. The conditions on Southern Thule were bad, but nothing which we weren't trained for. We had all the right kit and knew how to use it. And that is the vital point. All the expensive gear in the world won't help you to survive if you don't know how to use it and if you aren't exercising with it regularly.'

The marines had advanced part of the distance separating them from the Argentines when darkness fell. Their course took them across the top of the ice cap which flows down the side of the mountain. Under normal circumstances they would have been roped together. On this occasion, given the possibility of meeting enemy troops, Sergeant Napier opted to take the point on his own. There was the risk of falling down one of the numerous crevasses but, being a very experienced mountaineer, he accepted the hazard against the restraints of being literally tied to the other men. Like them, he was wearing specialist boots to which were fitted the crampons essential for traversing a surface such as this and, if necessary, climbing out of a crevasse. The patrol continued to advance through the dark hours. Sergeant Napier twice 'joined the Peter Pan Club' by falling down crevasses, but suffered no injury on either occasion. By 0600 he had reached a position overlooking Hewison Point. He and the other marines hacked at the snow and ice to make shallow trenches for themselves, and took shelter from the slanting squalls of snow sweeping over the mountain. The *Yarmouth* and the *Olmeda* were expected soon and dawn would bring fresh developments.

Captain Barker became increasingly concerned for the safety of the reconnaissance party. 'I had never experienced such cold, it really was exceptional. It was disturbing to think of Napier and his chaps being up on that mountain for eighteen hours in those conditions. One could not avoid drawing a comparison with the Fortuna Glacier episode. I called him several times during the night to make sure that he was coping and it was enormously reassuring to hear his rich Irish brogue saying, "Not to worry, Sir, 'tis cold but we're quite alright." The operation was not so much a contest with the Argentines as a battle with the elements.'

Further appeals to the Argentine base were broadcast by the *Endurance* during the night, again with no result. What the British could not know was that Martínez's men had spent the past few hours methodically destroying every item of equipment – scientific and military – in the base. Documents and weapons were taken to the shore and thrown into the sea. By their own reckoning, they destroyed property to a value in excess of one million pounds sterling. Included in this orgy of vandalism was their own radio installation and this precluded any chance of either hearing *Endurance* or talking to her.

The *Yarmouth* and the *Olmeda* joined HMS *Endurance* and the *Salvageman* at 0400. The morning hours were spent in completing final details and waiting for the weather to improve. Maps and written orders were flown to the two newly joined ships and the Blowpipe team transferred to the *Endurance* from the *Olmeda*. Low clouds obscured all but the lowest slopes of the 2,700 feet high Mount Larsen and the 3,500 feet of Mount Harmer (the highest point on Cook Island). The wind eased and, at 1140, Captain Barker decided to launch the operation before he lost the barely adequate visibility. 'Flying conditions were marginal, but

A few minutes before midday, and Sergeant Napier's party waits for the naval gunfire support programme to commence. Shortly afterwards the Argentines are seen to be showing a white flag. Operation Keyhole has achieved its purpose.

I couldn't afford to hang around any longer. I had *Salvageman* and the *Olmeda* four miles to the south-west, with *Yarmouth* to the east of them. I was in touch with Sergeant Napier and knew that his patrol was working its way forward – watched by thousands of penguins – to the flat strip of Hewison Point. I had still had no word from the Argentines so, with the intention of provoking some sort of response, I went in close and cruised along the shoreline of Ferguson Bay. This put the *Endurance* a few hundred yards off the Argentine base and nicely within range of my 20 mm pop-guns. One or two of my marines may have felt uncomfortable with this manoeuvre – bearing in mind what they had done to the *Guerrico* – but I needed to force the pace. We cruised very slowly past the buildings, my gunners very keyed up and ready to shoot if necessary. Nobody fired at us and there were no signs of activity ashore, so the next step was to put the NGS programme into operation.'

At 1140, Lieutenant-Commander Paul Miller lifted off from the small flight deck of HMS *Yarmouth* in Wasp 456. Occupying the observer's seat was LACMN Joe D'Souza. In the rear compartment were Brum Richards and his personal escort, Marine Mike Chin of the *Yarmouth*'s Royal Marines detachment, together with two radio sets. Richards recalls the flight: 'Within a minute of lift-off the left side of my face was completely numb. The chill factor in the cabin was about minus fifty because – to save weight – we were flying without doors. I had been through some strange episodes during the past two months but this was the strangest. The pilot set us down on the only flat piece of ground and then immediately went back to the ship. I set up my OP behind a rock and told Chin that we would be OK, but I was slightly nervous about the overhang of ice and rock two hundred yards to my rear. It was seven hundred and fifty feet high and extended along the shoreline to the southern edge of the island. I wondered how much of that stuff would come crashing down on us when the *Yarmouth*'s opening salvoes were fired. The radios were working perfectly and I told *Yarmouth* that I was ready to start spotting. Apart from the distant sound of the helicopters getting ready to embark the assault force, it was all very quiet. Then, with only six minutes to go before the gunfire was due to start, I was told that the shoot was cancelled. A white flag had been seen on Hewison Point.'

On board the *Olmeda*, the Sea King had been ranged on the flight deck at 1130. The rotors had been spread and both engines run to warm the whole aircraft. Take-off was scheduled for 1200. Ian McKenzie describes the atmosphere on the flight deck: 'Everyone was quite tense. The recommended all-up weight for the Sea King under icing conditions is 19,500 pounds – I would be lifting off with 22,700 pounds. However, we had a twenty-knot wind across the deck which helped to create lift and the engines loved the cold thick air. I took my most experienced crew – Spud Murphy as co-pilot, Peter Blackman as observer and Kim Alison as aircrewman. We took off using ninety per cent power and flew towards Hewison Point under a very low cloudbase. Visibility was about a thousand metres when we left the ship, but rapidly reduced as we neared the landing zone. Soon I was flying blind, relying on Peter Blackman's radar guidance. Spud set us into the wind about a mile from shore with an airspeed of sixty knots and an altitude of one hundred feet. I followed through on the controls and tried to peer through the murk. Occasionally I could see the sea below, but nothing ahead. As we neared the landing zone, speed was reduced to twenty-five knots, then ten knots at twenty feet. Peter talked us in, and I saw a stretch of beach passing close underneath. We were in almost whiteout conditions. Then we saw a blue smoke flare and I slowed down to a hover, just a few feet above the ground. Our downwash brought swirling snow up over the length of the aircraft, but I could see just enough to get the wheels down. As soon as I felt the contact I shouted "Go!" and the marines started to exit. Spud and I were both on instruments – with ninety-five per cent power on – and

Continuing to fly through one of the intermittent blizzards sweeping the area on 20 June, Ian McKenzie in Sea King 355 and Chris Blight in Wessex Whiskey Whiskey put more troops ashore and evacuate the small group of Argentine prisoners. The web of cables securing the tall radio pylons, almost impossible to see in these conditions, poses a serious hazard for the aviators.

A large tabular iceberg provides the background to yet another launch by Whiskey Whiskey. All five helicopters operated by Captain Barker's task group were hard worked during their brief deployment off Thule Island.

we had to reduce power very quickly to keep the wheels on the ground as 7,000 pounds of payload went out of the door in about five seconds. Alison reported "Last man!" and we applied full power again for a vertical take-off. I knew there was a very large mountain nearby so we went up to one thousand feet before pedal-turning the helicopter onto a safe heading out to sea. We picked up another load of troops and this time we had a clear run because the blizzard had cleared away. Flying inshore, I was startled to see just how close we had been flying to the enormous radio masts beside the Argentine base. Sometimes a little ignorance can be a good thing.'

Ewen Duncan's 9 Troop had been chosen to lead the main assault and was already airborne in the Sea King when Napier's report was passed to him. Duncan describes his feelings: 'We had been told that H-Hour was 1230, so we prepared for take-off at 1200. The guys were all very serious – – perhaps even a little apprehensive. When we ran out of the ship's hangar to board the helicopter there was an amazing number of people there to see us off. We must have looked pretty good – laden with weapons and ammunition, wearing our cam-whites [Arctic over-clothing], and the helo burning and turning on the flight deck. In flight I was seated at the door. We flew only a matter of feet above the water. I was listening in on a headset and I heard five minutes after take-off that a white flag was thought to have been raised. I didn't tell the guys, so they landed assuming that they would be assaulting. We took up all round defence on the ground, the snow driving hard, about a foot deep. This was a really bleak place. Somebody described it – with total accuracy – as "the icehole of the world". I saw Sergeant McLeman leading prisoners outside and Captain Nunn, who had just arrived, walking over to them.

Whiskey Whiskey, with Lieutenant-Commander Blight at the controls and Petty Officer Aircrewman Hogan seated at the door, delivers captured Argentine stores to the flight deck of HMS *Endurance*. In the background, the tug *Salvageman* displays a thick rime of frozen seawater on her bows.

Marines of 'M' Company adopt a pleasingly Victorian pose to celebrate their part in the ejection of Argentine forces from Thule Island. Their company flag is hoisted below the Union colours. To the right, Captain Chris Nunn, and, immediately under the larger flag, Warrant Officer Paul Juleff (holding a M16 rifle fitted with the M203 grenade launcher).

We took no chances and carefully cleared all the buildings with 7 Troop and the RM detachment from *Endurance* who came in after us.'

The prisoners – one civilian and nine military personnel – were searched and evacuated to the *Olmeda* by Ian McKenzie in the Sea King and by Paul Miller in Wasp 456. Having collected the disappointed NGS team from Cook Island, Miller was now faced with his first landing on snow in the gathering darkness. 'Never having operated under such conditions before, I had my mind well focussed on the job in hand. To avoid a whiteout, I made a zero zero landing – an approach where the pilot reduces height and speed at a constant rate so that the wheels make contact with the ground at zero groundspeed. All went well, and I made half a dozen ferry flights between the shore and the tanker. Next I was tasked to collect the captains from each of the other three ships and deliver them to the *Endurance* for the surrender document signing ceremony. Problems developed when I tried to pick up Captain Stockwell from the *Salvageman*. Joe D'Souza, my aircrewman, lowered the rescue strop and the Master had been

The main accommodation building on Hewison Point. The garrison has destroyed most of the scientific and radio equipment, but the buildings themselves are in excellent condition.

Royal Marines check the generator shed for weapons and explosives. Captain Nick Barker (left of picture) has come ashore to inspect the state of the buildings.

lifted a few feet off his deck when the winch motor jammed. I couldn't get him into the aircraft and couldn't safely get him back onto his ship, so I flew slowly across to the *Endurance* with my unfortunate passenger hanging underneath. He was frozen almost solid by the time we dropped him onto the flight deck – he disappeared into the warmth of the hangar very quickly indeed. We flew more than three and half hours that night and made thirty-two deck landings. With no doors on the cab it was exceptionally cold and draughty. The main worry was ice forming on the fuselage and the rotors. Flying at ninety knots through snow showers – with still-air temperatures well below freezing – you can quickly accumulate a load of ice which spoils the lift of the blades and makes the aircraft much heavier. There is the added risk of lumps of ice breaking away and entering the engine intake. Like most of the aircrew engaged in the South Georgia and Southern Thule operations, I was frequently flying on the outer limits of the operational safety rules.'

The captured Argentines were received and guarded on the *Olmeda* by marines of 8 Troop, 'M' Company. The tanker, accompanied by HMS *Yarmouth*, departed for South Georgia almost immediately. The *Endurance* and the *Salvageman* remained in the vicinity of Hewison Point overnight and, on the morning of 21 June, Lieutenant Keith Mills returned with 'the plum's' detachment to complete the search of the bright orange base buildings, ensuring that no explosives were concealed in the area. Mills and his men were disappointed that the affair had ended so tamely, but they had the satisfaction on this operation of being captors instead of prisoners. According to Lieutenant Mills: 'I wasn't happy when Captain Barker took us so close inshore – I knew the effect of infantry weapons at close range – but the white flag went up as soon as the people ashore saw the ship. The Wessex 5 was ready to go, so I quickly obtained permission to take the detachment ashore. Lieutenant-Commander Blight put us down by the Argentine base more or less simultaneously with the arrival of the Sea King. It was a great moment. After that it was a race between us and 'M' Company to see who could raise the Union Jack first.'

The accommodation buildings were found to contain a plentiful stock of cigars, wine and beer which made a welcome addition to the Keyhole ships' stores. Captain Barker was delighted to join in the celebrations. 'It was the *Endurance*'s people who had been forced to lower the Union Flag on South Georgia on 3 April, so it was clearly right that we should salute – with Argentine wine – the hoisting of the red, white and blue on Southern Thule.'

Once the buildings had been searched, they were sealed to make them weatherproof. These tasks consumed many hours, delaying the return of the *Endurance* and *Salvageman* until 22 June. In the interim, the *Yarmouth* was refuelled at sea by the *Olmeda*, the evolution being claimed as a new 'farthest south' record. A similar claim was made for the 'farthest south splicing the mainbrace' when the news came through on that day of the

His ceremonial sword laid on the table before him, Teniente de Corbeta Enrique Martínez signs the instrument of unconditional surrender. Seated opposite (left to right) are Captain Alan Stockwell (Master, *Salvageman*); Captain Chris Nunn, RM (Officer Commanding, 'M' Company); Captain Nick Barker, RN (Commander Task Group); Commander Tony Morton, RN (Officer Commanding, HMS *Yarmouth*); and Captain Gil Overbury, RFA (Master, RFA *Olmeda*).

birth of His Royal Highness Prince William.

By 24 June, HMS *Endurance* and the tug were back in Cumberland East Bay. HMS *Yarmouth* and the tanker were heading for the Falklands, 'M' Company was reunited at King Edward Point, and the Wessex 5, with its crew and maintainers, was restored to the RFA *Regent*. A few days later the satellite pictures showed pack-ice encrusting the shores of the Southern Thule islands. Operation Keyhole had been a close-run thing.

Describing his feelings after the operation, Nick Barker said: 'It was the most satisfying thing we did during the conflict. Southern Thule was the place where the whole sorry business had started, back in 1976. The Argentines had been allowed to get away with their occupation then and, once they were established, we were hamstrung in any attempt to get them to leave. Further, because of their policy of claiming wedges of Antarctic territory right down to the Pole from whatever offshore island they happened to occupy, chucking them off Thule Island signified that we were restoring their easternmost claimed boundary back to where it had been previously. It was not the island that was important, it was the political implication of their continental claims which really mattered. And that is the true worth – in historical terms – of Operation Keyhole. It was a very satisfying operation in other ways of course – nobody had been killed or injured. We were operating in the severest weather of the Antarctic winter and the risk of accidents – particularly for the aviators – was very high. Sergeant Napier's Recce Troop had the worst of it and I greatly admired their stamina in coping with twenty-four hours on the ice cap. But everyone put up a great performance and had every reason to feel proud of themselves.'

The final days of what can be described as 'the Paraquat period' were drawing to a close. Of those who had played a role, only 'M' Company, HMS *Antrim* and HMS *Endurance* still remained at South Georgia. The island was now a backwater. The turmoil in Argentina, and the collapse of the Galtieri regime on 17 June, suggested there was very little risk of further naval or military adventures coming from that country in the immediate future. For the British units who had fought in the South Atlantic, who had done their duty and buried their dead, it was time to think of home. The final days of June and the opening days of July saw the departure for warmer waters of most of the ships which had formed Sandy Woodward's Task Force and Brian Young's Task Group.

The first to depart from South Georgia was HMS *Antrim*. Of the seven warships sent south in early April from Woodward's original Exercise Springtrain force, she and HMS *Yarmouth* were the only two which had not either been sunk or ordered to sail for the United Kingdom. Brian Young was told that he, too, could head north – first for Port Howard and then for Portsmouth – on 26 June. The ship had not only provided South Georgia with its only anti-aircraft missile defence since 2 June, she had also conducted offshore patrols and had maintained – for the safety of other ships transitting the area – a detailed plot of the advancing icebergs. She had made good much of the damage inflicted by cannon-fire and a dud 1,000-pound bomb during the Argentine air raids on San Carlos Water on 21 May. The venerable 'Humphrey' had been punctured by splinters during the same attack, but her maintainers had patched her wounds and she was once again fully operational. Ian Stanley made one final flight – in fine weather – up to the Fortuna Glacier before the ship finally sailed. He collected some remaining items of kit from the two derelict Wessex before turning his back for the last time on a place which he could never forget.

Next to leave were Chris Nunn's 'M' Company marines. On 7 June they handed over their garrison duties to a half company of 2nd Bn Scots Guards, a unit which had fought magnificently on East Falkland, particularly in the battle to clear the daunting Mount Tumbledown. They were themselves due to return to England at the end of the month. They arrived at Grytviken aboard the Type 21 frigate HMS *Ambuscade* (Commander P. J. Mosse, RN) and she sailed again the same day for Ascension Island with 'M' Company's advance party embarked. The rest of the company sailed on the following day aboard the *Nordic Ferry*

HMS *Antrim*'s Log contains a simple but, to her ship's company, deeply important entry: '17 July. Arrived alongside in Portsmouth having been away for 121 days of which 110 were spent at sea. Tumultuous reception under blue skies.'

Early attempts to salvage the ARA *Santa Fe*. Lieutenant O'Connell's small diving team achieved surprisingly effective results when the technical difficulty of the task and the atrocious underwater conditions are taken into account.

(Captain R. Jenkins), a roll-on roll-off ferry requisitioned from Townsend Thoresen. She had arrived that morning from Port Stanley with the Scots Guards' stores and rear party.

Only the *Endurance* now remained. She had departed Portsmouth nine months earlier and was long overdue for her return, but there were various reasons why the Commander-in-Chief did not yet feel able to release her from her duties at South Georgia. One of these was the continuing need for a Royal Navy ship – experienced in local weather and navigation conditions – to shepherd the RFA and mercantile ships anchored in Cumberland East and Stromness Bays. They included the *Wimpey Seahorse*, *British Enterprise III*, *Scottish Eagle*, *Stena Inspector*, *Salvageman* and *Yorkshireman*.

Another reason was the unresolved problem of the *Santa Fe*. She was still resting on the bottom, blocking access to the Tijuca Jetty at Grytviken. Apart from being the best watering point on South Georgia, this berth was needed by other ships for the unloading of stores and fuel. During the second half of May, with the aid of technical advice and special equipment provided by the tugs *Salvageman* and *Typhoon*, the *Endurance*'s Clearance Diving Team had made a determined effort to refloat the wreck so that she could be towed away to a more distant point. That early attempt had failed (*vide* Appendix C), but a renewed salvage operation was now put in hand.

Commencing on 28 June, Dave O'Connell and his team started another intense period of diving, this time under the direction of Mr B. W. Vere-Stevens, Senior Mooring, Salvage and Boom Officer employed by the Directorate of Marine Services and seconded to the *Salvageman* for Operation Corporate. The operation called for a sustained period of highly technical and often dangerous work by all those serving in HMS *Endurance*, RMAS *Typhoon*, United Towing's *Salvageman* and *Yorkshire-*

Bearing an uncanny resemblance to a large whale lashed alongside a catcher of sixty years earlier, the *Santa Fe* is nursed carefully out of King Edward Cove on 15 July. Later that day she is grounded and abandoned in the shallows off Hestesletten, Moraine Fiord. She is destined to remain undisturbed in this lonely place for the following eight months.

man, and Stena Caribbean's *Stena Inspector*. It required more than two weeks to raise the submarine to the surface with sufficient buoyancy for a limited transit. On the afternoon of 15 July, the tug *Yorkshireman* took her sister *Salvageman* in tow, with the *Santa Fe* lashed securely alongside, and the trio moved slowly across the mouth of Moraine Fiord and around to Hestesletten where the submarine was beached. For the time being at least, she was safely out of the way.

HMS *Endurance* finally departed South Georgia on 16 July. Her first port of call was, naturally, Port Stanley. It was the first time the 'kelpers' had seen her since 21 March and she received an emotional welcome from the crowds gathered along the waterfront. There was a warm personal welcome from Rex Hunt, once more installed in Government House. He now had a knighthood and a new title – Civil Commissioner. He embarked in the ship for a short cruise to visit the isolated settlements at Fitzroy and Fox Bay and 'the plum' then sailed for Ascension Island.

Following the disruption of the *Endurance*'s seasonal programme, it had been decided to temporarily deploy another ship in her place. The choice fell on HMS *Hecate*, one of the Royal Navy's *Hecla* class survey ships. Although not equipped to duplicate the *Endurance*'s role in full, she did have an ice-strengthened hull and could continue the usual hydrographic work. The two ships met at Ascension Island on 5 August and Captain Barker handed over guardship duties to the *Hecate*'s captain, Commander Chris Gobey, RN.

It has been suggested that the *Endurance*'s return to the United Kingdom posed problems for the public relations people. The ship had been a political bone of contention for a long time. Her commanding officer had not endeared himself to everyone. Some critics of earlier government policies were claiming that they had

The changing of the guard. On 11 August, homeward bound up the west coast of Africa, HMS *Endurance* receives the greetings of three Royal Navy ships heading south for the Falkland Islands. Heading the column is the new aircraft carrier HMS *Illustrious* (Captain J. Slater, RN), taking over from HMS *Bristol* as South Atlantic Force Flagship. Following her are HMS *Battleaxe* (Captain D. Nolan, RN) and, nearest the camera, HMS *Amazon* (Commander I. Garnett, RN).

been vindicated by the war, that the decision to sell or scrap the *Endurance* had indeed been a fatal blow, in their view, to any hope of avoiding open hostility with Argentina. The sight of the now famous red hull coming up the English Channel was not only a symbol of British fighting grit, it was also an uncomfortable reminder of the arguments which had raged earlier in the year and throughout the second half of 1981. There seems to have been some uncertainty as to where and how she should make her landfall.

When it was announced that she would berth at Chatham, the Medway towns started to plan the biggest welcome they had ever extended to any ship. The *Endurance* was the only Chatham-based ship to have participated in the conflict. A civic spokesman told reporters: 'The Medway will not have seen anything like this since the Dutch invaded.' The joy was tinged with sadness. Her Majesty's Naval Base Chatham was another victim of the cuts in the Defence Budget. The dockyard was scheduled for closure and the winding down process had already started. This was the last time that HMS *Endurance* would come here for her annual refit.

The first visitor to the ship was Rear-Admiral Edward Anson, the new Chief of Staff to Admiral Fieldhouse. He flew aboard by helicopter while the ship was in mid-Channel. Another visitor was Lord Shackleton. He was accompanied by Viscount Trenchard from the Ministry of Defence who represented the government. In his speech to the assembled ship's company, Trenchard paid particular tribute to the Royal Marines Detachment for their defence of King Edward Point on 3 April and to the Wasp Flight for their part in the capture of the *Santa Fe*.

National coverage of her triumphant return began when the *Endurance* appeared in the Medway at 1400. More than eighty journalists were on hand to record the event. A crowd estimated at fifteen thousand waved from the Strand at Gillingham and thousands more cheered as she passed Grain and the river front at Hoo. Hundreds of yachts and pleasure boats trailed in her wake while, ahead, the way was cleared by the launches of the Queen's Harbourmaster, the Kent County Constabulary and the Ministry of Defence Police. Overhead circled helicopters and light aircraft chartered by the television companies. It was all in great contrast to the *Endurance*'s mundane departure from Portsmouth in October of the previous year.

At Bull Nose, where ships enter the naval dockyard through canal gates, the Mayors and other civic dignitaries – together with Mrs Elizabeth Barker and the wives of other members of the ship's company – stood with Rear-Admiral William Higgins, Flag Officer Medway, as he accepted the *Endurance*'s formal salute. When she came to rest at her berth and the brow was swung into place, wives, children and parents swarmed aboard and her hangar became a place of private joy and grateful celebration.

In the larger world outside, far from the tumult of Chatham, attention was concentrated upon the military defence and political future of the Falkland Islands. South Georgia, the setting for the Davidoff affair and all that flowed from it, sank back into the obscurity from which it had been dragged. Apart from a lonely company of Queen's Own Highlanders, the island was once again deserted by man. The only visible evidence of the human drama played out here, in April 1982, were the wrecks of three helicopters and, rusting on the foreshore of Moraine Fiord, the derelict victim of the Fleet Air Arm's unique 'search to destruction'. □

Epilogue

Looking back at the events of 1982, and indulging the questionably perfect vision of hindsight, it is possible to identify certain lessons and conclusions which emerge from Operation Paraquat. Some are more profound than others. Those set out below are not intended to be a definitive analysis of all aspects of the South Georgia story, but they do provide some food for thought.

POLITICS

The waging of war is a function of governments. The war of 1982 was started by the Argentine government and finished by the British government. A study of the preceding pages suggests that Buenos Aires was excessively impatient to start the conflict, London curiously reluctant to recognise its imminence.

The seeds of disaster were sown in 1942 and 1943 when the two navies commenced their joust in the white wilderness of Antarctica. For forty years, the shape and form of the final collision developed and matured. The encounter between Latin impetuosity and British stoicism produced results which did not serve the best interests of either country. With hindsight, the invasions of South Georgia and the Falkland Islands seem to carry with them a fateful aura of inevitability.

Given the long-running nature of the dispute regarding the South Atlantic and Antarctic territories, the impartial observer is bound to ask himself why neither government was able to recognise the peril in which it stood during those final months and weeks.

Argentina is not an isolated Third World country lacking cultural or intellectual ties with Europe. To the contrary, it is itself a modern European state in a South American setting. Given the one and a half centuries of peaceful contact with the United Kingdom, it is extraordinary that the Argentines failed to understand that their invasions would be met with a totally committed response. The vigour of the response was predictable not only in terms of national character but also in terms of the incumbent Prime Minister's own character. Mrs Thatcher had been not unflattered when the Russians had once described her as 'the iron lady'. General Galtieri and his colleagues could not have chosen a stronger adversary.

That said, the evidence does suggest that all was not well in the British camp prior to the outbreak of hostilities. Information and advices flowing from first the 'save the *Endurance*' campaign and then the Davidoff affair gave due notice that trouble was brewing, and brewing on a much larger scale than anything seen in previous Anglo-Argentine disagreements over the Falkland Islands and the Dependencies.

Storm cones were flying from the time of the illegal visit to Leith Harbour by the *Almirante Irizar* in December and following the abortive February talks between Richard Luce and Enrique Ros in New York. The intransigence of the new Argentine regime was underlined by its outright rejection, also in February, of the protest delivered by Ambassador Anthony Williams. By the time Captain Mitchell's reports – detailing the sailing of the Argentine navy – started to reach London, it was already too late to despatch an effective deterrent force. Even so, the evidence (*vide* Chapter III and Appendix A) suggests that the Foreign and Commonwealth Office accepted the gravity of the situation some time before the Ministry of Defence was forced to the same conclusion.

Bearing in mind that the United Kingdom's entire defence posture has been based for many years upon the supposition that the armed forces would have adequate time to respond to an identified escalating threat (which is precisely the type of threat which developed in the spring of 1982), the author has sought an explanation for the apparent lack of cohesion between the Foreign and Commonwealth Office on the one hand, and the Ministry of Defence apparatus on the other. He has failed. The explanation, however, may be as simple as that stated by one of his contributors: 'I can say, from nearly thirty years' experience as a civil servant, that some difficulties between the great departments of State are endemic in our system of government – as they are in some other Western countries such as France and the United States. It is a constant and permanent challenge to central co-ordinating agencies – such as the Cabinet Office and the Prime Minister's Office – to break down the instinctive insularity and self-interest of these great departments. Frequently the difficulties have to be resolved in Cabinet and this is not always faced – except in circumstances of dire national crisis.'

Shortly after the cessation of hostilities, the government ordered an enquiry into the events preceding the conflict. The Franks Committee interviewed many of the people who had been involved, in one way or another, in those events. It produced a long and comprehensive report which, given the short period of time in which it completed its work, was in some respects admirable. The Committee concluded that no fault could be found with any individual or any department of State in failing to prevent what was in effect Great Britain's first international war since 1945.

DIPLOMACY

Although the overwhelming majority of the British people were fully in accord with their government's decision to despatch a Task Force to the South Atlantic, and although the entire episode led to a powerful resurgence of national pride, there were some who disagreed with certain aspects of the campaign. One political opponent went so far as to accuse the Prime Minister of 'blood lust'. As far as Operation Paraquat was concerned, such an accusation is self-evidently without foundation. On the contrary, as the narrative makes plain, the desire for a bloodless solution to the crisis shaped the conduct of the operation in such a way that British lives were placed at greater risk than was militarily necessary. If there had not been a genuine and continuing wish to seek a diplomatic solution - leading to a voluntary withdrawal by the invaders from the Falkland Islands - Captain Young would have been able to follow a plan very different to the one adopted. Given a free hand, and if he had chosen to ignore normal humanitarian scruples, he could have simply laid off the Barff Peninsula and bombarded the Argentine garrison to destruction. The tiny force at Leith Harbour - comprising only sixteen armed troops - could have been battered rapidly into bloody submission. There would have been no need to insert the SAS and SBS reconnaissance parties, no need to hazard the lives of the naval aviators, and no resultant loss of two valuable helicopters. The 'kid glove' policy was chosen not only to sustain the sympathy of world opinion, but also to make it abundantly clear to Buenos Aires that the way was still open for them to change their minds without losing too much dignity. Operation Paraquat provided that opportunity. The Galtieri regime did not see fit to exploit it.

STRATEGIC PLANNING

The decision to mount Operation Paraquat as a precursor to Operation Corporate was based upon a large number of factors. Not everyone agrees upon the order of importance of those factors. Some contributors to the preceding narrative have emphasised the political and diplomatic value of the reoccupation. Others have underlined the military importance of 'clearing the left flank and creating a safe haven for our shipping'. In the event, all of these considerations proved to be well founded. The immediate impact of Captain Young's victory was indeed essentially moral and diplomatic. Another immediate benefit was the opportunity which the operation gave to various arms to put a final polish on their preparations for Operation Corporate. To quote Major Delves: 'South Georgia was, for us, a bitter experience, but it was important as a live-firing rehearsal for the Falklands. Some of the lessons learned were to pay off when we were planning the Pebble Island raid and other similar operations in May and June.'

The third benefit was undoubtedly greater than originally anticipated. The sheltered anchorages of Stromness Bay and Cumberland East Bay witnessed the arrival and departure of far more ships than the planners could have envisaged during the early days. Local waters were used for the delivery and cross-decking of thousands of men, thousands of tons of stores, and tens of thousands of tons of fuel. Some of this activity could have been conducted on the open ocean during rare fine spells, but not all of it. Referring specifically to the transfer of 5 Infantry Brigade from the *Queen Elizabeth II* to the *Canberra* and the *Norland*, it is hard to imagine how this reinforcement could have been carried through without the protection of South Georgia's mountain-ringed waters. Not a single life was lost during that transfer and it is stated that only one guardsman fell into the sea (he was quickly fished out again). The same operation, if attempted in the open sea, would have been totally dependent upon good weather (rarely encountered in that area and at that time of the year). The transfer would have taken much longer, consequently the ships would have been more vulnerable to detection by the enemy. In short, it would have been a vastly more hazardous undertaking.

Looking back at the operation, Admiral of the Fleet Sir Henry Leach has said: 'The island proved useful to us in practical terms in a number of ways, but it was the psychological value of the swift reoccupation which was most immediately apparent. It was necessary to offset the débâcle of having lost the islands in the first place. We needed a success. There was one alternative way of achieving a quick success and that would have been to send Captain Young and his ships down to Southern Thule instead of South Georgia. Throwing the Argentines off Thule Island would have been a positive gesture, but it would not have had the same popular or diplomatic impact as South Georgia. It would not have been internationally recognised. South Georgia was far more appropriate to our aim because it was demonstrably not a diversion from the main objective - the recovery of the Falkland Islands. Having said that, I must add that I had no illusions regarding the difficulties which Captain Young would be facing. His mission was almost impossible. He was required to somehow generate initiative under conditions which gave little scope for initiative. Framing a strategic objective is one thing, putting it into effect is another.'

The basic strategy of Paraquat proved to be sound and it achieved its objectives in every respect. Captain Young and Major Sheridan were able to meet the requirement of 'minimum loss of life and minimum damage to property' in full, and they did so within the time allotted.

TACTICAL PLANNING

The tactical plan for Paraquat fell into two distinct phases: initial reconnaissance by the SAS and SBS, then a follow-up exploitation by the 'M' Company Group. Both of the initial reconnaissances failed, and the 'M' Company Group was never employed as an assault force. On the face of it, therefore, the tactical plan was wrongly conceived and unworkable.

Curiously enough, even though the obstacles of weather and terrain proved to be almost insuperable, it is not easy with hindsight to suggest alternative tactical plans radically different to the one adopted. It is sometimes suggested that the Paraquat commanders were too cautious, that they should have landed the reconnaissance teams much closer to their targets, should have devoted less time to intelligence gathering, should have pressed forward more boldly, and so forth. One naval officer has said: 'The military wanted five days to count Argentine heads. I would have thought they could have done that in five minutes.' While understandable from a sailor's viewpoint, this remark illustrates the difference between combat on land and combat on the featureless sea. The sailor must accept his enemy as he comes over the horizon. The soldier needs time in which to assess his opponent's morale, professionalism, weaponry, and logistical organisation. Ideally, he needs to identify strongpoints, minefields, arcs of observation, arcs of fire, patrolling patterns, stand-to routines, and much else besides. Only when the fullest possible picture has been built up - within the time available - does a commander decide his lines of approach and plan of attack and consolidation. The losses suffered by the Argentines when they blundered ashore at King Edward Point on 3 April serve to prove the point.

If South Georgia's execrable weather during the key period of 21/24 April had not intervened, both reconnaissance insertions (the SAS on Fortuna Glacier and the SBS in Hound Bay) might have fulfilled their expectations. Similarly, if the Argentine Hercules reconnaissance flights had failed to find Captain Young's ships on 23 April, and if the *Santa Fe* had been sunk at sea instead of being able to stagger back to King Edward Point on 25 April, the Argentine garrison might have remained unaware of the nearness of the Paraquat ships and there would have been time for Captain Young to call forward the *Tidespring* and make use of the 'M' Company Group as first intended.

Commander Dennis Davis, commenting recently upon the tactics of Paraquat, has said: 'Yes, you could say it was a bad plan because it didn't work as anticipated. But the only reason it didn't work was the exceptionally foul weather at just the wrong time. Young was desperately unlucky in that respect. To a certain extent, there was a lack of appreciation at Northwood of just how bad it could be. But then look at how he reacted. When the "slowly slowly" approach over the Fortuna Glacier failed, he was fully prepared to accept the hazards of going straight into Stromness Bay for the Grass Island landing. Then, when the *Santa Fe* returned to Grytviken, and when Major Sheridan rightly stressed the tremendous impact this must be having on the morale of the troops ashore, he agreed to go straight in with a very small scratch force. Not an easy decision. Given all the restraints, his willingness to change tactics as circumstances and the needs of the military changed must be rated highly.'

Among the restraints indicated by Commander Davis, the following have been mentioned by various officers who were there: the unknown strength of the Argentine defence force, its willingness and ability to fight, the scale of reinforcement by the *Santa Fe*, the additional troops which might have been delivered by other (undetected) Argentine vessels during the previous three weeks, the possible presence of sea mines and stand-off weapons, the possibility of attack from the air, and the thought that the *Santa Fe*'s stern tubes were still pointing out towards Cumberland East Bay. The nature of the submarine's wounds were not known with certainty at that stage.

It is an interesting thought that Operation Paraquat might have had a very different outcome if the Task Group had *not* encountered adverse weather, if *less* time had been devoted to careful initial reconnaissance, and *if* the *Santa Fe* had remained undetected after leaving Mar del Plata on 9 April. Under these circumstances, Captain Young's ships might have been anchored snugly within Cumberland East Bay when Captain Bicain reached the island on 24 April. Even with the *Brilliant* conducting an anti-submarine patrol off the entrance to the Cumberland Bays, Bicain would have had an excellent chance of emulating Günther Prien who, in October 1939. stole quietly into the anchorage at Scapa Flow and sank HMS *Royal Oak*. The peculiarities of South Georgia's sea-bed and stratified deep waters would have made the relatively silent submarine very difficult to detect. The frustrations of the abortive Fortuna Glacier and Hound Bay insertions were possibly – with hindsight – a blessing in disguise.

COMMAND AND CONTROL

Two centuries ago a naval captain might have set sail from England with orders no more specific than 'make war with the King's enemies'. Those days are long gone. In an age of convoluted politics and uncertain international law, the captain of every ship carries a book of rules – drawn up by the Ministry of Defence on behalf of the government and frequently amended – which tell him how he should react to any given set of circumstances during times of tension or possible hostility. These are the Rules of Engagement. As the preceding account has shown, the Rules did from time to time appear to the participating officers and men to be inconsistent or even illogical during Operation Paraquat. It is perhaps sometimes difficult for a serviceman to patiently await the outcome of continuing peace initiatives if his own life stands in jeopardy in the meantime.

It was the early ban on shooting down Argentine reconnaissance aircraft which caused by far the greatest anxiety. The ban was perhaps not as foolish as it might seem. To quote Captain Young: 'Obviously I did not like being over-flown, especially when there was the likelihood of a submarine being guided towards me. But I could understand London's problem. What would have been the reaction of world opinion if, in those early days, I had shot down a 707 which later turned out to be a civilian aircraft full of Argentine schoolchildren on an educational flight to Antarctica, or a load of nuns allegedly being flown to a convent in South Africa, or something of that sort?'

According to Commander Davis (ideally placed to judge these matters in their larger context): 'The Task Group commander seldom had to ask for a change in the ROEs because his requests were usually anticipated. Everything had to be cleared at the highest level and it was marvellous that the Prime Minister and her advisors made these decisions without delay. It was a great morale booster for the military – not only in this particular war but as an indicator of the political-military relationship as a whole.'

Captain Coward is less sanguine: 'I think it is probably the ultimate in the modern trend in Whitehall to attempt to direct the minutiae of everything military, even at a great distance. It is an attempt to envisage every situation and lay down the predetermined response as one would like it to be. Ideally, what the commander on the spot instead needs is a forthright and honest explanation of the overall aims and requirements. There is probably only one thing which is certain, and which can be stated in advance of almost every military action, and that is that the circumstances encountered will not be those expected.'

Conversation with other senior officers reveals a healthy variety of personal opinions in respect of the whole concept of Rules of Engagement. Doubtless the debate will continue. As far as Operation Paraquat was concerned, and despite the very unexpected and extraordinary circumstances under which the operation was conducted, the concept seems to have worked well.

PUBLIC RELATIONS

No journalists or television camera crews had been authorised to travel to Ascension Island at the time of Captain Young's departure on 10 April. Even if they had, they would have been unaware of his orders or his destination. The entire operation was completed, therefore, without having been either reported or filmed by accompanying representatives of the media.

For the most compelling reasons, the Ministry of Defence attempted to impose a complete blackout on the operation for the entire period preceding the Argentine surrenders at King Edward Point and Leith Harbour. Thereafter, maximum instant publicity was required to prove to the world that the British government was prepared to back diplomacy with force. This must have been intensely frustrating for the men of Fleet Street because they were obliged to rely entirely upon official Press releases. These were based upon texts prepared by officers nominated by Captain Young who were already heavily involved in other duties. This is a standard practice but, given the exceptional circumstances around Cumberland East Bay between 25 and 28 April – with dozens of prisoners being processed, explosives being rendered harmless, defence plans being prepared, and ships being replenished for the coming battle for the Falkland Islands – the task was not an easy one. Looking back through the summaries forwarded to Whitehall, it is mildly surprising to see how complete and comprehensive they were. But in the nature of all naval signals – the Royal Navy being 'the silent service' – they inevitably lacked the colour and human interest which are the stuff of journalism. Sub-editors were obliged to add their own contributions.

An example of journalistic imagination was the 'Cindy and Annie story'. It was known that two attractive lady wildlife photographers had been overtaken by the rush of events and had been living in a tin hut somewhere on South Georgia. That they were alive and still at liberty was confirmed by the sparse messages reaching their families through the BAS network, but the details were all guesswork. That one of them was the daughter of a peer, Lord Buxton, and that their *Survival* documentary films had been seen by millions of television viewers, added to the journalistic value of their predicament. In Cindy Buxton's words: 'The amazing amount of publicity we received has always remained a bit of a mystery to both of us. I personally feel that it was all due to the fact that, in the early stages of the war, the media had nothing to report because nobody could get to either the Falklands or South Georgia. Then the media discovered that there were two females stuck in a very harsh environment, surrounded by a lot of men, ninety-five per cent of them the so-called enemy, and there they had a story. They loved it. They invented the stories that we had been taken prisoner, had repeatedly radioed for help, and were starving and freezing to death. It was all good stuff. The only problem was that when we got home and faced the media and told them that we were fine and nothing had happened to us, that we had not shot anybody, had not been captured, or raped, or starved and frozen, we were a bit of a disappointment!'

There is still a quiet residual resentment in some sections of Fleet Street that they were not given at least a background briefing during mid-April as to the existence and purpose of Operation Paraquat. If nothing else, the publication of this book should dispel any such sentiments. It was not a question of not trusting the discretion of editors – it was simply unthinkable that the success of the venture should be hazarded in the slightest way. Even Captain Ian Sutherland, the Director of Public Relations (Navy), had no details of the operation until the Argentine news agency TELAM started to report the presence of British ships off South Georgia on 23 April.

At that point, the BBC felt obliged to repeat the Argentine announcements on its World Service. It did so without comment and it quoted the possibly unreliable source. However, the reports triggered a burst of enquiries to the Ministry of Defence from British journalists seeking confirmation. The response was a denial that any British troops had landed on South Georgia. This was true – as far as an effective fighting force was concerned – and the statement did not further compromise the safety of the Paraquat ships. In the circumstances, it was probably the best way of handling an extremely sensitive story.

Several of the men who served on Operation Paraquat have since stated that their presence off South Georgia was first announced to the world by the BBC on 21 April, the very day when the first SAS and SBS patrols were being landed on the Fortuna Glacier and at Hound Bay. A meticulous search through the BBC archives has failed to uncover the transcript of such a broadcast. Similarly, a search through the index of *The Times* newspaper – for editions published on 22 and 23 April – reveals no mention of South Georgia. If the BBC had in fact exposed the existence of the Paraquat group in local waters – journalistically an exciting and very newsworthy story – one would have expected it be be pursued by Fleet Street. In the event, the earliest printed references to South Georgia appear in editions published during the night of 23/24 April, and these are plainly a follow-up to the BBC World Service broadcast late on 23 April.

It is true that the Task Group's existence was almost 'blown' on 21 April when Mr Francis Pym, the new Foreign Secretary, made a statement to the House of Commons. He stated, quite properly, that the despatch of the Corporate Task Force did not represent an abandonment of all hope by the British government of a peaceful solution to the crisis. When questioned by Members of the House, he confirmed that the armed forces were still planning to take early action if necessary. This well-intentioned statement was seized upon by some Members, and by those commentators who had weighed the options open to the Royal Navy, as an indication that South Georgia was an imminent target. Their suspicions were reinforced by a government statement that the civilians on the island had been contacted and that measures were in hand to secure their early release. This announcement referred in fact to current negotiations with the International Red Cross, but the straws flying in the wind suggested, again, that South Georgia was a likely scene of military action in the near future. Francis Pym left the House of Commons, having made his announcement, but returned an hour later to state that it would be wrong to draw any specific conclusions from his earlier statement. The speedy disclaimer served only to confirm the assessment of those engaged in the guessing game. Curiously, although Mr Pym's statements were reported verbatim in the British press on 22 April, they were not broadcast on the BBC World Service.

It has been suggested that the Argentine reconnaissance flights on 23 and 24 April were a direct result of alleged indiscretions by the British media, particularly the BBC. The evidence indicates that this was not the case. The first two overflights had already taken place by the time the BBC put out its first report. That report was a reflection of the announcements already broadcast to the world by TELAM. The fact that the flights were made on 23 April was, in the opinion of the author at least, an indication of the fact that the Argentine naval staff was just as capable of working out problems of time, distance and speed as its counterpart at Northwood. Having analysed the options open to the British, having created a scenario, the Argentines decided that, if the British were aiming for South Georgia, then 23 April would be a good day to go looking for them. Simultaneously, the aircraft could act as scouts for the *Santa Fe*. Captain Bicain would be approaching the island on that day and he would be as concerned by the possibility of encountering either icebergs or surface ships as Commander Wreford-Brown of the *Conqueror* had been. In short, and although they were giving the world an astonishingly detailed description of the strength, location and capability of Admiral Woodward's Corporate force, the British media appear to have given no culpable advance notice to the Argentine high command that South Georgia was about to be assaulted.

Apart from the lack of factual written information, the media were largely starved, during and immediately after the reoccupation, of visual material. Film shot by John Tippey and his crew was used in early April and, after the Argentine surrender, some of the film shot by Cindy Buxton and Annie Price was rushed back to London for inclusion in news programmes. Many of the 35 mm still photographs taken as a personal record by Royal Navy and Royal Marines personnel were retained – mainly on security grounds – and some of their pictures were later made available to the Press. Lieutenant Stewart Cooper's much-published photograph of SAS troopers being rescued from the Fortuna Glacier is a case in point.

The only official photographer in the Paraquat force was Leading Airman 'Peter' Allen, the resident 'phot' in HMS *Endurance*, and some of his excellent work has been used to illustrate this book, but he could not alone cover the entire operation. A further limitation was the lack of visibility encountered on those days and nights when some of the more dramatic events occurred. The aerial attacks on the *Santa Fe* were

Royal Marine Commandos exercising with the 66 mm HEAT weapon (high explosive anti-tank). It was while attempting to fire this short-range 'throw away' launcher that Acting Corporal Nigel Peters – the only British casualty during Operation Paraquat – was shot in the arm. All of Lieutenant Mills' weaponry was similarly lightweight. It is interesting to speculate what the fate of the *Guerrico* might have been if he had possessed one or more Milan launchers (the Milan having seven times the range and at least three times the explosive power of the 66 mm HEAT).

not photographed because none of the aircrew happened to have a camera with them on that day.

In short, Operation Paraquat was conducted without benefit of journalistic coverage, without any one officer being delegated as Task Group historian, and without coverage by a team of still photographers or video cameramen. This is not a criticism – it is simply a reflection of the speed and secrecy with which the operation was mounted.

The net outcome of all this was a rapid evaporation – after the initial euphoria – of media interest in the South Georgia episode. With Admiral Woodward's battle and carrier groups rapidly approaching the Falkland Islands – where 1,800 British nationals were suffering the outrage of foreign occupation – all eyes were turned in that direction. On British television, the endless debates between panels of retired senior officers concentrated upon the next phase: Would there be a sea battle? Would the troops land on East Falkland or West Falkland? Would the Argentines fight? How soon would victory come? With so much energy having been generated by those and subsequent debates, it is not surprising that Operation Paraquat – despite its extraordinary location and its brilliant success – has been neglected, even ignored, until now.

If the publisher and his author had not responded to Commander Paul Madge's suggestion by deciding to produce this book, it is very probable that the naval and military operations on and around South Georgia in 1982 would have remained largely unreported. Only Robert Headland, in his scholarly work *The Island of South Georgia* (Cambridge Univer-sity Press, 1984), and Cindy Buxton and Annie Price in their *Survival: South Atlantic* (Book Club Associates, 1983), have referred to these matters in any detail. For the rest, the Paraquat story would have been recorded only in official files and these will not be available to the general public until many years hence. The formal reports were in any case written in haste by officers who were about to plunge into the operations around the Falkland Islands and, as a consequence, they contain only the barest facts. Much of the information woven into the preceding narrative has been drawn, therefore, from the recollections of people who were there, or from the diaries which, in the teeth of standing instructions to the contrary, many of them fortunately maintained. There is an immediacy in these diaries – and a wealth of fine detail – for which there is no substitute.

One conclusion emerging from Operation Paraquat is that the armed forces might be well advised to appoint an 'archives officer' to every future task group. Ideally, he should be trained in the conservation and cataloguing of documents which could be later analysed, and should be someone with a sense of what might or might not be historically important. If he is allocated a photographer or video cameraman, so much the better. Events preserved in this way are of interest not only to aspiring authors but, more importantly, they permit detailed post-action analysis by everyone who seeks to learn from the operation.

MYTHS

Every battle and every campaign attracts its own mythology. A casual remark made in the heat of the moment, an imprecise newspaper report, the deliberate release of misinformation intended to mislead an enemy, any of these may subsequently be accepted as fact. Repeated often enough, the unlikeliest rumour can sometimes attain the stature of accepted truth. The war of 1982 was no different in this regard than any other military campaign. In particular, the battle at King Edward Point on 3 April generated a number of 'facts' which should now be laid to rest.

'Fact' one: when Captain Barker left his detachment of marines at King Edward Point on 31 March, he told Lieutenant Mills that, when the Argentines invaded, they should be met with 'token resistance'. The story continues by quoting Mills' response: 'Token resistance? No way! I'll squeeze their balls 'til their eyes water.'

This story is pure invention. Not only did Mills not make this remark, there was no reason why he should have even thought of it. At the time of the *Endurance*'s departure, the Argentines had not invaded the Falkland Islands, and few people in the ship really believed that they would go beyond the brink. Mills was working to the 'yellow card' rules and his main concern was the possibility of a confrontation with the Argentine party at Leith Harbour. Lieutenant Mills himself has no idea how this story originated.

'Fact' two: it is alleged by Argentine sources that Mills opened fire treacherously on their representatives who were coming ashore peacefully to accept his surrender.

This allegation is totally untrue. It is nothing more than an attempt to explain away the very poor tactics adopted by the Argentines. The unreliability of published Argentine accounts was exposed when Captain Trombetta's own story was published in *Boletin del Centro Naval* in June 1983. According to his version, the *Guerrico* never even entered King Edward Cove but remained in Cumberland East Bay and beyond the range of small-arms fire. If that had been so, one wonders how he could explain the damage she suffered from rifles, machine guns and short range anti-tank missiles? In his article, he admits to 'one dead and several wounded' in the *Guerrico*.

'Fact' three: another (and contradictory) Argentine version states that Captain Trombetta transmitted 'human rights guarantees in English' before attempting a landing on the island, and 'was told that there were no military personnel on it'. These phrases are quoted from an article published in the *Proceedings of the United States Naval Institute* in June 1982. It is evident that this otherwise highly respected organ accepted as truth that which the Argentines wished the publishers to believe. No more than a moment's reflection is required to understand that twenty-two men armed only with light weapons would be very unlikely to deny their own presence and thereby encourage a landing by such an overwhelming enemy force.

'Fact' four: during the battle of King Edward Point, on 3 April, the Carl Gustav rocket launched by Marines Stonestreet and Combes struck the sea fifteen yards short of the *Guerrico*'s hull. It then travelled in a straight line underwater and, acting as a torpedo, punched a hole deep below the waterline.

Stonestreet and Combes refute this story and the photographic evidence proves that it is untrue.

'Fact' five: the *Bahia Paraiso* was carrying three helicopters when she and the *Guerrico* attacked King Edward Point. Two of these – the Puma and an Alouette – were shot down by the Royal Marines.

The Puma was emphatically destroyed. However, there was only one Alouette and it survived with minor damage.

MORALE

Over a period of four decades, the armed forces of Great Britain had trained and worked in two overlapping but separate dimensions – to give aid to the civil power in colonies achieving independence, and simultaneously to maintain a posture of deterrence *vis-à-vis* a perceived threat from the Warsaw Pact countries. It had never been seriously contemplated that there might be an attack by an extreme right-wing regime. In a world of superpowers, such an event was almost unimaginable. Part of the shock of the Argentine invasions of the Falkland Islands and South Georgia devolved from the naïvety of the junta's attitudes. Their overt seizure of another nation's territory was an act of aggression which, in many ways, had become historically outmoded. It seemed almost as though Argentina's military rulers had failed to read a single history book or, if they had, had failed to understand what they were reading.

It was in the face of this alien situation that the British armed forces were ordered to mobilise. They were to travel to the other side of the world – ready to shoot and be shot at – but still in the expectation of a diplomatic settlement. To the military planners, the restraints imposed by the continuing negotiations were just one more complication to an already bizarre scenario. In such circumstances, it would have been not entirely surprising if the British commanders had wondered whether their men had been given sufficient time in which to mentally prepare themselves for war. The warrior needs to compose himself or, as the Bible has it, 'to gird up his loins'. Most of the men who took part in Operation Corporate – the reoccupation of the Falkland Islands – had nearly six weeks in which to adjust to the idea of the Argentines being 'the enemy'. Those who served in Operation Paraquat had less than three weeks.

It has never been the British way to attempt to indoctrinate the armed services with prepackaged attitudes towards other creeds or other political systems. Any such attempt would, in any event, most probably be met with derisory laughter. Morale is a quality easy to recognise, hard to define. The very high morale of the Paraquat force must in part have related to the quality of its leadership but, in the main, it reflected basic professional competence and a sense of rightness in a cause. Lieutenant-Colonel Keith Eve summarised his own impressions: 'There was a wonderful atmosphere in the ships, a sense of dedication to the task. Nobody had any doubts over what we were doing. It was a feeling of mutual confidence, of trust in the other chap and his ability to get on with his own job. And that is a battle winning factor.'

TRAINING

If there was a single lesson which emerged from Operation Paraquat, it was the importance and value of continuous in-depth training. At every stage and at every level, total familiarity with equipment and procedures shone through. There is no evidence to suggest that anyone fumbled the duties allotted to him. The quality of this superb training is underlined by the fact that it did not reduce any participant to the role of automaton. There was still room for individual commonsense and initiative. The temporary communications centre aboard HMS *Antrim* was a case in point. Likewise, the Fleet Air Arm maintainers in the *Antrim* threw away the rulebook when they manufactured a jury rig to mount a GPMG in their Wessex 3. The maintainers in HMS *Brilliant* used their initiative when they cut up two office swivel chairs to create jury GPMG mountings for the ship's Lynx helicopters. In the RFA *Tidespring*, Petty Officer Jim Humphries tried to overcome the ship's lack of defensive armament by inventing his own anti-aircraft weapon. He lashed two half pods of 2-inch rockets to a wooden pallet and fixed that to a helicopter towing arm. In an emergency, he intended to drag the contraption out onto the flight deck, point it over the stern and, from the shelter of the hangar, launch the rockets against incoming Argentine fighters by a home-made remote control electrical ignition system. He never had the opportunity to open fire, but the spirit was willing. Tony Ellerbeck and Tim Finding were similarly imaginative when they invented their own phantom anti-aircraft system – parking on high ground and preparing to fire AS12 rockets as 'spoof' surface-to-air missiles.

The *Brilliant*'s diving officer, Lieutenant Chris Sherman, tried to compensate for the lack of airborne weaponry by designing and building 'the Sherman bomb'. Consisting of a steel beer keg packed with plastic explosive and pieces of metal, it was fitted with two fifteen-second delay initiators and was intended to be dropped upon any suitable target which might present itself. Carried operationally by a *Brilliant* Lynx on at least one occasion, it was never employed in anger.

There are other instances which could be cited to show that the training of these officers and men was pitched at a level which did not prevent them from making their own individual contribution of ideas and *ad hoc* equipment. For a nation such as Great Britain, which will always have a relatively small number of men and women in the armed forces, imagination and drive are vital assets to be nurtured.

Lieutenant Nick Butler, one of the Lynx pilots serving in HMS *Brilliant*, summed up his own experiences: 'Paraquat was, for most of us, our first smell of gunsmoke, our first exposure to a shooting war. As aviators, we found ourselves doing few of the things we were trained to do, but doing a large number of things for which we had little or no formal training. In many respects, the rulebook simply went out of the window. The maintainers

and the aircrew were frequently operating in conditions which, in peacetime, would have been totally unacceptable, and using kit which was cobbled together as and when a need was identified. Although we could hardly believe it was all happening, everyone had tremendous determination to get involved. Such intense enthusiasm could have led to errors being made if the ingrained basic training – knowing what is workable and what is not – had not been as good as it was.'

After the return of HMS *Antrim* to the United Kingdom, her Wessex Flight 'Humphrey' was retired from active service to become a permanent exhibit at the Fleet Air Arm Museum, RNAS Yeovilton. In 1983, the gallant old warhorse's former aircrew and maintainers assembled at RNAS Portland for this formal photograph (staged in front of 'Humphrey's' replacement, Wessex XM328). *Standing (left to right)*: POACMN D. B. 'Fitz' Fitzgerald, AEM(M)1 S. V. 'Ozzy' Osment, LAEM(M) A. B. H. 'Tony G' Gigg, AEM(L)1 M. J. T. 'Mac' McGreal, AEM(R) Brian Kinsley, POAEM(L) B. P. 'Bungy' Williams. *Seated (left to right)*: POAEM(R) A. N. 'Tony' Davis, CAEA(M) D. J. 'Fritz' Heritier, Lieutenant C. J. Parry, RN, Lieutenant-Commander I. Stanley, DSO, RN, Sub-Lieutenant S. G. Cooper, RN, AEMN(R)1 J. E. 'Jim' McKee, POAEA(M) Lionel N. Kurn. Missing from this group is CPOAEA(L) T. 'Terry' Bullingham. He was severely wounded by shrapnel and splinters when HMS *Antrim* was attacked by Argentine Mirage aircraft in San Carlos Water. He received injuries to the arms and legs and lost the sight of both eyes. Petty Officer Fitzgerald and Lieutenant-Commander Stanley each received shrapnel wounds in the same attack. Also missing from the photograph are AEM(M)1 John Craven and LACMN Trevor M. Stapleton (who was loaned to the Flight by 826 NAS when Petty Officer Fitzgerald was wounded).

AVIATION

Helicopters were the one basic ingredient in Captain Young's planning without which he would have been unable to carry through the reoccupation of South Georgia. They were his principal vehicle for the cross-decking of men and stores, they were his principal eyes and ears, they were his principal anti-submarine weapon, and they were his sole practical means of delivering an assault force to the shore. All arms played an equal part in Operation Paraquat but, in the manner of the George Orwell story, some were more equal than others. The Fleet Air Arm helicopter was an irreplaceable component.

Despite the fully justified praise showered upon the Fleet Air Arm in the immediate aftermath of the South Atlantic conflict, it is a sad fact that it has been frequently a target for administrators seeking financial economies. As a consequence, the service has been obliged to fight a series of battles over the years for its very survival.

Today's Fleet Air Arm traces its roots back to the formation of the Kite and Balloon Sections in 1903 and the directive ordering the purchase of the airship *Mayfly* in 1909. It first acquired an identity of it own when, on 13 May 1912, it was established by Royal Warrant as the Naval Wing of the Royal Flying Corps. Two years later, on 1 July 1914, it became fully fledged as the Royal Naval Air Service. The RNAS became an effective and very large organisation during the Great War, but it was merged with the RFC on 1 April 1918 to form the Royal Air Force. For the next six years, naval aviation went into limbo.

The skills peculiar to flying over the sea, working with the Royal Navy, and operating against maritime targets, were all recognised by the formation of the Fleet Air Arm in April 1924. Further reorganisations during the 1920s and 1930s led to the strange arrangement whereby aircraft operating from naval bases and Royal Navy ships were flown and serviced by RAF personnel. It was an uneasy alliance, and the shortcomings were admitted when the Navy regained control of its own aeroplanes during the war of 1939–45. During that war, fifty per cent of all U-boats destroyed by the Allies were sunk by aircraft attacks. By 1945, the Fleet Air Arm was operating 1,300 aircraft with an establishment of 70,000 personnel, and the Navy was operating seventy-two aircraft carriers in a variety of classes.

The FAA led the world after the war with a series of revolutionary innovations – the steam catapult, pure jet fighters, mirror landing sights, the angled flight deck, and so forth. In 1947, the first naval helicopter squadron – number 705 – became operational in the 'plane guard' role. The helicopter was viewed at that time as being essentially a rescue vehicle for pilots whose fixed wing aircraft had fallen into the sea. The flexibility of the helicopter was displayed nine years later when Fleet Air Arm helicopters were a major component in the initial assault at Suez.

Trials showed that these versatile aircraft could carry a variety of weapons for attacking surface and sub-surface targets, and could also be fitted with sonar gear to track hostile submarines. The FAA developed into four principal groupings: fixed wing jet strike aircraft (epitomised by the Sea Harrier units which fought over the Falkland Islands), 'pinger' helicopter units (Naval Air Squadrons), 'jungley' helicopter units (Naval Air Commando Squadrons), and the Small Ships Flights of small strike helicopters which provide the eyes, ears and sting for the frigates and destroyers of the Royal Navy. Individual skills are today maintained at a high pitch by providing FAA aircrew with seaman branch appointments so that they may better comprehend the role of the mariner.

The experiences of the past seventy years have demonstrated time and again that a strong, well-equipped and highly trained Fleet Air Arm is a vital and integral element in the country's defences.

Some indication of the value of helicopters to the South Georgia operation can be gauged from the following letter, composed hastily on 27 April as the Task Group started to disperse, addressed to Lieutenant-Commander Stanley and the crew of Wessex 406 ('Humphrey'): 'Dear Ian and company – From all of us in "D" Squadron, many thanks to you all. The events of the last few days might have been lifted from a boys' adventure annual. Rescuing troops from a gale lashed glacier, from a dinghy swept out to sea, not to mention attacking an enemy submarine, all these have been your lot. Our indebtedness to you is great. There is no doubt that, without you, the successful outcome of the operation would have been soured by particularly high losses amongst the Squadron. So successful have you been that 406 was coming to be viewed as a cure-all. We wish you all good fortune – but take care. With luck we will meet again before too long. When we get back, and if it is possible, we would like to get you up to Hereford and thank you more appropriately. On behalf of us all, thanks. Cedric.'

WEAPONRY

The two weapons which won the battle for South Georgia were almost ludicrously old-fashioned. The free-falling depth charge – very similar to the Mark 11 which stopped the *Santa Fe* in her tracks – had first been used against German U-boats during the war of 1914–18. The breech-loading rifled-barrel gun – which so swiftly intimidated the Argentine garrison into surrender and of which the Mark 6 turret was basically no more than a semi-automatic refinement – had first entered general service with the Royal Navy in the 1880s. This is not to suggest that traditional weapons should remain in service for all time, or that high-technology weapons should be rejected. On the contrary, innovation has been a keynote over many decades in the history of the

The 105 mm recoilless rifle found on Hope Point after the Argentine surrender, with Warrant Officer Paul Juleff, RM, about to 'light the blue paper'. This weapon, and others like it, could have inflicted very serious damage on any British ship careless enough to come within its range.

Royal Navy and its aviation branch. However, and this applies particularly to ship-launched weapons, the 1982 conflict did demonstrate the continuing value of conventional naval gunfire in support of military operations on land. The rifled gun can deliver explosive to a given area with a combination of persistence and accuracy which, at the present state of the art, is not possible with the more expensive surface-to-surface missile.

On the Argentine side, one weapon which could have caused serious problems for the Paraquat ships was a 105 mm recoiless rifle. Said to have been an American design, it was found after the reoccupation on Hope Point, together with thirteen rounds of ammunition, by men of 'M' Company. This was a very potent weapon. Company Sergeant-Major Juleff could not resist the temptation to try it out. Having taken it down to the tip of King Edward Point, he pointed it across the Cove and fired off a round which impacted on the far shore with a large explosion. The rearward blast of the propellant blew in all the windows of a nearby building – much to the surprise of some marines working inside. As an observer commented: 'I'm glad they never used it against us. As far as we could judge, it had a range of 1,500 metres. The *Antrim* was stuffed with Sea Cat and Sea Slug missiles. If one of those things had come whacking through the side of her hull it would have made a hell of a mess.'

A cause for concern was the report that the Argentines might have laid sea mines in the approaches to King Edward Cove. The origins of this report are of some interest and they relate to the peace initiatives conducted by General Alexander Haig. During one of his visits to Buenos Aires on behalf of President Reagan, he had conversation with a former senior Argentine official who, referring specifically to South Georgia, boasted that 'the British will never reoccupy it – we have plenty of mines to block the bays'. This comment was passed to Downing Street. The Prime Minister recognised the significance of the information and it was incorporated in the initial operational directive for Paraquat. Northwood did not regard the presence of mines in South Georgia waters to be very likely because it was doubted that the Argentines would have had time – or possessed the means – to lay them. Nevertheless, the risk existed and the information was passed to Captain Young. It was found, in the event, that the Argentines had not managed to mine South Georgia waters, but they did lay mines off the Falkland Islands and at least twenty were swept up by the requisitioned trawlers of 11th Mine Countermeasures Squadron, Royal Navy.

The most powerful weapon in the Argentine armoury was the submarine *Santa Fe*, but she was never properly employed. When she departed Argentina, sixteen days earlier, the approaching British ships were still undetected. Initially, therefore, Bicain's role was simply that of transporting twenty marines as reinforcement for the King Edward Point garrison. Once it became known that Captain Young was in local waters, it might have been better if Bicain had deferred the disembarkation and instead concentrated upon the much more important objective of attacking the ships. By berthing at King Edward Point he compromised his boat's safety without materially influencing the military balance on shore. He further sealed his own fate when he failed to conceal his presence in South Georgia waters and proceeded to leave Cumberland East Bay on the surface in daylight.

With hindsight, it is interesting to consider what would have happened if he had succeeded in crippling or sinking a Paraquat ship before landing his passengers. After his capture, Captain Bicain claimed that he had had an opportunity to sink the *Endurance*, that he had seen her through his periscope off Hound Bay. When asked by Captain Barker why he had not attacked, Bicain replied: 'Ah, well, I remembered that cocktail party in Mar del Plata. I just couldn't do it.' The claim may or may not be based upon fact, but it lost some of its validity when Bicain subsequently claimed that he had also seen the *Tidespring* and the *Brambleleaf*. His stated reason for not firing torpedoes at these two prize targets was that 'they were zig-zagging at the time'.

EQUIPMENT

This is a very large topic. The author is not qualified to comment upon all aspects of the technology of Operation Paraquat, and some of the lessons learned are in any case subject to the restraints of national security. However, to touch lightly upon the subject, it is appropriate to refer to some of the more obvious points which have emerged in the preceding pages. First, it is ironic that the Paraquat ships were unable to make full use of many of the latest advances in electronic technology. The problem with both radar and radio is that, used incautiously, they reveal to the enemy your position, course and speed. Despite his twentieth-century assets, Captain Young was obliged to resort to the 'mark one eyeball' for much of the time. Admiral Horatio Nelson would have been no better served in this respect.

Then there was the sore subject of the outboard engines used by the SAS and the SBS to propel their Gemini inflatable boats. The four ships constituting the Paraquat force represented a capital investment of many millions of pounds, but the entire project almost ended in disaster for the lack of eight reliable motors costing a few thousand. In an age when a new weapons system may cost billions, there is still room for sound basic kit.

In the months following the conflict, considerable criticism was levelled at the designers of the Royal Navy's ships for having created 'one-hit' eggshells. The Ministry of Defence itself admitted to 'over two hundred desirable modifications' as a result of the conflict. This is a large and contentious topic. However, as far as Paraquat was concerned, the *Antrim* and the *Plymouth* both withstood sea conditions which their designers are unlikely to have ever contemplated. Due credit should be paid not only to the design teams but also to the yards who built them (Fairfields of Govan in the case of HMS *Antrim*, and Her Majesty's Dockyard, Plymouth, in the case of HMS *Plymouth*).

Helicopters were essential components in the South Georgia operation. Again, there is a touch of irony to the fact that the two types which played key roles – the Wasps (mainly from HMS *Endurance*) and the Wessex 3 (from HMS *Antrim*) were both verging on obsolescence. In their day they had been excellent machines. That they were able to perform as they did in 1982 is perhaps more a tribute to their pilots, crews and maintainers than to the aircraft themselves. The lack of modern navigation equipment in the two Wessex 5 lost on the Fortuna Glacier suggests, however, that courage and flying skills are not always enough.

COMMUNICATIONS

Secure radio communication is one field in which modern technology proved to have great value during the 1982 conflict. The development of satellites has revolutionised the speed and clarity with which messages can be passed over great distances. Senior officers can talk to each other in 'live voice' as though using a conventional telephone at home (and with less chance of a third party overhearing their conversation). The system permits the discussion of complex problems which, if transmitted via traditional types of radio equipment, would take longer to resolve. There is also the chance to hear inflexions of the voice, often an important factor in comprehending the other person's viewpoint.

A potential peril with communications satellites is that they could, if used unwisely, permit a senior commander, far removed from the scene of action, to ring up every couple of hours to enquire into progress. This did not happen during Operation Paraquat. When asked if Northwood had indulged in any 'back seat driving', Captain Young replied with an emphatic 'No.'

One lesson which did emerge was the need for a more comprehensive radio fit in the navy's destroyers and frigates so that they might be better able to handle a multiplicity of communications tasks. It is argued that there could be other occasions when Royal Navy ships might find themselves involved in a small-scale amphibious operation such as Paraquat, but without a preliminary period of several days during which 'lash up' arrangements could be made (as happened aboard HMS *Antrim*). This scenario need not necessarily involve actions against a hostile force – it could be associated with the rescue of British citizens from a civil war, or a rescue operation subsequent to a natural disaster such as a typhoon or earthquake.

'M' COMPANY, 42 COMMANDO ROYAL MARINES

'M' Company emerged from the Anglo-Argentine conflict with a strong sense of frustration. The first such unit to arrive in the South Atlantic, they never had the opportunity to exercise their skills in a direct confrontation with the Argentines. For any fighting man, to be left always on the outer fringe of great events is an unhappy arrangement. Captain Nunn and his people rendered excellent service as garrison force on South Georgia following the Argentine surrenders, and they did all that was asked of them during Operation Keyhole, but the greater glories eluded them.

The original plan had anticipated that it would be they who would seize South Georgia back from the invaders. The fact that 'D' Squadron 22 SAS and 2 SBS happened to be already in the South Atlantic, and happened by chance to be available for inclusion in Major Sheridan's land force, was for the marines sheer bad luck. If the *Fort Austin* had been engaged in other operations – perhaps around the Falkland Islands – 'M' Company would have had a clear run at South Georgia. Every man was Mountain and Arctic Warfare trained and well accustomed to surviving in an environment not unlike that found on South Georgia. The Reconnaissance Troop was specially trained for the types of initial clandestine insertion which, in the event, was taken over by 2 SBS and the SAS Mountain Troop. There are some Royal Marines who argue that the Reconnaissance Troop

While the bulk of 42 Commando RM was fighting on East Falkland, the frustrated men of 'M' Company were confined to the limited horizons of South Georgia's Cumberland East Bay. Daily patrols around King Edward Cove and Grytviken (where only the fin of the *Santa Fe* now remained above the surface) soon lost their novelty value.

was indeed better qualified than the SAS to tackle the perils of landing at a site such as the Fortuna Glacier. The Troop's performance during Operation Keyhole is offered as evidence to support this argument.

The obverse (left) and reverse (right) of the South Atlantic medal. Struck in cupro-nickel (instead of the more appropriate silver), its design, manufacture and distribution were influenced, it would seem, with economy as the primary consideration. This was at a time when the annual cost of garrisoning the Falkland Islands was calculated at nearly £500 million and the cost of the new Mount Pleasant airfield at £30 million. According to newspaper reports, the Prime Minister was dismayed when she discovered that medals sent to the next-of-kin of men killed in the campaign were being despatched in cheap brown envelopes accompanied by an impersonal and unsigned printed letter.

BRITISH ANTARCTIC SURVEY

From the details contained in Chapters IV and V, it can be seen that the BAS comes high on the 1982 list of unsung heroes. At Cambridge, in Antarctica, and on South Georgia itself, BAS personnel devoted thousands of man-hours to the collation of information which was – or could have been – useful to the Task Group commanders. The Royal Research Ships *John Biscoe* and *Bransfield* were important elements – the former as transport for the new Royal Marines Naval Party 8901 to Port Stanley in late March, the latter as a vital link in the BAS communications chain after the invasions. Often at considerable hazard to themselves, BAS personnel voluntarily provided information needed by the Northwood planners.

None of those who served in the operational area – either aboard the *Bransfield* or on South Georgia – received the South Atlantic medal. When representations were made to the Foreign Secretary, at that time Mr Francis Pym, he is quoted as having stated that 'there is no provision in the Royal Warrant for the award of this medal to civilians.' This, of course, is not quite so. A great number of medals were awarded to civilians who had served in RFA and STUFT ships. The medal was awarded to all the civilians who accompanied the Operation Corporate force in their professional capacity as journalists, television camera crews, photographers, and, in at least one instance, as war artist. Those who travelled below the latitude of 35 degrees south received the medal 'with rosette' (there being forty-eight such awards). Civilians employed on Ascension Island as Ministry of Defence public relations staff and as civil service engineers also received the medal, but 'without rosette' (there being fifteen such awards).

Dr Richard Laws was appointed a Commander of the British Empire in the 1983 New Year Honours List, presumably as recognition, at least in part, of his personal contribution and as a tribute to his colleagues. But it would have been a pleasing and inexpensive gesture if the government had struck a few extra medals for Captain Lawrence and his crew, and for the field parties who reconnoitred South Georgia's coastline just before and during the Argentine occupation. Now that their story has for the first time been told, perhaps a magnanimous government might see fit to reconsider their position.

The greatest mystery to have emerged from Operation Paraquat is the fate of the reports forwarded by the field parties to Cambridge after 5 April. From the detailed account in Chapter V, it can be seen that there was a regular flow of information from the field parties which related directly to the conduct – and lack of activity – of the Argentines at King Edward Point and Grytviken. It was not possible to observe the Stromness Bay area, but it was reasonable to assume that, if the former group were militarily slothful, then the latter might be similarly inactive.

None of this information ever reached the Paraquat commanders. The principal links between the BAS and the Ministry of Defence were Dr Laws, the BAS Director in Cambridge, and Admiral Tony Whetstone, the Assistant Director of Naval Operations (Sea) in Whitehall. The two men were in regular contact every day. Laws knew nothing of the secret Paraquat force, but the Navy's interest in the island made it plain that something might be afoot. Admiral Whetstone rightly regarded the BAS reports to be 'of considerable value'. He referred to them in telephone discussions with Vice-Admiral David Hallifax, Chief of Staff to Admiral Fieldhouse, and he also circulated the salient points in Dr Laws' reports to the officers concerned within the Ministry of Defence. That they were brought to the attention of the Commander-in-Chief is evident from his later comment that they were 'very useful'. What happened to them thereafter, within the Northwood structure, must be a matter for conjecture. However, apart from the initial estimate of 'about eighty men at Grytviken' – which was presumably the result of the reconnaissance conducted by Damien Sanders and Ian Barker on 5 April – no further reports were passed to the Task Group.

When first told – long after the end of the conflict – of the activities of the field parties, Captain Young, Major Sheridan, Major Delves and 'Clive' of the SBS all expressed surprise. They had each assumed that the field parties were confining their movements to the immediate vicinity of their huts while passively awaiting rescue. The significant patrols by the St Andrews Bay party – extending to the Barff Peninsula where the SBS intended to land – went unreported. 'If we had known that the Argentines were apparently all indoors making toast, we might have planned the insertions a bit differently,' was one comment.

A naval officer then employed at Northwood has said: 'We got lots of useful stuff about the natural characteristics of the island, but nothing specific about what the Argentines were up to. It was all negative information, and negative information isn't really very useful.' This is a debatable view. However, it is powerfully supported by Admiral Whetstone: 'My opinion remains that the [SAS/SBS] reconnaissances were not necessary to establish that there were large areas of the island where the BAS – and therefore our own forces – could wander at will, but to determine, without alerting the enemy, the detailed Argentine dispositions so that the final attack could be delivered effectively and with minimum casualties. It is worth making the point that the force allocated to Captain Young was not enormously superior to the believed size of the Argentine garrison. Had our estimates been incorrect, the need for detailed and accurate reconnaissance would have been even more important. Having said that, I can understand how it was that some of the available information did not get through to Captain Young. The Northwood planners gave his Task Group their immediate attention between 7 and 14 April because it would be the first force to get into action. However, once his operational orders had been approved by Admiral Fieldhouse and despatched by air for direct delivery, he was left to get on with it. Everyone's thoughts then became concentrated upon the main project – Operation Corporate. Young was a very senior and experienced captain. There was great confidence in his abilities and the quality of the people around him. It may – and one can only hypothesise – may have been felt that he knew all about the field parties and their activities, so it was not thought necessary to send him information presumably already in his possession. From personal experience, I know it is too easy to believe that subordinate units know more about the situation in the front line from direct contact with each other than is the actual case.'

The latter comment is an allusion to the ability of HMS *Endurance* to monitor the field parties' civilian radio traffic during the time she was in the vicinity of South Georgia. This was the case for only part of the time. Between 5 and 21 April the *Endurance* was away from the island, sailing more than 2,000 miles north to replenish from the *Fort Austin* and to join the *Antrim* group. It was during this period that the BAS made some of their most useful coastal surveys and obtained information which might have resolved some of the dilemmas facing Major Sheridan, Major Delves and 'Clive'. But not everyone shares this view. There are those who argue that the military plan could never have been influenced significantly by the BAS negative findings because they were not open to verification.

This line of reasoning is supported by Major Delves: 'Upon reflection, and although it would have been useful to know about the BAS before we made our final decisions, I think we would still have stuck to the adopted plan. We would have been obliged to treat any BAS negative reports with a high degree of caution. They could have been transmitted under duress, for all we would have known. They might have been saying "This stretch of coast is clear, and this and this" with, at the same time, an Argentine pistol pointed at the operator's head. We might have then walked smack into an ambush. So, yes, it is surprising that the information did not reach us, but perhaps it didn't make that much difference in the long run.'

For his part, Major Sheridan believes that even negative information, albeit transmitted via an insecure civilian radio net, would have been useful as general background to the tactical planning process. 'We were very much in the dark about the Argentines – just clutching at straws. Under those circumstances, you want all the information you can lay hands on. How you evaluate it, and how you use it, is another matter. It all depends upon the source and the quality. But I wonder if I would have agreed to the Fortuna Glacier insertion if I had had reason to believe that the Argentines were not actively patrolling the coast? Don't know. Impossible to say. Once again, its twenty-twenty vision.'

HMS ENDURANCE

Reference to Chapter II reveals some details of the long-running public debate regarding the fate of HMS *Endurance*. The Secretary of State for Defence, Mr John Nott, defended his decision to sell or scrap this venerable ship by stating, in a letter dated 12 January 1982, that she possessed 'only a limited military capability'. In terms of speed, armour and armament, this was a plain statement of fact which could not be disputed. However, her very important role in the events of the five months following the drafting of that letter demonstrate that maritime power does not depend exclusively upon the nuclear submarine, the modern frigate, or the new breed of aircraft carrier.

Captain Barker and his ship's company can fairly claim to have completed the following military (and naval) tasks with conspicuous success:

1. between 24–31 March, the clandestine surveillance of Leith Harbour following the illegal Argentine landings from the *Bahia Paraiso*.

2. the continuance of international radio monitoring services in the South Atlantic pending the arrival of the Operation Paraquat and Operation Corporate forces.

3. on 2 April, aiding in the defence of Port Stanley (when her Survey Team served alongside the Royal Marines of Naval Party 8901).

Battered but unbowed, the RFA *Tidespring* returns to Portsmouth on 22 July, much of her paintwork stripped by the South Altantic weather. The midships structural damage occurred when she struck the harbour wall at Gibraltar in early April.

4. on 3 April, destroying an Argentine Puma helicopter, damaging an Alouette helicopter and crippling a small but powerful enemy frigate, the *Guerrico*. The same action inflicted numerous casualties upon the South Georgia invasion force.

5. the provision of a forward operating base for the SAS and the SBS, and the insertion of the latter at Hound Bay.

6. the rescue of the abortive SBS patrol.

7. the early detection of the submarine *Santa Fe*.

8. participation in the helicopter attacks on the *Santa Fe* (her two Wasps making four separate sorties, launching seven AS12 missiles and scoring four hits).

9. participation in the action for the recovery of Leith Harbour.

10. acting as command ship for CTG 317.3 and – during a critical phase of the campaign – acting as unsupported guardship for South Georgia.

11. acting as command ship for Operation Keyhole.

12. carrying through the first phase in the operations for salvaging the *Santa Fe*.

The most critical period of Operation Paraquat was 23 April when Argentine air force reconnaissance flights first detected the Task Group, and the *Santa Fe* was identified as an immediate threat. The three other ships of the Paraquat force prudently withdrew 200 miles from the coast, but 'the plum' was judged well able to fend for herself.

On 14 October 1983, HMS *Endurance* was presented with the Wilkinson Sword of Peace for 1982. This prestigious award is given each year to those who, in the view of the donors, have made an outstanding contribution to international peace and goodwill. The award to the *Endurance* was self-evidently not a reflection of her involvement in Operation Paraquat but was instead a recognition of the ship's role, over a period of many years, in sustaining the welfare of the farming communities on the Falkland Islands and the many scientific bases in the Antarctic. The citation makes specific reference to the ship's contribution to the work of the British Antarctic Survey and the Scott Polar Research Institute, both of which are important members of the larger worldwide scientific community. One might conclude that HMS *Endurance* is remarkably effective in times of both war and peace. At the present time (1986), due to commence a major refit, costing £18 million, in Her Majesty's Dockyard, Devonport.

THE ROYAL FLEET AUXILIARY

The RFA is a civilian-manned fleet, formed in 1905, owned and operated by the Ministry of Defence. Its main task is to supply warships of the Royal Navy at sea with fuel, food, stores and ammunition which they need to remain operational while away from their land bases. The RFA also provides aviation support for the Navy, together with amphibious support and secure sea transport for the army.

The RFA employs 2,300 United Kingdom officers and ratings and 250 Hong Kong Chinese ratings. Although they serve under conditions of contract similar to those signed by Mercantile Marine employees, they are also bound (since 1982) by clauses which relate to the warlike hazards to which RFA ships are likely to be exposed.

At the present time (1985) there are thirty operational ships in the RFA fleet: fifteen fleet and support tankers, five dry cargo replenishment ships, five logistical landing ships, two roll-on roll-off logistical ships, one forward repair ship and two helicopter support ships.

Twenty-two RFA vessels were deployed in support of the Royal Navy and land forces during the South Atlantic campaign of 1982. Four of these, the *Fort Austin*, the *Tidespring*, the *Brambleleaf* and the *Olmeda*, were involved directly in Operations Paraquat and Keyhole.

The scale of this support is reflected in the activities of the RFA *Tidespring*. Without claiming to have been the busiest such vessel during the conflict, she amassed some impressive statistics during the period following her departure from Ascension Island until her eventual return to Rosyth. She conducted 42 replenishments

Captain Shane Redmond, Master of the RFA *Tidespring*. He had already taken part in one military campaign, having served in the Eastern Mediterranean in 1956 during the operations against Egypt. In 1982 his command was given the unusual combined roles of Operation Paraquat Task Group tanker, troop transport, amphibious assault ship and prison transport.

at sea, pumped over 6,050 tons of fuel and loaded 6,600 tons, delivered 6,800 gallons of lubricating oil, loaded and delivered 145 tons of ammunition, produced 1,508 tons of fresh water and delivered 395 tons, handled 266 helicopter landings on the flight deck, organised nearly 1,000 helicopter freight loads (vertreps), and served 19,843 meals. The radio room handled 2,300 signals (ten times the usual traffic) and used 300 miles of teleprinter paper. The engine room burned nearly 2,000 tons of fuel oil (at a cost of half a million pounds) to turn the propeller 3,558,090 times. To quote her Chief Officer, Alec Bilney: 'It was the most important period of my life. Everyone in the ship felt the same way. I hope nothing like it ever happens again but, while it lasted, it was exciting and very satisfying.'

The first British surface vessel to head south in March, 1982, was the RFA *Fort Austin*. Her Master, Commodore Sam Dunlop, MBE, had first gone to sea in 1939, and nearly forty years of his mercantile service was with the Royal Fleet Auxiliary. He is therefore uniquely qualified to assess the role of the RFA in the context of the 1982 campaign. 'You must understand that the RFA has undergone great changes since I first joined in 1943. At that time we were regarded primarily as freighters and station oilers. Replenishment at sea was, however, carried out, and RFA ships took part in practically every Navy operation, not least the Malta and North Russia convoys. It was a primitive and laborious evolution carried out with five-inch bronze pipes. Then we captured a German ship fitted with full replenishment at sea [RAS] equipment. We adopted the German system, the main advantage of which was the use of buoyant rubber hose instead of the old bronze pipe. The majority of RFA tankers were modified to operate the new gear and it was used with limited success by the Fleet Train to the British Pacific Fleet in 1944 and 1945. You could say that the foundation of the modern RFA came in 1955 with the advent of the *Tide* class tankers and the armament ships *Retainer* and *Resurgent*. Up until the 1960s, they and other RFA ships could defend themselves because they carried guns stowed below decks. In an emergency, these could be brought up and sited on the emplacements which were standard fitting on all RFA ships at that time. We lost that capability in the mid 1960s. The new official policy was: 'You remain firmly Merchant Navy.' This meant that, when *Fort Austin* sailed from Gibraltar in late March to support HMS *Endurance*, her only armament was a few line-throwing rifles and my personal 12-bore shotgun. It is to the never-ending credit of the RFA personnel that – in spite of not having been prepared either mentally or physically for that type of operation – their contribution to the 1982 campaign was so successful.'

There have been some major changes in RFA policy as a consequence of the 1982 campaign. The ships will no longer be registered as merchant vessels, special terms of service have been agreed for the civilian crews, point defence systems have been fitted on some ships, and there is a closer working relationship with the Royal Navy. A revitalised Royal Fleet Auxiliary is one of the positive results of the Anglo-Argentine conflict.

The hazards of service with the Royal Fleet Auxiliary in both wartime and peacetime were underlined in August 1984. Heading into the type of rough weather encountered so often in the South Atlantic, the RFA *Olwen* was hit by a freak wave seven hundred miles north of the Falkland Islands. Seven men working on the forecastle were swept down the length of the deck by a solid wall of water. Four were seriously injured and two killed. One of the latter was John Hocking, the carpenter who had done so much to assist the Fleet Air Arm aboard RFA *Tidespring* in April and May of 1982.

UNITED TOWING

Three ocean-going tugs were involved in operations off South Georgia and Thule Island. They were the *Salvageman* (Captain Alan Stockwell), the *Yorkshireman* (Captain Peter Rimmer), and the *Irishman* (Captain Tony Allen). All three were owned by United Towing Limited, a company established for over sixty years and a member of the North British Maritime Group. Based in Hull, North Humberside, United Towing already had a worldwide reputation prior to the events of 1982 for its expertise in towage, salvage and offshore oil industry operations. All of these skills were needed in full during the Anglo-Argentine conflict and for many months afterwards.

The *Salvageman*, the first commercial vessel to reach the war zone, entered Cumberland East Bay on 7 May. The most powerful tug operated by any British company, she displaced 1,600 tons and possessed an ice-strengthened hull. Her four Ruston main engines generated a combined 11,280 brake horsepower through twin propeller shafts, and she also carried various fixed and portable generators, compressors, pumps and fire-fighting equipment which made her a valuable component in logistical and salvage operations in the South Atlantic. She formed part of the Operation Keyhole force in June because, as Captain Nick Barker put it: 'It was a comfort to have her eleven thousand horsepower on hand when we were going down towards the pack-ice. I was counting on her to rescue the other ships if they got into difficulty.'

The *Yorkshireman* reached South Georgia on 7 June, via Tristan da Cunha and the Falkland Islands Maritime Exclusion Zone. She did much useful work in Cumberland East Bay and Stromness Bay, aiding merchant ships to berth and assisting with the transfer of personnel and equipment from ship to ship and ship to shore.

The 700-ton *Irishman* departed Portsmouth on 10 April and reached Tristan da Cunha a month later. Her services in Falkland

United Towing's biggest tug, *Salvageman*, shares an anchorage with HMS *Endurance* off Hewison Point on 20 June, while HMS *Yarmouth* patrols offshore. The intense cold is reflected in the fact that breaking waves have frozen solid on the rocky foreshore.

Island waters included a heroic forty-eight hours effort to take in tow and salvage the stricken *Atlantic Conveyor*. The tug moved to South Georgia on 7 June where she worked in tandem with her sister ship *Yorkshireman*. All three of these workhorse vessels played an important part in turning South Georgia – during the critical weeks of May and early June – into an efficient forward operating base.

HEAVYLIFT CARGO AIRLINES LIMITED

Helicopters were indispensable to the success of Operation Paraquat. The initial SAS and SBS reconnaissance patrols were inserted – and subsequently rescued – by helicopter. The principal threat to the Task Group – the *Santa Fe* – was destroyed by helicopters. The final assault at Hestesletten was delivered by helicopter. As shown in Chapter IV, the two Wessex 5 of 'B' Flight 845 Squadron were essential in the early stages – when the *Fort Austin* replenished the *Endurance* – and the two Wessex 5 of 'C' Flight would have been the keystone to the main landing if they had not been lost on the Fortuna Glacier.

As a matter of record, therefore, it is appropriate to mention that the four Wessex 5 would not have been available to take part in these operations without the swift collaboration of a civil airline, Heavylift Cargo Airlines Limited of Stansted, Essex. This company's load-lifting Belfast aircraft were delivering the first helicopters to Ascension Island within thirty-six hours of receiving a call from the Ministry of Defence. Other loads followed later, but the mounting of Operation Paraquat would have been delayed by many days if Heavylift and its small specialised fleet of big aeroplanes had not been available.

Heavylift was formed in 1978 as a joint venture between the Trafalgar House Group and Eurolatin Aviation. The company hoped to make use of the Belfast aircraft which Short Brothers had built originally to a military specification. Ten examples had been constructed and all were taken into service by the Royal Air Force. Operated by 53 Squadron RAF between 1966 and 1976, the Belfast was the first military aircraft in the world with a fully automatic landing system. It could carry 150 armed troops, a Chieftain tank, or two Saladin armoured cars. The type was withdrawn from RAF service following cuts in Defence expenditure and all ten examples were sold off. Five were purchased by Heavylift and, following exhaustive trials and proving flights, civil certification was obtained in 1980. By 1982, three were in regular freighting service and two were available in reserve.

Heavylift soon developed a market with oil and aerospace companies requiring the fast delivery of bulky high-value cargoes. The company also sought Ministry of Defence contracts between 1980 and 1982, but without success. The possibilities had not at the time been recognised in Whitehall. The operators were therefore delighted with the opportunity which presented itself during the conflict with Argentina to display its capabilities. Subsequently there has been a regular flow of additional contracts from the Ministry, particularly for the international movement of large helicopters.

Heavylift still own five Belfasts and a sixth is on permanent display at the Aerospace Museum, RAF Cosford. It wears the company's livery in recognition of the airline's contribution to the events of 1982. The four other examples of this impressive aeroplane have been broken up.

THE MERCANTILE MARINE

Although no merchant navy ships participated directly in Operation Paraquat, it can be seen from Chapter IX that a considerable number did make use of the anchorages in Stromness Bay and Cumberland East Bay during the weeks following its successful conclusion.

Forty-seven ships were requisitioned or chartered by the Ministry of Defence during the 1982 conflict. The National Union of Seamen estimates that approximately 3,000 of its members volunteered and served in these ships. All of those who entered the operational area were eligible to receive the South Atlantic medal. Six merchant navy personnel lost their lives in the conflict, all being killed when the *Atlantic Conveyor* was hit by an Exocet missile.

The long decline in the size of the British merchant fleet – attributed by some commentators to the lack of government financial support which leaves ships sailing under the Red Duster at a commercial disadvantage compared with the merchant fleets of many other nations – did not prevent British owners from responding to the government's call in April 1982. Commenting upon its role, Admiral Sir John Fieldhouse said that, without the merchant fleet's contribution, the campaign could not have been undertaken. 'I hope this message is clearly understood by the British nation,' he added.

The importance of the message is underlined by the following statistics of hulls available and suitable for defence purposes.

	1980	1981	1982	1983	1984
Tankers	118	100	95	84	78
Cargo ships	310	246	204	163	150
Passenger/ferry ships	170	158	145	144	145
Container ships	74	59	56	55	56
Trawlers (1,000–1,999 GRT)	37	35	23	18	16

There are several reasons to explain these figures, but 1982 did emphasise the fact that some of those ships which do in future enter British service should be designed and constructed in such a way that they can be swiftly converted to defence needs in time of crisis. Much effort is now being devoted to the study of ways and means of utilising tanker decks for Sea Harrier operations, installing helicopter flight decks on cargo and ferry ships, and creating a stock of point defence 'surface to air' missile packs for rapid installation on all ships bound for an operational area.

LINES OF COMMUNICATION

The two stepping stones which permitted the British to cross the Central and South Atlantic Oceans so swiftly during the early days of April were Dakar Airport and Wideawake Airfield. If the government of Senegal had not granted access to their big runway and refuelling facilities, it is difficult to see how the RAF VC10 and Hercules aircraft – and the contracted civil Belfast freighters of Heavylift Cargo Airlines Limited – could otherwise have made the rapid initial transfers of men and materiël described in the opening pages of Chapter VI. By the same token, if the American facilities on Ascension Island had not been made available, Operation Paraquat would not have taken place. Indeed it is questionable whether Operation Corporate itself could have been mounted.

The fact that the Senegalese and American governments were so responsive to the British dilemma suggests that somebody in the Foreign and Commonwealth Office was remarkably persuasive. There is an untold story here, and no doubt it will one day emerge.

Censored by the Ministry of Defence.

When considering the implications of the United Kingdom's triumph in the South Atlantic, it is appropriate to bear in mind the importance of retaining as many friends as possible in every corner of a rapidly shrinking world. Looking to the future, it is also worth remembering that Ascension Island is a cluster of currently dormant volcanoes. The explosion of Mount St Helens, Washington State, in 1983, was a reminder that Nature has the capability to interfere quite suddenly and comprehensively with British plans for the future of the Falkland Islands and the Dependencies.

THE GENEVA CONVENTIONS

The Conventions have their origins in the Second War of Italian Independence. In June of 1859 the Austrian Army, under Emperor Franz Josef, clashed with the French Army, led by Emperor Napoleon III, near the small town of Solferino, just south of Lake Garda. A quarter of a million men, with nearly 800 field guns, met in pitched battle. It was the first engagement in which modern artillery – rifled-barrel breech-loading pieces firing high explosive – was used with full force. The impact against infantry formations, still manoeuvering in the style of Waterloo, was catastrophic. At the end of the day, 40,000 soldiers were dead or wounded. The army surgeons could not cope with casualties on this scale and the majority of the injured were left to die – or be murdered by scavenging civilians – on the field of battle.

The human misery witnessed at Solferino shocked world opinion. It was evident that modern warfare called for a complete reassessment of the manner in which battle casualties were treated. The Swiss took the initiative by inviting other nations to attend a meeting in Geneva in 1864. It was agreed that the neutrality of the personnel and equipment of military hospitals would be guaranteed in time of war. Following that early step – the first of the Geneva Conventions – most of the world's governments have signed further agreements which seek to apply the same humanitarian principles to the treatment of sick and shipwrecked naval personnel, prisoners of war, and civilians caught up in acts of war. The current Conventions date from August 1949 and they grant a 'statutory right of initiative' to the International Committee of the Red Cross. The current English language edition of the Conventions runs to 245 pages – a fair indication of the scale and complexity of this subject.

As far as Operation Paraquat was concerned, the Conventions were relevant on a number of occasions. An obvious instance was the medical treatment given to the Argentine sailor whose leg was smashed in the *Santa Fe* attacks. The surgery and nursing which he received aboard HMS *Antrim* and the RFA *Tidespring* were precisely the same as if he had been a British sailor. The spirit of the Conventions was respected when Captain Bicain's crew were asked if they would help to move the submarine from King Edward Point to Grytviken. Having surrendered, they could not be ordered to place themselves in a position of jeopardy. However, they could be so invited, and they volunteered of their own free will. The same principle had applied when Corporal Daniels agreed to dismantle his mines and booby-traps on King Edward Point following Lieutenant Mills' surrender on 3 April.

Argentina and Great Britain are signatories to the 1949 Conventions and it is clear that both sides honoured the basic principles involved. How much of that respect was based upon commonsense and how much resulted from familiarity with the fine print of the 429 Articles is a matter open to debate. The evidence suggests that British officers do not receive a great deal of instruction regarding the Geneva Conventions during the course of their careers.

In the event, the only aspect of Paraquat which caused difficulty with regard to the Conventions was the disposal of the Argentines following their surrender. Technically, they were not 'prisoners of war'. In terms of the United Nations Charter, the two countries were not at war but were in 'a state of armed conflict'. Hence these men were simply 'prisoners'. However, for the British servicemen into whose hands they had fallen, the divide between a war and an armed conflict was so fine, under these circumstances, as to be indiscernible. Until such time as the captives could be evacuated to a safe place and then sent back to their homeland, they posed a constant threat. To all intents and

Glad to be going home. As the RFA *Tidespring* closes the final miles to Ascension Island, a photographer attached to the International Red Cross inspection team captures on film the scene in the ship's main hold. Despite the Spartan conditions, the Argentine prisoners appear relaxed and healthy.

purposes, they were prisoners of war and were so treated throughout their journey aboard the *Tidespring* to Ascension Island. It was to the Fleet Air Arm that the unusual duty of gaoler fell, and the previous pages have indicated the scale of the problem faced by Lieutenant Tidd and his Flight.

At an early stage in the voyage, a copy of the Geneva Conventions was handed to Captain Bicain – as senior officer and spokesman for the Argentines – and he soon discovered numerous Articles which appeared to cover his circumstances but which were totally impracticable in the overcrowded ship. Captain Redmond and Lieutenant Tidd made themselves available at a formal meeting each day to hear his complaints and to deal with them as best they could. An example of dissatisfaction was the fact that officers and men were housed in the same accommodation. Given the very wide social gulf between officers and other ranks in the Argentine armed forces, Bicain and his colleagues felt affronted that they should not be provided with separate cabins. This was quite impossible. Redmond and Tidd had had great difficulty in keeping apart the submariners (who were potentially only moderately dangerous), the noncombatant scrap workers (who were harmless), and Astiz's special forces (who were potentially very dangerous). There was simply no room in the ship for 'officer only' quarters.

Further rancour arose over British insistance that Argentine officers and men should eat the same food and should share it equitably. It was soon noticed that officers were taking the lion's share – pushing their way to the front of the queue and leaving the younger conscripts with nothing. This practice was quickly stamped upon by Mike Tidd, but the ban caused friction. It is evident that Argentine officers are not familiar with the axiom once taught to young British subalterns: 'first feed the horses, then feed your men, finally feed yourself.'

Representatives of the International Red Cross flew out to meet the *Tidespring* as she neared Ascension Island. Each Argentine was medically examined and questioned with regard to his treatment. Tidd and his Flight were questioned as to the manner in which they had guarded the prisoners. Captain Bicain lodged a complaint with the IRC, his basic objection being that the ship was an unsuitable vehicle for the numbers involved, and his comments were duly noted. However, and without compromising the impartiality to be expected of him, the senior IRC representative expressed satisfaction with the medical status and general welfare of the prisoners. Several of them shook hands with their former guards as they left the ship to be repatriated via Montevideo by IRC-chartered civil airliners.

The experience was summarised by Lieutenant Tidd: 'It was a totally unfamiliar situation, but I am immensely proud of the way in which the lads handled it. There was a language barrier, a food barrier, a cultural barrier, even a nationality barrier – three of the scrap men were Paraguayans and technically shouldn't have been under armed guard. We had every kind of problem with sleeping accommodation, sanitation and exercise, but we got through with no major incidents and we handed them over in better condition than when they first came to us. Not bad, for a bunch of Navy aviators!'

An unusual twist to the story of the prisoners was the fate of a 20-year-old Argentine marine conscript by the name of Riviera. He had won a place at Oxford University and was due to travel to England four months later. His prospects of obtaining a visa had slumped as a result of the war but, extremely fluent in English, he helped the guards as a very competent interpreter. Another prisoner, Teniente Richardson, was similarly saddened by the circumstances in which he found himself – his father had won the Distinguished Flying Cross while serving as a pilot with Bomber Command during the war of 1939–45. □

APPENDIX A

Diary of Events

THE FOLLOWING EPISODES are relevant to Operation Paraquat. For the sake of clarity and continuity, the events are arranged under two main groupings:

A - Argentine actions B - British actions

15 DECEMBER (1981)

A - Constantino Davidoff informs the British Embassy in Buenos Aires of his intention to visit South Georgia in furtherance of his contract with Christian Salvesen.

B - he is reminded of his obligations, under the terms of his contract and in accordance with international law, to report himself to the Magistrate at King Edward Point upon arrival.

16 DECEMBER

B - the House of Lords again debates the likely repercussions of the decision to sell or scrap HMS *Endurance*.

20 DECEMBER

A - Davidoff arrives at Leith Harbour aboard the Argentine naval icebreaker *Almirante Irizar*. He stays ashore for a few hours only, and the ship then departs. Neither he nor the ship's commanding officer, Captain Trombetta, make contact with the Magistrate, and the landing is therefore illegal under contractual, local and international laws.

23 DECEMBER

B - the Magistrate, Peter Witty, reports the visit to Governor Hunt in Port Stanley.

31 DECEMBER

B - Governor Hunt recommends to the Foreign & Commonwealth Office that a protest be made to the Argentine government and that Davidoff be prosecuted.

4 JANUARY (1982)

B - the FCO instructs Ambassador Sir Anthony Williams to deliver a strong protest and instructs Governor Hunt not to initiate legal proceedings against Davidoff.

6 JANUARY

B - Williams meets with officials of the Argentine Ministry of Foreign Affairs who deny any knowledge of the *Almirante Irizar* visit.

12 JANUARY

B - Colonel Stephen Love, Defence Attaché at the British Embassy, Buenos Aires, makes a voluntary eight-day visit to the Falkland Islands. He wishes to assess their military vulnerability in the light of his knowledge of the Argentine armed forces and the current situation on the mainland. His subsequent report provokes no response from the MOD or FCO.

22 JANUARY

B - Lord Carrington writes again to Secretary of State for Defence John Nott, reiterating the implications of his decision to withdraw HMS *Endurance*.

3 FEBRUARY

B - Mr Nott replies to Lord Carrington. He declines to reverse his earlier decision.

9 FEBRUARY

A - on or about this date, according to post-war reports, the junta secretly agree the proposal by Admiral Anaya for the seizure of the Falkland Islands. The plan anticipates a time-scale of several months ahead.

B - the Prime Minister endorses the decision to sell or scrap the *Endurance*. In Buenos Aires, Ambassador Williams delivers the British government's formal protest regarding the Davidoff landing in December.

17 FEBRUARY

B - Lord Carrington again appeals to John Nott to grant a reprieve for HMS *Endurance*.

18 FEBRUARY

A - the Argentine government rejects the British protest.

9 MARCH

A - Constantino Davidoff advises the British Embassy that he is now ready to commence work on South Georgia and intends to transport his workforce and equipment on board the naval auxiliary *Bahia Buen Suceso*. He is not told that he cannot proceed under the prevailing circumstances, nor is he asked to delay his departure. He is again instructed to observe normal Customs and Immigration procedures upon arrival. His declared plan of operations indicates that his men are expected to work through the worst of South Georgia's winter weather and nineteen-hour nights. This unlikely scenario does not arouse any traceable curiosity in London.

11 MARCH

A - a C-130 Hercules of the Argentine air force makes a reconnaissance flight down the length of the north-east coast of South Georgia.

B - the flight is observed by the BAS and the Joint Services Expedition and reported to Governor Hunt in Port Stanley. The FCO is informed and advised of the possible implications.

16 MARCH

B - Christian Salvesen inform the FCO that they have formally extended Davidoff's contract. Governor Hunt urges the FCO to find some way of preventing Davidoff's return to the island. HMS *Endurance* arrives at King Edward Point from Antarctica.

19 MARCH

B - HMS *Endurance* arrives at Port Stanley. The BAS discover the *Bahia Buen Suceso* at Leith Harbour where she is unloading large quantities of stores. Her commanding officer, Captain Briatore, has not reported his arrival and fails to seek Customs and Immigration clearance from the new Magistrate, Steve Martin. Governor Hunt is informed and he recommends to the FCO that the Argentines be told to leave immediately.

20 MARCH

B - the FCO instructs Hunt to give the Argentines the opportunity to reconsider their position and to observe normal procedures (if they are so inclined). Hunt drafts an appropriate note which is delivered to Captain Briatore by the BAS party at Leith Harbour. HMS *Endurance* is ordered to return to South Georgia from Port Stanley.

21 MARCH

B - the commanding officer of the *Endurance*, Captain N. J. Barker, RN, sends a radio request to Steve Martin, asking that the BAS keep watch from Jason Peak and report all activity in

Stromness Bay pending his arrival. The ship's Royal Marines Detachment rehearses for an eviction operation.

22 MARCH

A – the *Bahia Buen Suceso* departs Leith Harbour, leaving behind a party of 48 scrap metal workers.

B – the BAS continue to observe and report from Jason Peak.

23 MARCH

B – the FCO decides to follow Governor Hunt's advice. The Argentine government is informed that the Davidoff party will be evicted as a consequence of its non-compliance with legal requirements. It is decided to double (temporarily) the number of Royal Marines stationed at Port Stanley.

24 MARCH

B – Ambassador Williams and Colonel Love, Military Attaché, warn the MOD and FCO of the risks inherent in a direct confrontation with the Argentine navy under current circumstances. The FCO decides not to evict the Davidoff party. Captain Barker is instead instructed to divert to King Edward Cove and there await further orders. Lord Carrington warns the Cabinet Office and the MOD of the delicate situation on South Georgia and suggests the initiation of naval/military contingency planning. He again asks that the *Endurance* be left on station. The *Endurance* arrives in Cumberland East Bay and her Royal Marines take over from the BAS the task of watching Leith Harbour. Captain Mitchell, Naval Attaché, sends his first detailed report (based mainly upon local newspaper reports) informing London of the high state of readiness of the Argentine navy.

25 MARCH

A – the naval auxiliary *Bahia Paraiso* arrives at Leith Harbour from Southern Thule and illegally lands a party of marines with six months' stores for the scrap metal work-force.

B – Captain Mitchell informs London of the first sailings of the Argentine navy (ostensibly for anti-submarine exercises) and the departure for South Georgia of two small frigates. The British Embassy commences shredding its most sensitive military and naval files.

26 MARCH (FRIDAY)

B – Captain Mitchell sends a long and urgent signal to London, announcing that the Argentine naval bases are now almost empty of shipping and that at least one battalion of marines has been reported embarking with armoured vehicles. Taken in combination with information from other sources, his evidence indicates that the Argentine navy is unlikely to be deploying for anti-submarine exercises. The FCO advises Ambassador Williams to prepare for a worsening in the situation. John Nott agrees to leave HMS *Endurance* on station for the time being. The RFA *Fort Austin* is ordered to load stores at Gibraltar and to go to the support of the *Endurance*.

27 MARCH (SATURDAY)

B – in Buenos Aires, the shredding machines in the British Embassy Registry continue to operate throughout the weekend.

28 MARCH (SUNDAY)

A – the *Bahia Paraiso* departs Leith Harbour and proceeds to a loitering position approximately fifteen miles off the mouth of the Cumberland Bays.

29 MARCH (MONDAY)

A – the junta have confirmed the final details of Admiral Anaya's plans for the invasion of South Georgia and the Falkland Islands.

B – the RFA *Fort Austin* departs Gibraltar with replenishment stores for the *Endurance*. The first of three SSNs to be deployed, HMS *Spartan*, is ordered to load stores and war-scale ordnance at Gibraltar and to proceed to the South Atlantic with all despatch. The First Sea Lord informs CINCFLEET that further action may be required. In the House of Commons, during a long and acrimonious debate regarding the planned expenditure of £8,000 million for the purchase of the Trident II (D5) system, Mr Keith Speed, a former Navy Minister, asks why the MOD can no longer afford to keep the *Endurance* in service. The Secretary of State for Defence replies: 'I do not intend to get involved in a debate about the Falkland Islands now. These issues [Trident] are too important to be diverted into a discussion on HMS *Endurance*.'

30 MARCH (TUESDAY)

A – the *Bahia Paraiso* continues to patrol off the Cumberland Bays, shadowed by the *Endurance*. At least two Argentine frigates, a tanker, and possibly a submarine, are reported to be at sea somewhere between South Georgia and the Falkland Islands.

B – the British Embassy in Buenos Aires completes the task of destroying its naval and military files and starts to shred all other documents held in the Registry. In New York, Sir Anthony Parsons, Great Britain's representative at the United Nations, is investigating the possibilities of requesting an emergency debate in the Security Council. In Port Stanley, Governor Hunt is in regular radio contact with the Magistrate at King Edward Point but is unaware that the Argentine naval force now at sea has an amphibious landing capability.

31 MARCH (WEDNESDAY)

B – HMS *Endurance* is ordered back to Port Stanley. She disembarks her reinforced Royal Marines Detachment at King Edward Point before departing under cover of night. HMS *Spartan* departs Gibraltar. The first alert to home-based troops – the SBS at Poole – is issued by the MOD. The majority of other armed forces units prepare for Easter leave. The Joint Intelligence Committee – having now made its first assessment of the South Atlantic situation since July of the previous year – decides that imminent invasion of the Falkland Islands is a strong possibility. In the evening, the Prime Minister, having seen the new JIC assessment, consults with senior political advisors and the First Sea Lord. The decision is made, in principle, to mount a major Task Force, if that should prove to be necessary. In Port Stanley, there is a dinner party at Government House. Governor Hunt is unaware of developments in London, consequently he does not alert the Royal Marines at Port Stanley, or the BAS and Royal Marines personnel at King Edward Point, South Georgia. In Buenos Aires, Colonel Stephen Love, also uninformed by London of the latest JIC assessments, entertains to dinner General Mario Menendez, the new Argentine Governor-designate of the Falkland Islands.

1 APRIL (THURSDAY)

A – the Argentine invasion force is steering toward the eastern extremity of East Falkland. Off South Georgia, the *Bahia Paraiso* awaits the arrival of at least one A69 frigate. Two other warships *en route* are delayed by bad weather and mechanical problems aboard their supporting tanker.

B – HMS *Endurance* is approximately halfway between South Georgia and the Falkland Islands, her captain hoping to arrive at Port Stanley on Saturday. In England, shortly before midnight, Captain Lyn Middleton, commanding the aircraft carrier HMS *Hermes*, departs for Gibraltar on leave. Lacking any advice or orders to the contrary, thousands of other servicemen similarly depart for home or holiday. An exception are the Royal Marine Commandos – some are recalled. At Poole, 2 SBS and 6 SBS are ordered to start mobilising for service in the South Atlantic. HMS *Splendid* departs Faslane. At 1600 (London time), on his own initiative, the First Sea Lord brings the carriers HMS *Hermes* and HMS *Invincible* to forty-seven hours notice to sail. At

1630 (London time), the FCO telephones New York and instructs Anthony Parsons to request an emergency debate in the UN Security Council. The FCO sends a cable to Port Stanley, advising Governor Hunt that the town could be invaded by dawn the next day. In the evening, the Prime Minister meets again with John Nott, William Whitelaw and the First Sea Lord. The Prime Minister authorises the mounting of a large-scale Task Force. During the evening, in New York, the UN Security Council calls upon both sides to exercise restraint.

2 April (Friday)

A – at approximately 0400, Buzo Tactico (tactical diver) special forces come ashore at Mullet Creek, south of Port Stanley, and advance inland in two groups. At 0555, the western group attacks the empty Royal Marines barracks at Moody Brook. At 0600, the eastern group attacks Government House.

B – at 0227, HMS *Hermes* and HMS *Invincible* are brought to four hours notice to sail. Captain Middleton returns to Portsmouth from Gibraltar. Northwood starts to plan Operation Corporate. The Fleet Air Arm starts to mobilise. 148 (Meiktila) Commando Forward Observation Battery RA is recalled from leave. Port Stanley falls. At 1100 (London time), the proceedings of the House of Commons are interrupted by an announcement from the Lord Privy Seal, Mr Humphrey Atkins: 'There is now a real expectation that an Argentine attack against the Falkland Islands will take place very soon. It has become increasingly evident over the past few days [*sic*] that the Argentines had assembled a fleet which was operating in the vicinity of the Falkland Islands [*sic*].' The invasion is confirmed by the Leader of the House, Mr Francis Pym, at 1435 hours.

3 April (Saturday)

A – the *Bahia Paraiso* and the *Guerrico* attack King Edward Point. Lieutenant Mills and his marines surrender after a spirited defence.

B – HMS *Endurance* arrives off the Barff Peninsula just as the battle ends. The BAS reorganises its communications with South Georgia and Antarctica. The first RAF Hercules flights depart RNAS Yeovilton, transporting Fleet Air Arm Lynx helicopters to Ascension Island. Anthony Parsons addresses the UN Security Council. The UN adopts Resolution 502. Addressing the House of Commons, and in apparent contradiction of the announcement by the Lord Privy Seal on the previous morning, the Prime Minister states: 'Information about the Argentine fleet did not arrive until Wednesday.' During the same debate, Ted Rowlands, a former Minister at the Foreign Office, states: 'I shall make a disclosure. As well as trying to read the mind of the enemy, we have been reading its telegrams for many years. I am sure that many sources are available to the government and I do not understand how they failed to anticipate some of the dangers that suddenly loomed on the horizon.'

4 April (Sunday)

B – at Hereford, 'D' Squadron 22 SAS mobilises. HMS *Conqueror* departs Faslane with 6 SBS embarked. At RNAS Yeovilton, the first civil Belfast cargo flight departs with two Fleet Air Arm Wessex helicopters for delivery to Ascension Island.

5 April (Monday)

B – 'D' Squadron and 2 SBS depart RAF Lyneham for Ascension Island. On South Georgia, BAS personnel make their first reconnaissance (Grytviken and King Edward Point). The staff of the British Embassy in Buenos Aires start to leave the country. In London, Lord Carrington and Richard Luce both resign. The new Foreign Secretary is Mr Francis Pym. John Nott continues in office.

6 April

B – BAS personnel reconnoitre Hound Bay.

7 April

B – the secret decision is made to mount Operation Paraquat. Captain Young, in HMS *Antrim*, is ordered to detach from Admiral Woodward's force and to proceed at best speed, with HMS *Plymouth* in company, for Ascension Island. He is to join forces with the RFA *Tidespring* and, later, with HMS *Endurance*. His main assault force will be 'M' Company 42 Commando RM.

8 April

B – in pursuit of President Reagan's 'even handed' search for peace, Secretary of State General Alexander Haig visits Downing Street before flying on to visit General Galtieri in Buenos Aires.

9 April

A – the junta at last recognises the serious consequences of the invasions. Massive reinforcements are being sent to the Falkland Islands. A platoon-strength reinforcement departs for South Georgia aboard the *Santa Fe*.

B – the RFA *Fort Austin* embarks 2 SBS and 'D' Squadron at Ascension Island and sails south to rendezvous with HMS *Endurance*.

10 April

B – 'M' Company and 148 Battery arrive on Ascension Island.

11 April

B – BAS personnel reconnoitre Royal Bay. Peru proposes a 72-hour truce.

12 April

B – the RFA *Fort Austin* meets HMS *Endurance* in mid ocean and replenishes her dry stores.

13 April

B – early in the morning, the *Fort Austin* turns northward with three Troops of 'D' Squadron still aboard. HMS *Endurance* turns south, having embarked one SAS Troop and 2 SBS. Later in the morning, the *Antrim*, *Plymouth* and *Tidespring* rendezvous with the *Fort Austin* to replenish and to embark the other three SAS Troops. They then continue down the track of the *Endurance* while the *Fort Austin* continues northward to Ascension Island. The Organisation of American States (OAS) appeals to both countries for a peaceful solution to be found.

14 April

B – the *Antrim*, *Plymouth* and *Tidespring* overhaul the *Endurance* 600 miles south of Ascension Island. The Task Group forms up and steers for South Georgia. Operation Paraquat has started in earnest.

15 April

B – an RAF Nimrod aircraft drops Admiral Fieldhouse's operational order to Captain Young.

16 April

A – Argentina accuses Great Britain of ignoring the UN and Haig peace initiatives by mounting a Task Force. The junta are referring to Admiral Woodward's Operation Corporate force. They are unaware of the Paraquat group.

B – BAS personnel make a second visit to Hound Bay.

17 April

B – the RRS *Bransfield* is ordered to end her lonely watch on the 60th Parallel and to proceed north to Ascension Island. For the past two weeks, while the *Endurance* has sailed 2,000 miles north to join the Paraquat force, she has been an important link in the communications chain and the only British ship in the area.

19 April

B – HMS *Conqueror* arrives in South Georgia waters. BAS

personnel commence a third, and extended, reconnaissance of the Barff Peninsula.

20 April

B – HMS *Endurance*'s Royal Marines, together with Steve Martin, Bob Headland, and the other BAS personnel captured on 3 April, arrive back in the United Kingdom from Montevideo. RAF Victor aircraft mount a maritime reconnaissance over the northern approaches to South Georgia. The Paraquat force reaches a position 100 miles north of the island.

21 April

B – the SAS and SBS initial reconnaissance parties are inserted on the Fortuna Glacier and at Hound Bay respectively. The SBS encounter two BAS men at Hound Bay.

22 April

B – the SAS are evacuated from the Fortuna Glacier. HMS *Conqueror* is ordered to leave local waters and to steer north-west so that she can screen Task Group 317.8.

23 April

A – three C-130 Hercules make reconnaissance flights along the north-east coast of the island and detect the British ships. Argentine news agencies broadcast their sighting reports. The submarine *Santa Fe* approaches the island from the north-west.

B – Captain Young has entered Stromness Bay during the night and enabled the SAS Mountain Troop to make a landing on Grass Island. The ship's Wessex helicopter rescues three SAS soldiers from their drifting Gemini and all ships then concentrate 40 miles from the coast. The RFA *Brambleleaf* joins the Task Group and commences the pumpover of fuel. The Argentine overflights, plus reports of the *Santa Fe* being in the area, cause Captain Young to withdraw all ships but one to a new position 200 miles to the north-east. HMS *Endurance* returns alone to Hound Bay. HMS *Conqueror* is ordered back to South Georgia waters.

24 April

A – a military Boeing 707 makes a reconnaissance flight down to Hound Bay and identifies HMS *Endurance*.

B – HMS *Brilliant* joins the Task Group. The SBS reconnaissance party is evacuated by the *Endurance*'s Wasp helicopters from Cumberland East Bay.

25 April

A – the *Santa Fe* has entered Cumberland East Bay during the night and disembarked her passengers at King Edward Point. Captain Bicain prepares to sail at dawn, or earlier if possible.

B – the *Antrim*, *Plymouth* and *Brilliant* race in toward the coast. The first helicopter missions are launched. The *Santa Fe* is found, attacked and driven back to King Edward Point. A scratch military force is landed by helicopter at Hestesletten and the Argentine garrison at King Edward Point surrenders in the face of intimidatory naval gunfire.

26 April

A – in New York, the Argentine government lodges a protest at the United Nations.

B – the small Argentine garrison at Leith Harbour, together with the scrap workers who were one of the original causes of the conflict, surrender to Captain David Pentreath of HMS *Plymouth*. There is a similar formal surrender to Captain Young of HMS *Antrim* by the King Edward Point garrison. Operation Paraquat is successfully concluded. ☐

APPENDIX B

ARA *Santa Fe*: Chronology of Attacks, 25 April 1982

Time	Aircraft	Parent Ship	Weaponry	Remarks
0730	—	—	—	The radio office in HMS *Plymouth* hears the *Santa Fe* transmitting on Channel 16. The submarine is assessed to be leaving King Edward Point.
0810	Wessex 406	*Antrim*	2 × Mk11 depth charges 1 × GPMG	Launches from parent ship 70 miles east of the island and heads for the Cumberland Bays.
0815	Lynx 341	*Brilliant*	1 × Mk46 torpedo 1 × GPMG	Launches from parent ship 66 miles from the coast and heads for the northern coast of the island.
0840	Wessex 406	*Antrim*	—	Enters Cumberland East Bay and commences low-level search.
0850	Wasp 434	*Endurance*	2 × AS12	Launches from parent ship off Hound Bay.
0857	Wessex 406	*Antrim*	—	Sights and identifies the *Santa Fe* seven miles north of Barff Point, heading north-east.
0900	Wessex 406	*Antrim*	—	Drops both depth charges on target and transmits sighting report to other units. Attacks with GPMG. Commences 'dunking'.
0905	Lynx 341	*Brilliant*	—	Arrives from the north under the radar control of Wessex 406 and drops homing torpedo. Attacks with GPMG.
0906	Wessex 406	*Antrim*	—	Departs for parent ship to rearm and refuel.

Time	Aircraft	Parent Ship	Weaponry	Remarks
0908	Wasp 445	Plymouth	1 × AS12	Launches from parent ship 45 miles from the coast and approaches the Cumberland Bays under the radar control of Wessex 406.
0912	Wasp 434	Endurance	—	1st attack: enters Cumberland East Bay and attacks. First missile hits fin (delayed explosion), second missile falls 30 yards short. Aircraft departs for parent ship to rearm.
0920	Lynx 342	Brilliant	1 × GPMG	Launches 40 miles from the coast with intention of supporting Wessex 406 with MAD if target dives.
0920	Lynx 341	Brilliant	—	Aircraft continues to harass the submarine with intermittent GPMG fire as she makes her way back towards King Edward Point.
0940	Lynx 342	Brilliant	—	Arrives in Cumberland East Bay and finds the *Santa Fe* approaching King Edward Point. Attacks with GPMG.
0945	Wasp 445	Plymouth	—	Arrives and locates the target, half a mile from King Edward Point.
0950	Wasp 434	Endurance	2 × AS12	2nd attack: first missile launches out of control, second missile passes through submarine's fin without exploding.
0953	Wasp 445	Plymouth	—	Attacks with single AS12. Very near miss, port quarter.
0955	Wasp 435	Endurance	2 × AS12	Launches from parent off Hound Bay.
1000	—	—	—	Submarine is on final approach to King Edward Point under intermittent machine-gun attack from Lynx 341 and 342. Wasp 434 returns to parent to rearm again.
1005	Wasp 435	*Endurance*	—	Arrives over target and assesses the groundfire.
1010	Wasp 435	*Endurance*	—	Attacks with AS12 missiles. The first falls short by 50 yards, the second passes through the submarine's fin without exploding.
1012	Wessex 406	*Antrim*	1 × Mk11 depth charge	Returns to the scene of action. Pilot and observer assess the situation and decide not to repeat the earlier attack.
1015	Lynx 341/342	*Brilliant*	—	Circle the area at higher altitude, having exhausted their ammunition.
1015	Wessex 406	*Antrim*	—	Instructs other aircraft in radio contact to cease attacks and to withdraw.
1020	Wasp 445	*Plymouth*	2 × AS12	Starts to re-enter Cumberland East Bay. Told by Wessex 406 that *Santa Fe* has gone alongside the King Edward Point jetty and is disabled. Aircraft turns back to parent.
1020	Wasp 434	*Endurance*	2 × AS12	Returns and makes 3rd attack: first missile refuses to launch (later jettisoned), second hits fin (heavy explosion and structural damage).
1025	—	—	—	All remaining helicopters withdraw to parent ships.

Summary: six Fleet Air Arm helicopters took part in the action. They fired approximately 1,600 rounds of small-arms ammunition. They dropped two Mark 11 depth charges, one Mark 46 homing torpedo, and deployed nine AS12 missiles (two of these launches being abortive due to faulty electrical equipment).

THE COLOUR OF PARAQUAT

Persistently foul weather was a dominant feature of the operation. Just very occasionally, and for brief spells, the photographers were able to exploit the full potential of colour film. The results were often spectacular.

A helicopter which won fame and glory in the South Atlantic was the Wessex HAS3 ('Humphrey') operated by HMS *Antrim*. Here it is seen in three contrasting roles: delivering stores to HMS *Endurance* on 14 April (upper left), transporting military personnel along the South Georgia coast in June (upper right), and returning from an anti-submarine patrol during the passage south from Ascension Island in mid-April.

Anchored snugly in Stromness Bay (left), one of the big RFAs waits to resupply her next customer. It is early June, and South Georgia has become a major forward operating base.

King Edward Cove on a sunny summer day, early in 1985. The RMAS *Goosander* is moored over the wreck of the *Santa Fe*, with the tug *Salvageman* berthed at Grytviken. The picture suggests that the entry into this small enclosed stretch of water by the frigate *Guerrico*, on 3 April 1982, was a particularly ill-considered manoeuvre. If Lieutenant Mills and his marines had possessed heavier weaponry, the frigate might well have preceded the *Santa Fe* to the bottom of the Cove.

Annie Price and Cindy Buxton, as professional photographers, succeeded in producing some of the most breath-taking pictures of South Georgia's terrain and wildlife. A large number appear in their book 'Survival, South Atlantic'.

A Royal Air Force C-130 Hercules makes a low pass to drop mail to the post-conflict garrison at King Edward Point. Severe turbulence and strong down-drafts are always a hazard for aviators operating close to South Georgia's steep-sided mountains.

It is claimed that South Georgia is the only place in the world where 'all four seasons can happen in the space of one day'. The deceptive calm of Grytviken at sunset (upper right) is in total contrast to the huge seas which can build up offshore. Despite their great bulk, even the RFA ships were roughly handled by the South Atlantic swell during the campaign.

Long after the Argentines had returned to their homes, the British were still removing all traces of their temporary presence on British soil. In February 1983 the destruction of the hangar building on Hewison Point, Thule Island, was achieved by the simultaneous explosion of several naval mortar bombs and dozens of drums of captured diesel oil. Less dramatic, but technically far more difficult, was the raising and final scuttling of the submarine *Santa Fe* in February 1985. Operation Matchstick and Operation Okehampton set the seal on the 1982 Operation Paraquat.

APPENDIX C

ARA *Santa Fe*: Salvage Operations, 1982–85

REFERENCE HAS BEEN MADE (*vide* Chapter IX) to the early attempts to render safe the wreck of the Argentine submarine *Santa Fe*. More than three years were to pass before the British finally rid themselves of this relic of the Fleet Air Arm's unprecedented helicopter attack. Like the albatross in Coleridge's *Ancient Mariner*, this elderly submarine, built forty years earlier during the peak years of the United States of America's war with Japan, became a burden which could not be ignored.

On the morning of 26 April 1982, she was moved under the joint command of Captain John Coward, RN, and Capitan de Fragata Horacio Bicain from the BAS jetty at King Edward Point to the Tijuca jetty at Grytviken. There she gradually settled and grounded, semi-submerged, on the sea-bed. The saga of her salvage commenced a few days later.

PHASE 1 (EARLY MAY 1982)

A Clearance Diving Team, led by Lieutenant Dave O'Connell, RN, from HMS *Endurance* and assisted by technicians and engineering officers from the Royal Maritime Auxiliary Service ship *Typhoon*, made the first attempts to refloat the *Santa Fe* from her resting place at Grytviken. Some useful exploratory diving work was carried out during this time, but the task was too great for the resources available.

PHASE 2 (SECOND HALF OF MAY 1982)

The tug *Salvageman* was tasked to work with HMS *Endurance* and the RMAS *Typhoon* in further attempts to seal the leaks in the submarine's pressure hull and to pump her dry. Some headway was made in reducing the level of water within the hull between 18 and 22 May, but salvage operations were then suspended for two days as pumps broke down and the three ships were required for work connected with the arrival and departure of various merchantmen in Cumberland East Bay, Husvik Bay and Stromness Bay. Work was resumed by the *Typhoon* between 24 and 28 May when pumping operations were again halted by mechanical problems and by the arrival in Cumberland East Bay of the *Queen Elizabeth II*, *Canberra* and *Norland*. Small ships were again needed for cross-decking operations. During the delivery of 5 Infantry Brigade, RMAS *Typhoon* transported 1,500 troops, a full field medical unit, 1,000 kitbags and 100 tons of fresh water. She then departed for duties in the TRALA replenishment, repair and air drop zone 200 miles north-east of the Falkland Islands. The *Santa Fe* remained at Grytviken, once again flooded and resting on the bottom.

PHASE 3 (28 JUNE TO 15 JULY 1982)

After their return from Operation Keyhole, the *Endurance* and the *Salvageman* commenced a more ambitious attempt to raise the submarine. In charge of the operation was Mr B. W. Vere-Stevens, Senior Mooring, Salvage and Boom Officer employed by the Directorate of Marine Services (DMS) and seconded to the *Salvageman* for Operation Corporate. The diving team was again led by Lieutenant O'Connell with technical advice from Lieutenant-Commander David Ives and Lieutenant-Commander Arthur Ainslie from HMS *Endurance*.

It was a matter of urgency to move the *Santa Fe* to a place of safety. She was blocking access to the best water point on South Georgia and occupying a berth needed by other ships but, more importantly, she contained seven tons of explosives. Some of this material was of doubtful stability and the submarine was therefore a hazard to personnel working in the King Edward Cove area.

On 29 June, *Salvageman*'s two 100 tons per hour pumps started to lower the water level within the submarine. An abnormally high tide brought the sea pouring down through the fin hatch faster than it could be pumped out, and this necessitated two days of work to build a timber and concrete cofferdam on top of the fin. Pumping was then resumed and the Lower Control Room emptied of water. The submarine remained hard aground because the torpedo room, engine room and machinery spaces were still flooded and, lacking any working drawings of the boat, the salvage team were unable to blow all of the buoyancy and ballast tanks. Working conditions in and on the hull were extremely unpleasant, all surfaces being covered with a thick slime consisting of diesel oil, hydraulic and lubricating oils, rotting food, battery acid and excrement. Progress was painfully slow after one of the pump motors burned out. It could not be repaired until the cable ship *Iris* returned with spare parts and extra equipment on 13 July. All accommodation spaces were then pumped dry and the submarine floated free of the sea-bed.

The Clearance Diving Team aboard the *Iris* was led by Fleet Chief Petty Officer Jim Green who made a survey of the hull and the work achieved so far. He assessed that early action should be taken to dispose of the submarine. The TNT and detonators scattered about among the sludge and debris inside the hull were starting to dry out and would become increasingly hazardous. Most of the torpedoes were still primed and several could not be disarmed. There was an unlocated leak near the stern and it was evident that the boat could never be restored to seaworthy condition.

Captain Nick Barker chaired a meeting to discuss the next step and, after receiving approval from Northwood, the decision was made to tow the *Santa Fe* to a previously surveyed site on the Hestesletten shore of Moraine Fiord (where the British helicopter landings had been made on 25 April). This was done on 15 July after a final blow had brought the submarine to the surface.

The beaching of the *Santa Fe* at an isolated location – well away from buildings and regular anchorages – could not have been achieved without the expertise and special equipment supplied by the *Salvageman*, *Iris* and *Stena Seaspread*. Once again, however, the wide range of talents to be found aboard HMS *Endurance* had played a major part. Lieutenant O'Connell's four divers were fully trained and qualified, but salvage diving was certainly not their trade. Two were Able Seamen – M. Leach and J. Ferguson – and two were Royal Marines – S. Parsons and R. White. Both of the latter had fought in the original battle of King Edward Point on 3 April. Between them, these four accumulated 4,064 minutes of diving time on the *Santa Fe* in just four weeks (the record-holder being Knocker White with 726 minutes). Conditions inside the flooded hull had been far from ideal. Despite the use of underwater lights, visibility was minimal. Their work was not helped by the boat's heavy list to port or by the pancake ice and 'bergy bits' floating on the freezing surface of the Cove. Even with many layers of clothing under their dry suits, the divers could spend only a limited time under water during each session. Like the rest of the ship's company, they were delighted when the completion of their task signalled the end of the *Endurance*'s deployment in South Georgia waters and the beginning of her long voyage home.

PHASE 4 (9 MARCH TO 26 MARCH, 1983)

Concern had been expressed at the progressive deterioration of the explosives contained in the submarine. It was decided that these materials should not be disturbed to the extent necessary to remove them, and the only course, therefore, was to refloat her and scuttle her in very deep water. A salvage team headed once again by Mr B. W. Vere-Stevens was flown down from the United Kingdom. The twelve-man team arrived in King Edward Cove aboard the *Salvageman* (Captain J. Bold) on 9 March.

An initial survey showed that the *Santa Fe* appeared to be in much the same situation as when she was beached and abandoned nine months earlier. She was flooded throughout, only the fin and forward casing being above the surface of the water. She had settled in a bed of mud and sand with a twenty-degree list to port.

Two pontoons – carrying the pumps – were moored alongside the submarine while the *Salvageman* anchored 300 feet out from the beach. Pumping commenced the next day and, despite repeated technical problems, initial progress was rapid. The principal difficulty was the nature of the site. A heavy swell, combined with the rise and fall of the four-foot tide, made the wreck increasingly 'tender' as she was pumped out. On one occasion her list increased to forty-five degrees and she was hastily evacuated.

At 0700 on 16 March, the main ballast tanks were blown and the submarine became sufficiently buoyant to be warped out from the beach and secured, bow to stern, with a ten-degree list, alongside the *Salvageman*. It had not been possible to completely blow the saddle tanks – it had been intended to do this once the boat was secured alongside the tug. By early afternoon, however, a south-westerly gale began to create a heavy swell in Moraine Bay and it was decided to proceed into the sheltered waters of King Edward Cove for the night. There was still a great deal of work to be done before the submarine could safely be towed down the length of Cumberland East Bay to the open sea. A quantity of valuable salvage equipment was still inside her, and this would need to be removed before she was scuttled.

At 1450 the *Salvageman* – with her charge alongside – made fast to a buoy in King Edward Cove. All appeared to be well. Just three minutes later a loud crack was heard. The submarine rolled quickly to port and settled by the stern. The fore end and fin were still above the surface, but the hull assumed a trim of thirty degrees. Fleet Chief Petty Officer G. Biggs was in the lower control room at this time, and he hurriedly left the boat when he heard water pouring in aft.

Although the condition of the *Santa Fe* was alarming, it did not seem that all was yet lost. FCPO Biggs volunteered to go back into the control room via the forward hatch in an attempt to blow air into the stern and portside tanks. At great risk to himself, he made five entrances deep into the interior during the afternoon, accompanied each time by either Mr Vere-Stevens or Diver R. Williams for reassurance and assistance as required.

Air was pumped into the ballast tanks all afternoon, but to no effect. The trim of the submarine was so steep that the air was filling only the crowns of the tanks before spilling out through the free flood holes. Complete blowing was therefore impossible. The decision was made to risk towing the wreck out into the deepest part of Cumberland East Bay immediately and attempting to scuttle her there. Her condition was such that she could not be dragged alongside the *Salvageman* but instead would need to be towed.

By 1900 all was ready. The tug cast off and the tow commenced. Within a few minutes the submarine fouled the mooring of the buoy and came to a standstill. The towing bridle parted and the *Santa Fe* settled deep into the water with only the tip of the bow and the top of the attack periscope visible. In pitch dark, high winds and heavy rain, the *Salvageman* berthed at the King Edward Point jetty to consider the next move. At 2125 the *Santa Fe* disappeared entirely and sank in seventy feet of water.

On Thursday, 17 March, divers went down to examine the wreck and found her lying on her side, settling slowly into thick soft mud. A final survey, nine days later, showed that she had sunk so far into the mud that only the port side of her fin and a small part of her port bilge were visible to the divers. Ninety-five per cent of the wreck had been swallowed by the sea-bed. The *Santa Fe* was now a major salvage operation – far beyond the capabilities of the *Salvageman* team – and the project was abandoned. The tug sailed for Port Stanley on 27 March.

PHASE 5 (8 NOVEMBER 1984 TO 20 FEBRUARY 1985)

The time had come to lift the *Santa Fe* from the bottom of King Edward Cove and to complete the task of taking her out to sea. The Marine Services Directorate of the Ministry of Defence was about to undertake an operation unique in its remoteness and lack of base support. It was given the codename Okehampton, and the officer in charge was Captain Ferris Morton, the MOD's Chief Salvage Officer. Accompanying him was a seventeen-man team of divers and diving technicians.

It was known that the submarine was flooded throughout and had all internal transverse bulkhead doors and the conning hatch open. It was also known that the stern of the wreck lay embedded in deep mud and that the external fuel and ballast tanks had been damaged by depth charge, AS12 missile and small-arms attacks. With a wreck in this condition, seventy feet down in near freezing and almost tideless waters, a great deal of pre-departure planning was essential. It would require weeks to obtain any items of equipment inadvertently left behind in the United Kingdom. Furthermore, the operation involved the use of unusual items such as a Rimer-Alco oxygen concentrator. The team would need to be self-sufficient for the large quantities of oxygen required for underwater cutting. From Genflo Subsea Limited came a water-powered eductor pump capable of not only excavating mud from beneath the wreck but then pumping it 500 metres outside the Cove. Special heavy duty neoprene dry suits were tailor-made for each diver – they would be working for long periods in very cold conditions. The list of special requirements was a long one.

The two vessels deployed from the United Kingdom for this operation were the RMAS *Goosander* (Captain A. McGregor) and the by now familiar *Salvageman* (Captain A. Stockwell). The *Goosander* did not possess sufficient bunker capacity for such a long voyage and her maximum speed was eight knots. She was therefore towed all the way from Ascension Island to South Georgia, at 11 knots, by the *Salvageman*. This was not the longest tow on record, although it was possibly the longest 'north to south' tow. The expedition arrived in King Edward Cove on 4 November – after five weeks on passage – to find five inches of snow on the foreshore and half of the Cove frozen over with a thin layer of snow-covered ice.

An initial survey showed that the submarine lay with her stern buried in five metres of mud. Unusually in a submarine of this type and age, the pressure hull compartments were fitted with salvage air and spill pipes. Most of the connection points were hidden under the mud but, once exposed, they could be used for

King Edward Cove, 11 February 1982. The *Santa Fe* comes to the surface in a welter of muddy foam and a tangle of high-pressure air lines.

the blowing of compressed air directly into the boat's interior.

Air venting from the *Santa Fe*'s ballast tanks did not escape directly into the sea but came up through ducts to the underside of the deck casing. It was soon clear that the salvage team would be unable to fill the ballast tanks with air until they had first repaired the dozens of holes punched in these ducts by the machine gun and rifle bullets fired by Fitz Fitzgerald, Barry Bryant and Paul McKay (*vide* Chapter VIII). An even worse problem was the snorkel housing inside the fin. This had been severely damaged by the final AS12 – launched by Tony Ellerbeck and David Wells – when the submarine was alongside the BAS jetty. Until such time as a way could be found to seal it, any air pumped into the pressure hull would immediately escape through the snorkel.

The first four days were spent unloading stores at Grytviken and laying out four mooring anchors which would permit the *Goosander* to manouevre along the 94-metre length of the *Santa Fe*'s hidden hull. The *Goosander* was to remain on this mooring for the next ninety-four days, supplied with victuals, equipment, fuel and fresh water by the attendant *Salvageman*.

Over the following weeks the divers excavated the tons of glutinous mud gripping the stern, searched for bullet and splinter holes in the saddle fuel and ballast tanks, and located all the main salvage connections and venting controls. Given the lack of any detailed yard drawings, the latter task was both time-consuming and frustrating. Divers were obliged to enter through the fin hatch, feel their way down into the control room, make their way aft to the engine and motor rooms, and grope around in the tangle of pipework and detritus. Powerful lights and closed circuit television cameras were taken inside the hull, and each diver could communicate by voice with the controller in the *Goosander*, but the combination of cold and near zero visibility rendered the operation both difficult and dangerous. Throughout the operation, 868 separate dives were made and forty-two of these involved entries into the submarine's interior. It was a slow and painstaking business. In the words of their Director, Mr M. D. Walker: 'There were no accidents, diving incidents or decompression sickness, haste being sacrificed to safety throughout.'

The technical difficulties were overcome and, by early February, the wreck was ready to be brought to the surface. There was no question of attempting to lift the submarine's flooded deadweight of 2,500 tons by brute force. The task was achieved by a combination of air pumped into the pressure hull and saddle tanks, lifting bags attached to the bow, and wires passed under the hull and hauled in by the *Goosander*'s powerful winches.

Only two strops were used: one of 7-inch wire shackled around the fore end, abaft the diving planes, and another of 9-inch wire around the after section of the hull above the propeller shafts. To break the sea-bed suction, the after strop was gently winched up on 100-ton shackles while ten inflatable lifting bags, each of five tons lifting capability, were attached to the forward strop.

Over a period of several days the *Goosander* had moved progressively along the length of the submarine, paying out thirty-one compressed air hoses to the ingress points located or made by the divers along the sides of the tanks and pressure hull. During this time, the crew of the *Salvageman* were preparing moorings on the shore so that the submarine could be temporarily beached within the Cove.

The blowing of air to a carefully planned schedule commenced at 1050 on 8 February 1985 and, precisely seventy-two hours later, the *Santa Fe* reappeared in the light of day. After checking and consolidating the buoyancy, the *Salvageman* towed her to the beach where she took the ground with a heavy list to starboard.

Eight more days were required to generally prepare the submarine for the tow down the length of Cumberland East Bay to the open sea. It was planned to take her to a position well clear of the coast and to scuttle her at a depth of a thousand fathoms. At 0540 on 20 February, Captain Bold (who had routinely relieved Captain Stockwell shortly after Christmas) laid alongside the beached submarine, made fast, and very gently eased her off

Attended by the RMAS *Goosander*, the submarine is prepared for her final voyage. In the background is Mount Duse with, to the right, the white buildings of King Edward Point.

the beach and across King Edward Cove. More line was paid out until she was towing astern. The *Salvageman* proceeded slowly down the length of the Bay with the *Goosander* in attendance.

At 0835 it was seen that the submarine was starting to list to starboard and settling by the stern. The tow continued but, obstinate to the last, the *Santa Fe* continued to sink deeper into the water. By the time she and her custodian had cleared the entrance to the Cumberland Bays, it was evident to Ferris Morton that she was on the point of foundering. He and Captain Bold persisted until the last possible moment – by which time only the stem and the top of the fin were visible – before breaking the tow. At long last, after four decades of service, having survived nine helicopter attacks, having been three times stranded, once accidentally sunk and three times salvaged, the *Santa Fe* found her final resting place. She sank in 196 fathoms of water in position 54° 08.5′ South, 36° 19.8′ West.

Both salvage vessels returned to King Edward Cove, collected their stores and equipment, and departed for the United Kingdom five days later. The *Goosander* was towed north until she and the *Salvageman* parted company off Cape Finisterre. Captain Bold finally returned to Portsmouth on 25 March 1985, three years almost to the day from the *Santa Fe*'s original departure from her mainland base and her fateful voyage tc South Georgia. □

APPENDIX D

Exercise Green Skua

In Chapters I, VI and VII, reference has been made to the topographical features of South Georgia in general and the difficulties of traversing its many varied glaciers in particular. The decision by the SAS to land on the Fortuna Glacier is the one aspect of Operation Paraquat which elicits more dogmatic comment than any other. It is appropriate, therefore, to place on record a brief description of a military patrol conducted in 1985 by Captain Roger Morgan-Grenville and a small party of riflemen of 1st Bn The Royal Green Jackets.

Since the eviction of the Argentine invasion force in April 1982, South Georgia has been continuously garrisoned by a detached company from whichever battalion is currently serving as garrison force on the Falkland Islands. The main base is still King Edward Point, but foot patrols and fixed observation posts maintain a constant watch over the coastline. Despite frequent visits by supply ships, the garrison has a monotonous and isolated existence. Casting around for some means of exploiting the island's unique facilities as a training area, Morgan-Grenville selected the seventieth anniversary of Sir Ernest Shackleton's 1915 expedition as the theme for his patrol.

He sought authority to lead a twelve-man group from King Haakon Bay on the west coast to Stromness Bay on the east. Apart from himself, the group would consist of Captain Peter Gilbert, RAMC, (to provide medical support), Sergeant A. D. 'Tug' Wilson, RM, (a Mountain Leader and Arctic warfare instructor), and ten other ranks of The Royal Greenjackets. The patrol was intended to provide a test of physical endurance, to record and correct inaccuracies on the published maps, and to assess the probable route followed by Shackleton when he made his epic journey across the island in 1916.

The first consideration was safety at every stage of the transit. Approval was granted with the proviso that at least one support ship should be available throughout the two days of the exercise, and that this vessel should have a helicopter capable of lifting the patrol to safety in the event of an accident or physical exhaustion. By a superb stroke of irony, the ship which became available for this purpose was the RFA *Tidespring*. No longer operating Wessex helicopters, she carried a Sea King of 846 Naval Air Squadron.

At 0130 on 31 January, the patrol departed Peggotty Bluff –

The Royal Green Jackets party marches across the smooth dome of the Fortuna Glacier. Beyond the range of mountains is Stromness Bay, target for the SAS reconnaissance in April, 1982.

Captain Roger Morgan-Grenville, photographed on South Georgia shortly after the successful conclusion of Exercise Green Skua.

where Shackleton had left his sick companions under the upturned *James Caird* – and started to climb by torchlight up towards the heights of the Murray Snowfield. Reaching the 800-foot contour, Morgan-Grenville and his party entered thick cloud and driving rain. On higher ground they made faster progress and by midday had halted for lunch on the ridge line of The Trident, a series of rock peaks rising to 4,000 feet behind Possession Bay and Antarctic Bay. They were crossing an area of ice, snow and steep exposed scree. Crampons and abseil ropes were essential items of equipment when traversing this section of the route, but the patrol continued to make good progress despite the katabatic wind which had developed in the past hour. It increased in strength to Force 7 and the chill factor made it necessary for each man to have at least four layers of clothing under his outer waterproofs.

At 1330 the patrol reached the most difficult section, the severely contorted surface of the Crean Glacier. Visibility improved at this time, but the patrol could advance only slowly as they encountered at least three hundred crevasses over the next mile and a half. There were numerous halts as men fell through the snow bridges – in some cases leaving only their heads exposed above the surface of the ice – but their Bergen rucksacks, and the fact that they were roped to their companions, prevented them from disappearing into the bowels of the glacier.

Once clear of the Crean Glacier, the patrol was able to press forward at a steady two miles per hour over the sloping ice and snow-covered ridge which separates the Crean Glacier from the Fortuna Glacier. At 1730 they passed Peak 2749 on their left. This mountain provided one of the essential visual reference points for the helicopter pilots during the SAS insertion in 1982. A stark reminder of that dramatic episode is the hulk of Yankee Alpha, still stranded on the slopes below the mountain. The Greenjackets posed with the wreck for the obligatory photographs before setting off across the top of the Fortuna Glacier toward the head of Fortuna Bay.

According to Captain Morgan-Grenville: 'After the Crean, the Fortuna was pretty straightforward. Apart from the broken surface around one pressure ridge, we found few crevasses and we were marching at three and a half kilometres an hour. Our feet were cold and wet from the earlier rain, but the top of the glacier was good firm going. There was a fair amount of snow covering the ice and concealing some of the crevasses, but we had crossed the glacier and descended to sea-level at Fortuna Bay by 1930 hours. The descent was very bad – an eight hundred foot slope of loose scree. By then we had been marching for twenty hours – of which fourteen had been spent roped together – and we were becoming tired. Our other support ship – HMS *Berwick* – had radioed a warning that another gale was due, so I decided to bed down for the night rather than press on for Stromness.'

At 0500, on 2 February, the patrol broke camp and crossed the beach at the foot of the König Glacier. Here they encountered the wide and fast running stream which carries melt water down to the sea. It was forded with the aid of ropes and they then tackled the final section – the 2,000-foot ridge separating Fortuna Bay from Stromness Bay. By 0815 the patrol was on the crest, looking down at Stromness whaling station and the approaches to Leith Harbour. Morgan-Grenville was standing on the same ground which John Hamilton and his Mountain Troop had hoped to reach three years earlier. An hour later the Greenjackets were in Stromness itself, being greeted by a welcoming committee from the *Berwick* and the *Tidespring* with bottles of champagne and crates of beer.

Morgan-Grenville calculates that his patrol covered approximately 45 kilometres (including detours) between King Haakon Bay and Stromness. This was achieved in 32 hours and involved the transit of three glaciers. Two mild blizzards were experienced, and for much of the time feet and legs were soaked by rain-softened snow and, in the final stages, by immersion in the König Glacier melt stream. That part of the route which lays between Peak 2749 – where Tidd and Georgeson crashed in Yankee Foxtrot and Yankee Alpha – and Stromness, was covered in fifteen hours. The Fortuna Glacier itself – from Peak 2749 to the head of Fortuna Bay – was crossed in two hours.

From these facts, and bearing in mind that the Greenjackets are not specialist mountain troops, it can be argued that Cedric Delves and John Hamilton were not being over-ambitious when they settled upon their plan for the Fortuna Glacier insertion. It will be recalled that they were prepared to allow three days for the approach march and, given that they would have been moving forward cautiously toward a hostile force, this compares favourably with the fifteen hours taken by Morgan-Grenville's patrol.

Against this proposition it has been argued that Morgan-Grenville made his march in January and therefore at a time of the year (mid-summer) when conditions of weather and 'going' are much easier than those to be expected in late April (the onset of winter). The twelve men of the Green Jackets patrol would not necessarily agree that they encountered 'easy' conditions.

While it is true that the mean air temperature is somewhat higher in the summer, and while rather less snow will be encountered on high ground at that time of the year, sudden violent winds are liable to occur at any time. The 1964 Combined Services Expedition which followed the same route started their transit in high winds and torrential rain and completed it in a Force 11 storm accompanied by heavy snow. That was in late November (springtime). The middle section of the transit – over the Crean and Fortuna Glaciers – was achieved in reasonably settled weather and the Expedition's leader, Lieutenant-Commander Mike Burley, described the Fortuna Glacier crossing as 'uneventful'.

Ernest Shackleton and his two companions made their transit when the island was approaching the coldest part of the year, late May. Starving and exhausted, they covered the entire route in 36 hours – a truly remarkable achievement.

To the layman, it would appear that the prospects for anyone contemplating this route are determined not by the topography but by the rapidly changing weather patterns which pass over South Georgia throughout the year, winter and summer. Morgan-Grenville and Burley were quite lucky, Shackleton was very lucky, John Hamilton was totally out of luck. And that made all the difference. □

APPENDIX E

Operation Matchstick

In June 1982, as described in Chapter IX, the Argentine base on Thule Island (South Sandwich Islands) was the target for Operation Keyhole, commanded by Captain Nick Barker, RN. The garrison was arrested and evacuated, the buildings sealed against the weather, and the island left unattended and unguarded. In the general rush of events, any decision with regard to the future of the base was left in abeyance. And there the matter rested for several months.

Thule Island became once again a focus of British military activity at the end of the year. On Sunday, 19 December, an 'armed reconnaissance' was put ashore from HMS *Hecate* and Argentine colours were found to be again flying from the flagstaff on Hewison Point. The discovery was reported immediately to Port Stanley and London, and a curious chain reaction was thereby set in motion.

The presence of one of the Navy's survey ships in the Southern Ocean was itself unusual. Unlike HMS *Endurance*, the *Hecate* had an 'ice-protected' rather than 'ice-strengthened' hull, and she had never operated in Antarctica. However, with 'the plum' having been so delayed by the events of April, May and June, it was necessary to deploy a ship capable of sustaining at least part of her role as ice patrol ship and Falkland Islands Dependencies guardship. The *Hecate* was the vessel chosen as temporary replacement.

At the outbreak of the war, HMS *Hecate* had been undergoing a refit in HM Dockyard, Devonport. An intensive effort by the yard's work-force had her ready for sea and sailing south by 23 July. Initially it had been planned that she would conduct survey work in connection with Operation Corporate, but the Argentines had all gone home long before the *Hecate* could arrive on station. Rear-Admiral Derek Reffell, who had relieved Rear-Admiral 'Sandy' Woodward in the South Atlantic, then changed her tasking and she instead carried out various surveys on and around the Falkland Islands and South Georgia during the period August to December. These tasks included participation in the Joint Services feasibility studies for new airports and seaports at Mare Harbour, East Falkland, and in the Bay of Isles, South Georgia.

As the end of the year aproached, ice conditions surrounding the more southerly islands improved to the stage at which navigation by a ship of the *Hecate*'s class was feasible. She duly visited the British Antarctic Survey station on Signy Island (South Orkney Islands) on 15 December, and then moved on to make the Royal Navy's first inspection of Thule Island since Captain Barker's departure in June.

There was no reason to suppose that the Argentines had re-established themselves on the island, but the approach to a potentially hostile shore was planned with some care. The *Hecate*'s commanding officer, Commander Chris Gobey, RN, conferred with Captain Giles Noakes, RM, commanding the ship's ten-man Royal Marines detachment, and with Sub-Lieutenant John Knowles, Flight Commander of her single-aircraft Wasp Flight. On 18 December, as the *Hecate* pushed steadily through five-tenths pack ice, they settled the details of Operation Casino – the insertion of armed Royal Marines onto Hewison Point.

As the ship closed the coast early the following morning, it could be seen that the adjacent Ferguson Bay was choked with drift ice. A landing there by ship's boat was out of the question. Once again, a Fleet Air Arm helicopter was the best means of making an insertion. At 0430, Wasp 414 started to lift the marines ashore to a point where they could observe the base buildings. Noakes ordered the advance as soon as the insertion was completed and his men moved forward in tactical formation.

The buildings were searched and found to be unoccupied and undisturbed, but the Union Jack and the Blue Ensign left hoisted on the flagstaff by the Operation Keyhole visitors in June were discovered neatly folded and concealed under a nearby rock. They had been replaced by the blue and white of Argentina and this apparently sinister development was reported immediately to higher authority. The *Hecate* departed that evening and went on to visit other islands in the uninhabited South Sandwich group. Behind her she left a Union Flag once again flying over Hewison Point.

The first people to hear of Commander Gobey's discovery were Captain D. S. Dobson, RN, Senior Naval Officer, Falkland Islands, and Major-General David Thorne, CBE, Commander, British Forces, Falkland Islands. It was their decision to place a security blackout on the story while the implications were considered and a response formulated. During the final days of 1982, a stream of signals flowed between Port Stanley and London. The Foreign and Commonwealth Office was advised and consulted, but this was regarded as being essentially a matter for the Ministry of Defence. Not even Sir Rex Hunt, Her Majesty's Civil Commissioner for the Falkland Islands and the Dependencies, was told of the *Hecate*'s visit to Thule Island. As he commented later: 'It was one of the less satisfactory aspects of post-conflict civil/military relations that the Falkland Islands government was not kept informed of what the military and the FCO were doing, particularly with regard to the Dependencies. The FIG was the legitimate authority and so, in theory at least, its permission had to be sought before the military undertook any operation in the area. But wartime attitudes took a long time a-dying.'

Among those few who were aware of the episode, there seems to have been no doubt that the Argentines had made what was later described as a 'hit and run' raid on Thule Island. But to what purpose? Was the new regime in Buenos Aires simply cocking a snook at the British by replanting its flag as a claim to sovereignty, or was it actively planning to reoccupy the base in the near future? In the current political climate, neither possibility could be in any way acceptable to a British government.

The continuing tensions between Great Britain and Argentina were emphasised as the New Year opened. On 4 January, angry crowds demonstrated in the streets of Buenos Aires, making their protest at the 150th anniversary of the British settlement of the Falkland Islands. They burned the Union Jack and effigies of Mrs Thatcher. They demonstrated again a few days later when it was learned that the British Prime Minister, accompanied by her husband and Admiral Sir John Fieldhouse, had travelled south via the Royal Air Force's 'air bridge' to make a surprise visit to Port Stanley. She had come to show her government's continuing commitment to the islands, to reassure the local population, and to pay her personal respects at the Commonwealth War Graves Commission cemetery on the shores of San Carlos Water. It was during this visit – which aroused more furious protests in Argentina – that the subject of Thule Island was brought to her attention. What was to be done?

There were two puzzling features to the affair. Why, if they intended to reoccupy the island, would the Argentines give advance notice of their plans by running up their flag? If they did *not* intend to reoccupy, but simply wished to make a defiant gesture, why had they left the British flags neatly concealed under a rock instead of carrying them back to Argentina as a trophy?

Discussions continued throughout the first half of January, 1983. The Ministry of Defence contacted Dr Richard Laws, Director of the British Antarctic Survey, and asked if he could throw any light on the matter. Like all such Antarctic scientific organisations, the BAS was usually well informed regarding the movements of other nations' ships in the region. The *Almirante Irizar*, as an ice-breaker capable of forcing her way through the winter pack ice, had been for a while a prime suspect, but she was soon eliminated. It was suggested that perhaps a Soviet ice-breaker might have made a call at Thule, or perhaps some Eastern-bloc trawlermen had gone ashore there. Dr Laws was in no position, at that time, to comment on any of these possibilities. Apart from some biological, geological and cartographic surveys in earlier years, the British Antarctic Survey had never had a direct interest in the place and, since April 1982, had not been in a position to acquire any new information.

By the middle of January, the MOD was being forced to the conclusion that the alien flag had been raised on Hewison Point by the crew of an Argentine submarine. This somewhat unlikely scenario assumed that the Argentine navy had somehow re-activated the *Santa Fe*'s sister boat, the *Santiago del Estero*, previously reported to be unseaworthy, or had hazarded one of their two small German-built coastal submarines, the *Salta* or the *San Luis*. Why the Argentines should incur the navigational and operational risks of sending any of these three boats in order to run up a flag on an island nearly two thousand miles from their base at Mar del Plata was unclear, but, in the absence of any alternative explanation, this appeared to be the only possible answer.

The final assessment was placed before the Prime Minister (following her return to London) with the recommendation that the Hewison Point buildings be razed to the ground. It is clear that this was not an easy decision. There were several totally conflicting considerations. The Argentines had not formally ended hostilities and it would be politically disastrous if their base was left intact and they then re-established themselves on the island in strength. The big hangar had been already half converted into a barracks block and could be easily made to accommodate a company of soldiers or marines. Reinforcements and supplies could be delivered by sea or by air from the Marambio base. Once entrenched, a determined garrison would be difficult to overcome. Any British amphibious assault could be a bloody affair and, in the swiftly changing climate of world opinion, would almost certainly earn the condemnation of the United

Nations. British thinking was coloured by the fact that the Argentines had used Thule Island as a jumping-off point, in March 1982, for their assault on South Georgia. To quote Major-General Thorne: 'It was necessary to remove temptation from their path.'

All of these factors favoured the 'blow it up' decision. Conversely, however, the importance of the base as an internationally useful scientific facility was undeniable. The buildings were modern, well-built, and worth perhaps four or five million pounds. In a period of economic restraint, it would not be easy to find the money to replace them. Further, in the hostile environment of the sub-Antarctic, all buildings are potential life-saving refuges for the crews of any ships or aircraft obliged to make an emergency landing. Destruction of the base was certainly not compatible with the hopes of those who wished to see the Argentine and British governments starting to rebuild the former bridges of friendship between the two nations. By the same taken, it was not in keeping with the general spirit of international co-operation in the sub-Antarctic and Antarctic Treaty area. As Dr Laws has wryly observed: 'My opinion was against demolition as being contrary to accepted Antarctic practice.'

Having weighed the arguments, the Prime Minister authorised the mounting of a secret task group with orders to destroy all installations on Thule Island with the exception of one small refuge hut. The necessary order was drafted in late January by the staff of General Sir Edwin Bramall, GCB, OBE, MC, Chief of the Defence Staff, and despatched to Major-General Thorne in Port Stanley. Operation Matchstick was set in motion.

The two ships chosen to conduct the operation were HMS *Ariadne* (Captain David Evans, RN) and the by now familiar *Tidespring* (Captain John Wilkins, RFA). The Leander class frigate *Ariadne* was selected because, apart from being available in Falkland Island waters at the time, she was one of the five remaining ships of her class not yet converted to the missile carrying role. She still retained her 4.5-inch gun turret and, as Operations Paraquat and Keyhole had shown, the ability to employ naval gunfire support for amphibious operations was not simply useful, it was potentially vital. The *Tidespring*, by contrast, possessed no significant armament, but her big hangar and flight-deck could accommodate the Sea King helicopter. Gunfire and helicopter lift would be the keys to this little operation as they had been throughout the campaign of the previous year.

Orders to move south-east were received in the *Ariadne* and the *Tidespring* in the closing days of January. Additional stores were embarked and they started to make their way towards the South Sandwich Islands. Embarked in the big tanker were two Sea King HAS5 helicopters of 826 Naval Air Squadron. One was the *Tidespring*'s own Flight, the other loaned by the RFA *Fort Austin*. Commanding the combined Flight was Lieutenant-Commander Mike Wells, RN, who recalls the departure from Falkland Island waters: 'We left on 26 January, sailing under circumstances of great secrecy and remaining covert throughout the passage down to Thule. People in the ship felt ninety-five per cent certain that we would not encounter any Argentine ships or aircraft, but no chances were taken. We were in radio and radar silence. The main worry was that we might be spotted and reported by Russian trawlers working in the area, but in the event we saw no other ships and apparently nobody saw us. We needed four days to reach Thule, and during that time we planned and rehearsed the landing. Our military force was quite small – just the nine marines of *Ariadne*'s detachment and the RM detachment loaned to us by HMS *Antrim*. The main assumption was that we would find the place empty, and that we would then proceed to blow it up. For this purpose we had embarked a Sergeant, Royal Engineers, and a lot of explosives. The military commander was Lieutenant Charlie Wilson, RM, one of *Ariadne*'s watch-keeping officers. It was his intention to go ashore with a reconnaissance patrol at midnight on 30/31 January, find out whether the base was occupied, and then take whatever action seemed appropriate. In the event, it was decided to go straight in at first light on 30 January.'

The two ships had followed a circuitous approach route, sailing north and east of South Georgia. They met fewer icebergs and growlers than expected and made a fast passage. Planning and rehearsal for the operation was based in large part upon information obtained by the *Ariadne*'s Navigator, Lieutenant Ian Gibson, RN. Shortly before the ships' departure from Port Stanley, he had been flown down to Thule Island in a C-130 Hercules of the Royal Air Force and had returned with a set of excellent aerial photographs. These pictures tended to confirm that the Argentine base buildings were unoccupied and when, in brilliantly clear weather, HMS *Ariadne* arrived off Hewison Point, it was again apparent that the landing would probably not be opposed. This assessment was reported to the *Tidespring* – following some miles astern – and the order was given to launch her two Sea Kings with twenty-one Royal Marines embarked. At this point it was found that ZA129 was unserviceable. It was decided, therefore, to land the marines in two serials, both to be flown ashore in ZA136 piloted by Lieutenant-Commander Wells. He would make his approach from the east, flying low over the northern end of Hewison Point were it adjoins the rising ground of Mount Larsen. To protect his landing, to suppress any defensive fire which he might attract, HMS *Ariadne* would lay off the western side of the Point, near Ferguson Bay, ready to open fire at the shore if necessary. Overhead would be HMS *Ariadne*'s Wasp, flown by Lieutenant Chris Harrington, RN, and armed with a door-mounted general purpose machine gun (GPMG) manned by his aircrewman, 'Slinger' Wood.

Everything went according to plan. Wells put the Royal Marines down at the agreed point without difficulty. They advanced, checked the buildings, and found no signs of recent occupation. The next step, therefore, was to prepare the base for demolition. Pallets of explosive were lifted ashore by the Sea King during the remainder of the day and Sergeant Peter Chatfield of the Royal Engineers set about the placing of his charges.

Destruction of the buildings took much longer than anticipated. The Argentine designers and construction teams had done their work well. The task occupied four days and was not completed until 4 February. To save time, and to reduce the hazards of flying in sometimes adverse weather, the marines and Sergeant Chatfield remained ashore, living and sleeping in the large hangar which was deliberately left intact until all of the other buildings had been flattened. When the time came, the hangar was found to be an exceptionally strong structure and a charge consisting of several Mark 10 anti-submarine mortar bombs was needed to destroy it. One of these mortar bombs was placed beside a tracked Sno-cat found parked inside the hangar. After the explosion, the largest identifiable remnant of this vehicle was a small fragment of its gear-box casing. The building itself was literally blown to pieces. It would seem that a campaign in part triggered by the scrap metal of South Georgia had ended with the creation of yet more scrap metal on Thule Island. As Admiral of the Fleet Sir John Fieldhouse commented recently: 'Matchstick set the seal on the job.'

One unexpected problem encountered by the demolition party was the refusal by the large penguin colonies on Hewison Point to

Several of the Argentine base buildings are blown to pieces in one of a series of controlled explosions. The long building on the right side of the picture is the large helicopter hangar used by the British shore party as temporary accommodation and the last structure to be demolished.

remove themselves from danger. The Royal Marines spent much time trying to drive the birds away to a safe distance before each explosion but, inevitably, many shuffled back to their customary patch of ground at the last moment. There were numerous deaths and injuries as a result. Damaged birds were rounded up and humanely destroyed. According to Lieutenant John Gilbert, RN, of HMS *Ariadne*: 'This went on for several days and the penguins then got fed up with the bangs and they started to move up the slopes of the mountain. On the last day, as we were departing, we could see a big black U-shaped column on the snowfield. The leaders had waddled up several hundred feet, then changed their minds, and started to come down again. As they did so, they were passing the tail-enders who were still on their way up.'

The small task group departed Thule Island on 4 February and headed back to the Falkland Islands. Behind them they left intact one small refuge hut, packed with emergency food supplies and survival gear for the use of any distressed Antarctic travellers who might one day need shelter.

The secrecy of what had been done was maintained for several weeks, but it was inevitable that it would become common knowledge sooner or later. It seems likely that the destruction at Hewison Point was first reported by a Soviet fishing fleet working in the area, but there is no certainty. In any event, on 11 March 1983, *The Times* of London carried a report which stated: 'Argentina has made a hit and run raid on the British Dependency of Thule. A raiding party landed and ripped down the Union Flag and hoisted their own before retreating undetected.' *The Times* also quoted a Ministry of Defence spokesman as saying: 'We are not inclined to treat this incident as a serious indication of Argentina attempting to re-establish a presence on the Dependencies.'

This low-key MOD response to Press enquiries was clearly inconsistent with the decision to destroy the base. It was at variance also with the reaction in Buenos Aires where, for the first time, the authorities learned of the reports published in British newspapers. The Argentine Naval High Command 'denied all knowledge' of their flag having been hoisted on Thule, and the new President, Reynaldo Bignone, dismissed the stories as 'gossip'.

Suspecting that there was more to the story than met the eye, British journalists pressed Mrs Thatcher when, on the following day (13 March), she made a public appearance at the Port of Felixstowe. To quote *The Times* again: 'Asked by reporters if she knew that the Argentine flag had been taken down, she said, "Yes. It's a long way away and it must have been an Argentine ship calling there to put the flag up. We don't know. But we are on the alert the whole time."'

Further Press calls to the MOD elicited more details from a spokesman who appeared to be better briefed than his colleague of the previous day: 'The landing was discovered by HMS *Hecate* on 19 December. The Union Flag was folded and hidden under rocks. British forces were later sent to destroy all but one small building in view of the possibility that Argentina might seek to reoccupy the well-equipped base. We are obviously not encouraging a presence there and we have taken the necessary steps.'

On the face of it, that was the end of the story. Argentine newspapers carried the expected angry responses and accused the British of 'vandalism' and 'arrogance', but the episode was soon forgotten in the nation's drive to root out the guilty men of the 'dirty war' period and to re-establish democratic government. The British people also had other topics on their minds and Thule Island was, as the Prime Minister had observed, 'a long way away'. Only in the international scientific community did the affair continue to provoke interest.

As described in the main narrative and in the Epilogue, Antarctica is in many ways 'the international continent'. Politicians have drawn lines on the map and have laid claims to various wedges of territory leading down to the South Pole, but this nineteenth-century concept of territorial possession is fundamentally inappropriate to a region where the human race can never establish itself as it has in other parts of the world. The scientists of the eighteen nations working in Antarctica continue to maintain a friendly exchange of information, regardless of politics and boundaries. Very little can happen in the region without it being freely discussed on this international scientific network. One of the common denominators between these people is their perennial requirement for more money from their respective governments. Many were saddened to learn that a large, modern and expensive installation, a part of the meteorological/climatological research facilities in the deep south, was no more. They were curious about the circumstances leading to the British demolition visit, and they did not believe that the Argentines were culpable. So who was?

For a time it was suspected that a Soviet trawler or survey ship crew had replaced the Union Flag with Argentine colours on Hewison Point. There were various suggestions to explain why the Russians might have done such a thing – to embarrass the British, to show moral support for the defeated Argentines, or to generally sustain an atmosphere of tension and instability in the area. None of these arguments were very compelling, and the enquiries continued in other directions.

Each nation with scientific stations in the Antarctic and sub-Antarctic publishes an annual account of its current and planned activities. These reports detail the movements of each survey and base support vessel operating in the region. A study of these reports – dealing with the 1982/83 season – revealed which ships had made what visits and which, therefore, might have gone to Thule Island. More importantly, it was necessary to establish what visits might have been made during the few weeks between the time when the pack-ice started to break up, in early November, and the arrival of the *Hecate* on 19 December.

It was rumoured, for a while, that the South African naval survey ship, *Protea*, had called at Thule during the period under review. A *Hecla* class purpose-built hydrographic survey vessel purchased from a British yard, Yarrow (Shipbuilders) Limited, in 1969, she had for the past ten years played a valuable role in charting the waters of the Antarctic and Southern Oceans. She was owned and operated by a government which for many years had been (and still is) a highly respected participant in Antarctic research. It would have been very unwelcome news if any of her crew had been involved in an action likely to cause an international incident. Recent enquiries (1986) have in fact eliminated the *Protea* from the short-list of suspects. Throughout 1982 she was engaged in survey work off the coast of Namibia and South West Africa. On 27 October she went to Simonstown for her ten-year refit and was in dockyard hands for the following twelve months.

Attention turned next to another South African ship, the *Agulhas*. Built by Mitsubishi at Simonoseki, Japan, in 1977/78, this Antarctic survey and supply vessel of 3,000 tons was operated by the Department of Transport. It is acknowledged that she did indeed visit Thule Island, but not until 14 January 1983. The reason for her visit was a plan by the South African meteorological service to install an unmanned automatic weather station on Hewison Point. Thule Island is the most easterly island for thousands of miles at that latitude and the weather reports produced by the Argentines stationed there, prior to their removal by Captain Barker, had been of particular value to South African meteorologists. Pretoria hoped to obtain permission from London to take over the tasks previously performed by the Argentines (but without establishing a human presence on the island). We can only speculate on the full details of this proposal, but Pretoria must have felt reasonably confident that all would be well. There could have been no purpose in sending the *Agulhas* to Thule Island – with the new weather station in her cargo hold, ready to be installed on Hewison Point – if there had been any substantial doubts.

In the event, the *Agulhas* anchored in Ferguson Bay on 14 January and awaited formal clearance, from the FCO in London, via Pretoria, to send her installation team ashore. She remained at anchor all day but, receiving no response, moved on to carry out survey work elsewhere. London's reply was received shortly afterwards – permission to install the equipment was denied. The reason for the denial is not hard to imagine – the FCO already knew that Hewison Point would shortly become a very inappropriate location for sensitive scientific instruments.

With the South Africans absolved from responsibility, attention turned again to other vessels which might have visited Thule between early November and 19 December. All the usual known survey ships and warships were accounted for, so the visitors must have been officers or crewmen from a merchant vessel. There is insufficient evidence, at the present time, to enable the author to make a categoric statement. It is known, however, that certain Eastern-bloc ships were in the general area, and it is also a matter of record that a Western-bloc Antarctic cruise liner was within thirty miles of Thule during the key period.

Whoever it was who raised the Argentine flag on Hewison Point, he was almost certainly not an Argentine and he is very unlikely to have come ashore from any ship involved in the international Antarctic exploration and development programme. All of the available evidence points instead to a practical joke perpetrated by someone coming ashore for a casual sight-seeing visit from a commercial vessel. Bearing in mind the then still unresolved state of hostility between Argentina and the United Kingdom, and given the depth of passion aroused by the related question of whose flag should fly over the Falkland Islands, it was a remarkably irresponsible act and the joke soon back-fired.

It is always possible that, regardless of the *Hecate*'s December discovery, the British would have sooner or later decided to destroy the former Argentine base anyway, but the circumstances surrounding Operation Matchstick certainly provide a bizarre footnote to one of the most extraordinary campaigns ever fought by Her Majesty's armed forces. □

APPENDIX F

Honours and Awards

NO UNIFIED LIST of personnel receiving recognition for their part in Operation Paraquat has been published by the Ministry of Defence, and it is not difficult to understand why this should be. The nature of the 1982 campaign was such that certain ships and certain military units were engaged in more than one major operation and it is therefore impossible, in many cases, to decide whether an individual was honoured, at the conclusion of the campaign, for his contribution to Operation Paraquat, or to Operation Corporate, or to Operation Keyhole, or to a combination thereof.

There were only eleven 'immediate' awards for Operation Paraquat and these appeared in *The London Gazette* of 4 June 1982 (they are indicated in the following roll by an asterisk). However, many more names appeared later, in *The London Gazette* of 11 October 1982, and it is evident that some of those awards related in one way or another to services rendered on and around South Georgia and Thule Island. The same conclusion can be drawn from at least two awards which were announced in 1983.

It is with these difficulties in mind that the author has thought it proper to construct this roll according to a simple criterion: he has included *all* personnel who were honoured and who served in *any* of the ships or units deployed for Operation Paraquat and/or Operation Keyhole.

HMS *ANTRIM*

Distinguished Service Order
Captain B. G. Young, RN*
Lieutenant-Commander I. Stanley, RN*

Officer of the Order of the British Empire (Military Division)
Commander R. J. Sandford, RN

Distinguished Service Cross
Fleet Chief Petty Officer (Diver) M. G. Fellows[1]

British Empire Medal (Military)
Chief Air Engineering Artificer (M) D. J. Heritier

Mentioned in Despatches
Sub-Lieutenant S. G. Cooper, RN*
Marine Engineering Mechanic (M) 1st Class C. Crowley
Petty Officer Aircrewman D. B. Fitzgerald*
Sub-Lieutenant D. E. Graham, RN
Chief Petty Officer (Ops) (M) E. Graham
Marine Engineering Mechanician (M) 1st Class T. Miles
Lieutenant C. J. Parry, RN*
Sub-Lieutenant A. Pringle, RN

Commander-in-Chief's Commendation
Sub-Lieutenant M. A. Bonney, RN
Weapon Engineering Artificer 1st Class F. Latham
Petty Officer Weapon Engineering Mechanician (R) G. W. Lockwood
Petty Officer Weapon Engineering Mechanician (M) G. K. Robertshaw
Marine Engineering Mechanic (M) 1st Class D. I. Roberts
Able Seaman (M) C. W. Waterfall

Note:
1. FCPO Fellows was not a member of the ship's company but was tasked to HMS *Antrim* on 21 May to investigate an unexploded Argentine bomb lodged in her Seaslug missile magazine. He and his team removed the bomb after working for ten hours in very hazardous conditions.

HMS *BRILLIANT*

Distinguished Service Order
Captain J. F. Coward, RN

British Empire Medal (Civil)
Mr Shik-Ming Kang[1]

Mentioned in Despatches
Lieutenant N. A. M. Butler, RN
Lieutenant-Commander B. W. Bryant, RN
Lieutenant-Commander J. N. Clark, RN
Leading Seaman (Missile) R. M. Gould
Lieutenant-Commander L. S. G. Hulme, RN
Weapon Engineering Mechanician (R) 1st Class J. R. Jesson
Marine M. A. Neat, RM

Commander-in-Chief's Commendation
Lieutenant P. A. McKay, RN
Air Engineering Mechanic (M) 1st Class P. J. O'Hara

Note:
1. Mr Kang was the head laundryman in HMS *Brilliant* and also the contractor for laundry services in several other ships deployed in the South Atlantic. His powers of persuasion and cheerful example contributed significantly to the fact that his compatriots stayed at their posts in difficult and dangerous circumstances, thereby helping to sustain hygiene and morale in Paraquat and Corporate ships.

HMS *CONQUEROR*

Distinguished Service Order
Commander C. L. Wreford-Brown, RN

Distinguished Service Medal
Acting Petty Officer (Sonar SM) G. J. R. Libby

Mentioned in Despatches
Weapon Engineering Artificer 2nd Class J. M. C. Foy
Marine Engineering Artificer (M) 2nd Class D. J. Leaning
Marine Engineering Artificer (M) 2nd Class S. D. Mitchell

Commander-in-Chief's Commendation
Radio Supervisor (SM) J. M. Bysouth
Lieutenant-Commander J. M. R. Garland
Lieutenant-Commander D. A. Hall
Lieutenant-Commander T. P. McClement

HMS *ENDURANCE*

Commander of the Order of the British Empire (Military Division)
Captain N. J. Barker, RN

Member of the Order of the British Empire (Military Division)
Lieutenant D. C. W. O'Connell, RN

British Empire Medal (Military)
Petty Officer Medical Assistant K. Adams
Chief Mechanical Engineering Mechanician (M) G. S. Cox
Petty Officer D. W. Craig
Chief Petty Officer D. Hamilton

Distinguished Service Cross
Lieutenant-Commander J. A. Ellerbeck, RN*
Acting Lieutenant K. P. Mills, RM*

Distinguished Service Medal
Sergeant P. J. Leach, RM*

Mentioned in Despatches
Marine D. S. Combes, RM
Marine L. Daniels, RM
Lieutenant-Commander W. E. Hurst, RN*
Able Seaman (R) M. S. Leach
Petty Officer (R) J. Pearson*
Marine J. Stonestreet, RM
Lieutenant C. Todhunter, RN
Lieutenant D. A. M. Wells, RN*

Commander-in-Chief's Commendation
Lieutenant-Commander A. A. Ainslie, RN
Petty Officer (NAAFI Manager) D. R. Atkinson
Leading Radio Operator (G) S. P. Brereton
Leading Radio Operator (G) I. D. Campbell
Able Seaman J. W. Ferguson
Lieutenant T. S. Finding, RN
Lieutenant D. Hutchings, RN
Leading Aircrewman R. B. Nadin
Marine S. J. Parsons, RM
Leading Seaman D. J. Snell
Marine R. K. White, RM

HMS *PLYMOUTH*

Distinguished Service Order
Captain D. Pentreath, RN

British Empire Medal (Military)
Petty Officer (Missile) J. J. T. Waterfield

Mentioned in Despatches
Radio Operator (Tactical) 1st Class R. J. Ash
Leading Aircrewman J. A. Harper
Leading Radio Operator (T) R. J. Hutcheson
Lieutenant-Commander K. M. Napier, RN
Acting Petty Officer Marine Engineering Mechanic (M) D. M. K. Ross

Commander-in-Chief's Commendation
Able Seaman (M) P. M. Orr
Lieutenant-Commander D. J. Sayer, RN
Leading Seaman A. Whitehead

The Distinguished Service Cross

The Distinguished Service Order

HMS *YARMOUTH*

Distinguished Service Cross
Commander A. Morton, RN

Mentioned in Despatches
Sergeant I. D. Fisk, RM
Lieutenant-Commander P. G. Miller, RN

RFA *FORT AUSTIN*

Distinguished Service Order
Commodore S. C. Dunlop, CBE, RFA[1]

Member of the Order of the British Empire (Civil Division)
James Quirk, Esq, Supply & Transport Officer (Naval)

Note:
1. Commodore Dunlop had been appointed MBE some years earlier. He was elevated to CBE in the Queen's Birthday Honours List on 12 June 1982 in recognition of his services as Commodore, RFA. He received both decorations, CBE and DSO, from the hands of Her Majesty upon his return to the United Kingdom.

RFA *OLMEDA*

Officer of the Order of the British Empire (Military Division)
Captain G. P. Overbury, RFA

RFA *TIDESPRING*

Officer of the Order of the British Empire (Military Division)
Captain S. Redmond, RFA

'M' COMPANY GROUP, 42 COMMANDO ROYAL MARINES

Officer of the Order of the British Empire (Military Division)
Major J. M. G. Sheridan, RM

Mentioned in Despatches
Sergeant J. Napier, RM

Commander-in-Chief's Commendation
Captain C. J. Nunn, RM

148 (MEIKTILA) COMMANDO FORWARD OBSERVATION BATTERY, ROYAL ARTILLERY

Military Cross
Captain W. A. McCracken, RA

Mentioned in Despatches
Captain C. C. Brown, RA
Lieutenant-Colonel K. R. H. Eve, RA
Bombardier N. R. Jackson
Captain J. H. McManners, RA[1]
Warrant Officer 2nd Class M. D. Richards, RA

Note:
1. Captain McManners was honoured for his services on East Falkland.

'D' SQUADRON, 22 SPECIAL AIR SERVICE REGIMENT

Distinguished Service Order
Major C. N. G. Delves

Military Cross
Captain G. J. Hamilton (posthumous)[1]
'Tim' (Boat Troop)

Mentioned in Despatches
Eight officers and other ranks

Note:
1. Captain Hamilton was killed in action on East Falkland under circumstances of great gallantry.

No 2 SECTION, SPECIAL BOAT SQUADRON, ROYAL MARINES

Mentioned in Despatches
'Scobes'
'Willie'

Commander-in-Chief's Commendation
'Clive'
'Thommo'

57 SQUADRON, ROYAL AIR FORCE

Mentioned in Despatches
Squadron Leader J. G. Elliott, RAF

Note: although his award was not related directly to Operations Paraquat, Keyhole or Matchstick, it is appropriate to mention that Sub-Lieutenant J. M. Knowles, RN, was awarded the Air Force Cross as a direct result of his aviation services in the Antarctic. In 1982–83, he was Flight Commander in HMS *Hecate*, temporarily deployed in place of HMS *Endurance*. Subsequently he was appointed in a similar role in HMS *Herald*. According to Commander C. S. Gobey, commanding officer of HMS *Hecate*, John Knowles 'displayed outstanding and extended airmanship' in situations in which no other helicopter was available on standby to assist him in the event of an emergency. During his time in the south, he flew each month an average of seventy-five hours 'over glaciers, ice and freezing seas'.

The British Empire Medal The Distinguished Service Medal

APPENDIX G

Biographical Notes

THE FOLLOWING NOTES refer to some of the personnel mentioned by name in the preceding narrative. The ranks, appointments and decorations are those held in April 1982.

Captain Nicholas BARKER, RN
Commanding Officer, HMS *Endurance*

Born Malta, 1933. Son of a serving naval officer, educated at Canford School, he obtained a regular commission in 1953. Subsequently commanded seven ships and held a number of staff appointments. Deputy Commodore Contract Built Ships, 1973–75. Stood by and commanded HMS *Arrow*, 1975–79. Assumed command of HMS *Endurance* in 1980. He is a Fellow of the Royal Geographical Society, a Fellow and Committee Member of the British Institute of Management, a Member of the Nautical Institute, and a Younger Brother of Trinity House. Subsequently awarded The 1983–84 Defence Fellowship at Churchill College, Cambridge University. Married, with four children, his family home is in Gillingham, Dorset. Appointed a Commander of the Order of the British Empire (Military Division).

Lieutenant-Commander Christopher BLIGHT, RN
Flight Commander, 'A' Flight 848 Naval Air Squadron, RFA *Regent*

Born 1948, Torquay, Devon. Educated Torquay Grammar School. Joined the Birmingham City Police as a Cadet in 1965 but then entered the Royal Navy as a Sub Lieutenant in 1967. Qualified as a fixed wing and helicopter pilot. Over the next nine years he served with 845, 771 and 707 NAS in HM Ships *Albion*, *Bulwark*, *Hermes*, and *Invincible*, and in the RFAs *Engadine*, *Resource*, *Regent*, *Olwyn*, *Reliant* and *Olmeda*. Qualified helicopter instructor, 1979. He was working in a staff appointment at RNAS Yeovilton at the time of the Argentine invasions and was loaned to 848 NAS for the duration of the conflict.

Lieutenant-Commander Barry BRYANT, RN
Flight Commander, HMS *Brilliant* Lynx Flight

Born Dorchester, Dorset, 1949. Educated at Wymondham College, Norfolk. He joined the Royal Navy as a prospective pilot, having chosen his career at the age of four (after witnessing a Dragonfly helicopter demonstration during a Devonport Navy Day). Qualified as Observer after initial flying training. Served with 824 NAS in HMS *Ark Royal*, 1970, flying in the Sea King, then returned to the Wessex. Served in HMS *Fife* and as an Instructor at RNAS Portland. Operations Officer in HMS *Tiger*. Principal Warfare Officer, HMS *Achilles*, 1977–78. After a staff appointment, he converted to the Lynx and was serving in HMS *Avenger* at the outbreak of the 1982 campaign. A year later he was badly injured when his Lynx ditched off Oman, but later returned to normal duties. Married, two daughters, resident of Sydling St Nicholas, Dorset.

Lieutenant Nicholas BUTLER, RN
Pilot, HMS *Brilliant* Lynx Flight

Born Eastbourne, 1955, and educated at King George V School, in Hong Kong, and at Tunbridge Wells Grammar School. Entered BRNC Dartmouth in 1973. Sea service in HM Ships *Falmouth*, *Bacchante* and *Scylla*. Commenced flying training in 1978. Joined 820 NAS as a Sea King pilot and served in HM Ships *Blake*, *Hermes* and *Invincible* before converting to the Lynx in 1981. Like his friend Barry Bryant, he was serving in HMS *Avenger*, as Flight pilot, at the commencement of the conflict, and was loaned to HMS *Brilliant* for its duration. Married, he lives in Dorset with his wife and two young sons.

Above left: Capt Nicholas Barker, RN.
Above right: Lieut-Cdr Christopher Blight, RN.
Centre left: Lieut-Cdr Barry Bryant, RN.
Centre right: Lieut Nicholas Butler, RN.
Left: Lieut-Cdr John Clark, BSc, RN.

Lieutenant-Commander John CLARK, BSc, RN
Pilot, HMS *Brilliant* Lynx Flight

Born 1949, Sunderland, educated at Bede Grammar School. Entered Britannia RNC 1968. Served as a Midshipman in HM Ships *Diana* and *Minerva*. Promoted Sub Lieutenant in 1970 and entered the Royal Naval Engineering College, Manadon, Plymouth, to read Mechanical Engineering. Promoted Lieutenant 1974 and joined HMS *Daedalus* as Air Engineering Officer. Selected for pilot training and qualified as Sea King pilot in 1977. Served with 826 NAS in HMS *Tiger*. Converted to the Lynx and

Above left: Sub-Lieut Stewart Cooper, RN.
Above right: Capt John Coward, RN.
Centre left: Lieut-Cdr John Dransfield, RN.
Centre right: Cdre Sam Dunlop, MBE, RFA.
Left: Lieut-Cdr Anthony Ellerbeck, RN.

commanded HMS *Arrow*'s ship's flight, 1979–81. Promoted to Lieutenant-Commander in 1981 and appointed as instructor, Lynx Flight Simulator Unit. Loaned to HMS *Brilliant* as Lynx pilot for the duration of the South Atlantic conflict. Retired from the Navy in 1983 and joined Bristow Helicopters Ltd. Mentioned in despatches.

Sub-Lieutenant Stewart COOPER, RN
Pilot, HMS *Antrim* Wessex 3 Flight

Born in Aberdeen, 1954, and attended Robert Gordon's College until 1971. Joined the Royal Navy in 1974 as Radio Electrical Mechanic (Air) and served with 737 NAS. In 1976 he was drafted to RAF Lossiemouth and worked for eighteen months in radio workshops and the Radar Simulator before joining 849 NAS in HMS *Ark Royal*. Accepted for officer training in 1977 and entered BRNC *Britannia*. Left the Navy briefly in 1979, then re-entered and commenced flying training. Appointed to HMS *Antrim* in December, 1981. Married, he resides with his family in Weymouth, Dorset. Mentioned in despatches.

Captain John COWARD, RN
Commanding Officer, HMS *Brilliant*

Born in 1937 and educated at Downside School. After initial training at BRNC Dartmouth, he volunteered for service in the submarine branch in 1960. After a number of appointments, he became First Lieutenant of HMS *Warspite* in 1969 (his Commanding Officer being Commander Sandy Woodward). Subsequently appointed to command HMS *Oracle* and then the nuclear submarine HMS *Valiant*. Promoted Captain in 1978, he served as Naval Assistant to the First Sea Lord before assuming command of HMS *Brilliant* in 1980. Following the conflict, he was appointed Commodore, Director of Operational Requirements (Navy). Married, with two sons, he resides in Cornwall. Admitted to the Distinguished Service Order.

Major Cedric DELVES
Officer Commanding, 'D' Squadron,
22 Special Air Service Regiment

Born 1947, son of a serving army officer, entered the RMA Sandhurst in 1966. Played rugby, learned to parachute. Commissioned into the Devon & Dorset Regiment, 1966, and served with them in Germany, Malta and Ulster. Encouraged by his Adjutant, Captain 'H' Jones, to apply for secondment to the SAS. Accepted 1973, and subsequently served as a Troop Commander in Dhofar and elsewhere. He returned to his parent regiment as a Company Commander, serving again in Germany and Ulster. Selected for Staff College Camberley, 1979, passed 1980. Returned to 22 SAS in 1981 as 'D' Squadron commander. Married, three children. Admitted to the Distinguished Service Order.

Lieutenant-Commander John DRANSFIELD, RN
Flight Commander, HMS *Plymouth* Wasp Flight

Born 1945, entered the Royal Navy in 1965. Qualified as ASW pilot. Flew Wessex 1 and Wessex 3 with 826 and 737 NAS in the Far East and at RNAS Portland, 1967–70. Seconded to Royal Australian Navy (725 Sqn, RANAS Nowra). Commanded Lee-on-Solent SAR Flight (Whirlwind 9), 1972–74. Converted to the Sea King and served in HMS *Hermes* and *Ark Royal*, 1977–79. In charge of the Sea King flight simulator unit at RNAS Culdrose, 1979–81. Flight Commander Ship's Wasp Flight, HMS *Ariadne* 1981, joined HMS *Plymouth* in April, 1982.

Commodore Sam DUNLOP, MBE, RFA
Master, RFA *Fort Austin*

Born 1922 in Ulster, educated at Belfast Royal Academy, he joined Andrew Weir & Co as an apprentice in April, 1939. Transferred to the RFA in 1943. Served extensively in the North Atlantic, Mediterranean and Pacific Ocean during the 1939–45 war. Appointed a Member of the Order of the British Empire. Continued to serve with the RFA throughout his career, becoming the most senior officer afloat and Commodore. He was host to HM the Queen and HRH the Prince Philip when they went aboard the RFA *Fort Austin* during the celebrations to mark the 70th anniversary of the RFA. Appointed Commander of the Order of the British Empire in the 1982 Birthday Honours List and admitted to the Distinguished Service Order for his services during operations off the Falkland Islands.

Lieutenant-Commander Anthony ELLERBECK, RN
Flight Commander, HMS *Endurance* Wasp Flight

Born in Harrogate, Yorkshire, in 1943, he attended Leeds Grammar School before joining the Cunard-Brocklebank Line as a Deck Officer Cadet in 1959. Served in various cargo liners and tankers and promoted to Chief Officer before leaving the Merchant Navy to join the Royal Navy as a Cadet in 1967 with the specific intent of learning to fly. After leaving BRNC Dartmouth, he qualified as a pilot and joined 845 NACS in HMS *Albion*. Flight Commander, HMS *Sirius*, 1974. Seconded to the Royal New Zealand Air Force as an Instructor, 1976–78.

Appointed Maintenance Test Pilot at RNAS Yeovilton, 1978–80, before joining HMS *Endurance* in May 1980 as Flight Commander. Married, with three children, he resides in Somerset, where much of his leave periods are devoted to deer stalking and fishing. Awarded the Distinguished Service Cross.

Lieutenant-Colonel Keith EVE, RA
Royal Artillery Liaison Officer for Naval Gunfire Support

Born 1928 in Srinagar, Kashmir, son of an Indian Army officer. Raised in India, Hong Kong, France and England, he completed his education at Blundell's, Tiverton. Entered the army as a National Serviceman in 1946, commissioned in 1948. Played rugby for Combined Services (Scotland) before serving in Hong Kong and Germany. Promoted Captain in 1955. Qualified parachutist. Attached briefly to the French Army before taking part in anti-EOKA operations in Cyprus, 1956, and operations in the Yemen, 1957. Passed Commando course 1964. Commanded 79 Cdo Light Battery RA on operations in Aden and The Chestnut Troop RHA in Radfan, 1966–67. Appointed RALONGS and promoted Lieutenant-Colonel, 1967. He became responsible for all aspects of NGS training and documentation at MOD and sub-unit levels, and for NGS training and planning with NATO and Commonwealth forces. Commanded the NGS element in Operation Paraquat and Operation Tornado, served as NGS staff officer for Operation Sutton. Mentioned in despatches.

Captain Michael FARLEY, RFA
Master, RFA *Brambleleaf*

Born 1936, Torquay, Devon. Educated at Torquay Grammar School and Plymouth School of Navigation. He entered the Merchant Navy in 1954 and served with J. & C. Harrisons Ltd until joining the RFA in 1961 as a Second Officer. He has served in a large number of RFA ships, receiving his first command in 1977. Married, resident of Paignton, Devon.

Lieutenant Timothy FINDING, RN
Senior Pilot, HMS *Endurance* Flight

Born 1959, son of a former RAF pilot. Entered Britannia RNC 1977, qualified as a helicopter pilot 1979. Served with 819 NAS (Sea King), 1979–81. Converted to Wasps in 1981 and immediately joined HMS *Endurance*.

Petty Officer Aircrewman David 'Fitz' FITZGERALD
Aircrewman, HMS *Antrim* Wessex Flight

Born 1948, educated at Redbridge School. Entered the Royal Navy, HMS *Raleigh*, in 1965, and trained as a seaman. Specialised in torpedo and anti-submarine warfare. Served in HM Ships *Tiger*, *Brighton*, *Arethusa* and *Diomede*. Commenced aircrew training in 1974 and subsequently served in ships' helicopter flights in HM Ships *Tiger*, *London* and *Antrim*. Married, one son, resident of Southampton. Mentioned in despatches.

Lieutenant Ian GEORGESON, RN
Pilot, 'C' Flight 845 Naval Air Commando Squadron, RFA *Tidespring*

Born in Latheron, Caithness, in the north of Scotland in 1959. Educated at Latheron Primary and North Berwick High Schools. He entered the Royal Navy in 1976 and qualified as a Wasp pilot. Subsequently served in HMS *Galatea* and, as Flight Commander, in HMS *Ashanti* and HMS *Tartar*. Qualifying on the Wessex, he served a tour of duty in Northern Ireland before moving to 846 NAS flying the Sea King. During this time he became one of the very few people to qualify for the 'Lebanon' clasp on his Campaign Service Medal.

Above left: Lieut-Col Keith Eve, RA. Above right: Capt Michael Farley, RFA. Centre left: Lieut Timothy Finding, RN. Centre right: POACMN David 'Fitz' Fitzgerald. Below left: Lieut Ian Georgeson, RN. Below right: Sgt Peter Leach, RM.

Sergeant Peter LEACH, RM
Detachment Sergeant-Major, HMS *Endurance*

He joined the Corps in 1963 from the Royal Marines Reserve. Completed two tours in Borneo during the Confrontation with Indonesia and saw action as a leading scout. Served aboard HMS *Zest* in 1966–67, qualified as an Observer on Sioux and Scout helicopters with the Brigade Air Squadron in 1970. Completed nine years' service in 1972 and joined the Liverpool & Bootle Constabulary as a Police Constable. Resigned fourteen months later and rejoined the Corps. Again promoted Corporal, served in Ulster and Cyprus, and qualified as air photography instructor and small boat coxswain. Promoted Sergeant in 1976 and qualified as a sniper marksman. Employed at HMS *Raleigh* 1978–81 as seamanship instructor before joining HMS *Endurance*. Awarded Distinguished Service Medal.

Lieutenant Paul McKAY, RN
Observer, HMS *Brilliant* Lynx Flight

Born 1953 in Warrington, Lancashire. Educated at Boteler Grammar School, Warrington, he entered BRNC Dartmouth in 1973 and commenced flying training shortly afterwards. Qualifying as an ASW Observer, he served successively with 824 NAS in HMS *Ark Royal* (1974), 819 NAS at HMS *Gannet*, Prestwick (1975), and with 737 NAS at RNAS Portland (1976). During this time he became a Helicopter Warfare Instructor. In 1979 he moved to 829 NAS and was appointed to HMS *Endurance* as Senior Observer for the next two years. During this period he became thoroughly conversant with the South Atlantic and Antarctica, and this background experience was invaluable during the 1982 campaign. Married, with one son, he lives in Weymouth, Dorset.

Above left: Lieut Paul McKay, RN. Above right: Lieut-Cdr Ian McKenzie, RN. Centre left: Lieut-Cdr Paul Miller, RN. Centre right: Cdr Antony Morton, RN. Below left: Capt Christopher Nunn, RM. Below right: Lieut-Cdr Christopher Parry, MA, RN.

Lieutenant-Commander Ian McKENZIE, RN
Flight Commander, 'A' Flight 824 Naval Air Squadron, RFA *Olmeda*

Born 1950, educated at Morrison's Academy, Crieff, and entered the Royal Navy in 1968. Served in HM Ships *Abdiel* and *Tiger* before transferring to flying training in 1975. Flew Wessex 3 in HMS *Devonshire*, then converted to the Sea King in 1978. Served with 814 NAS in HM Ships *Hermes* and *Bulwark*. To 706 NAS in 1980 as instructor at RNAS Culdrose and in the RFA *Engadine*. Assumed command of 'A' Flight 824 NAS in June 1981. Appointed a Member of the Order of the British Empire.

Lieutenant-Commander Paul MILLER, RN
Flight Commander, HMS *Yarmouth* Wasp Flight

Born in London, 1952. Educated in the United Kingdom and Singapore. Entered the BRNC Dartmouth as a Cadet in 1971. Qualified as an Observer in 1972. Served with 819 NAS in Scotland before joining 771 NAS at RNAS Culdrose. Transferred back to Sea King helicopters with 826 NAS in 1976, serving aboard HMS *Tiger*, and then 824 NAS for the final deployment of HMS *Ark Royal*. In 1978 he commenced pilot training and, when awarded his 'wings', also received the Louis Newmark Trophy for his performance on the course. In 1979 and 1980, he flew as a Wasp pilot with the ships' flights of HMS *Ajax* and HMS *Ariadne*. Was awarded the Admiral Sir Dudley Pound Prize for 1979. Commanded the Wasp flight in HMS *Berwick* (1980–81) before joining HMS *Yarmouth*, again as Flight Commander. Married, with two young sons, he resides near Weymouth, Dorset. Mentioned in despatches (for gallantry under fire during the landings in San Carlos Water).

Commander Antony MORTON, RN
Commanding Officer, HMS *Yarmouth*

A Yorkshireman, he was educated in Cornwall and entered Britannia RNC in January, 1960. After service in HMS *Leverton*, he qualified as a fixed wing pilot and served with several Naval Air Squadrons flying Hunter and Buccaneer aircraft. Qualified as air warfare instructor, 1968. Commanded HMS *Walkerton*, 1971. Attended the Canadian Forces Staff College, Toronto. Commanded 809 NAS (Buccaneers) in HMS *Ark Royal*. Promoted Commander, 1978. Employed for eighteen months at the RAF Central Tactics and Trials Organisation before assuming command of HMS *Yarmouth*. Awarded the Distinguished Service Cross.

Captain Christopher NUNN, RM
Officer Commanding, 'M' Company, 42 Commando RM

He joined the Royal Marines in 1969 and received the Bruce Lumsden Award upon completion of the Commando Course. Served as a Troop Commander in Northern Ireland, Malta, Cyprus, Turkey, Sardinia and the United States, 1970–72, learning to speak Greek. Recruit Squad Officer, CTCRM Lympstone, 1973, and Physical Training Officer, Britannia RNC, 1974–76. Attended the Army Air Corps Centre, Middle Wallop, 1976–77, and qualified as a fixed wing and Gazelle helicopter pilot. Second in command of 45 Commando Group Flight, 1977–78 and completed Arctic Warfare Course. Appointed Flight Commander of 3 Commando Brigade Air Squadron RM in 1978, serving three more tours in Northern Ireland. Awarded the Queen's Commendation for Valuable Service in the Air, July 1980. Appointed Company Commander of 'M' Company, 42 Commando RM, in 1981 after attending Combat Team Commanders Course at the School of Infantry, Warminster. Married, with two young sons, he resides in Bracknell, Berkshire. Awarded the Commander-in-Chief's Commendation for his services during Operations Paraquat and Keyhole.

Lieutenant-Commander Christopher PARRY, MA, RN
Observer, HMS *Antrim* Wessex 3 Flight

Born 1953, educated at Portsmouth College and Jesus College, Oxford (Honours Degree in Modern History). Entered the Royal Navy as a University Cadet Seaman Officer in 1972. Joined HMS *Ark Royal* after post-University training at BRNC Dartmouth. Served subsequently in HM Ships *Alfriston* and *London* before

Above left: Capt David Pentreath, RN.
Above right: Mne Nigel Peters, RM.
Centre left: Flt Lt Andrew Pulford, RAF.
Centre right: Capt Shane Redmond, RFA.
Left: Maj Guy Sheridan, RM.

sub-specialising as an Observer in 1979. Joined 826 NAS, embarked in HMS *Bulwark* until 1981, then appointed to HMS *Antrim* as Flight Observer. Married with one daughter, he resides in Farlington, Hampshire. Mentioned in despatches.

Captain David PENTREATH, RN
Commanding Officer, HMS *Plymouth*

Educated at Haileybury College, he entered the Royal Navy in 1951. As a Lieutenant, he served in HM Ships *Ark Royal* and *Victorious* as a Seahawk and Scimitar pilot. Promoted Lieutenant-Commander 1963, he served in staff appointments and commanded HM Ships *Kemerton* and *Daring* during the next five years. Promoted Commander in 1970, he had further staff appointments, commanded HMS *Brighton*, and attended the National Defence College, Latimer. Promoted Captain in 1977, appointed Deputy Director of Naval Recruiting and attended the Royal College of Defence Studies, London, in 1980. He was appointed Commanding Officer of HMS *Plymouth* in December 1980, and became Captain Sixth Frigate Squadron in April 1981. Appointed as Director of the Royal Naval Staff College, Greenwich, upon his return from the South Atlantic. Married, with three sons, he resides in Upwey, Dorset. Admitted to the Distinguished Service Order.

Marine 1st Class Nigel PETERS, RM
Acting Corporal, HMS *Endurance*'s Royal Marines Detachment

Born Letchworth, Hertfordshire, 1961. Educated at Norton Road School, Letchworth, and joined the Royal Marines at the age of 16. Served in Malta during the final year of the British presence on that island. Served with 41 Commando RM in Cyprus and Northern Ireland before joining the RM Detachment aboard HMS *Endurance* in 1981. Commanded the Jason Peak observation point. He was the only British serviceman to become a casualty during operations on and around South Georgia. Subsequently joined the Surrey Constabulary. Married, resident in Camberley, Surrey. Mentioned in despatches.

Flight Lieutenant Andrew PULFORD, RAF
Pilot, 'C' Flight 845 Naval Air Commando Squadron

Born in Newark, Notts, in 1958. Educated Magnus Grammar School, Newark. Entered RAF Henlow in 1976, commissioned 1977, and qualified as a fixed wing pilot in the same year. Helicopter pilot training 1978, served with 72 Sqn RAF (Wessex HC2) and promoted Flying Officer. Seconded to the Fleet Air Arm (707 NAS) in 1980 to fly the Wessex HU5. Promoted Flight Lieutenant 1981 while serving with 845 NACS at RNAS Yeovilton as a 'jungley' pilot.

Captain Shane REDMOND, RFA
Master, RFA *Tidespring*

Born Hatch End, Middlesex, 1938. Entered HMS *Conway* as an Officer Cadet, age 14. Joined the RFA in 1955. Subsequently served in most parts of the world, including the operations off Suez in 1956. Appointed to his first command in 1975 at age 36. Married, with two sons and a daughter, resident of Winchelsea, East Sussex. Appointed Officer of the Order of the British Empire (Military Division).

Major Guy SHERIDAN, RM
Commander Land Forces, Task Group 317.9

Born in Cairo, Egypt, 1942, son of an Indian Army officer. Educated at Downside School and entered the Royal Marines in 1960. Promoted Lieutenant, 1963, after serving with 45 Commando RM in Aden. Transferred to 40 Commando RM and served with them in Borneo and Malaya. In 1968 he was seconded to The Muscat Regiment in the Sultan of Oman's Army. During his two years of service in the Gulf area, he was awarded the Sultan of Oman's Commendation for Gallantry. Staff training at Camberley followed in 1974. In 1975 he moved to Iran and was senior advisor for the formation of a Commando for the Imperial Navy. Between 1978 and 1981 he held a staff appointment in Portsmouth and then joined 42 Commando RM as Second-in-Command. Apart from the ski-ing and mountaineering achievements already noted in Chapter VI, he also made a 600km ski traverse of the Zagros Mountains, Iran, in 1977, and participated in the Royal Marines expedition which made the first ascent of Mount Menthosa, 21,140 feet, in the Himalayas. His mountaineering activities have taken him to most of the world's continents. In the winter of 1985 he made a ski traverse of 800km of the McKenzie Mountains, Yukon, North West Canada. Married, with one daughter, he resides in Bovey Tracey, Devon. Appointed an Officer of the Order of the British Empire (Military Division).

Lieutenant-Commander Ian STANLEY, RN
Flight Commander, HMS *Antrim* Wessex Flight.

Born 1946, Edinburgh. Educated at Weymouth Grammar School. Entered the Royal Navy in 1963 as an artificer apprentice. Obtained indentures in 1966 and entered Britannia RNC as Acting Sub Lieutenant Air Cadet. Commissioned 1967, then trained as a helicopter pilot. Second pilot, HMS *London* Flight, 1969–71. Staff pilot, 737 NAS, 1971–72. Flying instructor, 705 NAS, 1973–75. Flight Commander, HMS *Ajax*, 1975–77, during which time the ship made a round-the-world cruise and made a three-month visit to Canada. Training Officer, 723 Squadron, Royal Australian Naval Air Station, Nowra, 1979. Promoted Lieutenant-Commander and appointed Senior Pilot, 771 NAS. Joined HMS *Antrim* November 1981. Married, two children, resident in Mullion, Cornwall.

Lieutenant Michael TIDD, RN
Flight Commander, 'C' Flight 845 Naval Air Commando Squadron, RFA *Tidespring*

Born Guernsey, 1955. Educated at Elizabeth College, Guernsey, and entered the Royal Navy in 1975. First tour served as a commando pilot with 845 NACS in Ulster and Norway, flying the Wessex 5, 1977–79. Converted to the Wasp and served with 829 NAS as Flight Commander, HMS *Naiad*, in the Far East. Rejoined 845 NACS as helicopter warfare instructor in 1981. At the time of the Argentine invasions, he was commanding the Detachment Cadre in Northern Ireland. He was rushed back to the United Kingdom and placed on an RAF VC10 for Ascension Island. Upon arrival he was given command of 'C' Flight and departed immediately for Operation Paraquat.

Lieutenant David WELLS, RN
Observer, HMS *Endurance* Wasp Flight

Born 1951, educated at Eastbourne College, he entered the Royal Navy in September 1976 and completed training as an Observer in 1978. Between June 1978 and March 1980 he served in 824 NAS (Sea King helicopters) in HMS *Ark Royal* and the RFAs *Fort Grange* and *Olwen*. After a conversion course to the Wasp, he joined HMS *Endurance*'s ship's flight in October 1980. Most of his Antarctic and South Georgia missions were flown as Observer to Lieutenant-Commander Tony Ellerbeck in Wasp 434 and he launched most of the AS12 missiles during the series of attacks against the *Santa Fe* on 25 April 1982. Married, he lives in St Mawgan, Cornwall. Mentioned in despatches.

Commander Christopher WREFORD-BROWN, RN
Commanding Officer, HMS *Conqueror*

He joined the Royal Navy from Rugby School in 1963. Qualified as a submariner in 1969. After various appointments in a number of submarines, and having qualified as a Navigator and Principal Warfare Officer, he passed the Commanding Officer's Qualifying Course and took command of HMS *Opossum* in 1976. Subsequently was Second-in-Command of HMS *Courageous* and Staff Officer Operations, Second Submarine Squadron. Promoted Commander in 1980, he attended the National Defence College before being appointed to HMS *Conqueror*. Married, resident in Devon. Admitted to the Distinguished Service Order.

Captain Brian YOUNG, RN
Commanding Officer, HMS *Antrim*
Commander Task Group, Operation Paraquat

Entered the Royal Navy in 1944 as a Cadet at Eaton Hall, Chester, and at Dartmouth. Served as Midshipman and Sub Lieutenant in HM Ships *King George V*, *Theseus* and *Wren*. Promoted Lieutenant in 1952 and sent to the United States for fixed wing pilot training. Served with 803 and 804 NAS flying

Above left: Lieut-Cdr Ian Stanley, RN.
Above right: Lieut Michael Tidd, RN.
Centre left: Lieut David Wells, RN.
Centre right: Cdr Christopher Wreford-Brown, RN.
Left: Capt Brian Young, RN.

Sea Hawks from HM Ships *Albion*, *Centaur*, *Bulwark* and *Ark Royal*, 1954–58. Took part in ground attack missions over Egypt during the Suez operation, flying with 804 NAS from HMS *Bulwark*. Exchange posting with the RAF, 1958–60, as a Hunter instructor, before becoming senior pilot of 804 NAS (Scimitars) in HMS *Hermes*. Promoted Lieutenant-Commander, 1960, attended RN Staff College. After a staff appointment, he commanded 892 NAS (Sea Vixens) aboard HMS *Centaur* during the confrontation with Indonesia in 1964–65. In 1965–66 he commanded 766 NAS (Sea Vixens), Naval Air Fighter School, RNAS Yeovilton, and was promoted Commander. Appointed to command HMS *Wiston*, Senior Officer 9th Mine Countermeasures Squadron based on Bahrein, Arabian Gulf. Attended Joint Services Staff College 1968, was Commander (Air) RNAS Lossiemouth 1968–70, Commander (Air) HMS *Eagle* 1970–72, Staff Aviation Officer to Flag Officer Carriers and Amphibious Ships 1972–74, commanded HMS *Danae* 1974–75, and attended Senior Officers' War Course, RNC Greenwich 1976. He was appointed Assistant Chief of Staff (Warfare) to CINCFLEET in 1976 and Chief of Staff to FONAC (as Commodore) in 1978. His final appointment was to HMS *Antrim* in 1981 and he retired from the service two years later. Married, resident in Penn, High Wycombe. Admitted to the Distinguished Service Order. □

APPENDIX H

Fleet Air Arm Helicopters deployed for Operations Paraquat and Keyhole

(with details of aircrew who flew on operations described in the main narrative)

737 Naval Air Squadron

HMS Antrim *Ship's Flight (Operation Paraquat)*

Wessex HAS3, serial number XP142, side number 406
Flight Commander, Lieutenant-Commander I. Stanley, RN
Second Pilot, Sub Lieutenant S. G. Cooper, RN
Flight Observer, Lieutenant C. Parry, RN
Aircrewman, POACMN D. Fitzgerald
Senior Maintenance Rating, CAEA(M) D. J. Heritier

815 Naval Air Squadron

HMS Brilliant *Ship's Flight (Operation Paraquat)*

Lynx HAS2, serial number XZ725, side number 341
Lynx HAS2, serial number XZ729, side number 342
Flight Commander, Lieutenant-Commander B. W. Bryant, RN
Flight Observer, Lieutenant N. A. M. Butler, RN
Senior Pilot, Lieutenant-Commander J. N. Clark, RN
Senior Observer, Lieutenant P. A. McKay, RN
Senior Maintenance Rating, AEMN(M)1 P. J. O'Hara

824 Naval Air Squadron

'A' Flight, RFA Olmeda *(Operation Keyhole)*

Sea King HAS2A, serial number XV649, side number 355
Flight Commander, Lieutenant-Commander I. S. McKenzie, RN
Senior Pilot, Lieutenant D. P. Murphy, RN
Senior Observer, Lieutenant P. Blackman, RN
Aircrewman, LACMN K. J. Alison
Senior Maintenance Rating, AEA(M)1 P. S. Leeper

829 Naval Air Squadron

HMS Endurance *Ship's Flight (Operations Paraquat and Keyhole)*

Wasp HAS1, serial number XS527, side number 434
Wasp HAS1, serial number XS539, side number 435
Flight Commander, Lieutenant-Commander J. A. Ellerbeck, RN
Senior Observer, Lieutenant D. A. M. Wells, RN
Senior Pilot, Lieutenant T. S. Finding, RN
Observer, Sub Lieutenant R. J. F. Buckland, RN
Aircrewman, LACMN R. B. Nadin
Senior Maintenance Rating, CAEMN(M) K. Williams

HMS Plymouth *Ship's Flight (Operation Paraquat)*

Wasp HAS1, serial number XT429, side number 445
Flight Commander, Lieutenant-Commander J. R. Dransfield, RN
Aircrewman, LACMN J. Harper
Senior Maintenance Rating, AEA(L)1 N. N. P. Pollard

HMS Yarmouth *Ship's Flight (Operation Keyhole)*

Wasp HAS1, serial number XV624, side number 456
Flight Commander, Lieutenant-Commander P. G. Miller, RN
Aircrewman, LACMN J. D'Souza
Senior Maintenance Rating, AEMN(L)1 D. McCaughy

845 Naval Air Commando Squadron

'C' Flight, RFA Tidespring *(Operation Paraquat)*

Wessex HU5, serial number XT464, call-sign Yankee Alpha
Wessex HU5, serial number XT473, call-sign Yankee Foxtrot
Flight Commander, Lieutenant M. J. Tidd, RN
Pilot, Lieutenant I. Georgeson, RN
Pilot, Flight Lieutenant A. Pulford, RAF
Aircrewman, LACMN R. Wilson
Aircrewman, LACMN A. Lomas
Senior Maintenance Rating, AEMN(R)1 W. T. Raines

848 Naval Air Commando Squadron

'A' Flight, HMS Endurance *(Operation Keyhole)*

Wessex HU5, serial number XS486, call-sign Whiskey Whiskey
Flight Commander, Lieutenant-Commander C. J. Blight, RN
Second Pilot, Lieutenant T. Mason, RN
Aircrewman, POACMN M. P. Hogan
Senior Maintenance Rating, CPOAEA(L) E. J. Cole

APPENDIX I

Nominal Roll (Royal Marines) for the Battle of King Edward Point, 3 April 1982

Lieutenant	K. P. (Keith) Mills	NO25861B	Awarded Distinguished Service Cross
Marine	R. C. (Bob) Ashton	PO34358C	
Marine	S. P. (Steve) Chubb	PO37464G	NP8901
Marine	L. M. (Laurie) Church	PO36081Y	NP8901
Marine	D. S. (Dave) Combes	PO37746K	Mentioned in despatches
Marine	L. (Les) Daniels	PO38702V	NP8901
Marine	J. W. (Brasso) Hare	PO35634C	NP8901
Marine	S. (Steve) Holding	PO37997A	NP8901
Marine	D. (Jock) Hunter	PO38186N	NP8901
Marine	A. D. (Jesse) James	PO38468R	NP8901
Corporal	A. (Al) Larkin	PO28516V	
Marine	A. M. (Andy) Lee	PO38827E	
Sergeant	P. J. (Pete) Leach	PO31491S	Awarded Distinguished Service Medal
Marine	I. D. (Paddy) McCallion	PO32876H	
Marine	S. J. (Steve) Parsons	PO34440X	Commander-in-Chief's Commendation
Acting Corporal	N. A. (Nige) Peters	PO36756H	Wounded (gunshot)
Marine	M. R. (Spike) Poole	PO33908G	
Marine	R. S. (Paddy) Porter	PO37092N	NP8901
Marine	J. (John) Stonestreet	PO35576L	Mentioned in despatches
Lance-Corporal	S. (George) Thomsen	PO38258M	NP8901
Marine	J. (Jock) Thomson	PO33317W	
Marine	R. K. (Knocker) White	PO38389Y	Commander-in-Chief's Commendation

APPENDIX J

Operation Blackbull One

As described at the end of Chapter VI, the Royal Air Force was unable to play a major role in Operation Paraquat because the target area, South Georgia, was simply too far from its normal operating bases. On 20 April, however, and by the imaginative use of improvised techniques, 55 and 57 Squadrons mounted from Ascension Island what was at that time (for the RAF) a record-breaking maritime reconnaissance operation to survey the waters off the northern and eastern approaches to the island. It was given the local and unofficial code-name of Operation Blackbull, so named because a bull is the central figure in the emblem of RAF Marham, the home base of these two Squadrons. Listed below are the Victor aircraft and pilots who took part in the operation.

XL192	55 Sqn	Squadron Leader John G. Elliott (Primary)
XL163	55 Sqn	Squadron Leader R. 'Bob' Tuxford (Short leg out)
XM715	57 Sqn	Squadron Leader Frank Milligan (Short leg out)
XL189	55 Sqn	Flight Lieutenant A. M. 'Al' Skelton (Long leg out)
XH671	57 Sqn	Squadron Leader Martin D. Todd (Airborne top-up)
XL164	57 Sqn	Squadron Leader Martin D. Todd (Brazilian coast RV)
XL163	55 Sqn	Squadron Leader Frank Milligan (Short leg inward)
XL188	55 Sqn	Squadron Leader R. 'Bob' Tuxford (Short inward reserve)
XL511	55 Sqn	Wing Commander Colin C. B. Seymour (Long inward reserve)

55 and 57 Squadrons were subsequently involved in two other Blackbull operations (22/23 April, Argentine coast, and 24/25 April, Falkland Islands). They took part also in rehearsals for the Operation Blackbuck bombing missions by RAF Vulcan aircraft over Stanley Airfield.

APPENDIX K

British Antarctic Survey Personnel

THE FOLLOWING MEN were stationed on South Georgia at the time of the Argentine invasion on 3 April 1982.

BALES, Christopher

Age 22, from Brentwood. Meteorologist and physicist. Captured at King Edward Point. He photographed the *Guerrico* manoeuvring in King Edward Cove shortly before the ship opened fire.

BANNER, Robert

Age 24, from Ballinluig (Tayside). Electrician. Captured at King Edward Point. He was a member of the party which discovered the *Bahia Buen Suceso* at Leith Harbour on 19 March, and he took clandestine photographs of the Argentines unloading their ship.

BARKER, Ian

Age 21, from Halifax. General assistant (fieldwork and mountaineering). Had been working at the Lyell Glacier hut for ten days prior to the Argentine attack. He was one of the two-man secret reconnaissance team which observed King Edward Point from the slopes of Mount Hodges on 5 April. Their report gave London and Northwood the first indication of the strength of the Argentine garrison. With his three colleagues, he remained at Lyell Glacier throughout the occupation. Presented himself to British forces at King Edward Point on 29 April.

BAYMAN, Peter

Age 30, from East Leake. Radio operator. Captured at King Edward Point. He worked long hours, day and night, transmitting and receiving teleprint, voice and cryptographic materials. He continued to use a portable radio in Grytviken church during the battle, monitoring Argentine communications. After his capture, he supplied the Argentines with details of the field parties' radio frequencies in the hope that the invaders would make contact with the Britons still on the island and ensure their safety.

DOIDGE, William

Age 28, from Mount St Hilaire, Quebec, Canada. Biologist (specialising in seals). He was working at Schlieper Bay at the time of the invasion and remained there throughout the occupation. With his colleague, Paul Martin, he maintained observation of the local coastline and continued to confirm the lack of any Argentine activity in the area. Evacuated by HMS *Endurance* on 1 May.

DRAGE, Clifford

Age 33, from Royston. Electrician. Captured at King Edward Point. Assisted the Royal Marines when the defensive positions were being prepared on 2 April.

EDWARDS, Trefor

Age 37, from Cambridge. Carpenter. He had spent the summer working on the new building at Bird Island. On 19 March he was with the field party which discovered the *Bahia Buen Suceso* at Leith Harbour and it was he who delivered Governor Hunt's message to Captain Briatore on the following morning. On 2 April, he departed King Edward Point on foot for Lyell Glacier where he remained throughout the occupation. Rejoined British forces on 29 April at King Edward Point.

GEMMELL, Campbell

Age 23, from Stirling. Glaciologist. He was already working from the Lyell Glacier hut at the time of the invasion and remained there throughout the occupation. Presented himself to British forces at King Edward Point on 29 April.

GOODALL-COPESTAKE, Paul

Age 24, from Malvern. Biologist (ornithologist). Officer-in-Charge of the Bird Island field station. He and the three other men at Bird Island observed and reported Argentine overflights before and during the occupation. They also coordinated radio reports from the other three field parties and acted as the main link with Signy Island. Evacuated by HMS *Endurance* on 1 May.

HEADLAND, Robert

Age 37, from Cambridge. Biologist. Had already completed two seasons on the island. Captured at King Edward Point. He established and managed the BAS observation point on the Busen Peninsula, 21/23 March, and prepared the reports of Argentine activity at Leith Harbour. Surrendered the Grytviken church party to the Argentines on 3 April, and arranged with his captors the security of BAS scientific equipment on King Edward Point. Interviewed by Astiz aboard the *Bahia Paraiso* and by the tribunal at Puerto Belgrano. Extensively debriefed by the MOD at Montevideo and on the flight north to RAF Brize Norton. Continued to provide MOD with charts and other details regarding South Georgia following his return to Cambridge.

HECTOR, Julian

Age 23, from Bedford. Biologist (ornithologist). Remained at Bird Island throughout the occupation. Shared the tasks listed for Paul Goodall-Copestake. Evacuated by HMS *Endurance* on 1 May.

HOPKINS, Dermot

Age 33, from Birmingham. Diesel mechanic. Captured at King Edward Point. He greatly assisted the Royal Marines on 2 April, moving equipment and materials with the BAS tractor. Throughout the crisis period, he travelled in the launch *Albatros* during all her voyages to Carlita Bay, Jason Harbour and the Sörling Valley. On 3 April he drove the tractor and trailer between Grytviken and King Edward Point, saving much of his colleagues' personal kit but thereby having no time to salvage any of his own.

HUMPHRIES, Paul

Age 35, from Yeovil. Carpenter. He had worked throughout the summer at Bird Island, and was still there at the time of the invasion. Assisted in keeping watch for Argentine ships and aircraft off the north-west coast of the island. Evacuated by HMS *Endurance* on 1 May.

LOCKWOOD, Brian

Age 24, from Thornaby. Meteorologist and physicist. Had already completed a season on the island. Captured at King Edward Point. He accompanied Trefor Edwards to Leith Harbour on 19 March and took some of the clandestine photographs.

MARTIN, Paul

Age 21, from Carlisle. Biologist (specialising in seals). He was working with Bill Doidge at Schlieper Bay prior to the invasion and remained there until evacuated by HMS *Endurance* on 1 May. Maintained the local coastal watch and continued to transmit negative sighting reports.

MARTIN, Stephen

Age 26, from Ashbourne. Boatman, and winter Base Commander. Appointed Magistrate on 4 March. Had already completed a season on the island. Captured at King Edward Point. Martin was the key figure in dealing with the illegal Argentine landing at Leith Harbour and with all the communications prompted by that event during the closing days of March. He was responsible for BAS liaison with HMS *Endurance* and, following her departure on 31 March, with Lieutenant Mills and his detachment. He handled the radio traffic with the *Bahia Paraiso* on 2 and 3 April, assisted the Royal Marines with their defensive arrangements, and ensured the safety of his own people. Interviewed by Astiz and the Argentine enquiry tribunal at Puerto Belgrano. Extensively debriefed by MOD in Montevideo and on the return flight to RAF Brize Norton.

MORRIS, David

Age 29, from Portishead. Marine Biologist attached to the Offshore Biological Programme. Captured at King Edward Point. He accompanied the launch *Albatros* on some of her coastal voyages during the crisis period.

NORTH, Anthony

Age 28, from Leicester. Marine biologist, attached to the Offshore Biological Programme. With Peter Stark and Myles Plant, he was dispersed to St Andrews Bay by the Base Commander on 2 April. Either alone or in company with Myles Plant, he made four extended reconnaissances along the coast and across the Barff Peninsula during the Argentine occupation. In the event of an emergency, there would have been little hope of support or rescue. His hard-won reports concerning the lack of Argentine activity sadly did not reach the Operation Paraquat commanders. Evacuated from St Andrews Bay by HMS *Endurance* on 30 April.

PINHEIRO-TORRES, Alexandro

Age 21, from Cardiff. Meteorologist and physicist. Portuguese national, resident in Wales. Captured at King Edward Point. He assisted with monitoring and translating Argentine radio communications before and during the battle.

PLANT, Myles

Age 32, from Lincoln. Builder and carpenter. Dispersed to St Andrews Bay on 2 April after assisting the Royal Marines with their defensive arrangements. He accompanied Tony North on three of the coastal foot patrols conducted during the occupation, and shared fully in the risks thereof. With North, he made the first (unplanned) contact with the SBS at Hound Bay on the night of 21/22 April and briefed the Royal Marines regarding the local situation.

SANDERS, Damien

Age 25, from Haywards Heath. Diver. Dispersed to Lyell Glacier by Base Commander on 2 April. He accompanied Ian Barker on 5 April to reconnoitre King Edward Point. Presented himself to British forces on 29 April and assisted in the task of salvaging BAS equipment and records at King Edward Point.

SHAW, Neil

Age 25, from Abergele. Plumber. Had been working on the new building at Bird Island. Captured at King Edward Point. He was one of the men who kept watch on Stromness Bay from the Busen Peninsula, 21/23 March. He assisted in preparing the Royal Marines defences on 2 April.

STANSWOOD, Brian

Age 23, from Southampton. Radio operator. Captured at King Edward Point. He and Peter Bayman had worked long hours, around the clock, maintaining contact with Port Stanley and HMS *Endurance*.

STARK, Peter

Age 24, from London. Meteorologist and physicist. Had already completed two winters on the island. He was one of the observation team on the Busen Peninsula, 21/23 March. Dispersed to St Andrews Bay by Base Commander on 2 April. He kept watch on the local coastline and supported the lady film producers during the absences on patrol of Tony North and Myles Plant. Was taken aboard HMS *Endurance* (21–24 April). He briefed the SAS regarding the layout of King Edward Point and its approaches.

WHEELER, Barry

Age 28, from Rochester. Carpenter. Had worked over the summer at Bird Island. Remained there throughout the occupation and assisted in keeping watch for Argentine sea and air movements. Evacuated by HMS *Endurance* on 1 May. He was killed in a road traffic accident shortly after his return to England.

WHITAKER, Terrence

Age 34, from Cambridge. Marine biologist, attached to the Offshore Biological Programme. Captured at King Edward Point. Accompanied and assisted several coastal voyages by the launch *Albatros* prior to the invasion. Assisted the Royal Marines on 2 April. He took a leading role in organising Grytviken church as a BAS refuge.

Note: it was deemed by Authority that none of the men listed above were entitled to the South Atlantic medal. No other forms of recognition for their services have been (to date) forthcoming.

Glossary

APC	Armoured Personnel Carrier
ASE	Automatic Stabilisation Equipment
ASW	Anti-submarine Warfare
BAS	British Antarctic Survey
BAT	British Antarctic Territory
CINCFLEET	Commander-in-Chief, Fleet
CLF	Commander Land Forces
CTF	Commander Task Force
CTG	Commander Task Group
DMS	Directorate of Marine Services
FCO	Foreign & Commonwealth Office
FCS	Flight Control System
FIBS	Falkland Islands Broadcasting Service
FIC	Falkland Islands Company
FIDS	Falkland Islands Dependencies Survey
FIG	Falkland Islands Government
FIGAS	Falkland Islands Government Aviation Service
GPMG	General Purpose Machine Gun
HAS	Helicopter, Anti-submarine
HF	High Frequency (radio)
HMY	Her Majesty's Yacht
HU	Helicopter, Utility
JIC	Joint Intelligence Committee
LADE	Argentine military air transport service
LMG	Light Machine Gun
L42	British sniper rifle
ML	Mountain Leader (Royal Marines)
MOD	Ministry of Defence
MTS	Maritime Tactical School
NGFO	Naval Gunfire Forward Observer
NGLO	Naval Gunfire Liaison Officer
NGS	Naval Gunfire Support
Northwood	HMS *Warrior*, operational headquarters
OP	Observation Post
The 'plum'	HMS *Endurance*
P&O	Peninsular & Oriental
RAS	Replenishment At Sea
RFA	Royal Fleet Auxiliary (Service)
RM	Royal Marines
RMAS	Royal Maritime Auxiliary Service
RN	Royal Navy
RNAS	Royal Naval Air Station
ROE	Rule of Engagement
SAS	Special Air Service
SBS	Special Boat Squadron
SLR	Self-loading Rifle
SMG	Sub Machine Gun
SSN	Attack Submarine, nuclear powered
STUFT	Ship(s) Taken Up From Trade
SWAG	South West Atlantic Group
TELAM	Argentine information service
UHF	Ultra High Frequency (radio)
VERTREP	Vertical Replenishment (by helicopter)
VHF	Very High Frequency (radio)

British operational code-names:

CASINO	Armed insertion, Thule (19.12.82)
CORPORATE	Recovery of Falkland Islands
DANDELION	Initial landings, South Georgia (21/23.4.82)
EAGLE	The Hestesletten landing site (25.4.82)
FLOATER	SBS insertion, Hound Bay
FOXHOUND	SAS insertion, Fortuna Glacier
KEYHOLE	Recovery of Thule
MATCHSTICK	Destruction of Thule base
OKEHAMPTON	Salvage of the *Santa Fe* (1984/85)
PARAQUAT	Recovery of South Georgia (originally PARAQUET)
SUTTON	The San Carlos (Falkland Islands) landings (21.5.82)
TORNADO	Initial observation and intimidation of Argentine garrison on Falkland Islands (1/21.5.82)

Argentine operational code-names:

ROSARIO	Invasion of Falkland Islands (2.4.82)
TG60	King Edward Point attack force (3.4.82)

Index

THE PEOPLE

THE SHIPS

Photographic Sources & Acknowledgments

This lists, by page number, the photographs which illustrate OPERATION PARAQUAT. The author is grateful to all those organisations and individuals who assisted him in tracing these pictures and in making them available for publication.

CHAPTER I

2, 3 (lower), 6 (both), 7 (lower), 13: Denis Hayward, Army.
3 (upper): Michael Parker, National Geographic Society.
7 (upper): R. K. Headland, Scott Polar Research Institute.
8 (left): J. Webber, National Portrait Gallery.
8 (upper right), 8 (lower right), 10, 11 (lower): Christian Salvesen plc.
11 (upper): R. G. Eves, National Portrait Gallery.
12: *Daily Star* Picture Library.

CHAPTER II

15, 22 (all), 23: Camera Press Limited.
16: Michael Montgomery.
17 (upper): Cyril Blakeman.
17 (lower), 18, 30 (upper), 31 (lower): R. K. Headland, Scott Polar Research Institute.
20, 31 (upper), 33 (upper), 35: British Antarctic Survey.
28: George McVay, Christian Salvesen plc.
29 (upper): Frank H. Mason, Christian Salvesen plc.
29 (lower): John Adams, Camera Press Limited.
30 (lower): Christian Salvesen plc.
32 (left): *Jane's Defence Weekly*.
32 (upper right): Ministry of Defence.
32 (lower), 33 (lower): Forces Photographic Section, Falkland Islands.
34: Denis Hayward, Army.

CHAPTER III

37 (upper): LS(Phot) T. Harding.
37 (lower), 38: *Daily Star* Picture Library.
40 (large), Forces Photographic Section, Falkland Islands.
40 (inset), 42, 43, 44 (both), 45, 53, 56: LA(Phot) N. R. Allen.
41: Photographic Section, HMS *Osprey*.
46: *Jane's Defence Weekly*.
47, 55: R. K. Headland, Scott Polar Research Institute.
48, 49: Robert Banner, BAS.
51 (both): Photographic Section, HMS *Hermes*.
57: Photographic Section, HMS *Herald*.

CHAPTER IV

60, 62, 63: Fleet Air Arm PR, RNAS Yeovilton.
61, 67 (lower), 73, 76: Camera Press Limited.
64, 67 (upper): Forces Photographic Section, Falkland Islands.
65: Nigel Thomas, Navy.
68 (both): Royal Marines PR, MOD London.
69, 86: Lieut K. P. Mills, DSC, RM.
71: R. K. Headland, Scott Polar Research Institute.
74 (upper), 75 (both): Nigel Peters.
74 (upper): author.
78: Survival Anglia Limited.
79, 81: Denis Hayward, Army.
85: P&O Steam Navigation Co Limited.
87: *Jane's Defence Weekly*.

CHAPTER V

89 (upper): LA(Phot) N. R. Allen.
89 (lower): C. J. Gilbert, BAS.
90: R. K. Headland, Scott Polar Research Institute.
91 (upper), 93: British Antarctic Survey.
91 (lower): C. J. Collinge, BAS.
95: Survival Anglia Limited.
97: Fleet Air Arm.
99: *Daily Star* Picture Library.
100: Royal Marines Museum, Eastney.

CHAPTER VI

103: the late Steve Piercey.
104, 105: Photographic Section, HMS *Heron*.
106, 108, 109: Lieut Kevin White, RN.
107, 115, 117: Fleet Air Arm.
110: Lieut-Col Guy Sheridan, OBE, RM.
111, 123, 127: Ministry of Defence.
112, 113: Royal Marines Museum, Eastney.
114: L. Cherrault, Camera Press Limited.
118: Chief Officer Alec Bilney, RFA.
119: Royal Maritime Auxiliary Service.
120, 121: Lieut Ewen Duncan, RM.
124, 125: J. Bishop, BAS.
130: Photographic Section, RAF Marham.

CHAPTER VII

132, 134, 137, 139 (upper): Lieut Stewart Cooper, RN.
133, 135, 141, 151: Lieut Kevin White, RN.
139 (lower): Ministry of Defence.
142: 'Clive'.
143: Anglia Survival Limited.
145, 152: Denis Hayward, Army.
149: Lieut-Col Keith Eve, RA.

CHAPTER VIII

155: Photographic Section, HMS *Osprey*.
156, 157: Ministry of Defence.
159: Navy PR, MOD.
161 (both): Photographic Section, HMS *Neptune*.
162: Lieut-Cdr Chris Parry, RN.
163, 180: Lieut-Cdr John Dransfield, RN.
164: Photographic Section, HMY *Britannia*.
165, 170, 171, 173, 174, 176: Lieut Kevin White, RN.
169: FPU, HMS *Excellent*.
172: Photographic Dept, RNAS Culdrose.
175: Royal Marines Museum, Eastney.
177 (left): 'Clive'.
177 (right): Lieut Ewen Duncan, RM.
179: LA(Phot) N. R. Allen.

CHAPTER IX

182: Lieut Stewart Cooper, RN.
183, 188: Lieut Ewen Duncan, RM.
186, 189, 190, 194 (upper left), 195 (upper), 197, 199, 203 (upper): Fleet Air Arm.
187 (upper), 203 (lower), 204, 205, 206, 207 (lower), 208, 209: LA(Phot) N. R. Allen.
187 (lower): Chief Officer Alec Bilney, RFA.
191, 192, 193, 200, 201, 202: Royal Marines Museum, Eastney.
194 (upper right), 194 (lower): Lieut Kevin White, RN.
195 (lower): Colour Sergeant Peter Leach, DSM, RM.
198: Lieut-Cdr Paul Miller, RN.
207 (upper): Portsmouth & Sunderland Newspapers.

EPILOGUE

214: author.
216: Fleet Air Arm.
217, 219 (upper): Royal Marines Museum, Eastney.
219 (lower): Robert Scarlett.
221: Danny du Feu, Ministry of Defence.
222: Capt Shane Redmond, OBE, RFA.
223: United Towing.
225: Press Association.

APPENDIXES

233 (all): United Towing.
234: Royal Maritime Auxiliary Service.
235 (upper): Capt Roger Morgan-Grenville, RGJ.
235 (lower): Ministry of Defence.
239 (upper): Lieut John Gilbert, RN.
239 (lower): LA(Phot) N. R. Allen.
242, 243: Robert Scarlett.

THE COLOUR OF PARAQUAT

The author is indebted to the following individuals and organisations for their contribution to the colour section: Lieut Kevin White, RN, Chief Offr Alec Bilney, RFA, United Towing, Survival Anglia, Royal Fleet Auxiliary Service, Lieut Charles Wilson, RM, Royal Fleet Auxiliary Service, A. J. Glen, Esq, Fleet Air Arm.

The following notice appears on instructions from the Ministry of Defence

The Fleet Air Arm is pleased to have commissioned the writing of this book and to have helped to bring together many of the participants in Operation Paraquat to assist the author.

However the Fleet Air Arm is aware that a large part of the subject matter is opinionative and speculative by nature and therefore takes no responsibility for the views expressed or the accuracy of the information herein.

For his part the author acknowledges that, because of restraints which the reader will appreciate, a lack of precision and clarity marks those areas of the narrative which pertain to matters of national security. The author also acknowledges that though he has received help from many of the participants in the Operation the responsibility for the accuracy of the information in the book and for its interpretation is entirely his.

Production Manager, Mike Wicks
Production Assistant, Adrian Hillier
Dust-jacket design, Brian Ainsworth and Adrian Hillier
Original canvas of 'Santa Fe and Humphrey', Michael Turner
Original pencil sketch (endpapers), Michael Turner
Maps, charts and other graphic design, Brian Ainsworth
Typescript, Christine Kirk
Typesetting and page make-up, Charles Innes and Veronica Pinniger
Technical Editor, Commander Paul Madge, RN
Picture research, Roger Perkins